Advances in Science, Technology & Innovation

IEREK Interdisciplinary Series for Sustainable Development

Advances in Science, Technology & Innovation (ASTI) is a series of peer-reviewed books based on important emerging research that redefines the current disciplinary boundaries in science, technology and innovation (STI) in order to develop integrated concepts for sustainable development. It not only discusses the progress made towards securing more resources, allocating smarter solutions, and rebalancing the relationship between nature and people, but also provides in-depth insights from comprehensive research that addresses the **17 sustainable development goals (SDGs)** as set out by the UN for 2030.

The series draws on the best research papers from various IEREK and other international conferences to promote the creation and development of viable solutions for a **sustainable future and a positive societal** transformation with the help of integrated and innovative science-based approaches. Including interdisciplinary contributions, it presents innovative approaches and highlights how they can best support both economic and sustainable development, through better use of data, more effective institutions, and global, local and individual action, for the welfare of all societies.

The series particularly features conceptual and empirical contributions from various interrelated fields of science, technology and innovation, with an emphasis on digital transformation, that focus on providing practical solutions to **ensure food, water and energy security to achieve the SDGs.** It also presents new case studies offering concrete examples of how to resolve sustainable urbanization and environmental issues in different regions of the world. The series is intended for professionals in research and teaching, consultancies and industry, and government and international organizations. Published in collaboration with IEREK, the Springer ASTI series will acquaint readers with essential new studies in STI for sustainable development.

ASTI series has now been accepted for Scopus (September 2020). All content published in this series will start appearing on the Scopus site in early 2021.

Sandeep Panda · Srabani Mishra · Ata Akcil ·
Eric D. Van Hullebusch
Editors

Biotechnological Innovations in the Mineral-Metal Industry

Springer

Editors
Sandeep Panda
Department of Industrial Biotechnology
Gujarat Biotechnology University
Gandhinagar, Gujarat, India

Srabani Mishra
Department of Mining Engineering
Suleyman Demirel University
Isparta, Turkey

Ata Akcil
Department of Mining Engineering
Suleyman Demirel University
Isparta, Turkey

Eric D. Van Hullebusch
Institut de Physique du Globe de Paris
Paris, France

ISSN 2522-8714 ISSN 2522-8722 (electronic)
Advances in Science, Technology & Innovation
IEREK Interdisciplinary Series for Sustainable Development
ISBN 978-3-031-43624-6 ISBN 978-3-031-43625-3 (eBook)
https://doi.org/10.1007/978-3-031-43625-3

This Springer imprint is published by the registered company Springer Nature Switzerland AG
The registered company address is: Gewerbestrasse 11, 6330 Cham, Switzerland

Preface

With the advent of technology, the demand for metals in increasing. Natural resources of metals are getting depleted, and there has been a huge concern to meet this ever-growing demand. To this aspect, the secondary resources, for example, the urban mines (basically referring to the electronic wastes), have been quite attractive to process for metal recovery. It is important to note that electronic wastes, for example, have been seen to contain an array of metals in a concentration higher than obtained from the primary resources (ores from mines). Over several decades, metallurgical strategies, e.g., pyrometallurgical and hydrometallurgical applications, have been significantly contributing to the supply of metals to meet the demand. However, the mining industry is increasingly facing to several challenges, for example, dumping of low-grade ores and/or drainage of acidic mine effluents are a matter of concern. Similarly, is the case while dumping or burning the end-of-life (EOL) electronic items. Therefore, processing of such ores and wastes has become highly essential in view of industrial waste management and meeting the circular economy guidelines.

It is quite encouraging to note that research has been progressing well to understand the fundamental as well as applied aspects of biotechnology to provide an economic and eco-friendly answer to such problems. It not only aims at looking forward to providing solutions for the environment but also focuses at providing value-added products that can be used in the mineral-metal industry or other industries/sectors where such metals or products can be used. This is the main source of inspiration behind coming up with a book devoted to providing the recent biotechnological innovations and applications related to the treatment of varied mineral-metal wastes. Herein, several key chapters are included that cover the main theme of the book. These chapters have been written by well-known experts in the field, and the contributions have been received from several parts of the globe. We thank them for such wonderful contributions. In addition, we are extremely thankful to our expert reviewers who have given their valuable time and feedback that allowed all the contributors to significantly improve their submissions. It has been a great collective effort, and we believe this book will provide the much-needed valuable information to all students/researchers working in the subject.

Gandhinagar, India — Sandeep Panda
Isparta, Turkey — Srabani Mishra
Isparta, Turkey — Ata Akcil
Paris, France — Eric D. Van Hullebusch

Contents

About the Editors

Dr. Sandeep Panda (Lead Editor) received B.Sc. with honors in Zoology (2006) from Utkal University, M.Sc. in Biotechnology (2008) from Ravenshaw University, and Ph.D. in Life Sciences (2015) from North Orissa University, India. He is working as Assistant Professor at the Department of Industrial Biotechnology, Gujarat Biotechnology University, India. His main research areas include: bio-hydrometallurgical approaches for metal extraction from primary and secondary resources; bio-desulphurization; bio and chemical approaches for mine water treatment and application of eco-friendly approaches for sustainable mineral-metal waste recycling and management.

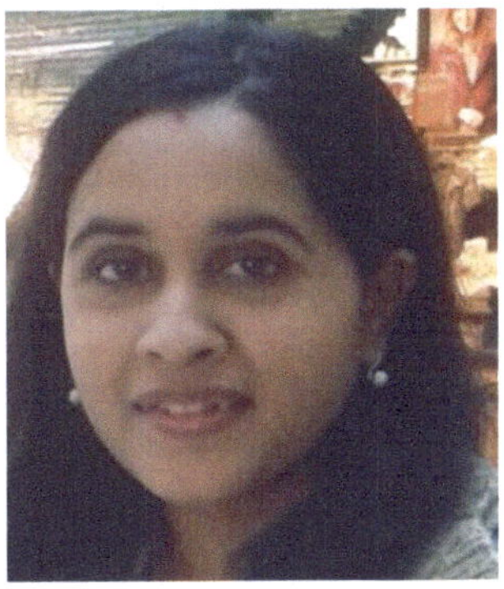

Dr. Srabani Mishra received B.Sc. with honors in Microbiology (2007) from Orissa University of Agriculture Technology, M.Sc. in Applied Microbiology (2009) from Vellore Institute of Technology, and Ph.D. in Biological Sciences (2018) from Academy of Scientific and Innovative Research, India. She is working as Assistant Professor at the Department of Mining Engineering, Suleyman Demirel University, Turkey. Her main research interests include: bio-mineral processing; bio-hydrometallurgical applications for metal extraction from both primary and secondary mineral resources; bio-desulphurization of coal and pet-coke and nano-technological applications in environmental remediation.

Prof. Dr. Ata Akcil holds a B.Sc. in Mining Engineering (1990), M.Sc. in Mineral Processing (1994), and Ph.D. in Mineral Processing (1999) from Dokuz Eylul University, Izmir, Turkey. He is working as a full professor at the Suleyman Demirel University (SDU), Turkey. He is also the head of Mineral-Metal Recovery and Recycling (MMR&R) Research Group, Department of Mining Engineering at SDU. His main research areas include: industrial waste management/recycling and hydrometallurgical/bio-hydrometallurgical process applications for metal recovery.

Prof. Dr. Eric D. Van Hullebusch holds a B.Sc. in Chemistry (1997), M.Sc. in Aquatic Chemistry and Microbiology (1999), and Ph.D. in Aquatic Chemistry and Microbiology (2002) from Université de Limoges, France. He is working as full professor at the Institut de Physique du Globe de Paris/Université de Paris in Biogeochemistry of engineered ecosystems. His main research areas include: bio-weathering of natural and industrial materials, bio/hydrometallurgy process applications for resource recovery from secondary waste streams, and soil bio/remediation.

Microbes, Metal(Loid)s and Microbe–Metal(Loid) Interactions in the Context of Mining Industry

G. Kiran Kumar Reddy and Y. V. Nancharaiah

Abstract

Mining activities generate large quantities of wastes which pose threat to aquatic life, environment and human health, if left untreated. The generation of large quantities of wastewater with low concentrations of metals (for example: 10 mg/L Te(IV) in used solar cell leachate; 5–30 mg/L Cr, 0.14–21 mg/L Cu, 0.2–30 mg/L Ni, and 0.2–28 mg/L Zn in electroplating wastewater) makes the biological treatment processes more attractive, over physicochemical processes. Often, these wastes are laden with critical, scarce metals and are considered as resource due to their limited availability, cost and intended applications. Metal–microbe interactions through various redox reactions and acid-producing metabolism allow the extraction of base metals from ores. In the industrial biomining and bioleaching processes, microbes are employed for metal extraction from ores and the same can be applied for extraction of precious metals from low-grade ores, solid wastes and mine tailings. This book chapter presents an overview of metal–microbe interactions for potential biotechnological applications in the treatment of metal laden wastes generated in mining activities. These metal-microbe interactions include mobilisation (biomining) and immobilisation (biosorption, bioaccumulation, bioreduction and bioprecipitation) of metal ions present in different forms of wastes. Up to date studies on different microorganisms and biofilm systems employed for successful treatment and recovery of metals from wastes have been included along with underlying biochemical process. The chapter ends with a discussion on sustainable technological platforms such as bioelectrochemical systems, wherein the oxidation of organic contaminants in wastewater is coupled to the removal and recovery of metal ions from metal-bearing wastewater. Important studies on bioelectrochemical systems for the recovery of metal ions and associated removal mechanisms are provided.

Keywords

Biofilms · Bioremediation · Metal–microbe interactions · Mining wastes · Biomining · Bioelectrochemical systems

1 Introduction

In metal mining, ores are extracted through open pit, placer or underground mining processes. Extraction and beneficiation of metals from the earth are common to active mining operations (Dudka & Adriano, 1997). A continuous uptrend is seen in the production of metals (ferrous and non-ferrous) including precious and scarce metals (Agboola et al., 2020) due to the development of innovative technologies, industrialisation, increased gross domestic product (GDP), high disposable income, world population explosion, urbanisation and high living standards (Langkau & Tercero Espinoza, 2018). The annual production of metals in the world is about 3.3 billion tonnes amounting to £2500 billion (Smith & Wentworth, 2022). The availability, supply and prices of metals in international markets are influencing factors for the economy of many countries in terms of exports and employment (Hennebel et al., 2015). Demand for metals in developing technologies such as renewable energy generation, batteries and electric vehicles can spur increased mining operations.

Sites of mining operations for metallic ores have negative social and environmental impacts. The large land usage for mining operations cause irreversible damage

G. Kiran Kumar Reddy · Y. V. Nancharaiah (✉)
Biofouling and Biofilm Processes Section, WSCD, Bhabha Atomic Research Centre, Kalpakkam, Tamil Nadu 603102, India
e-mail: yvn@igcar.gov.in; venkatany@gmail.com

G. Kiran Kumar Reddy · Y. V. Nancharaiah
Homi Bhabha National Institute, Anushakti Nagar, Mumbai 400 094, India

S. Panda et al. (eds.), *Biotechnological Innovations in the Mineral-Metal Industry*,
Advances in Science, Technology & Innovation, https://doi.org/10.1007/978-3-031-43625-3_1

to surrounding ecosystem, landscape and biodiversity (Simmons et al., 2008). Generation of enormous quantities of overburden or waste rock with toxic substances at mining sites is a concern. The fugitive dust generated during ore extraction deteriorates air quality and is a health hazard to surrounding biota. Wastewater generated during mining and metal processing can contaminate surrounding water bodies if discharged without proper treatment (Cidu, 2011; Mudd, 2008). Grinding (milling) of the metal ore to generate very fine particles followed by physicochemical leaching processes of metal concentration or beneficiation is used for extracting metals form non-metallic ores. These processes generate high volumes of toxic waste called tailings which are stored in tailings pond (Kossoff et al., 2014). Leachates from the mining pits, tailing ponds and other mining wastes contain toxic metals and can be potentially toxic to living organisms (Kossoff et al., 2014). Acid mine drainage (AMD) is generated by the interaction between the mining wastes such as waste rock, tailings and mine structures of metal-sulphide mines and air (oxygen) and water. AMD is typically characterised with low pH (<3) and higher concentrations of heavy metals including Cd, Cu, Fe, Mn, Pb and Zn. Thus, AMD poses a severe environmental and health hazard (Park et al., 2019).

In addition to mining and metallurgical operations, waste streams containing heavy metals are generated from other anthropogenic activities such as manufacturing of paper, plastic, fertilisers, pesticides, cement, burning of fossil fuels, tanneries, electroplating, corrosion inhibitors, batteries, photovoltaic devices, dyes, paints and pigments (Fu & Wang, 2011; Nancharaiah et al., 2015). Based on the associated health and environmental hazards, the US Environmental protection agency (US EPA) has listed 13 metal(loid)s (Ag, As, Be, Cd, Cr, Cu, Hg, Ni, Pb, Sb, Se, Tl and Zn) in the priority pollutants' list with stringent discharge limits. Since metals cannot be biodegraded unlike organic contaminants, they accumulate in biota across trophic levels due to bioavailability (Barakat, 2011). Therefore, treatment of metal-laden wastewater is necessary for safe disposal of treated wastewater to environment.

Different physical, chemical and biological methods are developed with varying degrees of efficiency and feasibility for treating metal-contaminated water (Alalwan et al., 2020; Fu & Wang, 2011). Physicochemical processes such as chemical precipitation, ion exchange, chemical oxidation/reduction, reverse osmosis, electrodialysis and ultrafiltration have limitations such as lower efficiency, sensitive to operating conditions and production of secondary sludge (Alalwan et al., 2020; Wang & Ren, 2014). Often, metal-laden wastewater is characterised by large quantities of wastewater with low concentrations of metals (Wang & Ren, 2014). Thus, treatment of such wastes requires efficient technologies. The interactions of microbes with diverse metals, metalloids and radionuclides are well known. In natural environments, biogeochemical cycles of metals are controlled through diverse metal–microbe interactions involving oxidation, reduction, methylation, precipitation and chelation. These metal–microbe interactions can either mobilise or immobilise metals. Hence, microbiological processes can be utilised in engineered environments to treat metal-laden solid and liquid wastes (Jing & Kjellerup, 2018). Mobilisation of metals by microbial leaching aids in extraction of precious/scarce metals from low-grade ores and wastes. In contrast, metal immobilisation by microbes through sorption, accumulation or transformation is beneficial for bioremediation of contaminated soil/metal-laden wastewaters. Along with treatment, biorecovery of metals from waste is promising for critical and scarce metals. Due to scarce availability, unequal global distribution, and enormous consumption has led to rapid dwindling of resources for critical and scarce metals. This necessitates lookout for alternative sources. Solid e-waste, process streams, mining wastewater, mine leachates and waste streams from metallurgical process are potential secondary sources for critical and scarce metals (Nancharaiah et al., 2016). Metal–microbe interactions can be utilized for developing novel biotechnologies for treatment of metal-laden wastes coupled to resource recovery. This book chapter presents a detailed account of microbial–metal(loid) interactions and biogeochemical transformations of metal(loid)s as biotechnologies for treatment of metal-laden wastewaters. Different classes of microorganisms, biofilm systems and mixed microbial consortia for mobilisation and/or immobilisation of metals for process development for managing mining wastes have been discussed. Finally, a detailed account of bioelectrochemical systems (BES) was presented in the context of prospective technologies for biorecovery of metal(loid)s.

2 Metal–Microbe Interactions

Biogeochemical cycles of several metals in environmental settings are governed by microorganisms (Gadd, 2010). Microbial interactions with metals can either decrease or increase (mobilisation) the mobility of metals. These activities are dependent on metal type, microorganism type and environmental conditions. Diverse biochemical processes employed by microbes influence the bioavailability of metals as these processes can alter solubility and mobility in natural settings. Figure 1 depicts common microbe–metal(loid) interactions. These microbe–metal interactions are exploited for developing microbial biotechnologies that are sustainable and aimed at metal extraction, recovery and/or bioremediation. The microbial interactions with metals have been primarily studied with a perspective to

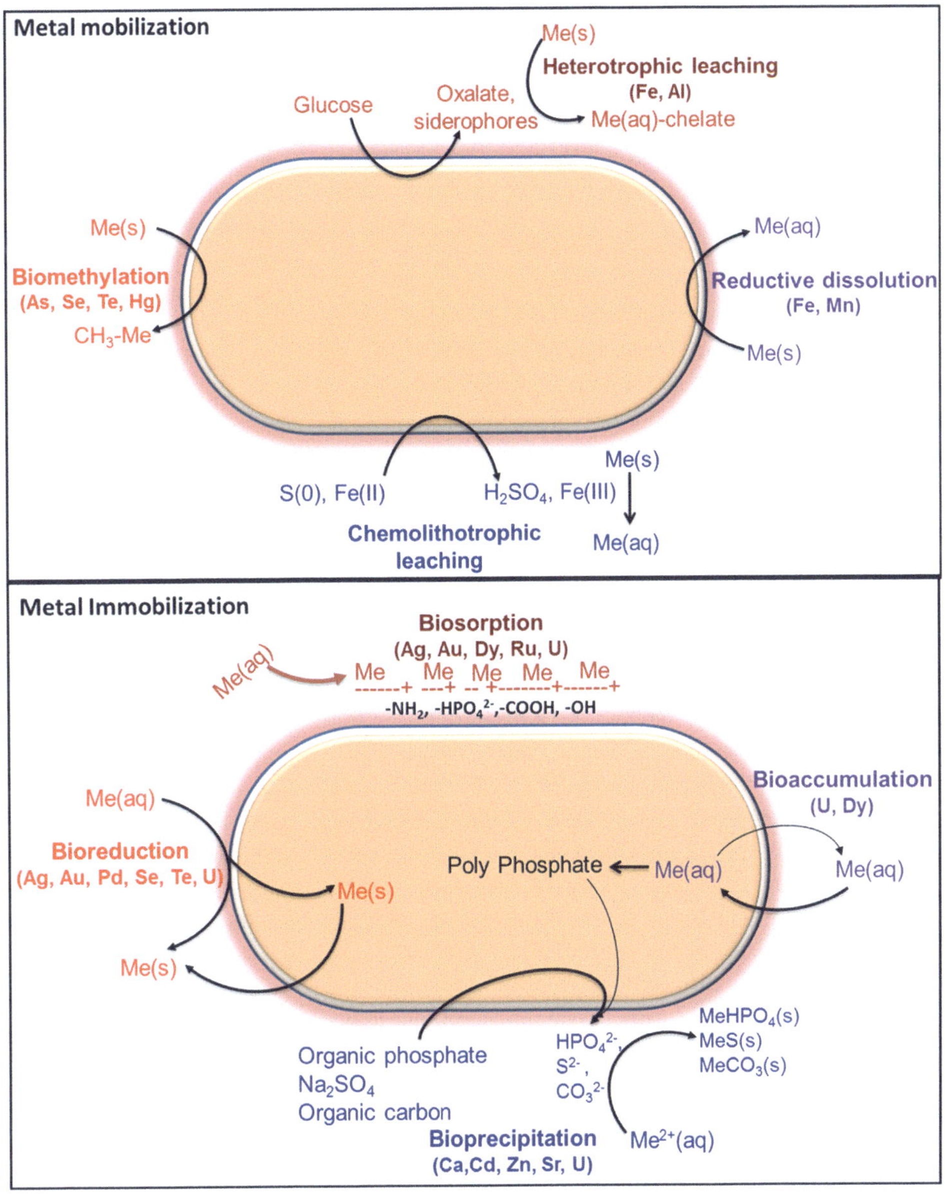

Fig. 1 Illustration of microbe–metal interactions involved in mobilisation and immobilisation of metals

develop sustainable biotechnological processes for metal recovery/removal processes through solubilisation of metals from low-grade ores/secondary sources and to immobilise the metal contaminants from soil and wastewaters (Nancharaiah et al., 2016).

Mobilisation of metals is achieved by microbes through different biochemical routes such as heterotrophic, chemolithotrophic, bioreduction and methylation of metals (Gadd, 2010). In heterotrophic metabolism, microbes release organic acids and metal chelating molecules into surrounding environments that aid in metal mobilisation (Farkas et al., 2021). Microbes produce organic acids including oxalic acid and citric acid which facilitate favourable microenvironments for metal leaching by providing protons and by forming soluble metal complexes. For example, oxalic acid forms stable and mobile complexes with iron and aluminium. Similarly, citrate forms soluble and stable complexes with different metal ions. Siderophores are biogenic chelating agents which aid in metal extraction due to high affinity towards metals to form metal complexes (Williamson et al., 2021). In chemolithotrophic leaching (or autotrophic leaching), microbes obtain energy and electrons through ferrous iron [Fe(II)] oxidation and from reduced sulphur substrates generating ferric iron [Fe(III)] and sulphuric acid, respectively. The acidic environment created by this autotrophic metabolism helps in solubilising metal sulphides/other metal compounds (Korehi et al., 2013). Microbe-mediated reduction and methylation increase the solubility of some metal(loid)s leading to their mobilisation. For example, increased solubility of Fe and Mn is achieved by the microbial reduction of Fe(III) to Fe(II) and Mn(IV) to Mn(II). It is well known that certain metal(loid)s can serve as terminal electron acceptors (TEAs) in anaerobic respiration of microbes. Microbial enzyme-mediated methylation of metal(loid)s also influences the corresponding metal(loid) solubility and mobility. For example, methylated metal(loid)s such as methylarsenicals, dimethyl selenide, dimethyl diselenide and dimethyl telluride compounds are volatile and generated by microbial biomethylation processes (Gadd, 2010). The detailed understanding of underlying biochemical mechanisms in mobilisation of metal(loid)s is explored to develop environmentally benign biotechnological processes for leaching and recovering valuable metals from solid wastes and low-grade ores.

Immobilisation of metal(loid)s by microbes involves physicochemical and metabolic processes. In microbial biosorption, both organic and inorganic metal species are immobilised on cell surfaces by physicochemical sorption processes. These interactions include electrostatic interactions, ion/proton displacement, chelation or complexation (Diep et al., 2018). Phosphoryl-, carboxyl-, hydroxy-, amino-, imidazole-, sulphate- and sulfhydryl-groups present on microbial cells serve as adsorption sites (Aryal, 2020). In contrast, immobilisation of metal(loid)s by bioaccumulation is an energy-intensive process requiring metabolically active cells for accumulating metals inside the cells. Through various membrane transport mechanisms, cells take up the metal ions from external environment and accumulate intracellularly (Diep et al., 2018). In some cases, metals biosorbed on the cell surface are transported and bioaccumulated inside the cells. To minimise cellular toxicity due to the accumulated heavy metals, microbes employ sequestration strategies like metal-binding proteins (MBPs), peptides and other polymers. Polyphosphate, a polymer of phosphate anions can sequester metals through complex formation and aids in compartmentalisation inside the cells. Many bacteria induce reductive transformations of metal(loid)s to decrease the solubility and to immobilise them. In dissimilatory metal reduction by anaerobic bacteria and archaea, metal(loid)s serve as TEAs to thrive and to produce energy (Lloyd, 2003). The anaerobic metabolism of microorganisms influences the change in redox status of metal(loid)s and mobility (Benzerara et al., 2011). Bioprecipitation is another method by which microbes can immobilise the metals as mineral forms. Active microbial metabolism produces substrates such as sulphide, carbonate, phosphate or hydroxide. These react with soluble metals in the surrounding environment to form metal precipitates (Benzerara et al., 2011). Metal immobilisation by microbes has been extensively studied and employed as a potential biotechnological tool for remediating contaminated soil, water and wastewater.

3 Biotechnological Applications of Metal–Microbe Interactions for Removal and Recovery of Metals

Direct and indirect interactions of microorganisms with metal(loid)s have been successfully explored for developing different biotechnological processes. These approaches are extremely beneficial for metal extraction from wastes, low-grade ores and removal and/or recovery of metals from metal-laden wastewaters (Nancharaiah et al., 2016). Mobilisation of metals through microbial bioleaching processes from ores with low concentrations of metals, solid wastes and immobilisation of metals from wastewaters through various microbial bioremoval processes are discussed in this section. Additionally, bioelectrochemical systems were discussed in detail with potential utility in biorecovery of metals.

3.1 Bioleaching: Mobilisation of Metals from Low-Grade Ores and Solid Wastes

Large quantities of wastes are generated in mining and metallurgical operations which need to be managed properly to avoid environmental problems (Simmons et al., 2008). But, some of these wastes are a potential resource for critical and scarce metal(loid)s. Recovery of scarce and critical elements from alternative sources like end-of-use equipment, mining wastes and low-grade ores is needed for improving environmental and material sustainability. Electronic wastes (e-wastes) is a valuable secondary source for precious and scarce metal(loid)s (Nancharaiah et al., 2016; Sethurajan et al., 2018). Extraction followed by recovery of these metal(loid)s from the secondary sources would help in resource management and environmental protection. Bioleaching is a biohydrometallurgical process employs one or more microorganisms for leaching metals from solids. Compared to chemical hydrometallurgy, pyrometallurgy and electrometallurgical processes, biohydrometallurgy is economical, eco-friendly, less energy intensive and associated with lower greenhouse gas emissions, minimal toxic sludge production and does not require hazardous chemicals (Massianaie et al., 2006). Moreover, biohydrometallurgy allows in situ leaching and is suitable for extracting metals from wastes or low-grade ores (Johnson, 2013). Thus, bioleaching can be considered for low-grade ores, mining wastes, metallurgical wastes and e-wastes.

Bioleaching is classified as direct or indirect depending on microbial contact with metals (Mishra et al., 2005). Direct bioleaching involves a close contact between microbes and metal minerals and involves oxidation through enzymatically catalysed reactions. Oxidation of pyrite by direct bioleaching is shown in Eq. 1.

$$2\,FeS_2 + 7\,O_2 + 2\,H_2O \rightarrow 2\,FeSO_4 + 2\,H_2SO_4. \quad (1)$$

Indirect bioleaching does not involve any direct contact of microbes with metal minerals. However, oxidised intermediates or leaching agents produced by microbes aid in metal solubilisation. Copper leaching from chalcopyrite by *Acidithiobacillus ferrooxidans* involves oxidation of ferrous to ferric ions (Eq. 2), which in turn oxidises chalcopyrite (Eq. 3) to indirectly leach metals.

$$4\,Fe^{+2} + O_2 + 4\,H^+ \rightarrow 4\,Fe^{+3} + 2\,H_2O, \quad (2)$$

$$CuFeS_2 + 4\,Fe^{+3} \rightarrow 5\,Fe^{+2} + Cu^{+2} + 2\,S^0. \quad (3)$$

According to the involved mechanism, bioleaching is classified as redoxolysis, acidolysis or complexolysis (Mishra et al., 2005). In redoxolysis, transfer of electrons from metal minerals to microorganisms is coupled with oxidation of ferrous to ferric ions to cause metal leaching. Dissolution of metals from insoluble to soluble species is influenced by biogenic acids (proton) produced by microbes is involved in acidolysis. Organic acids or carbonic acid produced through microbial metabolism can acidify the surrounding environment and aid in solubilisation of metals. In complexolysis, microbial metabolites such as siderophores, cyanide can chelate the metals to form stable metal complexes. Besides the type of microorganisms, diverse factors such as pH, temperature, particle size of minerals, mineralogy, oxygen and iron concentrations, pulp density and redox potential influence bioleaching, either independently or in combination (Massianaie et al., 2006; Sethurajan et al., 2018).

Low-grade ores can be a potential source for scarce or critical metals owing to their large quantities. Bioleaching can be considered for metal extraction from wastes or low-grade ores in an environmentally benign manner. Microbial bioleaching using bacteria and fungi was studied for leaching of low-grade copper-sulphide ores. Bacterial cultures such as *Acidithiobacillus ferrooxidans* and *At. thiooxidans* were used for copper leaching from low-grade copper-sulphide ore (5% Cu, 34.5% Fe and 22% S). Both pulp density and microbial inoculum were critical in improving leaching efficiency. Leaching of up to 45% of copper was achieved during 45 days experiment with an inoculum density of 2.2×10^7 cfu mL^{-1} and 10% pulp density (Wang et al., 2014). Higher copper-leaching efficiencies of up to 68% were achieved from a low-grade ore (0.72% Cu, 2.6% Fe, 0.02% Zn, 0.002% Ni) using fungi *Aspergillus niger*. Organic acid metabolites produced by fermentation of sucrose by fungi were responsible for achieving bioleaching (Mulligan et al., 2004).

Microbial metal leaching from reduced minerals such as chalcocite and chalcopyrite is achieved by redoxolysis or ferric iron-mediated oxidative dissolution and it has been successfully implemented across the world. Waste flue dusts collected from smelter or furnace exhausts contain metals at significant concentrations (Sethurajan et al., 2018). For example, copper flue dusts from copper smelters contain 20–40% of Cu as reduced minerals such as chalcocite, chalcopyrite, bornite and covellite (Massianaie et al., 2006). Electric arc furnaces and blast furnaces of steel industries generate metallic dusts with varied concentrations of metals. As the metals in these flue dusts are in reduced sulphidic mineral phases, bioleaching can be considered as an alternative economic method over pyrometallurgical or chemical leaching procedures. *At. ferrooxidans, At. thiooxidans* and *Leptospirillum ferrooxidans* have been investigated either as pure cultures or mixed cultures for bioleaching of flue dusts (Bakhtiari et al., 2010; Bayat et al., 2009). Bioleaching from flue dusts was demonstrated in large volume bioreactors (Massianaie et al., 2006). Details of some important bioleaching processes proposed for extracting metals from flue dusts are mentioned in Table 1. Tailings are wastes generated in large quantities during ore beneficiation and contain reduced minerals such as pyrite, chalcopyrite, chalcocite, arsenopyrite and sphalerite (Sethurajan et al., 2018). Tailings contain different metals (Fe, Zn, Cu, Pb and As) in significant concentrations (Lee et al., 2015; Liu et al., 2008). Oxidative dissolution of metals from these reduced wastes employing either single or mixed cultures of *At. ferrooxidans*, *At. thiooxidans* and uncultured acidophilic iron-oxidising and sulphur-oxidising consortia have been reported to leach metals as in the case of flue dusts (Lee et al., 2015; Liu et al., 2008; Nguyen et al., 2015). The details of bioleaching of tailings waste are summarised in Table 1.

Oxidative dissolution using acidophilic iron- and sulphur-oxidising bacteria applied for bioleaching of reduced minerals is not applicable for extracting metals from oxidised mineral sources (Johnson et al., 2013). But, metals can be leached from oxidised mineral ores and wastes using organic acid-producing fungi and/or bacteria (Johnson, 2013). Some studies have reported that metals can be extracted from oxidised minerals (oxidised lateritic ores) by chemolithotrophs and anaerobic reductive dissolution. Studies on metal bioleaching from oxidised metallurgical wastes (e.g. mineral sludge, residues, red mud and slags) are given in Table 1. Sludges are generated as waste by-products in diverse ferrous and non-ferrous industries. Sludges from furnaces, converters, plating and metal industries are produced during metallurgical operations. Iron industries, especially during steel production, generate large quantities of sludge (up to 4 tonnes sludge per 1 tonne steel produced) (Das et al., 2006). Nature of the processed ore

Table 1 Bioleaching of metals from low-grade ores and solid wastes

Metal source	Metal content (%)	Organisms	Process conditions	Leaching yield	References
Low-grade copper ore	Cu—5, Fe—34.5	*At. ferrooxidans, At. Thiooxidans*	Inoculum: 2.2 × 10^7 cfu/mL; pulp density: 10% w/v; pH: 2	45% Cu was leached in 45 days	Wang et al. (2014)
Low-grade metal ore	Cu—0.72, Fe—2.6, Zn—0.02, Ni—0.002	*A. niger*	Carbon sources: sucrose, molasses, potato peels, sawdust, leaves, potato chip waste, corn kernels and husk; pH: 2–4.5	68% Cu was leached in sucrose medium in 13 days	Mulligan et al. (2004)
Electric arc furnace dust	Zn—1.2%,Fe—54.7%	*At. ferrooxidans*	Particle size: 0.038 mm; pulp density: 1% w/v; pH: 1.3	35% Zn and 33% Fe were leached in 21 days	Bayat et al. (2009)
Copper smelter flue dust	Zn—1.67, Cu—22.2, Fe—5.9, Pb—1.5	Mixed consortia of *At. ferrooxidans, At. thiooxidans, L. ferrooxidans*	Pulp density: 2.7% w/v	85.5% Cu was leached in 23 days	Bakhtiari et al. (2010)
Shuikoushan mine tailing, China	Zn—0.12, Cu—0.13, Pb—0.02, Fe—0.47	Enriched sulphur-oxidising bacteria indigenous to site	Sulphur addition: 2% w/v, pulp density: 1% w/v; pH: 2	95% Zn, 95% Cu and 44% Pb were leached in 13 days	Liu et al. (2008)
Janggun mine tailing, South Korea	As—3.3, Zn—2.8, Pb—0.98, Cu—0.27, Fe—13.3	*At. thiooxidans*	Pulp density: 0.5% w/v; pH: 1.8	50% As, 35% Zn and 16% Cu were leached in 25 days	Lee et al. (2015)
San Marcelino tailings, Philippines	Zn—0.01, Mn—0.13, Cu—0.13, Fe—3.61	Pure and binary cultures of *At. thiooxidans and At. ferrooxidans*	Pulp density: 5% w/v; pH: 2	Efficient solubilisation by mixed consortia; 95% Mn, 75% Zn and 40% Cu were leached in 25 days	Nguyen et al. (2015)
Pyritic sludge	Zn—0.13, Fe—44	Bacteria (*At. ferrooxidans, At. caldus*), archaea (*S. metallicus*)	Pulp density: 2.5% w/v; pH: 2; temperature: 65 °C; microbe: *S. metallicus*	90% Zn was leached in 10 days	Hita et al. (2008)
Copper mining residues	Cu—0.72, Fe—2.64, Zn—0.02	*A. niger*	Pulp density: 5% w/v; pre-treatment: ultrasound pre-treatment prior to bioleaching	60% Cu from raw residue and 80% Cu from pre-treated residue in 10 days	Mulligan et al. (1999)
Zinc leach residue	Zn—2.51, Fe—11.5, Pb—2.35	*At. thiooxidans*	Pulp density: 2.1% w/v; sulphur addition: 25 g/l; pH: 3.3	70% Zn in 45 days	Sethurajan et al. (2017a, 2017b)
Red mud	Al—9.4, Ti—2.9, Fe—13.74	*A. niger* (PTCC 5210)	Pulp density: 1% w/v; pH: 2.5	69% Al, 60% Ti, 25% Fe by spent medium in 30 days	Vakilchap et al. (2016)
Copper slag	Zn—1.7, Cu—0.35, Fe—40.7	Mixed cultures of *At. ferrooxidans, At. thiooxidans, At. caldus, L. ferrooxidans, Sulfobacillus thermotolerans*	Pulp density: 5% w/v; sulphur addition: 1% w/v; pH: 1	62% Cu, 35% Zn in 29 days	Kaksonen et al. (2011)
Lead/zinc smelting slag	Zn—14.1, Cd—0.88, Pb—32.4, In—0.09	Pure, binary cultures of *At. thiooxidans*, and *L. ferriphilum*	Pulp density: 10% w/v; pH: 1	90% Zn, 86% Cd and 71% in 9 days	Wang et al. (2015)
Copper smelting slag	Zn—2.7, Cu—0.38, Pb—0.60	Mesophilic mixed cultures of *At. thiooxidans, At. ferrooxidans* and *L. ferriphilum*	Pulp density: 10% w/v; iron addition: 0.45% Fe(II); sulphur addition: 1% w/v; pH: 2.1–2.2	70% Cu, 7% of Zn in 25 days	Kaksonen et al. (2016)

determines the mineralogical and elemental composition of sludge produced. For example, during steel production, the generated sludge has the mineral phases such as magnetite, hematite and wustite. Apart from Fe, most of the metallurgical sludges contain significant levels of Zn, Pb, Cd and As (Mansfeldt & Dohrmann, 2004). Different hydrometallurgical and microwave-assisted processes have proposed for improving the efficiency and economics of extraction of metals from these sludge wastes. Bioleaching is not a preferred method due to the oxidative nature of minerals in these sludges. Studies on the bioleaching of oxidised minerals present in the sludges are mentioned in Table 1. Bioleaching of Zn, Fe and As from sulphidic minerals such as pyritic sludge with iron-oxidising bacteria and archaea was reported through oxidative dissolution process (Hita et al., 2008). Mixed species cultures of *Acidithiobacillus* spp. and *Leptospirillum* spp. were shown to bioleach Zn from steel plant sludge through biogenic acid (sulphuric acid) production (Nguyen et al., 2015). Other oxidised mineral wastes such as residues are generated during different metallurgical processes. These include leach residues (from filtration of acid leached slurry) and purification residues (generated during purification processes like cementation) (Sethurajan et al., 2018). Metallic wastes co-generated during Zn production are termed as Zn plant residues (ZPR). Production of one tonne Zn generates 0.5–0.9 tonne ZPR, and these residues contains high concentrations of zinc along with iron and other impurities (Creedy et al., 2013). Organic acid (gluconic acid and citric acid)-producing *A. niger* was able to bioleach Cu from copper residues (Mulligan et al., 1999). Biogenic sulphuric acid-producing *At. thiooxidans* was able to bioleach Zn from oxidised minerals in ZPR (Sethurajan et al., 2017a, 2017b).

Red mud is another potential waste source for various metal ions. It is alkaline and saline in nature and metals are present in oxidised forms (Vakilchap et al., 2016). Alkaline extraction of alumina from bauxite ore generates large quantities of red mud. Approximately, per tonne of aluminium extraction generates two tonnes of red mud (Vachon et al., 1994). Red mud is poly-metallic in nature (Al, Fe, Si, Ti) and commonly contains different rare earth elements. *Acidithiobacilli* was adapted to grow at red mud pH of 9 for bioleaching. This approach has facilitated release of up to 47% Al from the red mud. Biogenic sulphuric acid produced from *T. thioparus* and *At. thiooxidans* was shown to leach Al from red mud (Vachon et al., 1994). Compared to bacterial leaching, Al leaching from red mud by fungal cultures (*Penicillium simplicissimum* and *A. niger*) through complexolysis was efficient. Additionally, cell-free supernatant of fungal cultures was shown to bioleach metals efficiently from red mud over metal leaching by growing or pre-grown fungal cells (Vakilchap et al., 2016). Smelting operations generate huge quantities of slags (each tonne of Cu produced generates 2.2 tonne salgs) as waste by-products which are commonly piled in heaps or tailings dams (Gorai et al., 2003). Metal and mineral phases of slags are variable and depend on ore type and employed metallurgical processing. Slags are being poly-metallic in nature, and they contain metal(loid)s such as Fe, Al, Ca, Cu, Zn, Pb, Cd and As in reasonable amounts (Gorai et al., 2003; Wang et al., 2015). Pyro- or hydro-metallurgical processes are suitable for metal leaching from slags but require hazardous chemicals and energy (Gorai et al., 2003). Bioleaching by bacterial and fungal cultures is an environmentally benign approach for dissolution of metals from slags. Oxidative bioleaching of slags by chemolithotrophs such as sulphur (*Acidithiobacillus thiooxidans)* and iron-oxidising bacteria (*Leptospirillum ferriphilum*) has been reported (Wang et al., 2015). *Aspergillus niger*-produced organic acids were shown to leach metals from slags (Sukla et al., 1995). Indigenous thermophilic and heterotrophic bacteria such as *Bacillus* spp., *Pseudomonas* spp. and *Sporosarcina* spp. were reported to efficiently bioleach metals from slags over mesophilic heterotrophs (Cheng et al., 2009; Potysz et al., 2015). Bioleaching of metals from slags was demonstrated in lab-scale continuously stirred tank reactors (CSTRs) with a comparable efficiency to that of flask studies under shaking conditions (Kaksonen et al., 2011).

3.2 Biological Treatment, Removal and Recovery of Metals from Contaminated Aqueous Solutions

Mining, mineral processing and metallurgical operations generate various types of metal-laden wastewaters that differ in metal composition and concentration of individual metals. Besides these, metal-contaminated wastewaters are generated from other anthropogenic activities involving batteries, pesticides, rayon industry, tanning industry, textile industry, petrochemicals, manufacturing of paper and electrolysis applications (Qasem et al., 2021). Metal(loid) s can enter the environments through natural weathering, discharging improperly treated wastewater and through accidental leaks is a serious environmental concern. These wastewaters can contaminate the soil, surface, sub-surface aquatic environments and can damage the ecosystem (Kossoff et al., 2014). Immobilisation of metals from these contaminated aqueous streams is essential for environmental protection. Additionally, metal(loid)s recovery from these wastewaters and e-waste leachates is highly desired in the current era of dwindling resources, especially for achieving material sustainability of scarce and critical metals including rare earth elements (Nancharaiah et al., 2016). For achieving removal and recovery of metal(loid)s from wastewaters, diverse physicochemical, electrochemical and filtration processes are proposed (Barakat, 2011). These methods are energy intensive, expensive, require hazardous chemicals and generate secondary wastes. Employing microorganisms for removal and recovery of metal(loid)s from waste streams can be more economic, sustainable and environmentally benign (Wang et al., 2021). As presented in Fig. 1, microbes immobilise metal(loid)s through different mechanisms such as bioprecipitation, biosorption, bioreduction and bioaccumulation. Microbial immobilisation of metals is highly sustainable for the removal and recovery of critical and scarce metals from wastewaters (Fig. 2). Besides heavy metals, more emphasis is given on the removal and recovery of precious, rare earth and scarce elements in this section.

Microbial biosorption of metals employing bacteria, fungi and microalgae has been well studied for biological treatment of leachates and industrial wastewaters. Microbial biosorption efficiency is mainly governed by different parameters such as presence of functional groups, biosorbent (microorganisms) dosage, pH, temperature and retention time (Beni & Esmaeili, 2020). Waste biomass of *A. niger* and immobilised acidophilic heterotrophic bacterial consortium onto shell sand, biopolymers, waste biomasses were found to remove most of the leachate metals at mine waste disposal sites (Petrisor et al., 2002). Near-complete metal removal from AMD was achieved using a calcined eggshell and a microalgae hybrid system as shown in Table 2 (Choi & Lee, 2015). Column bioreactor containing *Candida lipolytica* and sewage sludge was used for treatment of electroplating wastewater with efficient removal of Cr, Cu, Ni and Zn (Ye et al., 2010). Hybrid treatment system comprising chemical precipitation and biosorption was used for treating tannery wastewater. Chemical precipitation and biosorption by *Kitasatospora* sp. were sequentially used for removing Cr(III) and Cr(VI), respectively, for treating tannery wastewater (Ahmed et al., 2016). Besides heavy metals, removal and recovery of rare transition metal such as Ru from industrial effluents through microbial biosorption were explored. Ru-laden industrial wastewater (metal-plating and acetic acid production industries) was subjected to treatment using *Rhodopseudomonas palustris* and *Corynebacterium glutamicum* bacteria. Acid pre-treatment of *R. palustris* cells with acid has significantly improved Ru loading onto bacteria (from 86 to 145 mg/g dry cells) and the preferential Ru selectivity over Ni and Zn in the metal-plating effluents. Precious metal catalysts (Ru, Ir and Rh) are used in commercial production of acetic acid. Composite fibres of polyethylenimine (PEI)-coated *C. glutamicum*/chitosan exhibited 16.5-fold higher Ru loading of 110 mg/g over commercial ion-exchange resin from acetic acid solutions (Kwak et al., 2013; Won et al., 2014). Biosorption followed by biorecovery of precious metals from leach liquors of waste electrical and electronic equipment using microbial biomass has been systematically reviewed by Ilyas and Lee (2014).

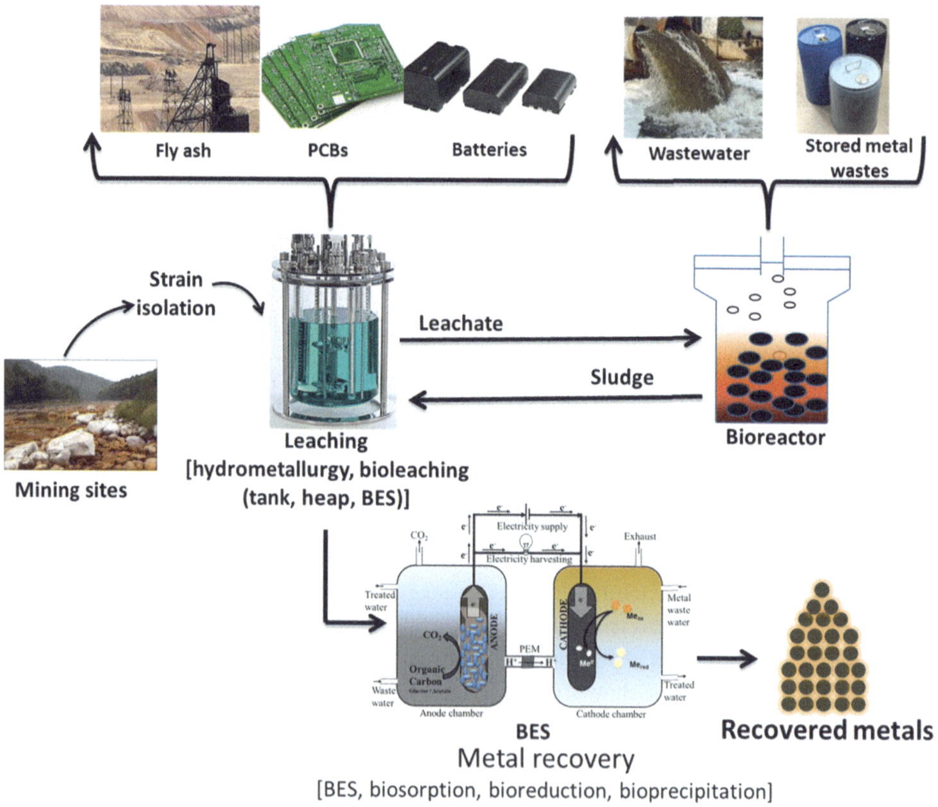

Fig. 2 Integrated processes for extraction and recovery of metal(loid)s from metal-laden solid and liquid wastes

Microbial biosorption for the recovery of rare earth elements such as Dy, La, Nd, Sc, Eu and Yb using pure cultures of *Pseudomonas* sp., *Myxococcus* sp., *Bacillus subtilis*, *E. coli*, *Saccharomyces cerevisiae*, *Aspergillus terreus* and *Rhizopus arrhizus* has been reported (Ilyas & Lee, 2014).

Conversion of toxic, soluble and mobile metal species into less or non-toxic, insoluble forms can be achieved through microbial reduction. Bioreduction of many redox-active metal(loid)s including heavy metals, precious metals and metalloids is summarised in Table 2. Microbial reduction of metals is catalysed by various enzymatic and non-enzymatic processes (Suja et al., 2014; Tripathi & Garg, 2014). Chromium (Cr)-containing waste streams are generated in diverse industrial processes, such as manufacture of stainless steel, electroplating, wood processing, leather tanneries, and in dye production. In natural environments, Cr predominantly exists in hexavalent form (Cr(VI), chromate, soluble and highly toxic) and trivalent form (Cr(III), insoluble and less toxic). Conversion of Cr(VI) to Cr(III) is a potential strategy for treatment of contaminated soils and waters (Reddy & Nancharaiah, 2018). Different single and mixed culture microbes are reported for bioreduction of Cr(VI) (Wang et al., 2021). Self-immobilised biofilms in the form of granular sludge were shown to tolerate and remove elevated Cr(IV) concentrations even in presence of other toxic co-contaminants. Efficient reduction associated with recovery of nearly 92% of supplied 0.75 mM Cr(VI) as Cr(III) precipitates on biomass was reported (Reddy & Nancharaiah, 2018). Granular sludge consisting of mixed microbial consortium was able to reduce highly mobile Co(III) complex to less mobile Co(II) (Reddy et al., 2016). Immobilised cells of *Bacillus cereus* and *Pseudomonas putida* were shown to remove 42% of Cr(VI) from real tannery effluent (Tripathi & Garg, 2014). Besides toxic metals, microbial reduction followed by recovery of precious metals such as Pd(II) as bio-Pd(0) was studied from catalyst leachates of automobiles (Yong et al., 2003). Biogenic nanopalladium on granular sludge (bio-Pd) generated from the reduction of Pd(II) was shown to catalyse p-Nitrophenol and Cr(VI) reduction (Suja et al., 2014). *Desulfovibrio alaskensis* and *Desulfovibrio* spp. were shown to produce biogenic metallic nanoparticles of precious metals (Pt and Pd) (Capeness et al., 2015).

Recovery of scarce metal(loid)s such as selenium and tellurium from process streams and wastewaters is important for pollution mitigation and improving material sustainability (Wadgaonkar et al., 2018). Selenium wastewaters containing selenate (Se(VI)) or selenite (Se(IV)) oxyanions are generated in various industrial activities such as mining activities, metal refining, coal combustion, oil refining along with manufacturing of glass, electronics, steel, semiconductors and photoelectric cells (Nancharaiah & Lens, 2015a, 2015b). Due to high bioaccumulation potential, discharge limit for Se is regulated by the US EPA at 5 μg/L to aquatic bodies (Nancharaiah et al., 2018). Hence, it is necessary to apply efficient treatment methods for treating selenium-contaminated wastewaters. Microbial reduction of soluble forms of Se (selenate or selenite) to insoluble metallic Se is commonly employed for the treatment of Se-containing wastewaters. Biofilm-based full-scale bioreactor (ABMet™) is able to completely remove soluble forms of Se and other wastewater pollutants to safe discharge levels. Microbial reduction of Se oxyanions using bioreactors with suspended and attached microbial growth was previously reviewed (Nancharaiah & Lens, 2015a, 2015b; Nancharaiah et al., 2016). Mixed microbial communities as aerobic granular sludge and anaerobic granular sludge were shown to efficiently reduce and recover Se oxyanions (Nancharaiah et al., 2018; Wadgaonkar et al., 2018). Abundance of tellurium is at 0.00001% in the Earth's crust which makes it a scarce element. Tellurium and its composites are used in solar cells, petroleum refining, electronics, electroplating, optics and glass and sensor industries (Reddy et al., 2023). Leachates and waste streams from these operations generate tellurium-containing wastewaters. Soluble forms of Te (Te oxyanions: tellurite (TeO_3^{2-}) and tellurate (TeO_4^{2-})) are toxic, whereas insoluble metallic form of Te (as, Te(0)) is less toxic. Thus, reduction of Te oxyanions to metallic tellurium is a treatment process for remediation of contaminated wastewaters. With the aid of bacteria and fungi, an integrated biosorption–bioreduction process was applied for recovering Te from used CdTe solar cells. Biosorption of Ag and Cd was achieved by fungal cells, whereas *Pseudomonas mendocina* has reduced and recovered Te as precipitates of Te(0) from CdTe solar cell acid leachates (Rajwade & Paknikar, 2003). Reduction followed by recovery of Te oxyanions as Te(0) nanostructures on anaerobic granular sludge was reported by Wadgaonkar et al. (2018).

In bioaccumulation, metabolically active microbes uptake the metals and accumulate intracellularly. Structural, physiological and genetic factors of microbes determine bioaccumulation potential besides speciation, bioavailability of metals. Bioaccumulation of Cu and Zn by bacteria isolated from wastewater-irrigated soils has been reported by Ahmed and Malik (2011). Continuous exposure of metals in wastewater-enriched metal accumulating bacteria in the soil and the isolated strains was shown to have enhanced tolerance to different heavy metals (Ahmed & Malik, 2011). Similarly, different actinomycetes strains (*Streptomyces* sp.) isolated from abandoned mines were shown to tolerate heavy metals (Pb, Cr, Zn and Cu) and were able to accumulate Pb (El Baz et al., 2015). Apart from toxic heavy metals, specific bioaccumulation of rare earth elements such as dysprosium (Dy) by acidophilic fungal strain, *Penidiells* sp. was isolated from

Table 2 Different microorganisms (bacteria, microalgae, fungi) and microbial processes investigated for removing metals from wastewaters

Metal (mg/l or mM)	Metal source	Microbe/mixed culture/composites	Mechanisms	Removal efficiency	References
Bacteria-based treatment					
Fe(II) (5000), As(III) (1)	Synthetic acid mine drainage	*A. ferrooxidans*	Bioprecipitation	53.6% Fe, 17.2% As	Song et al. (2018)
Cr(VI) (625), Total Cr (2250)	Tannery wastewater	Combined chemical-biological treatment; *Kitasatospora* sp.	Chemical precipitation, biosorption	99.3% total Cr, 98.4% Cr(VI)	Ahmed et al. (2016)
Ru (1823)	Ru-acetate	Bacterial biomass/chitosan composite fibres	Biosorption	Loading of 110.5 mg/g for polyethylenimine-modified composite	Kwak et al. (2013)
La (100)	$LaCl_3$	Immobilised *P. aeruginosa*	Biosorption	Loading of 1 mM La/g biomass	Philip et al. (2000)
Cr(VI) (108)	Tannery effluent	Immobilised co-cultures of *P. putida* and *B. cereus*	Bioreduction	52% Cr(VI)	Tripathi and Garg (2014)
Co(III) (up to 5 mM)	$[Co(III)\text{-}EDTA]^-$	Denitrifying granular sludge	Bioreduction	More than 99% in 5 days	Reddy et al. (2016)
Pd(II) (1 mM)	Na_2PdCl_4	Aerobic granular sludge	Bioreduction	100% in 10 min by biogenic H_2 of sludge	Suja et al. (2014)
Pt(IV) (2 mM)	$PtCl_4$	*Desulfovibrio alaskensis* G20	Bioreduction	Pt(IV) reduction to Pt(0)	Capeness et al. (2015)
Se(IV) (12.7)	Na_2SeO_3	Aerobic granular sludge bioreactor	Bioreduction	About 82–98.6% removal of Se(IV)	Nancharaiah et al. (2018)
Te(IV) (10)	Acid leachate of used CdTe solar cell	*Pseudomonas mendocina MCM B-180*	Bioreduction	Complete reduction of tellurite and > 99% recovery as tellurium	Rajwde and Paknikar (2003)
Zn (3.2 mM), Cu (5.8 mM)	$ZnCl_2$; $CuSO_4$	*Pseudomonas* sp.	Bioaccumulation	Up to 26 mg Zn/g dry cells Up to 30 mg Cu/g dry cells	Ahmed and Malik (2011)
Pb (500)	$Pb(NO_3)_2$	*Streptomyces* sp. BN3	Bioaccumulation	615 mg Pb/g dry cells	El Baz et al. (2015)
U	$UO_2(NO_3)$, uranyl-carbonate	*Bacillus* sp., *Rahnella* sp., *Citrobacter* sp., *Pseudomonas* sp., *Serratia* sp., *Deinococcus radiodurans*	Bioprecipitation	U was bioprecipitated as hydrogen uranyl phosphate by diverse microbes in acidic and alkaline conditions at varying removal efficiencies	Newsome et al. (2014), Paterson-Beedle et al. (2010) and Kulkarni et al. (2013)
Microalgae-based treatment					
Fe, Cu, Zn, Mn, As, Cd	Acid mine drainage	Calcined eggshell and *Chlorella vulgaris* hybrid system	Biosorption	99.7% Fe, 99.5% Cu, 99.9% Zn, 99.8% Mn, 100% As, 100% Cd in 6 d	Choi and Lee (2015)
Fungi-based treatment					
Cr (5–30), Cu (0.14–21), Ni (0.2–30), Zn (0.2–28)	Electroplating wastewater	Mixture *of Candida lipolytica* and sewage sludge	Bioreduction, biosorption	96–100% Cr, 76–100% Cu, 67–100% Ni, 88–100% Zn	Ye et al. (2010)
Dy	$DyCl_3$	*Penidiella* sp. T9	Bioaccumulation	0.91 mg Dy/mg dry cells	Horiike and Yamashita (2015)

abandoned mine through microbial enrichment. This strain accumulated most of the REEs except Sc, even in the presence of divalent and trivalent metal cations. Interestingly, this stain showed very high Dy accumulation capacity of 0.91 mg Dy/mg dry cells at an optimum pH of 2.5 making it suitable for Dy recovery (Horiike & Yamashita, 2015).

Microbial bioprecipitation of metals as sulphides, hydroxides, phosphates and carbonates is one of the conventional methods employed for treating metal-laden wastewaters. Removal and recovery of metals as metal precipitates serve the treatment as well as resource recovery from mine drainages and industrial wastewaters (Sethurajan et al., 2018). Biogenic sulphide produced through the metabolism (assimilatory/dissimilatory sulphate reduction) of sulphate reducing bacteria can precipitate heavy metals present in the wastewater and remove them from solution (Thauer et al., 2007). Metal sulphide precipitation (MSP) occurs through two different steps involving production of biogenic H_2S followed by precipitation as metal sulphides. Different factors such as type of electron donor, concentration of electron donor, source of sulphate, electron donor-to-sulphate ratio, pH and metal-to-sulphide ratio influence MSP (Esposito et al., 2006). Metal precipitation through biogenic sulphide has been reported for the treatment of synthetic bioleachate and AMD (Cao et al., 2009; Song et al., 2018), industrial wastewater (Kosinska & Miskiewicz, 2012), Zn leach residues (Sethurajan et al., 2017a, 2017b) and C metallurgical slags of Cu (Kaksonen et al., 2011). As solution pH determines the metal sulphide formation, a selective and sequential precipitation of metals in a mixed-metal wastewater can be achieved by varying the system pH (Sethurajan et al., 2018). Microbial mediated bioprecipitation of soluble uranium (U(VI)) as insoluble hydrogen uranyl phosphate has received the attention of many researchers for U removal and recovery from wastewaters. Different bacterial strains as mentioned in Table 2 are reported for U bioprecipitation (Kulkarni et al., 2013; Newsome et al., 2014; Paterson-Beedle et al., 2010).

3.3 Bioelectrochemical Systems for Removal and Recovery of Metals

In recent decades, bioelectrochemical systems (BES) have emerged as most sustainable systems for coupling wastewater treatment to resource recovery (Nancharaiah et al., 2015). In BES, microbial oxidation of organic substrate present in water or wastewater generates energy in the form of reducing equivalents (e^- and H^+). These microbiologically released electrons facilitate development of potential difference across external electrodes (anode/cathode) which leads to bioelectrogenic activity. Depending on BES configuration and purpose, BESs are classified into microbial fuel cells (MFCs) for electricity generation, bioelectrochemical treatment (BET) for waste treatment and microbial electrolysis cells (MECs) for synthesis of value-added products/hydrogen generation (Venkata Mohan et al., 2014a, 2014b). Although they share similarities in anode, classification of BESs is driven by different cathodic reactions. In literature, all these configurations are interchangeably used under different names such as BES, BET, MEC, microbial electrochemical technology (MET) (Nancharaiah et al., 2015; Venkata Mohan et al., 2014a, 2014b).

The commonly built BES configurations consist single- or two-chamber systems. In dual-chamber BES, partition between anode and the cathode compartments is achieved by proton-exchange membrane (PEM or cation-exchange membrane (CEM)). Anaerobic oxidation of organics by anaerobic bacteria present in wastewater generates e^- and H^+ in anodic compartment. These electrons are shunted to anode, and through external circuit, electrons are carried to cathode. In the cathode chamber, these electrons are ultimately consumed by the TEAs. In a typical MFC, oxygen serves as the TEA. Besides, wastewater can contain other oxidised TEAs such as nitrates, sulphates, redox metals or dyes (Nancharaiah et al., 2015). Distinct cathode compartment does not exist in single-chamber BES, and PEM may or may not be included. From the sidewall of anode compartment, TEA (oxygen from atmosphere) as well as protons passes through the porous cathode (Reddy & Nancharaiah, 2019). Two-chamber BESs under different configurations have been considered for metal removal and recovery of metals in the cathode chamber (Fig. 3).

The configuration of anode/bioanode of a BES is similar in performing oxidation of organic substrates and transferring electrons to the solid anode. However, anodes of different BESs differ in terms of microbial community depending on substrate type, substrate concentration, other pollutants and operating conditions. In BES configuration with biocathode, microbes accept the electrons from the cathode and transfer to other TEAs present in cathode chamber. However, electrons transferred to microbes from the cathode do not have any energy advantage as these electrons may not serve in gaining energy (Rosenbaum et al., 2011). Bacteria employ different mechanisms of extracellular electron transport (EET) for transferring electrons to the anode and to accept electrons from the cathode. EET mechanisms include (i) direct contact, in which c-type cytochromes in outer cell membrane mediate shunting electrons to the solid anode, (ii) electron shuttle-mediated (soluble electron shuttles or redox mediators (flavins, phenazines, quinines, etc.)) shunt electrons between cells and electrodes and/or (iii) conductive pili-mediated (conductive matrix of biofilm also called nanowires shunt electrons to solid electrodes). On biocathodes, outer membrane c-type cytochromes and electron shuttles are proposed for mediating EET from cathode

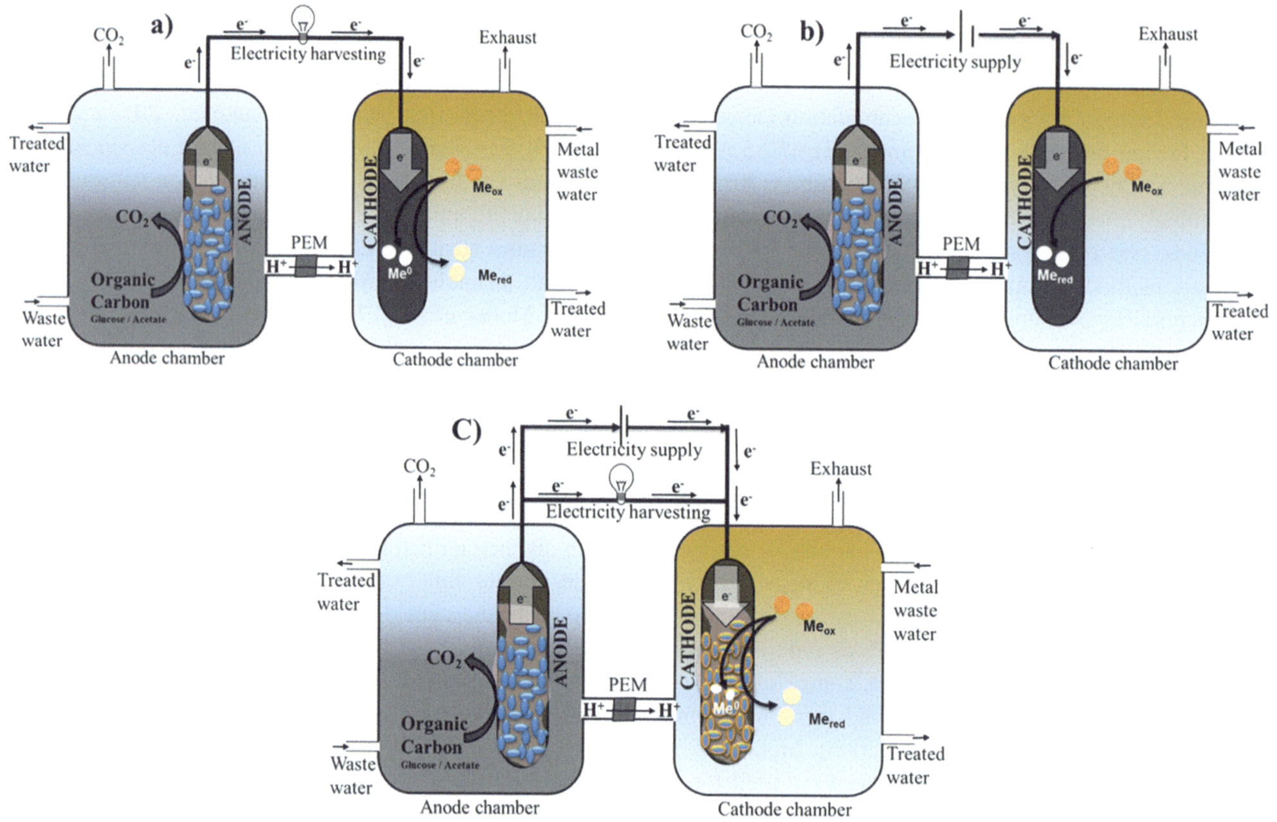

Fig. 3 Different configurations of two-chamber bioelectrochemical systems used for metal removal and recovery. **a** Microbial fuel cell, **b** microbial electrolysis cells, **c** microbial fuel cells and microbial electrochemical cells with biocathodes

to microbial cells (Nancharaiah et al., 2015). Organisms use any of these mechanisms either singly or in combination for achieving EET. Different EET mechanisms operating in BES are depicted in Fig. 4.

The sum of half-cell reactions at the anode and the cathode determines net redox potential which drives the feasibility of metal transformations and subsequent removal or recovery in BESs. MFCs with net positive redox potential (mainly due to positive redox potential at cathode), spontaneous flow of electrons from the anode to the cathode are thermodynamically favourable and associated with electricity generation. In half-cell cathode reactions with lower redox potential (generally, negative redox potential), the spontaneous flow of electrons is thermodynamically not favourable and requires an additional input of electricity to force the flow of electrons as in MEC. Generally, acetate or glucose (to simulate sewage or organic laden wastewater) is used as the model organic substrate for anaerobic microbial oxidation at the anode and has a redox potential of −0.289 and −0.41 V, respectively. Oxidation of these organic substrates in the anode chamber can be coupled to the reduction of metal(loid)s such as Co(III), Cr(VI), Au(III), Pd(II), Hg(II), Ag(I), Se(IV) and Cu(II) at the cathode in MFCs. Therefore, the reductive transformation of these metal(loid)s can be done without additional power input as these transformations have positive redox potential and the net sum of two half-cell reactions is positive. However, reduction of certain metals requires input of electricity and requires operation of BES in MEC mode. For example, reduction of metals such as U(VI), Pb(II), Co(II), Ni(II), Cd(II) and Zn(II) has negative redox potential which requires the supply of electricity to continue the flow of electrons (Amanze et al., 2022; Nancharaiah et al., 2015). A compilation of half-cell reactions along with redox potentials for the metals studied in BESs is summarised by Nancharaiah et al. (2015).

Table 3 provides a summary of various studies that evaluated the potential of different configurations of BESs for removing various metal(loid)s from aqueous solution. Reductive precipitation of Cr(VI) has been well studied in different configurations of BES and under different operating conditions. The positive redox potential of +1.33 V (versus standard hydrogen electrode, SHE) for Cr(VI)/Cr(III) couple makes the reduction favourable in MFCs with current generation. Complete removal of 100 mg/l Cr(VI) coupled to anaerobic oxidation of acetate in two-chamber MFC was reported for the first time by Wang et al (2008). Compared to abiotic cathode, enhanced removal of

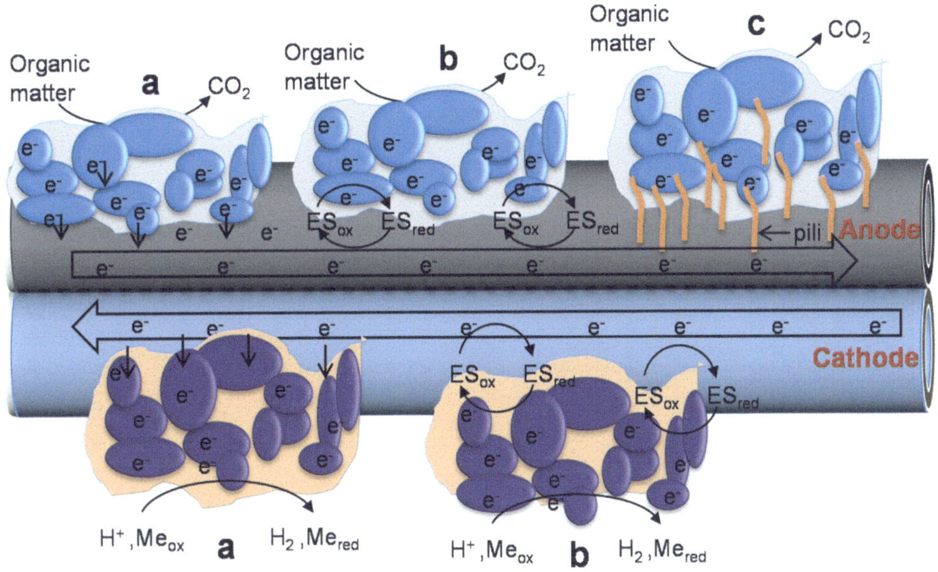

Fig. 4 Depiction of various extracellular electron transport mechanisms operating between microbes and electrodes (modified and redrawn from Nancharaiah et al., 2015)

Cr(VI) by a factor of 2.9 was reported when acclimatised biocathode with exoelectrogenic biofilm was enriched on the electrode (Wu et al., 2015). Influence of different cathodes and Cr(VI) concentrations on net power density along with concomitant Cr(VI) removal in MFC was studied (Li et al., 2008). Compared to carbon brush and carbon felt electrodes, MFCs with carbon cloth cathode displayed better performance for bioelectricity generation and Cr(VI) removal (as $Cr(OH)_3$). Maximum power density of 1221.91 mW/m^2 was achieved in MFC operated with 120 mg/l Cr(VI) and was higher by 1.1 and 1.4 times than that of MFCs operated with 80 mg/l and 50 mg/l Cr(VI), respectively.

The rare metal, Co, is produced as a by-product of Cu and Ni mining. It is a critical metal with a potential supply risk. It finds different strategic, military and industrial applications. Co is used in alloys of aircrafts' engines and orthopedics. Lithium cobalt oxide ($LiCoO_2$)-based lithium ion batteries (LIBs) are commonly used in many of the electronic gadgets which constitute 5–10% of Co. Used LIBs are a potential secondary source for Co. The redox potential of half-cell reaction for Co(III)/Co(II) is +1.61 V which makes Co(III) as an efficient TEA for MFC operation. A MFC with acetate as electron donor in the anode chamber and $LiCoO_2$ deposited carbon felt as the cathode was constructed to monitor Co(III) reduction followed by leaching as Co(II) (Huang et al., 2013). Under these MFC conditions, anaerobic oxidation of acetate at the anode was effectively coupled to the reduction of Co(III) on the cathode to Co(II) followed by leaching into catholyte. Compared to chemical leaching, Co(II) leaching in MFC was higher by about 3.4 times. The leached Co(II) can be recovered by further reduction into metallic Co(0) form. However, Co(II)/Co(0) has negative redox potential, and spontaneous electron transfer is not possible. To overcome this, MECs are studied by applying external voltage and to recover Co(0) (Jiang et al., 2014). The above two systems are integrated to make self-driven MFC-MEC system for the complete recovery of Co from $LiCoO_2$ (Huang et al., 2014).

Copper (Cu) is a micronutrient for living organisms. However, it is potentially toxic at slightly higher concentrations. Mining and metallurgical operations, wire and copper polishing industries produce copper-containing waste streams. Physicochemical processes such as adsorption, precipitation and electrochemical reduction are commonly practised to recover Cu from Cu-rich waste streams. The positive redox potential for Cu(II)/Cu(0) half-cell reactions spontaneously drives the cathodic reduction when coupled to acetate/glucose oxidation in the anode chamber. For Cu(II) reduction, catholytes such as copper chloride, copper sulphate and fly ash leachate were used, and the solution pH was adjusted and maintained at acidic conditions to avoid precipitation of Cu(II) ions. Under MFC conditions, a bipolar membrane was used between the anode and the cathode compartments to restrict the entry of protons into the anode compartment and to keep pH differences intact. With $CuCl_2$ catholyte having 1000 mg/l Cu(II), the MFC was able to remove near-complete Cu(II) at pH 3 over a period of 6–7 d of operation. Reduced Cu(II) was deposited as metallic Cu(0) on the cathode (ter Heijne et al., 2010). With $CuSO_4$ as the catholyte (50–6412 mg/l Cu(II)) at pH 4.7, the glucose oxidation-driven MFC was able to remove > 99% of Cu(II) at a concentration of 196 mg/l (Tao et al., 2011). When fly ash leachate was used as catholyte, 36 h MFC operation yielded more than 97% Cu(II) removal efficiency with no effect on removal of other fly ash metals such as Pb(II) and Zn(II) (Tao et al., 2014). Modified cathodes and their influence of Cu(II) reduction, power output

Table 3 Metal(loid) removal and recovery in different configurations of bioelectrochemical systems

Metal(loid)	BES configuration, electrode materials	Electron donor, concentration	Metal salt; metal concentration; catholyte pH	Metal removal efficiency	Maximum power output (W/m^2) in MFC or voltage applied in MEC	References
Chromium	tMFC, graphite plates for anode and cathode	Sodium acetate, 2.64 g/l	$K_2Cr_2O_7$; 25, 50, 100 and 200 mg/l Cr(VI); pH 2–6	100% from 100 mg/l in 150 h	0.15 W/m^2 at 200 mg/l Cr(VI) and pH 2	Wang et al. (2008)
Chromium	tMFC, graphite felt for anode and cathode. Biocathode with exoelectrogenic biofilm	Glucose, 1 g/l	$K_2Cr_2O_7$; 6–20 mg/l Cr(VI); pH 7	79.3% in 24 h	–	Wu et al. (2015)
Chromium	tMFCs, carbon felt as anode and three different cathodes (carbon cloth, carbon brush, carbon felt)	0.75 g/l acetate	$K_2Cr_2O_7$; 50–120 mg/l Cr(VI); pH 2–7	100% from 120 mg/l Cr(VI) at pH 2 in 132 h with carbon cloth cathode	1.221 W/m^2 at 120 mg/l Cr(VI) with carbon cloth cathode at pH 2	Li et al. (2008)
Chromium	tMFC; carbon cloth anode and cathode	Sodium acetate, 1 g/l	$K_2Cr_2O_7$, 1000 mg/l Cr(VI); pH 3	98% in 8 days	0.138 W/m^2	Aiyer (2020)
Cobalt	tMFC, graphite felt for anode and cathode	Sodium acetate, 0.38 COD/l	$LiCoO_2$ particles; solid/ liquid ratios of 50–1000 mg/l (w/v); pH 1–3	99.1% from solid/liquid ratio of 50 mg/l in 48 h	–	Huang et al. (2013)
Cobalt	tMEC, graphite brush anode, graphite felt cathode	Sodium acetate, 1 g/l	$CoCl_2$; 874 µM Co(II); pH 3.8–6.2	92% in 6 h	0.3–0.5 V	Jiang et al. (2014)
Cobalt	Coupled tMFC-tMEC, graphite felt anodes and cathodes except carbon rod cathode in MEC	Sodium acetate, 1 g/l in each of anodes	MFC: Co(II); 50 mg/l Co(II); pH 6, MEC: $CoCl_2$; 20 mg/l Co(II); pH 6.2	MFC: 7 mg/l/h Cu(II); MEC: 88% in 6 h	Co(II) leaching MFC and 0.2 V input in MEC	Huang et al. (2014)
Copper	tMFC, graphite plate for anode, graphite foil pressed on Ti plate for cathode	Sodium acetate, 1.64 g/l	$CuCl_2$; 1 g/l Cu(II); pH 3	99.88% in anaerobic cathode in 6 d; 99.95% in aerobic cathode in 7 d; residual Cu below 1.2 mg/l Cu	0.43 and 0.8 W/m^2 under anaerobic and aerobic cathodes, respectively	ter Heijne et al. (2010)
Copper	tMFC, graphite plates for anode and cathode	Glucose, 5 g/l	$CuSO_4$; 50, 200, 500 or 1000 mg/l Cu(II); pH 4.7	>96% from 200 mg/l Cu(II) in 264 h	0.339 W/m^3 at 6412 mg/l Cu(II), pH 4.7	Tao et al. (2011)
Copper	tMFC; carboncloth anode and cathode	Sodium acetate, 1 g/l	$CuCl_2$; 1000 mg/l Cu(II); pH 3	98% in 8 days	0.108 W/m^2	Aiyer (2020)
Copper	tMFCs, graphite anode, three different cathodes (graphite, polypyrrole-graphite, graphene oxide/polypyrrole composite graphite)	Glucose, 2 g/l	50, 500, 1000 mg/l Cu(II); pH 3–4	98.88% at 1000 mg/l with graphene oxide/polypyrrole-modified graphite in 200 h	600.25 mW/m^3 at 1000 mg/l with graphene oxide/polypyrrole-modified graphite	Rikame et al. (2021)
Gold	tMFC, carbon brush anode, carbon cloth cathode	Sodium acetate, 1 g/l	$AuCl_4^-$; 100–200 mg/l Au(III); pH 2	99.8% from 200 mg/l; residual 0.22 mg/l Au(III)	6.58 mW/m^2 for 2000 mg/l Au(III) at pH 2	Choi and Hu (2013)
Silver	tMFC, carbon brush anode, carbon cloth cathode	Sodium acetate, 1 g/l	$AgNO_3$; 50, 100, 200 mg/l Ag(I); pH 7	99.9% of 50 mg/l in 8 h; 0.05 mg/l remaining	4.25 mW/m^2 at 1000 mg/l, pH 7	Choi and Cui (2012)

(continued)

Table 3 (continued)

Metal(loid)	BES configuration, electrode materials	Electron donor, concentration	Metal salt; metal concentration; catholyte pH	Metal removal efficiency	Maximum power output (W/m^2) in MFC or voltage applied in MEC	References
Silver	tMFC, graphite plate anode, graphite felt cathode	Sodium acetate, 1.28 g/l	$AgNO_3$; 50, 100 or 200 mg/l Ag(I); pH 2, 4, 6.6; $[AgS_2O_3]^-$	95% in 36 h	0.109 W/m^2	Tao et al. (2012)
Silver	tMFC, graphite fibre brush anode, graphite felt cathode	Sodium acetate, 1 g/l	$AgNO_3$; 50–500 mg/l	83% removal and 67.8% recovery at 500 mg/l in 72 h	3006 mW/m^3	Ali et al. (2019)
Selenium	sMFC, carbon cloth anode, coated carbon cloth cathode	Acetate or glucose	SeO_3^{2-}; 50–200 mg/l Se(IV); pH 7	99% of 75 mg/l in 48 h in acetate-fed MFC	2.90 W/m^2 at 25 mg/l Se(IV)	Catal et al. (2009)
Selenium	tMFC, graphite anode and cathode	Glucose, 3 g/l	SeO_3^{2-}; 25, 51, 102 mg/l Se(IV); pH 7	73.6% Se(IV) removal and 26.4% Se(0) recovery at 102 mg/l Se(IV) in 72 h	0.034 W/m^2 at 25 mg/l Se(IV)	Sravan et al. (2020)
Tellurium	tMFC, graphite anode and cathode	Glucose, 3 g/l	TeO_3^{2-}; 11, 22, 44 mg/l Te(IV); pH 7	54.7% Te(IV) removal and 45.3% Te(0) recovery at 44 mg/l Te(IV) in 72 h	0.044 W/m^2 at 44 mg/l Te(IV)	Sravan and Mohan (2022)
Mercury	tMFC, graphite felt anode, carbon paper cathode	Sodium acetate, 0.82 g/l	$HgCl_2$; 25, 50, 100 mg/l Hg(II); pH 2	98.2–99.5% in 10 h; 0.44–0.69 mg/l	0.433 W/m^2 at 100 mg/l, pH 2	Wang et al. (2011)
Vanadium	sMFC coupled to BER, carbon felt electrodes except plain carbon paper cathode in sMFC	Glucose, 0.75 g/l	V(V); 75 mg/l	93.6% in 12 h	0.543 W/m^2	Hao et al. (2015)
Vanadium	tMFC, carbon fibre felt for anode and cathode	Glucose, 0.81 g/l; sulphide, 100 mg/l	$NaVO_3$; 500 mg/l V(V); pH 2	25% in 72 h	0.572 W/m^2 at 500 mg/l, pH 7	Zhang et al. (2009)
Vanadium	tMFC; carbon cloth anode and cathode	Sodium acetate, 1 g/l	$NaVO_3$; 1000 mg/l V(V); pH 3	Reduced to around 20 mg/l in 8 days	0.094 W/m^2	Aiyer (2020)
Cadmium	tMEC, carbon brush anode, carbon cloth cathode	Sodium acetate, 1 g/l	$CdSO_4$; 50, 100, 200 mg/l Cd(II); pH 6	93.6% from 50 mg/l Cd(II) in 60 h	Cr(VI) reducing MFC	Choi et al. (2014)
Nickel	tMEC, carbon felt anode, stainless steel cathode	Sodium acetate, 1 g/l	$NiSO_4$; 50–1000 mg/l Ni(II); pH 5	99% of 50 mg/l Ni(II) in 20 h	0.9 V	Qin et al. (2012)

tMFC two-chamber microbial fuel cell, *sMFC* single-chamber microbial fuel cell, *tMEC* two-chamber microbial electrolysis cell, *BER* bioelectrical reactor in MFC with glucose substrate as anodic fuel were studied by Rikame et al. (2021). Compared to bare graphite and polypyrrole-modified graphite cathode, graphene oxide/polypyrrole composite graphite as the cathode showed enhanced Cu(II) removal and high-power output.

Precious metals (e.g. Au, Ag, Pt and Pd) are commonly used in jewellery and medical applications. Scrap and waste streams are the potential secondary sources for removal and recovery of these precious metals. With acetate as the fuel in the anode chamber, Au(III) reduction and removal as Au(0) in the cathode chamber from tetrachloroaurate were studied in a dual-chamber MFC. Au(III)/Au(0) has a positive redox potential of +1.002 V to favour the spontaneous flow of electrons from the anode to the cathode. Relatively fast and efficient reduction of Au(III) followed by deposition as Au(0) on the cathode at tested concentrations of 50 and 100 mg/l was reported. At these respective concentrations, the removal efficiencies were 97.8% and 94.6% in 12 h of operation. Simultaneously, maximum power output of 0.89 W/m^2 was achieved in MFC at 100 mg/l Au(III) (Choi & Hu, 2013). Ag(I) is a precious metal and used in electronic and photographic applications besides its common use in ornaments. As the half-cell reaction of Ag(I)/Ag(0) has positive redox potential (+0.799 V), Ag(I) can be a potential TEA at the cathode in MFC. With $AgNO_3$ as the Ag(I) source, >99% removal was observed in a dual-chamber MFC operated at 50–200 mg/l in 12 h with acetate as the anodic fuel. Reduced Ag ions were deposited as Au(0) metallic crystals on the cathode (Choi & cui, 2012). Similarly, recovering Ag from silver thiosulphate complex (simulated for photographic wastewater) was feasible in MFC using acetate as organic substrate. Varying catholyte pH from 2 to 6.6 did not influence Ag(I) reduction. Silver sulphite reduction in the cathode chamber indicated deposition of metallic Ag(0) along with traces of acanthite (Ag_2S), indicating thiosulphate reduction in MFC (Tao et al., 2012). In another study, silver-laden synthetic wastewater was tested for recovering Ag as metallic silver nanoflakes in MFC using acetate as organic substrate. In 72 h of MFC operation, 83% removal and 67.8% recovery at 500 mg/l Ag(I) were observed in the cathode chamber. Silver nanoflakes were deposited on cathode which inhibited growth of bacterial strains and biofilm formation from activated sludge indicating potential antimicrobial and antibiofouling activities (Ali et al., 2019).

The metalloids selenium and tellurium are used in diverse industrial applications, and their oxyanion forms are potentially toxic. Removal of these elements from process streams and wastewaters is a treatment approach. Recovery of these metalloids from wastewaters is highly sustainable owing to their scarcity and unequal global distribution (Nancharaiah & Lens, 2015a, 2015b). The positive redox potential of +0.41 V for Se(IV)/Se(0) couple indicates the potential of Se(IV) as TEA in MFC. Selenite removal in a single-chamber air cathode MFC was studied by Catal et al. (2009) employing acetate or glucose as electron donor. MFC operation of 48 and 72 h was able to completely reduce respective selenite concentrations of 75 and 200 mg/l. Precipitates in bright red colour were observed on cathode, anode and in solution indicated the successful reduction of selenite in MFC (Catal et al., 2009). In another study, a dual-chamber MFC was operated with biotic anode and abiotic cathode for selenite removal and recovery (Sravan et al., 2020). With glucose as the anodic fuel for electrons, reduction and removal of selenite were observed in cathodic chamber. With increase in concentrations from 25 to 103 mg/l selenite, increased reduction and removal were observed. However, there was a reduction in power output with increase in concentration of Se(IV) in MFC which was reported. The similar setup was also studied for removal and recovery of tellurite oxyanions, a toxic form of tellurite metal(loid). Tellurite (TeO_3^{2-}) as the TEA for its reduction and removal in cathode chamber was evaluated in a two-chamber MFC with glucose as the electron donor in anode chamber (Sravan & Mohan, 2022). With increase in concentration of Te(IV) from 11 to 44 mg/l, increased tellurite removal and recovery as Te(0) were observed along with concomitant increase in power density. The results indicated the possibility of MFCs for the treatment and recovery of selenite and tellurite-contaminated wastewaters.

Mercury is an another trace element which finds diverse applications. Hg-laden wastewaters are produced in burning of fossil fuels, use of mercury-based antifungals and use as catalysts. Many microbes are known to reduce soluble Hg(II) to insoluble metallic Hg(0), and microbial reduction is a remediation strategy to remove and immobilise Hg in contaminated wastewaters. Redox potential of Hg(II)/Hg(0) couple is positive at +0.91 V and can be a potential TEA in MFC. Two-chamber MFC with acetate as the anodic fuel and $HgCl_2$ as catholyte, rapidly removed Hg(II) up to 100 mg/l under anaerobic conditions. The reduced mercury was deposited as Hg(0) on cathode and also as Hg_2Cl_2 in cathode chamber (Wang et al., 2011).

Vanadium is extensively used as a catalyst in petroleum refining, metallurgy and in production of phthalic anhydrides. Vanadium toxicity to biota necessitates the treatment of contaminated water. In wastewaters, vanadium mainly exists in pentavalent state (V(V)) which is soluble and toxic. Reduction of V(V) to V(IV) results in precipitation and reduces the mobility and toxicity (Hao et al., 2015). The reduction of V(V) to V(IV) was studied with sulphide as electron donor and V(V) as TEA in a two-chamber MFC. Simultaneous removal of two pollutants (sulphide and vanadium) was achieved in this MFC and the respective removal efficiencies were 82% (sulphide) and 26% (vanadium) (Zhang et al., 2009). In another study, a single-chamber

MFC powered by glucose was coupled to V(V) fed bioelectrical reactor (BER). The energy and electrons produced in MFC were coupled to efficient reduction of V(V) in the BER. This kind of designs can be used for in situ treatment of V(V) containing ground water with the bioelectricity produced from MFCs (Hao et al., 2015).

For removing metals such as cadmium and nickel, BES has to be operated in MEC mode. For example, Cd(II)/Cd(0) and Ni(II)/Ni(0) couples have a negative redox potential of -0.40 V and −0.25 V, respectively. The movement of electrons from high potential anode to low potential cathode is aided by a small amount of external power input (Nancharaiah et al., 2015). In a study by Luo et al. (2014), 0.7 V was applied to remove Ni(II) in MEC. From a catholyte solution having 500 mg/l Ni(II), 94% was removed in 40 h. Another study with 500 mg/l Ni(II) at pH 5 showed 51–67% removal efficiencies at an applied voltage of 0.5–1.1 V (Qin et al., 2012). However, modification of existing MFCs and use of biocathodes are explored to reduce and recover these metals without any external power input (Choi et al., 2014; Singh & Kaushik, 2022).

Table 4 provides a summary of studies that evaluated standalone BES or integrated BES for recovering metal(loid)s from multi-metal solutions or real waste streams. Studies on BES are extended for removing and recovering metals from simulated multi-metal wastewaters and real wastewaters (Ai et al., 2020; Amanze et al., 2022; Li et al., 2008; Lim et al., 2021; Zhang et al., 2020). Some of the BESs are integrated with other electrolysis reactors or thermoelectric generators for achieving complete and efficient recovery of multiple metals present in wastewater (Ai et al., 2020; Tao et al., 2014). BES coupled with electrolysis reactor was shown to recover metals (Zn, Pb and Cu) from fly ash leachate (Tao et al., 2014). With the input of external power, near-complete (97%) removal of Cu(II) was observed in BES. The effluent from BES having Zn and Pb is fed to a conventional electrolysis reactor with the energy input to remove 95.4% of Zn(II) and 98.1% of Pb(II). Integration of thermoelectric generator with MFC was attempted for sequential recovery of Cu, Cd and Co from synthetic smelting wastewater on cathodic surfaces (Ai et al., 2020). Cu(II) was bioelectrochemically recovered as Cu(0). While Cd(II) was recovered as $Cd(OH)_2$, $CdCO_3$ and Co(II) were recovered as $Co(OH)_2$ by electrodeposition on cathode surface. BES alternatingly used as both MFC and MEC was tested for removal and recovery of mixed-metals having copper, chromium, cadmium. The MFC was stabilised with three different strains of *Castellaniella* sp. which are shown to be strong exoelectrogenic organisms for efficient anodic activity. Compared to independent experiments with these three strains, use of mixed culture in anode chamber had better electrogenic activity and efficient metal removal. Cr(II) and Cu(II) were completely removed in BES operated as MEC and the remaining Cd(II) in catholyte was removed with an applied voltage of 1.2 V (Amanze et al., 2022). These studies revealed the flexibility of using single BES as MFC and MEC systems for simultaneous removal of mixed-metals present in wastewater. In a bioelectrochemical assisted electrodeposition system, selective and separate recovery of mixed-metals (Pb and Zn) onto different electrodes in cathode chamber by varying the cathodic potential was reported (Zhang et al., 2020).

In addition to removal of metals in multi-metal studies, BES is also studied for treatment of real industrial wastewater. A dual-chamber MFC fuelled by acetate was operated to remove Cr(VI) from real electroplating wastewater along with simultaneous electricity production. With 204 mg/l Cr(VI) in wastewater, almost complete removal of Cr(VI) and 66% recovery of total Cr were observed (Li et al., 2008). In another study with real electroplating wastewater having very high concentrations of Cr(VI) (around 2070 mg/l), MFC was operated to study removal of Cr(VI). Wastewater was diluted from 5 to 50% to get Cr(VI) concentrations in the range of 93–998 mg/l. With increase in concentrations of Cr(VI), high-power output was observed. However, the Cr(VI) removal efficiencies were reduced to 53% at 998 mg/l as compared to 100% at 183 mg/l (Kim et al., 2017). In a recent study, real industrial wastewaters with Zn(II) concentrations ranging from 1.2 to 1.8 mM along with other minor metal(loid)s such as Na(I), Mg(II), Ca(II), Pb(II) and Si(IV) were effectively treated using a acetate-powered MFC. Near-complete removal of Zn(II) was reported with a recovery of 42% as zinc precipitates (Lim et al., 2021). These studies suggest that BES is a promising sustainable option for removal and recovery of metals from multi-metal-laden industrial wastewaters generated from mining, metallurgical operations and process streams.

4 Concluding Remarks and Future Perspectives

Microbe–metal(loid) interactions play critical roles in biogeochemical cycling of various metal ions in natural environments. These interactions are explored for developing novel microbial biotechnologies for bioremediation and sustainable management of metal-laden wastes. The beneficial role of microbes has been exploited for large-scale applications in bioleaching, bioremediation and wastewater treatment. Microbial biotechnologies are already applied at industrial scale for biomining of sulphide ores and for treating wastewaters. Advancements in this field are needed for coupling treatment to recovering precious/scarce elements as part of material sustainability. More research is needed on understanding metal–microbe interactions in case of

Table 4 Standalone BES or integrated BES for recovering metals from multi-metal containing wastewater and treatment of real wastewaters

Metal(loid)	BES configuration, Electrode materials	Electron donor, concentration	Metal salt; metal concentration; pH	Metal removal efficiency	Maximum power output (W/m^2) in MFC or voltage applied in MEC/ER	References
Zinc, lead, copper	tMFC-ER; graphite felt anode, graphite plate with titanium wire as cathode	Sodium acetate, 1 g/l	Fly ash leachate; 52.1 mg/l Cu(II); 100 mg/l Pb(II); 180 mg/l Zn(II) at pH 2	Cu(II): 97% in 36 h in BES; 95% of Zn(II) and 98% of Pb(II) in 10 h in ER	MEC: 0.080 V; ER: 6	Tao et al. (2014)
Copper, Iron, Nickel	tMEC, graphite brush anode, carbon cloth with platinum layer as cathode	Sodium acetate, 1 g/l	$CuSO_4$, $FeSO_4$, $NiSO_4$; 9 mM Fe(II) + 5 mM Cu(II) + 5 mM Ni(II), pH 2.8	Cu(II) was first recovered, followed by Ni(II) and then Fe(II); Some amount of Fe was also recovered during Ni(II) recovery stage	1.0 V	Luo et al. (2014)
Copper, chromium, cadmium	tMFC and tMEC; carbon brush anode, carbon cloth cathode	Yeast extract, 1.25 g/l; 2.5 g/l peptone	$K_2Cr_2O_7$, $CdCl_2$, $CuSO_4$; 134 mg/l Cr(VI), 103 mg/L Cd(II), 130 mg/l Cu(II) at pH 1.8	Cr(VI): 99.6% and Cu(II): 99.9% in MFC; Cd(II): 99.9%	MFC: 0.32 W/m^2; MEC: 1.2 V	Amanze et al. (2022)
Chromium	tMFC, carbon felt anode, graphite paper cathode	Sodium acetate, 1 g/l	$K_2Cr_2O_7$; Cr(VI) in real electroplating wastewater; 204 mg/l; pH 2.5	99.5% in 25 h from 204 mg/l;	1.6 W/m^2 at 204 mg/l Cr(VI), pH 2.5	Li et al. (2008)
Chromium	tMFC with bipolar membrane; carbon felt anode, cathode	Sodium acetate, 1 g/l	5%–50% diluted real electroplating wastewater; 93–998 mg/l Cr(VI); pH 1.8	100% at 93 and 183 mg/l; 63% at 573 mg/l; 53% at 998 mg/l	0. 15 W/m^2 at 998 mg/l Cr(VI) at pH 1.8	Kim et al. (2017)
Zinc	tMFC; plain carbon felt anode, cathode	Sodium acetate, 1 g/l	Industrial wastewater; 1.55 ± 0.35 mM Zn(II)	96–99%	0.10 ± 0.05 A/m^2	Lim et al. (2021)

ER electrolysis reactor

strategic, critical and rare earth elements. Developments in biological and bioelectrochemical techniques and their integration with hydro-metallurgy would help in novel biotechnologies for better management of wastes coupled to recovering scarce/critical metals.

References

Agboola, O., Babatunde, D. E., Fayomi, O. S. I., Sadiku, E. R., Popoola, P., Moropeng, L., Yahaya, A., & Mamudu, O.-A. (2020). A review on the impact of mining operation: Monitoring, assessment and management. *Results in Engineering, 8*, 100181.

Ahemad, M., & Malik, A. (2011). Bioaccumulation of heavy metals by zinc resistant bacteria isolated from agricultural soils irrigated with wastewater. *Journal of Bacteriology, 2*, 12–21.

Ahmed, E., Abdulla, H. M., Mohamed, A. H., & El-Bassuony, A. D. (2016). Remediation and recycling of chromium from tannery wastewater using combined chemical–biological treatment system. *Process Safety and Environment Protection, 104*, 1–10.

Ai, C., Yan, Z., Hou, S., Huo, Q., Chai, L., Qiu, G., & Zeng, W. (2020). Sequentially recover heavy metals from smelting wastewater using bioelectrochemical system coupled with thermoelectric generators. *Ecotoxicology and Environment Safety, 205*, 111174.

Aiyer, K. S. (2020). Recovery of chromium, copper and vanadium combined with electricity generation in two-chambered microbial fuel cells. *FEMS Microbiology Letters, 367*, fnaa129.

Alalwan, H. A., Kadhom, M. A., & Alminshid, A. H. (2020). Removal of heavy metals from wastewater using agricultural by products. *Journal of Water Supply: Research Technology—AQUA, 69*, 99–112.

Ali, J., Wang, L., Waseem, H., Sharif, H. M. A., Djellabi, R., Zhang, C., & Pan, G. (2019). Bioelectrochemical recovery of silver from wastewater with sustainable power generation and its reuse for biofouling mitigation. *Journal of Cleaner Production, 235*, 1425–1437.

Amanze, C., Zheng, X., Man, M., Yu, Z., Ai, C., Wu, X., et al. (2022). Recovery of heavy metals from industrial wastewater using bioelectrochemical system inoculated with novel *Castellaniella* species. *Environmental Research, 205*, 112467.

Aryal, M. (2020). A comprehensive study on the bacterial biosorption of heavy metals: Materials, performances, mechanisms, and mathematical modelling. *Reviews in Chemical Engineering*, 20190016.

Bakhtiari, F., Zivdar, M., Atashi, H., & Seyed, B. S. (2010). Continuous bioleaching of copper from copper smelters dust with a mixed culture of mesophilic bacteria. *Journal of Chemical Society of Pakistan, 32*, 215–222.

Barakat, M. A. (2011). New trends in removing heavy metals from industrial wastewater. *Arabian Journal of Chemistry, 4*, 361–377.

Bayat, O., Sever, E., Bayat, B., Arslan, V., & Poole, C. (2009). Bioleaching of zinc and iron from steel plant waste using *Acidithiobacillus ferrooxidans*. *Applied Biochemistry and Biotechnology, 152*, 117–126.

Beni, A. A., & Esmaeili, A. (2020). Biosorption, an efficient method for removing heavy metals from industrial effluents: A Review. *Environmental Technology and Innovation, 17*, 100503.

Benzerara, K., Miot, J., Morin, G., Ona-Nguema, G., Skouri-Panet, F., & Ferard, C. (2011). Significance, mechanisms and environmental implications of microbial biomineralization. *Comptes Rendus Geoscience, 343*, 160–167.

Cao, J., Zhang, G., Mao, Z., Fang, Z., & Yang, C. (2009). Precipitation of valuable metals from bioleaching solution by biogenic sulfides. *Minerals Engineering, 22*(3), 289–295.

Capeness, M. J., Edmundson, M. C., & Horsfall, L. E. (2015). Nickel and platinum group metal nanoparticle production by *Desulfovibrio alaskensis* G20. *New Biotechnology, 32*, 727–731.

Catal, T., Bermek, H., & Liu, H. (2009). Removal of selenite from wastewater using microbial fuel cells. *Biotechnology Letters, 31*, 1211–1216.

Cheng, Y., Guo, Z., Liu, X., Yin, H., Qiu, G., Pan, F., & Liu, H. (2009). The bioleaching feasibility for Pb/Zn smelting slag and community characteristics of indigenous moderate-thermophilic bacteria. *Bioresource Technology, 100*, 2737–2740.

Choi, C., & Cui, Y. (2012). Recovery of silver from wastewater coupled with power generation using a microbial fuel cell. *Bioresource Technology, 107*, 522–525.

Choi, C., & Hu, N. (2013). The modeling of gold recovery from tetrachloroaurate wastewater using a microbial fuel cell. *Bioresource Technology, 133*, 589–598.

Choi, C., Hu, N., & Lim, B. (2014). Cadmium recovery by coupling double microbial fuel cells. *Bioresource Technology, 170*, 361–369.

Choi, H. J., & Lee, S. M. (2015). Heavy metal removal from acid mine drainage by calcined eggshell and microalgae hybrid system. *Environmental Science and Pollution Research, 22*, 13404–13411.

Cidu, R. (2011). Mobility of aqueous contaminants at abandoned mining sites: Insights from case studies in Sardinia with implications for remediation. *Environment and Earth Science, 64*, 503–512.

Creedy, S., Glinin, A., Matusewicz, R., Hughes, S., & Reuter, M. (2013). Ausmelt technology for treating zinc residues. *World Metall-Erzmetall, 66*, 230–235.

Das, B., Prakash, S., Reddy, P. S. R., & Misra, V. N. (2006). An overview of utilization of slag and sludge from steel industries. *Resources, Conservation and Recycling, 50*, 40–57.

Diep, P., Mahadevan, R., & Yakunin, A. F. (2018). Heavy metal removal by bioaccumulation using genetically engineered microorganisms. *Frontiers in Bioengineering and Biotechnology, 6*, 157.

Dudka, S., & Adriano, D. C. (1997). Environmental impacts of metal ore mining and processing: A review. *Journal of Environmental Quality, 26*, 590–602.

El Baz, S., Baz, M., Barakate, M., Hassani, L., El Gharmali, A., & Imziln, B. (2015). Resistance to and accumulation of heavy metals by actinobacteria isolated from abandoned mining areas. *Scientific World Journal, 761834*, 1–14.

Esposito, G., Veeken, A., Weijma, J., & Lens, P. N. L. (2006). Use of biogenic sulfide for ZnS precipitation. *Separation and Purification Technology, 51*, 31–39.

Farkas, B., Bujdoš, M., Polák, F., Matulová, M., Cesnek, M., Duborská, E., Zvěřina, O., Kim, H., Danko, M., Kisová, Z., Matúš, P., Urík, M. (2021). Bioleaching of manganese oxides at different oxidation states by filamentous fungus *Aspergillus niger*. *Journal of Fungi (Basel), 7*, 808.

Fu, F., & Wang, Q. (2011). Removal of heavy metal ions from wastewaters: A review. *Journal of Environmental Management, 92*, 407–418.

Gadd, M. (2010). Metals, minerals and microbes: Geomicrobiology and bioremediation. *Microbiology, 156*, 603–643.

Gorai, B., Jana, R. K., Premchand. (2003). Characteristics and utilization of copper slags-a review. *Resources, Conservation and Recycling, 39*, 299–313.

Hao, L., Zhang, B., Tian, C., Liu, Y., Shi, C., Cheng, M., & Feng, C. (2015). Enhanced microbial reduction of vanadium (V) in groundwater with bioelectricity from microbial fuel cells. *Journal of Power Sources, 287*, 43–49.

Hennebel, T., Boon, N., Maes, S., & Lenz, M. (2015). Biotechnologies for critical raw material recovery from primary and secondary sources: R&D priorities and future perspectives. *New Biotechnology, 32*, 121–127.

Hita, R., Wang, H., Bigham, J. M., Torrent, J., & Tuovinen, O. H. (2008). Bioleaching of a pyritic sludge from the Aznalcollar (Spain) mine spillage at ambient and elevated temperatures. *Hydrometallurgy, 93*, 76–79.

Horiike, T., & Yamashita, M. (2015). A new fungal isolate, *Penidiella* sp. strain T9, accumulates the rare earth element dysprosium. *Applied and Environment Microbiology, 81*, 3062–3068.

Huang, L., Li, T., Liu, C., Quan, X., Chen, L., Wang, A., & Chen, G. (2013). Synergetic interactions improve cobalt leaching from lithium cobalt oxide in microbial fuel cells. *Bioresource Technology, 128*, 539–546.

Huang, L., Yao, B., Wu, D., & Quan, X. (2014). Complete cobalt recovery from lithium cobalt oxide in self-driven microbial fuel cell-microbial electrolysis cell systems. *Journal of Power Sources, 259*, 54–64.

Ilyas, S., & Lee, J. C. (2014). Biometallurgical recovery of metals from waste electrical and electronic equipment: A review. *ChemBioEng Reviews, 1*, 148–169.

Jiang, L., Huang, L., & Sun, Y. (2014). Recovery of flakey cobalt from aqueous Co(II) with simultaneous hydrogen production in microbial electrolysis cells. *International Journal of Hydrogen Energy, 39*, 654–663.

Jing, R., & Kjellerup, B. V. (2018). Biogeochemical cycling of metals impacting by microbial mobilization and immobilization. *Journal of Environmental Sciences (china), 66*, 146–154.

Johnson, D. B. (2013). Development and application of biotechnologies in the metal mining industry. *Environmental Science and Pollution Research International, 20*(11), 7768–7776.

Johnson, D. B., Grail, B. M., & Hallberg, K. B. (2013). A new direction for biomining: Extraction of metals by reductive dissolution of oxidized ores. *Minerals, 3*, 49–58.

Kaksonen, A. H., Lavonen, L., Kuusenaho, M., Kolli, A., N€arhi, H., Vestola, E., Puhakka, J. A., Tuovinen, O. H. (2011). Bioleaching and recovery of metals from final slag waste of the copper smelting industry. *Minerals Engineering, 24*, 1113–1121.

Kaksonen, A. H., Särkijärvi, S., Puhakka, J. A., Peuraniemi, E., Junnikkala, S., & Tuovinen, O. H. (2016). Chemical and bacterial leaching of metals from a smelter slag in acid solutions. *Hydrometallurgy, 159*, 46–53.

Kim, C., Lee, C. R., Song, Y. E., Heo, J., Choi, S. M., Lim, D.-H., Cho, J., Park, C., Jang, M., & Kim, J. R. (2017). Hexavalent chromium as a cathodic electron acceptor in a bipolar membrane microbial fuel cell with the simultaneous treatment of electroplating wastewater. *Chemical Engineering Journal, 2017*(328), 703–707.

Korehi, H., Blothe, M., Sitnikova, M. A., Dold, B., & Schippers, A. (2013). Metal mobilization by iron-and sulfur-oxidizing bacteria in a multiple extreme mine tailings in the Atacama Desert, Chile. *Environmental Science and Technology, 47*, 2189–2196.

Kosinska, K., & Miskiewicz, T. (2012). Precipitation of heavy metals from industrial wastewater by *Desulfovibrio desulfuricans*. *Environmental Protection Engineering, 38*(2), 51–60.

Kossoff, D., Dubbin, W. E., Alfredsson, M., Edwards, S. J., Macklin, M. G., & Hudson-Edwards, K. A. (2014). Mine tailings dams: Characteristics, failure, environmental impacts, and remedi- ation. *Applied Geochemistry, 51*, 229–245.

Kulkarni, S., Ballal, A., & Apte, S. K. (2013). Bioprecipitation of uranium from alkaline waste solutions using recombinant *Deinococcus radiodurans*. *Journal of Hazardous Materials, 262*, 853–861.

Kwak, I. S., Won, S. W., Chung, Y. S., & Yun, Y. S. (2013). Ruthenium recovery from acetic acid waste water through sorption with bacterial biosorbent fibers. *Bioresource Technology, 128*, 30–35.

Langkau, S., & Tercero Espinoza, L. A. (2018). Technological change and metal demand over time: What can we learn from the past? *Sustainable Materials and Technologies, 16*, 54–59.

Lee, E., Han, Y., Park, J., Hong, J., Silva, R. A., Kim, S., & Kim, H. (2015). Bioleaching of arsenic from highly contaminated mine tailings using *Acidithiobacillus thiooxidans*. *Journal of Environmental Management, 147*, 124–131.

Li, Z. J., Zhang, X. W., & Lei, L. C. (2008). Electricity production during the treatment of real electroplating wastewater containing Cr6+ using microbial fuel cell. *Process Biochemistry, 43*, 1352–1358.

Lim, S. S., Fontmorin, J. M., Pham, H. T., Milner, E., Abdul, P. M., Scott, K., Head, I., & Yu, E. H. (2021). Zinc removal and recovery from industrial wastewater with a microbial fuel cell: Experimental investigation and theoretical prediction. *Science of the Total Environment, 776*, 145934.

Liu, Y. G., Zhou, M., Zeng, G. M., Wang, X., Li, X., Fan, T., & Xu, W. H. (2008). Bioleaching of heavy metals from mine tailings by indigenous sulfur-oxidizing bacteria: Effects of substrate concentration. *Bioresource Technology, 99*, 4124–4129.

Lloyd, J. R. (2003). Microbial reduction of metals and radionuclides. *FEMS Microbiology Reviews, 27*, 411–425.

Luo, H., Liu, G., Zhang, R., Bai, Y., Fu, S., & Hou, Y. (2014). Heavy metal recovery combined with H_2 production from artificial acid mine drainage using the microbial electrolysis cell. *Journal of Hazardous Materials, 270*, 153–1159.

Mansfeldt, T., & Dohrmann, R. (2004). Chemical and mineralogical characterization of blast furnace sludge from an abandoned landfill. *Environmental Science and Technology, 38*, 5977–5984.

Massianaie, M., Oliazadeh, M., & Bagheri, A. S. (2006). Biological copper extraction from melting furnace dust of Sarcheshmeh copper mine. *International Journal of Mineral Processing, 81*, 58–62.

Mishra, D., Kim, D. J., Ahn, J. G., & Rhee, Y. H. (2005). Bioleaching: A microbial process of metal recovery; A review. *Metals and Materials International, 11*, 249–256.

Mudd, G. M. (2008). Sustainability reporting and water resources: A preliminary assessment of embodied water and sustainable mining. *Mine Water and the Environment, 27*, 136–144.

Mulligan, C. N., Galvez-Cloutier, R., & Renaud, N. (1999). Biological leaching of copper mine residues by *Aspergillus Niger*. *Process Metall., 9*, 453–461.

Mulligan, C. N., Kamali, M., & Gibbs, B. F. (2004). Bioleaching of heavy metals from a low-grade mining ore using *Aspergillus niger*. *Journal of Hazardous Materials, 110*, 77–84.

Nancharaiah, Y. V., & Lens, P. N. L. (2015a). Ecology and biotechnology of selenium respiring bacteria. *Microbiology and Molecular Biology Reviews, 79*, 61–80.

Nancharaiah, Y. V., & Lens, P. N. L. (2015b). Selenium biomineralization for biotechnological applications. *Trends in Biotechnology, 33*, 323–330.

Nancharaiah, Y. V., Sarvajith, M., Lens, P. N. L. (2018). Selenite reduction and ammoniacal nitrogen removal in an aerobic granular sludge sequencing batch reactor. *Water Research, 131*, 131–141.

Nancharaiah, Y. V., Venkata Mohan, S., & Lens, P. N. L. (2015). Metals removal and recovery in bioelectrochemical systems: A review. *Bioresource Technology, 195*, 102–114.

Nancharaiah, Y. V., Venkata Mohan, S., & Lens, P. N. L. (2016). Biological and bioelectrochemical recovery of critical and scarce metals. *Trends in Biotechnology, 34*, 137–155.

Newsome, L., Morris, K., & Lloyd, J. R. (2014). The biogeochemistry and bioremediation of uranium and other priority radionuclides. *Chemical Geology, 363*, 164–184.

Nguyen, V. K., Lee, M. H., Park, H. J., & Lee, J. U. (2015). Bioleaching of arsenic and heavy metals from mine tailings by pure and mixed cultures of *Acidithiobacillus* spp. *Journal of Industrial and Engineering Chemistry, 21*, 451–458.

Park, I., Tabelin, C. B., Jeon, S., Li, X., Seno, K., Ito, M., & Hiroyoshi, N. (2019). A review of recent strategies for acid mine drainage prevention and mine tailings recycling. *Chemosphere, 219*, 588–606.

Paterson-Beedle, M., Readman, J. E., Hriljac, J. A., & Macaskie, L. E. (2010). Biorecovery of uranium from aqueous solutions at the expense of phytic acid. *Hydrometallurgy, 104*, 524–528.

Petrisor, I. G., Komnitsas, K., Lazar, I., Voicu, A., Dobrota, S., & Stefanescu, M. (2002). Biosorption of heavy metals from leachates generated at mine waste disposal sites. *European Journal of Mineral Processing Environmental Protection, 2*, 158–167.

Philip, L., Iyengar, L., & Venkobachar, C. (2000). Biosorption of U, La, Pr, Nd, Eu and Dy by *Pseudomonas aeruginosa*. *Journal of Industrial Microbiology and Biotechnology, 25*, 1–7.

Potysz, A., van Hullebusch, E. D., Kierczak, J., Grybos, M., Lens, P. N. L., & Guibaud, G. (2015). Copper metallurgical slags-current knowledge and fate: A review. *Critical Reviews in Environment Science and Technology, 45*, 2424–2488.

Qasem, N. A. A., Mohammed, R. H., & Lawal, D. U. (2021). Removal of heavy metal ions from wastewater: A comprehensive and critical review. *NPJ Clean Water, 4*, 36.

Qin, B., Luo, H., Liu, G., Zhang, R., Chen, S., Hou, Y., & Luo, Y. (2012). Nickel ion removal from wastewater using the microbial electrolysis cell. *Bioresource Technology, 121*, 458–461.

Rajwade, J. M., & Paknikar, K. M. (2003). Bioreduction of tellurite to elemental tellurium by *Pseudomonas mendocina* MCM B-180 and its practical application. *Hydrometallurgy, 71*, 243–248.

Reddy, G. K. K., & Nancharaiah, Y. V. (2018). Sustainable bioreduction of toxic levels of chromate in a denitrifying granular sludge reactor. *Environmental Science and Pollution Research, 25*, 1969–1979.

Reddy, G. K. K., & Nancharaiah, Y. V. (2019). *Biofilm and granular sludge bioreactors for textile wastewater treatment*. Microbial Biofilms in Bioremediation and Wastewater Treatment. CRC Press.

Reddy, G. K. K., Nancharaiah, Y. V., & Mohan, T. V. K. (2016). Bioreduction of [Co(III)-EDTA]—By denitrifying granular sludge biofilms. *Chemical Engineering and Technology, 39*, 1669–1675.

Reddy, G. K. K., Pathak, S., & Nancharaiah, Y. V. (2023). Aerobic reduction of selenite and tellurite to elemental selenium and tellurium nanostructures by *Alteromonas* sp. under saline conditions. *International Biodeterioration and Biodegradation, 179*, 105571.

Rikame, S. S., Mungray, A. A., & Mungray, A. K. (2021). Electrochemical recovery of metal copper in microbial fuel cell using graphene oxide/polypyrrole cathode catalyst. *International Journal of Energy Research, 45*, 6863–6875.

Rosenbaum, M., Aulenta, F., Villano, M., & Angenent, L. T. (2011). Cathodes as electron donors for microbial metabolism: Which extracellular electron transfer mechanisms are involved? *Bioresource Technology, 102*, 324–333.

Sethurajan, M., Lens, P. N. L., Rene, E. R., van de Vossenberg, J., Huguenot, D., Horn, H. A., Figueiredo, L. H., & van Hullebusch, E. D. (2017a). Bioleaching and selective biorecovery of zinc from zinc metallurgical leach residues from the Tres Marias zinc plant (Minas Gerais, Brazil). *Journal of Chemical Technology and Biotechnology, 3*(92), 512–521.

Sethurajan, M., Lens, P. N. L., Rene, E. R., van de Vossenberg, J., Huguenot, D., Horn, H. A., Figueiredo, L. H., & van Hullebusch, E. D. (2017b). Bioleaching and selective biorecovery of zinc from zinc metallurgical leach residues from the Tres Marias zinc plant (Minas Gerais, Brazil). *Journal of Chemical Technology and Biotechnology, 3*(92), 512–521.

Sethurajan, M., van Hullebusch, E. D., & Nancharaiah, Y. V. (2018). Biotechnology in the management and resource recovery from metal bearing solid wastes: Recent advances. *Journal of Environmental Management, 211*, 138–153.

Simmons, J. A., Currie, W. S., Eshleman, K. N., Kuers, K., Monteleone, S., Negley, T. L., Pohlad, B. R., & Thomas, C. L. (2008). Forest to reclaimed mine land use change leads to altered ecosystem structure and function. *Ecological Applications, 18*, 104–118.

Singh, A., & Kaushik, A. (2022). Removal of Cd and Ni with enhanced energy generation using biocathode microbial fuel cell: Insights from molecular characterization of biofilm communities. *Journal of Cleaner Production, 315*, 127940.

Smith, D., & Wentworth, J. (2022). *POST (Parliamentary Office of Science and Technology), POSTbrief 45, Mining and the sustainability of metals*. UK Parliament.

Song, Y. W., Wang, H. R., Yang, J., Zhou, L. X., Zhou, J. C., & Cao, Y. X. (2018). Evaluation and optimization of a new microbial enhancement plug-flow ditch system for the pretreatment of acid mine drainage: Semi-pilot test. *RSC Advances, 8*, 1039–1046.

Sravan, J. S., & Mohan, S. V. (2022). Bioelectrocatalytic reduction of tellurium oxyanions toward their cathodic recovery: Concentration dependence and anodic electrogenic activity. *ACS ES&T Water, 2*, 40–51.

Sravan, J. S., Nancharaiah, Y. V., Lens, P. N. L., & Mohan, S. V. (2020). Cathodic selenium recovery in bioelectrochemical system: Regulatory influence on anodic electrogenic activity. *Journal of Hazardous Materials, 399*, 122843.

Suja, E., Nancharaiah, Y. V., & Venugopalan, V. P. (2014). Biogenic nanopalladium production by self-immobilized granular biomass: Application for contaminant remediation. *Water Research, 65*, 395–401.

Sukla, L. B., Kar, R. N., & Panchanadikar, V. (1995). Bioleaching of copper converter slags using *Aspergillus niger* isolated from lateritic nickel ore. *International Journal of Environmental Studies, 47*, 81–86.

Tao, H.-C., Gao, Z.-Y., Ding, H., Xu, N., & Wu, W.-M. (2012). Recovery of silver from silver(I)-containing solutions in bioelectrochemical reactors. *Bioresource Technology, 111*, 92–97.

Tao, H.-C., Lei, T., Shi, G., Sun, X.-N., Wei, X.-Y., Zhang, L.-J., & Wu, W.-M. (2014). Removal of heavy metals from fly ash leachate using combined bioelectrochemical systems and electrolysis. *Journal of Hazardous Materials, 264*, 1–7.

Tao, H.-C., Liang, M., Li, W., Zhang, L.-J., Ni, J.-R., & Wu, W.-M. (2011). Removal of copper from aqueous solution by electrodeposition in cathode chamber of microbial fuel cell. *Journal of Hazardous Materials, 189*, 186–192.

Ter Heijne, A., Liu, F., Weijden, R. V., Weijma, J., Buisman, C. J., & Hamelers, H. V. (2010). Copper recovery combined with electricity production in a microbial fuel cell. *Environmental Science and Technology, 44*, 4376–4381.

Thauer, R. K., Stackebrandt, E., & Hamilton, W. A. (2007). Energy metabolism and phylogenetic diversity of sulphate-reducing bacteria. In *Sulphate-reducing bacteria: Environmental and engineered systems* (1st ed., pp. 1–37). Cambridge: Cambridge University Press.

Tripathi, M., & Garg, S. K. (2014). Dechlorination of chloroorganics, decolorization, and simultaneous bioremediation of Cr^{6+} from real tannery effluent employing indigenous *Bacillus cereus* isolate. *Environmental Science and Pollution Research, 21*, 5227–5241.

US EPA, Priority pollutants. Appendix A to 40 CFR Part 423. Available from: http://water.epa.gov/scitech/methods/cwa/pollutants.cfm

Vachon, P., Tyagi, R. D., Auclair, J. C., & Wilkinson, K. J. (1994). Chemical and biological leaching of aluminum from red mud. *Environmental Science and Technology, 28*, 26–30.

Vakilchap, F., Mousavi, S. M., & Shojaosadati, S. A. (2016). Role of *Aspergillus Niger* in recovery enhancement of valuable metals from produced red mud in Bayer process. *Bioresource Technology, 218*, 991–998.

Venkata Mohan, S., Velvizhi, G., Annie Modestra, J., & Srikanth, S. (2014a). Microbial fuel cell: Critical factors regulating bio-catalyzed electrochemical process and recent advances. *Renewable and Sustainable Energy Reviews, 40*, 779–797.

Venkata Mohan, S., Velvizhi, G., Vamshi Krishna, K., & Lenin Babu, M. (2014b). Microbial catalyzed electrochemical systems: A biofactory with multi-facet applications. *Bioresource Technology, 165*, 355–364.

Wadgaonkar, S. L., Mal, J., Nancharaiah, Y. V., Maheshwari, N. O., Esposito, G., & Lens, P. N. L. (2018). Formation of Se(0), Te(0), and Se(0)–Te(0) nanostructures during simultaneous bioreduction of selenite and tellurite in a UASB reactor. *Applied Microbiology and Biotechnology, 102*, 2899–2911.

Wang, G., Huang, L., & Zhang, Y. (2008). Cathodic reduction of hexavalent chromium [Cr(VI)] coupled with electricity generation in microbial fuel cells. *Biotechnology Letters, 30*, 1959–1966.

Wang, H., & Ren, Z. J. (2014). Bioelectrochemical metal recovery from wastewater: A review. *Water Research, 66*, 219–232.

Wang, J., Huang, Q., Li, T., Xin, B., Chen, S., Guo, X., Liu, C., & Li, Y. (2015). Bioleaching mechanism of Zn, Pb, In, Ag, Cd and as from Pb/Zn smelting slag by autotrophic bacteria. *Journal of Environmental Management, 159*, 11–17.

Wang, J., Zhu, S., Zhang, Y., Zhao, H., Hu, M., Yang, C., Qin, W., & Qiu, G. (2014). Bioleaching of low-grade copper sulfide ores by *Acidithiobacillus ferrooxidans* and *Acidithiobacillus thiooxidans*. *Journal of Central South University, 21*, 728–734.

Wang, S., Liu, T., Xiao, X., & Luo, S. (2021). Advances in microbial remediation for heavy metal treatment: A mini review. *Journal of Leather Science and Engineering, 3*, 1.

Wang, Z., Lim, B., & Choi, C. (2011). Removal of Hg2+ as an electron acceptor coupled with power generation using a microbial fuel cell. *Bioresource Technology, 102*, 6304–6307.

Williamson, A., Folens, K., Matthijs, S., Paz Cortes, Y., Varia, J., Du Laing, G., Boon, N., & Hennebel, T. (2021). Selective metal extraction by biologically produced siderophores during bioleaching from low-grade primary and secondary mineral resources. *Minerals Engineering, 163*, 106774.

Won, S. W., Kwak, I. S., & Yun, Y. S. (2014). The role of biomass in polyethylenimine-coated chitosan/bacterial biomass composite biosorbent fiber for removal of Ru from acetic acid waste solution. *Bioresource Technology, 160*, 93–97.

Wu, X., Zhu, X., Song, T., Zhang, L., Jia, H., & Wei, P. (2015). Effect of acclimatization on hexavalent chromium reduction in a biocathode microbial fuel cell. *Bioresource Technology, 180*, 185–191.

Ye, J., Yin, H., Mai, B., Peng, H., Qin, H., He, B., & Zhang, N. (2010). Biosorption of chromium from aqueous solution and electroplating wastewater using mixture of *Candida lipolytica* and dewatered sewage sludge. *Bioresource Technology, 101*, 3893–3902.

Yong, P., Rowson, N. A., Farr, J. P., Harris, I. R., & Macaskie, L. E. (2003). A novel electrobiotechnology for the recovery of precious metals from spent automotive catalysts. *Environmental Technology, 24*, 289–297.

Zhang, B., Zhao, H., Shi, C., Zhou, S., & Ni, J. (2009). Simultaneous removal of sulfide and organics with vanadium (V) reduction in microbial fuel cells. *Journal of Chemical Technology and Biotechnology, 84*, 1780–1786.

Zhang, L., Xu, Z., & He, X. (2020). Selective recovery of lead and zinc through controlling cathodic potential in a bioelectrochemically-assisted electrodeposition system. *Journal of Hazardous Materials, 386*, 121941.

Chalcopyrite Dissolution: Challenges

Denise Bevilaqua, Ailton Guilherme Rissoni Toledo, Laíze Guimarães Crocco, Riberto Nunes Peres, Rachel Biancalana da Costa, Assis Vicente Benedetti and Olli H. Tuovinen

Abstract

Chalcopyrite is the main source of copper in the world, amounting to nearly 70% of the copper reserves. Nonetheless, chalcopyrite is highly recalcitrant to chemical and biological processing for copper extraction. Concentration by flotation and Cu recovery by pyrometallurgical techniques are still the main route for processing chalcopyrite concentrates, although they are unfeasible for copper extraction from low-grade ores that make up the most copper reserves. Acid bioleaching is a promising technique for extracting copper from low-grade copper ores, and the technology has been studied for decades, but there is still no commercial-scale bioleaching application for copper recovery from chalcopyrite concentrates. Bioleaching is practiced with low-grade chalcopyrite ores in heap leaching processes with ores of multiple sulfide minerals. Research in this area has probed electrochemical reactions, biological activities, and interactions with microbes and mineral surfaces to integrate operational models for chalcopyrite bioleaching. The purpose of this chapter is to review the evolution in the understanding of the chemical leaching and bioleaching of chalcopyrite in the last 20 years, and the progress achieved so far.

D. Bevilaqua (✉) · A. G. R. Toledo · L. G. Crocco · R. B. da Costa
Department of Biochemistry and Organic Chemistry, Institute of Chemistry, São Paulo State University (UNESP), Araraquara, SP 14800-060, Brazil
e-mail: denise.bevilaqua@unesp.br

R. N. Peres · A. V. Benedetti
Department of Analytical, Physical-Chemical and Inorganic Chemistry, Institute of Chemistry, São Paulo State University (UNESP), Araraquara, SP 14800-060, Brazil

O. H. Tuovinen
Department of Microbiology, Ohio State University, 484 West 12th Avenue, Columbus, OH 43210, USA

Keywords

Chalcopyrite bioleaching · Electrochemical techniques · Galvanic interactions · Redox potential control · Polarization

1 Introduction

Copper ranks as the third most consumed metal in the world, and its consumption has been increasing consistently with the industry and technology. Copper has a wide range of industrial and consumer applications due to its high thermal and electrical conductivity and the propensity to form metallic alloys with many other metals. Global demand for copper continues to increase because it plays an indispensable role in modern technologies including applications in renewable energy areas. About 70% of the global copper reserves are chalcopyrite ($CuFeS_2$) ores, and 70–80% of copper production comes from this mineral (Nyembwe et al., 2018).

Chalcopyrite is a recalcitrant mineral in hydrometallurgy, characterized by slow dissolution kinetics. Flotation and pyrometallurgical techniques are still the main route for processing chalcopyrite, although they are economically unfeasible for copper extraction from low-grade ores, which constitute the most copper reserves. Several chemical and biological strategies have been proposed to improve copper solubilization from chalcopyrite. Chemical alternatives in hydrometallurgy include for example process adjustments with catalytic ions, controlled redox potential, and use of ferrous and ferric iron and chloride to modify the leach solution. Hydrogen peroxide, Na-nitrate, Cr^{6+}, and Cu^{2+} have also been tested as chemical oxidants (Li et al., 2013), but their best use may be in the elucidation of the mechanisms and steps on chalcopyrite oxidation. As predicted from the Van't Hoff equation, temperature in the range of ambient to 120 °C in autoclaves is efficient in enhancing

S. Panda et al. (eds.), *Biotechnological Innovations in the Mineral-Metal Industry*, Advances in Science, Technology & Innovation, https://doi.org/10.1007/978-3-031-43625-3_2

the kinetics of chemical leaching of chalcopyrite. Many advances in biological strategies were achieved in the late twentieth century, leading to greatly increased knowledge of physiological capabilities of acidophilic microorganisms for chalcopyrite solubilization. Some processes with chalcopyrite concentrates were advanced to pilot scale and also demonstrated at a large scale (Watling, 2013), but commercialization has not materialized. Thus, bioleaching processes are yet to develop to economically competitive technology for copper extraction from chalcopyrite concentrates. Heap leaching applications of low-grade ores that contain chalcopyrite typically also contain secondary copper and other metal sulfides, which are more readily subjected to dissolution. An example of polymetal sulfide ore extraction is the Terrafame heap leaching operation (www.terrafame.com) in NE Finland. The operation recovers Zn, Ni, Co, Cu, and U in the acid leach cycle, and a second leach cycle with extended residence time is practiced accommodating the slow dissolution of Cu from chalcopyrite and U from thucholite in secondary bioleaching heaps.

This chapter addresses the refractoriness of chalcopyrite dissolution and the role of microorganisms in chalcopyrite bioleaching. The chapter emphasizes electrochemical techniques inasmuch as they help to elucidate the mechanisms of chalcopyrite dissolution. The redox potential as a strategy to promote chalcopyrite dissolution is also appraised in this chapter.

2 Chalcopyrite Properties and Models to Explain the Refractoriness

The lattice energy of chalcopyrite is close to 17,000 kJ mol^{-1}, and the standard enthalpy of formation ΔH_f and the standard Gibbs free energy of formation ΔG_f° values are −193.6 and −190.6 kJ mol^{-1}, respectively. These ΔH_f and ΔG_f° values are in the common range for sulfide minerals and do not explain the recalcitrance and unfavorable kinetics of chalcopyrite dissolution (Li et al., 2013).

Slow dissolution of chalcopyrite has been recognized through the years (e.g., Dutrizac, 1978, 1991; Nicol et al., 2017; Nicol, 2017a, 2017b; Viramontes-Gamboa et al., 2006, 2007, 2010). The slow dissolution has been mainly attributed to the formation of passive, metal-depleted layers on chalcopyrite surface and to chalcopyrite properties as semiconductor material (Crundwell, 1988; Liu & Li, 2011; Weisener et al., 2003; Yu et al., 1973).

Potential measurements have been used to analyze chalcopyrite passivation. The increase in the current with positive potentials is considered to signal a transpassive phenomenon. Transpassivation is a phenomenon, in which a passivated surface (metal, metal alloys, or mineral) begins to dissolve fast when the electrode potential becomes too positive and surpasses the breakdown potential of the passive film. According to Nicol (2017b), the formation of a transpassive region and the oxidation of chalcopyrite in sulfate solutions occur at lower potentials than in chloride solutions. Nicol (2017a) reported that the capacitance values vary with the applied potential, decreasing between 0.40 and 0.95 V_{SHE} in 0.3 mol L^{-1} sulfuric acid and sharply increasing with potentials > 1 V_{SHE}, thus manifesting a trend similar to the measured current. These results support the findings that intermediate solid-phase products are formed on chalcopyrite between 0.4 and 0.95 V_{SHE}, causing the formation of a passivation layer, which hinders further dissolution.

Both Fe^{3+} and Cu^{2+} act as oxidants in the leaching of chalcopyrite in chloride-containing solutions (Nicol & Zhang, 2017). Potentiostatic measurements (current vs. time) showed that the potential region between 0.65 and 0.80 $V_{Ag|AgCl|KCl(3\ mol\ L^{-1})}$ is important in the electrochemical leaching process. The authors also reported slow dissolution of chalcopyrite even after 24 h. Different concentrations of chloride did not influence chalcopyrite leaching, whereas the pH and specific potentials (0.65–0.80 V) in chloride-containing solutions had major effects (Nicol & Zhang, 2017). In contrast, Bevilaqua et al. (2013) demonstrated that the addition of 0.10–0.20 mol L^{-1} NaCl enhanced both the chemical leaching and bacterial leaching of chalcopyrite in shake flasks and stirred tank bioreactor conditions at mesophilic temperatures. The highest rate chalcopyrite leaching occurred at < 0.45 $V_{AglAgCllKClsat}$ redox potential (Bevilaqua et al., 2013). Chloride ions inhibit the formation of secondary solid phases from chalcopyrite, decreasing the precipitation of ferric iron as jarosite-type secondary phases (Vakylabad et al., 2022). Secondary Cu-sulfides and S^0 are also decreased, sometimes below the detection by X-ray diffraction analysis of solid residues (Bevilaqua et al., 2013). Thus, chloride ions decrease the passivation of chalcopyrite (Martins et al., 2019, Martin & Leão, 2023). Chloride toxicity at the > 0.2–0.3 mol L^{-1} range may, however, impede the bioleaching action depending on the microbial culture (Akcil et al., 2013; Dopson et al., 2017; Huynh et al., 2019). Several halotolerant iron- and sulfur-oxidizers have been described, some tolerating chloride concentrations above the ~ 0.5 mol L^{-1} in seawater (Zammit et al., 2012; Khaleque et al., 2018; Martins et al., 2019, Martin & Leão, 2023). Wang et al. (2022) reviewed prospects of using seawater-based lixiviants in heap leaching systems. The toxicity of seawater, ~ 0.5 mol L^{-1} Cl^- with ~ 35‰ salinity, in heap bioleaching systems necessitates testing, selection, and acclimatization for salt-tolerant microbial communities.

Chalcopyrite oxidation in alkaline solutions involves mineral activation at low potentials, followed by passivation and transpassivation at high potentials and decreasing oxidation at > 18 h of contact (Nicol, 2019). In mixed chloride and sulfate solutions containing Fe^{3+} and Cu^{2+}, the solution potential of 0.2 $V_{Ag|AgCl|KCl(3\ mol\ L^{-1})}$ varied, and potentiostatic measurements confirmed that a specific mixed potential region supports oxidative dissolution of copper and iron from chalcopyrite (Nicol, 2021).

Zhao et al. (2019) reviewed the dissolution and passivation mechanism of chalcopyrite in the bioleaching process, pointing out that several secondary products are formed that can cause chalcopyrite passivation during the leaching process. The three main passivating products formed during contact of chalcopyrite in leach solution are S_n^{2-} (polysulfides), S^0 (elemental sulfur), and $XFe(SO_4)(OH)_6$ (jarosite-type precipitates), where X is usually mixtures of K^+, NH_4^+, H_3O^+, and Na^+. The polysulfide film formed on chalcopyrite can have a thickness of up to 1 μm and its formation is due to the dissolution of Fe^{2+}, thus leading to Fe-deficient copper polysulfides (CuS_n), which are unstable and readily converted to other Cu-sulfides (Zhao et al., 2019). Some studies suggest that polysulfides are not the main passivating agents of chalcopyrite because they are oxidized increasingly to form elemental sulfur at redox potentials of>0.9 V_{SHE} (Klauber, 2008; Parker et al., 2003; Zhao et al., 2019). The formation of S^0 is considered as the main passivating agent in sulfate-rich and bioleaching solutions and is further oxidized to sulfate over time (Dutrizac, 1989; Khoshkhoo et al., 2014; Nava et al., 2008). Other studies indicate that the S^0-layer on chalcopyrite is porous and does not hinder the dissolution of chalcopyrite (Klauber, 2008; Klauber et al., 2001; Sasaki et al., 2012). Thus, the formation of a S^0-layer and its passivating effect on the acid leaching of chalcopyrite is a controversial subject because opposite effects have been reported. Some of the different interpretations emanate from the initial phase of chalcopyrite oxidation, which forms a layer on mineral surface of oxidic Fe(III) and metastable sulfide phases of unoxidized S and Cu, leading to passivation. As the leaching reactions continue, more Fe is extracted from chalcopyrite surface layers, and solid-state diffusion becomes increasingly rate controlling. Intermediate Cu-sulfide phases of the stoichiometry of chalcocite (CuS_2) and idaite (Cu_5FeS_6) have been detected on passivated chalcopyrite surfaces (Varotsis et al., 2022). Several studies suggest that jarosite-type precipitates are the main secondary phase that passivates chalcopyrite leaching: i.e., precipitation of poorly soluble Fe(III)-sulfates together with enrichment of phases with Cu–S bonds (Sandström et al., 2005; Zhao et al., 2019).

The importance of the secondary phases hindering chalcopyrite dissolution is on debate. The slow dissolution of chalcopyrite may be linked to other factors such as the structure of chalcopyrite and its semiconductor behavior. O'Connor and Eksteen (2020) have expressed strong criticism about the use of the "passivation" term. "Passivation" is not universally accepted, most likely due to its superficial resemblance, if any, to the well-known passivation behavior of metals and metal alloys, and the lack of a clearly identifiable surface layer. Several reports use the term to explain the slow chalcopyrite dissolution without the necessary scientific and analytical basis. The products formed on the mineral surface are variations of metal-deficient phases such as Fe-deficient polysulfides or sulfides, but all with the same structure and behavior (O'Connor & Eksteen, 2020).

Passivation is normally defined for metals and alloys, as discussed and explained in the literature (Fontana, 1987; Sedriks, 1996; Uhlig, 1978). The term and its derivatives in mineral studies were adopted from corrosion research and were not originally defined for minerals. A passive surface presents an appreciable and nobler potential than a non-passive surface and has a low current density due to the proximity of anodic and cathodic potentials for heterogeneous materials as well as a significant potential region with very small current up to the passive film breakdown, typically as shown in Fig. 1a. Figure 1a is different from the polarization curves observed for minerals as shown in Fig. 1b. As mentioned by O'Connor and Eksteen (2020), there are differences

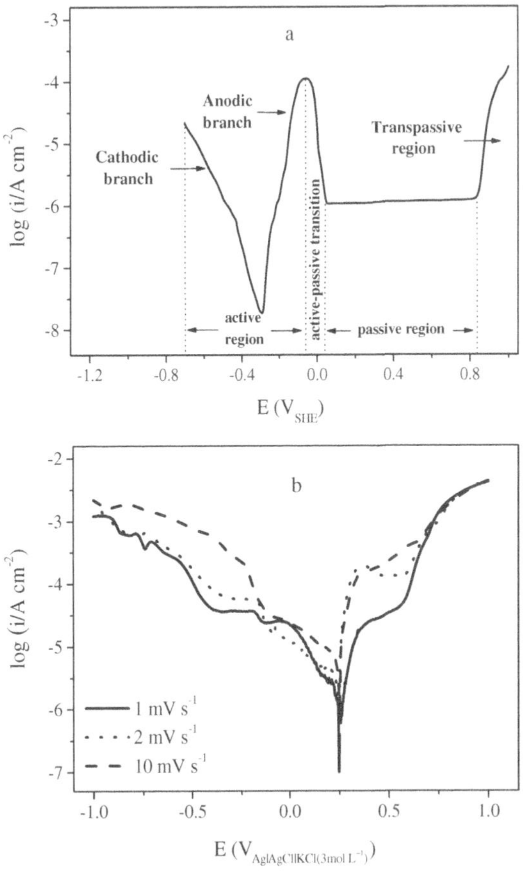

Fig. 1 Schematic representation of polarization curves: (**a**) a generic curve with characteristic parameters and specific regions; (**b**) chalcopyrite in salt leach solution (ionic strength = 0.08 mol L^{-1}) scanned at different scan rates

in passivation characteristics between metals/alloys and chalcopyrite. The mineral behavior is different from the passivation of metals, and the active–passive–transpassive regions are generally not obvious (O'Connor & Eksteen, 2020). This can be seen at different scan rates as shown in Fig. 1b. The tests performed at low scan rate (0.1 up to 10 mV s^{-1}) do not show a passivation region with low dissolution current. Sulfur species on polarized, metal-deficient chalcopyrite surface assent with interfacial mineral-leach solution species (Ram et al., 2020). Passivation and critical current density for metals tend to decrease at low pH values, but this is not observed for chalcopyrite, because its passivation can also be related to its electrochemical properties.

Mikhlin et al. (2017) analyzed the near-surface region of an oxidized chalcopyrite sample that was conditioned in acid ferric sulfate solution for 30 min at 50 °C. The group used X-ray photoelectron spectroscopy (XPS), X-ray absorption near-edge spectroscopy (XANES), and density functional theory (DFT) calculations to describe the oxidized, metal-depleted regions in the layer. Mikhlin et al. (2004, 2017) analyzed three metal-depleted layers: (i) a thin (1–4 nm) outmost layer containing polysulfide species, (ii) a ~ 20 nm thick, highly metal-deficient layer rich in disulfides but negligible in polysulfides, and (iii) a defective and near-stoichiometric underlayer of about 100 nm thickness. The slow chalcopyrite dissolution was attributed to metal depletion on chalcopyrite surface and slow diffusion of copper and iron species from the bulk solid to the mineral surface (Mikhlin et al., 2004, 2017).

Mikhlin et al. (2004) determined the capacity of non-stoichiometric sulfides and intermediates from chalcopyrite, bornite (Cu_5FeS_4), and chalcocite (Cu_2S) to passivate their surfaces. Covellite (CuS) was not formed on the oxidation of chalcopyrite, whereas the formation of non-stoichiometric sulfides ($Cu_{1-x}Fe_{1-y}S_{2-z}$, Cu_xS) was confirmed (Mikhlin et al., 2004). Chalcopyrite oxidation yielded copper and iron in the solution phase and the formation of S–S bonds on the mineral surface. The sulfur/metal and copper/iron ratios in the aqueous phase were dependent of the potential that was applied. The non-stoichiometric layers on chalcopyrite surface were not the cause of passivation. Mikhlin et al. (2004) concluded that the low chalcopyrite dissolution was due to the extremely slow diffusion of copper and iron from the bulk solid to the chalcopyrite surface (i.e., metal depletion). Thus, the slow chalcopyrite dissolution was associated with the semiconductor behavior or the formation of passive layer on chalcopyrite surface.

Nasluzov et al. (2019) demonstrated in DFT + *U* (*U* = Hubbard-type correction parameter) simulation and chalcopyrite XPS studies that the crystal structure comprises centers with tri- or pentasulfide or tri- and disulfide complex anions, with a negative energy formation of 1.2–1.5 eV for each Fe atom extracted from the structure. The XPS data suggested initial depletion of iron, but not copper, on chalcopyrite surface and the presence of sulfides and polysulfide anions (S_n^{2-} with $n>5$). During chalcopyrite oxidation, Cu was depleted as S–S chains were formed. The stability of the polysulfide centers was considered responsible for the delayed oxidation and leaching of chalcopyrite (Nasluzov et al., 2019).

Contrary to the conclusions suggested by Mikhlin et al. (2004), Zhao et al. (2015a, 2019) reported bornite and covellite as the main intermediates associated with chalcopyrite dissolution. Bornite formation represented a reductive step, which was believed to be a rate-limiting reaction in the overall chalcopyrite dissolution. Covellite formation from bornite was an oxidation step and not a cause of the rate limitation (Zhao et al., 2019).

The contrasting interpretations may be attributed to the differences in chalcopyrite surface characterization, although the surfaces were characterized in cyclic voltammetry experiments in both studies. It is possible that due to experimental differences, atypical secondary solid phases with variable stability were formed in the two studies. This controversy shows, however, that multiple experimental and analytical approaches should be used to interpret the formation of intermediate solid phases during the time course of chalcopyrite dissolution.

In the semiconductor model of chalcopyrite and leaching, the path of electron transfer between chalcopyrite and a redox pair in the solution depends on their respective energy levels (Memming, 2015). It is necessary that the energy level of the redox pairs in the electrolyte (E_{redox}) approaches the energy of the edge of the conduction band (E_C) or the valence band (E_V) of the semiconductor chalcopyrite (Crundwell, 1988; Osseo-Asare, 1992). If this condition is established, but the E_{redox} and the semiconductor Fermi level (E_F) are not at the same energy level, a charge transfer arises between the semiconductor and the redox pairs in solution in order to establish the equilibrium (Bott, 1998).

Electron transfer makes the phases negatively or positively charged. This affects the density of the states of the redox pairs, and in the case of the semiconductor, the excess or lack of charge is distributed within the solid up to a distance of about 10–1000 nm (Bott, 1998; Crundwell, 2015); this zone is called the space charge region. The space charge regions with a lack or excess of major charge carriers (electrons for *n*-type semiconductors and holes for *p*-type) are also called the depletion and accumulation regions, respectively (Bott, 1998).

The charge transfer of a semiconductor is a function of the concentration of major charge carriers. The semiconductor behaves like a metal in an accumulation condition, as there are excess charge carriers available for charge transfer. Slow reactions are expected in a depletion layer situation (Bott, 1998; Crundwell, 2015). Therefore, according to the chalcopyrite semiconductor model, efficient leaching under accumulation conditions is expected.

O'Connor and Eksteen (2020) argued that many studies claim that chalcopyrite leaching in acidic solutions leads to passivation. However, other studies have shown no passivating effect in alkaline solutions with a complexing agent, although purportedly the same passivating species are formed on the mineral surface (O'Connor & Eksteen, 2020). If copper- or iron-oxides are naturally formed on the chalcopyrite surface, they are dissolved when immersed into acid solution, and eventually, the protection of the surface is destroyed. It is accepted that the Fe-S bond is broken more readily than the Cu–S bond and the direct decomposition of sulfide surface has a very slow rate. Surface analyses indicate that polysulfide chains of different sizes, elemental sulfur, and intermediate Cu-sulfides are formed. The *n*-type semiconductor character of chalcopyrite as determining its slow dissolution was also criticized by Nicol (2017a), attributing the observed effects to improper operation of laboratory equipment or experimental artifacts. Mikhlin et al. (2017) and Nasluzov et al. (2019) argued that the semiconductor character disappears after the modification of the first layers of the mineral surface. Ozone treatment after the bioleaching step was shown to oxidize reduced-S-containing complexes on chalcopyrite surface (Lv et al., 2021). Measurements of corrosion current densities (Tafel curves) and open-circuit potentials indicated increased reactivity of chalcopyrite after ozone treatment.

An alternative to the different approaches on chalcopyrite surface oxidation/dissolution is the reductive/dissolution route (Biegler & Horne, 1985; Hiroyoshi et al., 1997, 2000, 2001, 2004, 2008; Sandström et al., 2005; Gu et al., 2013; Zhao et al., 2015c), which is based on simultaneous electrochemical reactions occurring spontaneously under moderate acid leaching conditions. Based on this approach, Toledo et al. (2022) explained the high yield of copper recovery from a chalcopyrite concentrate sample under abiotic conditions at 1 atm and at 65 °C. The experimental variables were the initial concentration of Fe^{2+} ions and pulp density (ρ_{pulp}). Relatively high copper extractions were obtained at the initial $[Fe^{2+}]/\rho_{pulp}$ ratio of about 80, with an optimal range of solution potential maintained during almost all the time courses of 28 days. A response surface with statistical confidence of 0.997 was obtained using a central composite factorial design, allowing to reach the optimal condition with > 90% of chalcopyrite dissolution.

3 Galvanic Interaction

When dissimilar metals are immersed in a corrosive or conductive solution, there is a potential difference between the metals. If these metals are brought into contact or electrically connected, this potential difference produces a flow of charges. The corrosion of the less resistant metal increases and that of the more resistant metal decreases, compared to their behavior when they are electrically separated. Thus, the metal more resistant to corrosion acts as a cathode and the less resistant as an anode generating a galvanic cell. This behavior can be extended to other materials, for example, to ores with multiple sulfide minerals.

The electrochemical action produced by different metal reactions through a path leading to electrons and electrolytes generates a difference in potentials between the involved phases. The galvanic interaction that occurs between two minerals is caused by the different rest potentials, which lead to different electrochemical reactivities (Peters, 1977). In the case of sulfide minerals, when they come into contact with each other, they can form a galvanic cell, and thus, oxidation–reduction reactions occur, caused by the difference in the resting potentials of the mineral phases (Peters, 1977). The rest potential difference distinguishes between cathodically protected and anodically sacrificed minerals.

The rest potential indicates the electrochemical reactivity. Therefore, in the galvanic cell, the mineral with the higher resting potential is considered the nobler mineral, and the mineral with the lower resting potential is actively dissolved. Classification of common sulfide minerals in terms of their rest potentials and their mineral reactivity has been presented in the literature (Peters, 1977; Tanne & Shippers, 2021). According to Zhao et al. (2015a), the pyrite addition for chalcopyrite dissolution generates a catalytic effect that is mainly attributed to the increase in galvanic current. In this galvanic cell, pyrite is the cathodic phase, while chalcopyrite acts as the anodic phase leading to preferential dissolution (Zhao et al., 2015a). The galvanic effect of pyrite has been reported in many studies of chemical leaching and bioleaching of chalcopyrite (Li et al., 2017; Natarajan & Kumari, 2014; Olvera et al., 2014; Zheng et al., 2021).

In leaching tests with chalcopyrite and bornite in different ratios, the copper extraction was greater than 90% after 30 days in all tests containing the two minerals in a mixture (Zhao et al., 2015a). Copper dissolution was 48% at the most when the two Cu-sulfide minerals were tested separately. In the presence of *Leptospirillum ferriphilum*, the leaching of copper greatly increased when the redox potential exceeded 380 $mV_{Ag|AgCl|KClsat}$, and the dissolution almost ceased when the redox potential exceeded 480 $mV_{Ag|AgCl|KClsat}$. If the redox potential was maintained in the range (380–480 $mV_{Ag|AgCl|KClsat}$) during the time course, the bioleaching of chalcopyrite and bornite admixtures yielded high copper extraction. Similar results were obtained with chalcopyrite and pyrite mixtures, with the best results with 4:1 pyrite:chalcopyrite by weight ratios. Zhao et al. (2015a) attributed these results to the optimal redox potential range for chalcopyrite dissolution (380–480 $mV_{Ag|AgCl|KClsat.}$), while the galvanic effect by pyrite was not considered important in this case. Chalcopyrite dissolution and pyrite activation effect were dependent on the mixing ratio.

Tanne and Shippers (2021) monitored the evolution of electric current in galvanic cells with chalcopyrite–chalcopyrite, pyrite–pyrite, and chalcopyrite–pyrite mixtures in different proportions without current disturbance (ZRA—Zero Resistance Ammeter mode). The results showed that the galvanic effect on the coupling between chalcopyrite and pyrite was relatively small. The authors speculated that chalcopyrite was too recalcitrant to dissolve during the experiment because the difference in the resting potentials between the two minerals was too small. Bioleaching results confirmed that the concentrate was leached much faster and to a greater extent when assisted galvanic bioleaching was applied. In the bioleaching assays, acidophilic microorganisms played a key role in keeping the redox potential of the solution high and may have partially oxidized electrically insulating sulfur layers. The increase in pyrite content in the mixed mineral system resulted in higher Cu recovery (Tanne & Shippers, 2021).

Hiroyoshi et al. (2000) reported that the dissolution of chalcopyrite was accelerated when the redox potential was controlled at a relatively low value. Under these conditions, chalcopyrite was reduced to Cu_2S, and its subsequent rapid dissolution yielded a high copper extraction. Thus, there are also major conflicts over topics concerning the effects of bornite or pyrite on the bioleaching of chalcopyrite. The enhancement of chalcopyrite leaching is related to these interactions between the different minerals, and the optimal region of the redox potential in mixtures of chalcopyrite and other sulfide minerals also improves the leaching.

Pathak et al. (2017) reviewed several catalysts that have been shown to enhance the bioleaching of chalcopyrite. These catalysts include metals, most notably silver (e.g., as $AgNO_3$), which precipitates as Ag_2S and Ag^0 on chalcopyrite surface, enhancing the semiconductor properties of the mineral and reducing the formation of the passivating S^0 layer on the mineral surface. Silver catalyst in the bioleaching is in flux between the solution and solid phases because Ag_2S is readily oxidized by Fe^{3+} to Ag^+ and S^0. Ag_2S also acts cathodically in contact with chalcopyrite, thus assisting the galvanic coupling effect (Yang et al., 2019; Zhao et al., 2022).

Activated carbon has also been shown to enhance the bioleaching of chalcopyrite (Ahmadi et al., 2013; Méndez et al., 2022). This effect is attributed to the activated carbon sorption of sulfur intermediates, thereby partially alleviating passivation effect caused by S^0 formation. Activated carbon is cathodic with respect to chalcopyrite, and this galvanic interaction has a positive effect in the bioleaching. It is conceivable that, as shown with biochar treatment, excessive activated carbon results in the formation of a passivating surface layer with sulfur-laden carbon intermixed with jarosite-type precipitates. Yang et al. (2017) demonstrated that visible light and 0.1% graphene accelerated Fe^{2+} oxidation during chalcopyrite bioleaching. The effect also involved jarosite precipitation on graphene particle surfaces, thus reducing its formation on chalcopyrite surface. Cyclic voltammetry results were consistent with these effects.

4 Bioleaching Microorganisms

Several bacteria and archaea capable of oxidizing Fe- and S-compounds produce acid leaching conditions for the dissolution of chalcopyrite (Latorre et al., 2016; Sadeghieh et al., 2020). They produce protons from sulfur oxidation (Eq. 1) under acidic conditions and regenerate Fe^{3+} as the chemical oxidant (Eq. 2) in the bioleaching process.

$$2S^0 + 3O_2 + 2H_2O \rightarrow 2SO_4^{2-} + 4H^+, \quad (1)$$

$$4Fe^{2+} + O_2 + 4H^+ \rightarrow 4Fe^{3+} + 2H_2O. \quad (2)$$

Rather than pure cultures, mixed cultures containing S- and Fe-oxidizers are recognized to be more efficient in the bioleaching. Many of these microorganisms are commonly found in acid mine drainage and sediments. They vary in their temperature requirements, responses to pH, and tolerance to high concentrations of metals. Several reviews have been published in the last decade on the diversity of these bacteria and archaea (e.g., Johnson & Quatrini, 2020; Mahmoud et al., 2017; Moya-Beltrán et al., 2021; Nuñez et al., 2017; Quatrini & Johnson, 2018; Wang et al., 2020; Zhang et al., 2019).

Acidithiobacillus ferrooxidans is the most studied acidophile active in the bioleaching, with more than 9000 papers in the Web of Science. Its genome (NBCI txid920) was the first to be sequenced among bioleaching microorganisms. The biological leaching of Cu from chalcopyrite has always been only partial, reaching a maximum of 60–80% over 3–4 weeks in bench-scale studies depending on the specific experimental conditions. Limited copper dissolution under mesophilic conditions is a consequence of the increased redox potential of the leach solution, which is associated with the high ferric/ferrous ratio, leading to formation of Fe(III) precipitates (Li et al., 2013; Tian et al., 2021; Zhao et al., 2019). These changes in the leach solution are known to hinder the dissolution of chalcopyrite.

Moderately thermophilic (approx. 45–60 °C) and extremely thermophilic (approx. 60–80 °C) microorganisms have faster oxidation rates and thereby bring about improvement in chalcopyrite bioleaching. Some examples of chalcopyrite bioleaching and microbial diversity include the isolation and characterization of a novel, extremely thermoacidophilic, obligately chemolithotrophic *Acidianus sulfidivorans* (Plumb et al., 2007). This archaeon grows optimally at 74 °C and is active at a pH range of 0.4–2.2. Vilcáez et al. (2008) evaluated chalcopyrite bioleaching with three thermophiles at 65, 70, 75, and 80 °C. *Acidianus brierleyi* was the least active Fe oxidizer, suppressing the redox potential of the leach solution near the critical value of 450 $mV_{Ag/AgCl}$, thus favoring chalcopyrite leaching. *Sulfolobus metallicus* and *Metallosphaera sedula* oxidized Fe^{2+} at faster rates, thus promoting higher redox potential of the leach solution but causing lower efficiencies of chalcopyrite leaching. Iron oxidation by these thermophiles also resulted in ferric iron precipitation, which on the one hand suppresses chalcopyrite leaching and on the other hand decreases the solution redox potential, thus favoring chalcopyrite leaching. The results also demonstrated that a threshold concentration of either Fe^{2+} or Fe^{3+} is required to initiate the bioleaching of chalcopyrite.

Castro and Donati (2016) characterized a thermophilic archaeon, *Acidianus copahuensis*, which also had a low iron oxidation capacity. In bioleaching experiments, iron released from the chalcopyrite matrix remained mainly in the ferrous form because of the low oxidation activity, thus contributing to low redox potential of the leach solution. Safar et al. (2020) tested *Ac. copahuensis* further in chalcopyrite bioleaching experiments and demonstrated that initial cell adhesion on the mineral particles combined with low iron oxidation activity achieved high copper leaching, which was attributed to a low redox potential.

Liu et al. (2017) evaluated the bioleaching of a chalcopyrite concentrate sample with mixed cultures of mesophilic, moderately thermophilic, and extremely thermophilic microorganisms. Secondary covellite, chalcocite, and bornite were formed during the time course. The formation of bornite and chalcocite was observed at solution redox potentials $< 500\ mV_{SHE}$. At > 550 mV, covellite was formed, but bornite and chalcocite were not detected. In addition, elemental S and jarosite-type precipitates were also formed; they did not appear to hinder chalcopyrite oxidation. The yields of copper leaching increased with the temperature of incubation: 59% in about 20 days at 30 °C, 78% in 16 days at 45 °C, and 85% at 65 °C in 10 days (Liu et al., 2017). Correspondingly, Hedrich et al. (2018) tested a moderately thermophilic mixed culture of *L. ferriphilum*, *Acidithiobacillus caldus*, and *Sulfobacillus* spp. for the bioleaching of a chalcopyrite concentrate sample in stirred tank temperature-controlled bioreactors. The yields of copper leaching increased with the temperature and with the lower redox potential of the leach solution.

Combinations of microbes expand the metabolic range in the bioleaching process and have been invariably more efficient than pure cultures of microbes. Pure culture work has been, however, important in elucidating oxidation steps of sulfide minerals and pathways of sulfur oxidation, iron redox shuttling, and coupling of the leaching process with biochemical energy transduction, respiratory chain activity, and physiological traits (Dopson & Okibe, 2023). Pure and mixed culture work has provided fundamental principles and premises on the limits of environmental pH, temperature, and responses to potentially toxic metals and solutes in acid leach solutions. Molecular-level and genetic studies have revealed potential to modify and manipulate properties of these microorganisms (Jung et al., 2021). These approaches are now widely used in characterization of acidophilic iron- and sulfur-oxidizing microbes, but they have yet to be integrated for commercialization of bioleaching processes.

5 Electrochemical Approaches for Chalcopyrite Dissolution

There have been ongoing discussion and interpretation of the interaction between microorganisms and the chalcopyrite substratum in the bioleaching process. Electrochemical techniques have been applied in endeavors to unravel mechanistic information and thermodynamics on bacteria–mineral interactions as it pertains to the bioleaching of chalcopyrite.

Biofilm formation impacts chalcopyrite surface and over time contributes to a passivation effect. Biofilms are typically composed of microbial cells and their extracellular polymeric substances (EPS), which can sequester metals, trap nano-size particles, and possibly also nucleate Fe(III) precipitation. Biofilms are invariably formed on mineral surfaces in bioleaching processes, causing changes in the electrochemical properties of the system (García-Meza et al., 2013; Lara et al., 2013; Bobadilla-Fazzini & Poblete-Castro, 2021). Zhao et al. (2019) and Zeng et al. (2023) elaborated on the biofilm aspects and discussed factors that are involved in bacterial attachment on chalcopyrite. Bobadilla-Fazzini and Poblete-Castro (2012) reported that biofilms of *Acidithiobacillus thiooxidans* and *Leptospirillum* spp. were not formed on chalcopyrite in stirred and laminar flow conditions. This may be a unique feature of the mixed culture as numerous other studies have proven that cell adhesion and attachment leading to biofilm layers are integral parts of the bioleaching of chalcopyrite and other sulfide minerals.

Electrochemical impedance spectroscopy (EIS) has proven a useful tool in detailing these effects. Figure 2 shows three equivalent electrical circuits (EEC) used to adjust the EIS data (Bevilaqua et al., 2004).

In the circuit A (Fig. 2a), Rs is the solution resistance and $Q_{act}(R_{ct}W)$ is connected to the electrode active region, where Q is the constant phase element composed by the admittance

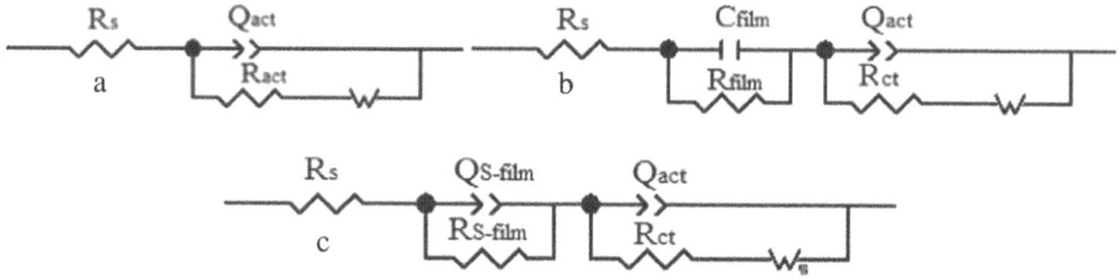

Fig. 2 Equivalent electrical circuit for EIS data: **a** 7 and 24 h in the absence and 7 h in the presence of bacteria; **b** 71 and 120 h in the absence and 24–120 h in the presence of bacteria; **c** for times longer than 120 h (Bevilaqua et al., 2004), slightly modified by the authors. Elements of the circuit: Rs, solution resistance; Q_{act}, constant phase element of the electrode active region; R_{act}, charge transfer resistance of the electrode active region; W, Warburg element that means semi-infinite linear diffusion; C_{film}, capacitance of biofilm or sulfur film partially covering the surface; R_{film}, resistance of biofilm or sulfur film partially covering the surface; $Q_{S\text{-}film}$, constant phase element of the film covering all electrode surfaces; $R_{S\text{-}film}$, resistance of the film covering almost all electrode surfaces; R_{ct}, charge transfer resistance; W_s, Warburg element, diffusion through the finite layer thickness

Yo and the exponent n of the electric double layer, both of them independent on the frequency. R_{ct} is the charge transfer resistance and W is the Warburg associated with a semi-infinite linear diffusion process. This circuit was able to describe the experimental conditions: 7 and 24 h of incubation in the absence and 7 h in the presence of bacteria. For longer times, modified circuits were proposed to represent mineral surface in part and were almost entirely covered by an adherent porous layer by compounds containing sulfur, hydroxides, biomolecules, and biomass. For immersion times between 71 and 120 h in the absence and 24–120 h in the presence of bacteria, circuit B [$Rs(R_{film}C_{film})(Q_{act}(R_{ct}W))$] (Fig. 2b) was used where the association $R\text{-}C_{film}$ represents the resistance and capacitance of a biofilm or sulfur film formed on part of the mineral surface. For times longer than 120 h, circuit C [$(Rs(Q_{S\text{-}film}R_{S\text{-}film})(Q_{act}(R_{ct}W_s))$] is shown in Fig. 2c, in which the $Q_{S\text{-}film}R_{S\text{-}film}$ sub-circuit was associated with the sulfur or biofilm covering almost entirely the electrode surface with the mass transport limited by diffusion through the film, where W_s represents a finite diffusion element (Bevilaqua et al., 2004).

Electrochemical noise analysis (ENA) has also been used to characterize bacteria–mineral interactions (Bevilaqua et al., 2006, 2007, 2011). The addition of chloride and silver ions to the bioleaching system was shown with ENA to influence the electrochemical behavior of chalcopyrite. Chloride caused an increase in the electrochemical potential and the current response of the system, thus enhancing the dissolution of chalcopyrite. Silver ions caused considerable variations in the noise resistance (R_n) values, indicating signal instability and a lack of steadiness in the leaching of chalcopyrite (Horta et al., 2009a).

The use of carbon paste electrodes (CPE) with minerals instead of massive mineral electrodes has improved reproducibility in electrochemical tests. Massive mineral electrodes are subjected to signal variability due to polishing and fracture and the lack of homogeneity (Horta et al., 2009b). The use of CPE showed a linear correlation between the mass of chalcopyrite and the charge obtained from cyclic voltammetry assays when the proportion of mineral in the CPE electrodes was between 20 and 80 wt%. By converting the noise resistance to admittance, the results of CPE studies suggested that chalcopyrite passivation was associated with slow chalcopyrite dissolution (Horta et al., 2009a). In the presence of *A. ferrooxidans* and additional chloride in the electrochemical cell, bacteria adhered to the electrode, resulting in an activated state with the influence of the two components (bacteria + Cl^- ions). This increased the admittance with a greater dispersion of the admittance points, showing a synergism between the bacteria and chloride ions in solution. Further results with CPE manufactured with solid residues from bioleaching experiments showed that chloride addition inhibited the secondary solid-phase formation on chalcopyrite. Thus, chalcopyrite dissolution was less hindered, indicating a less resistive behavior and more susceptible mineral dissolution, showing synergism between bacteria and chloride in chalcopyrite bioleaching.

The composition, structure, and other properties of chalcopyrite electrodes change during contact in the leaching solution due to precipitation, dissolution, and biomass accumulation (Kinnunen et al., 2006; Varotsis et al., 2022). The electrochemical potential of chalcopyrite electrodes decreases during these reactions; thus, it can be used to monitor the time course of the leaching process. In general, mineral electrodes with specific electrochemical potentials are used to optimize and follow mineral leaching processes.

Impedance experiments showed that the addition of Fe^{2+} decreased the capacitive arcs, suggesting the removal or absence of precipitates blocking of the electrode (Arena et al., 2016). Hydrodynamic impedance tests indicated a resistance decay of the system, which was attributed to the improved mass and ion transport, avoiding the accumulation of passivating precipitates on the electrode surface (Fig. 3). Semiconductor behavior (Crundwell, 2015) and slow diffusion of metal ions from the interior to the chalcopyrite surface (Mikhlin et al., 2017) have also been shown to explain the slow dissolution of chalcopyrite.

The optimal redox potential range of 380–430 $mV_{Ag|AgCl|KCl(3\ mol\ L^{-1})}$ minimizes the accumulation of surface layers, because under reducing conditions, the extraction of Fe from the crystal structure and the formation of less refractory, Fe-deficient Cu-sulfides can lead to almost complete chalcopyrite dissolution (Third et al., 2000, 2002; Vilcáez & Inoue, 2009). At low redox potentials, chalcopyrite is reduced in several steps to form Cu_2S, which is relatively readily dissolved, thus improving the dissolution kinetics (Hiroyoshi et al., 2000; Gu et al., 2013; Zhao et al., 2015b, 2015d, 2017).

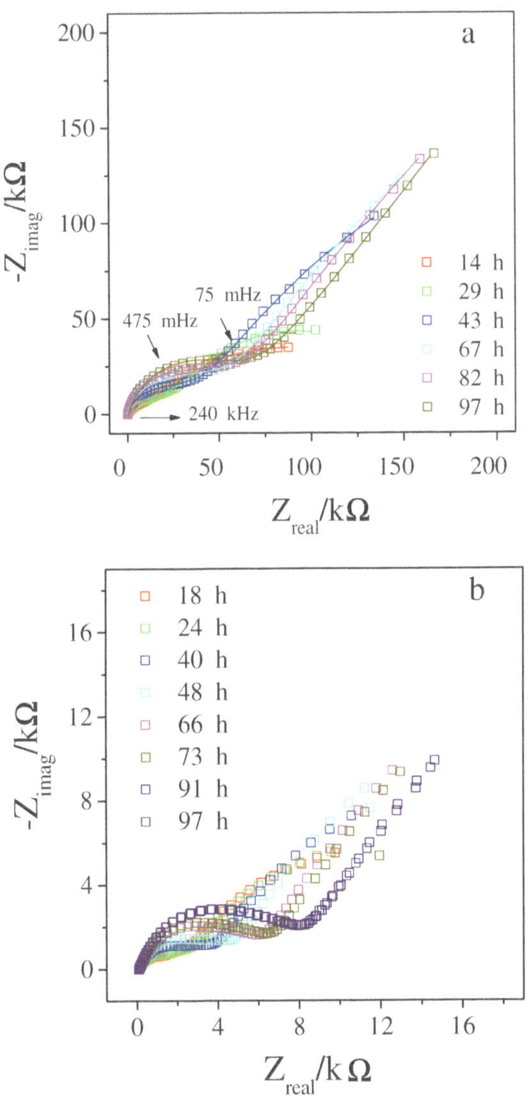

Fig. 3 Nyquist diagrams for carbon paste electrodes modified with chalcopyrite in natural aerated salts solution with 0.10 mol L^{-1} ferrous ions (**a**) (Arena et al, 2016) steady electrode (**b**) hydrodynamic electrode (1500 rpm)

Ghahremaninezhad et al. (2010, 2013) studied the chalcopyrite behavior in sulfuric acid solution in the absence of additional adding Fe^{2+} or Cu^{2+} and extended this study using XPS analysis. Based on potentiodynamic and EIS measurements, they proposed the formation of a passive layer due to the formation of metal-deficient sulfides ($Cu_{1-x}Fe_{1-y}S_2$, $y>>x$) at low potentials, and a second, copper sulfide-rich layer ($Cu_{1-x-z}S_2$), which was formed at higher potentials on the first layer as it was being dissolved. The two layers showed passive characteristics, but they started to dissolve in the range of 0.74–0.86 $V_{Ag|AgCl|KCl(3\ mol\ L^{-1})}$ leading to chalcopyrite dissolution (Ghahremaninezhad et al., 2010). The authors also studied the kinetics of Fe^{3+}/Fe^{2+} ions on anodically passivated chalcopyrite in sulfuric acid. They observed an increase in the dissolution of passivated chalcopyrite by ferric iron, which was reduced to ferrous iron and thereby lowered the redox potential (Ghahremaninezhad et al., 2012).

Processes occurring on the chalcopyrite/solution interface can be defined in relation to electrochemical mineral surface features and solution composition at specific electrochemical potentials. Mineral surfaces strongly influence the rates of chalcopyrite dissolution and the formation of intermediates and products (Lara et al., 2013; Zhao et al., 2015b). The formation of Fe(III)-phosphate on the chalcopyrite surface due to the presence of phosphate ions in the electrolyte has been noted (Lara et al., 2015). Yang et al. (2014) detected phosphorus on the surface of chalcopyrite bioleaching residues by X-ray photoelectron spectroscopy (XPS), probably as Fe(III)-phosphate precipitates at 0.60–0.65 $V_{Ag|AgCl|KClsat}$. Phosphate is used invariably in bioleach solutions as it is an important nutrient for microorganisms. Electrochemical experiments with chalcopyrite and mesophilic acidophiles and the corresponding XPS surface analyses support the formation of fractions of disulfides ($S_2^{2}-$), monosulfide (S^{2-}), polysulfides ($S_n^{2}-$) and elemental sulfur (S^0) in different proportions (Zhao et al., 2015b, 2015c). Zhao et al. (2015b) also found covellite as the main intermediate of chalcopyrite dissolution. Intermediary covellite formation during chalcopyrite oxidation has been disputed (Arce & González, 2002; Mikhlin et al., 2017). Microbial cell and EPS distribution, intermediate chemical species, and secondary solid phases on leached chalcopyrite surfaces vary spatially as well as in composition (García-Meza et al., 2013; Varotsis et al., 2022).

6 Redox Potential Control

Several reports emphasize the importance of the solution potential of the redox pairs in determining the kinetics of chalcopyrite reactions. Various hypotheses have been presented to explain this effect, and strategies for controlling the potential have been discussed in the literature (Li et al., 2013; Zhao et al., 2019). Tian et al. (2021) summarized the effects of redox potential on the chemical leaching and bioleaching of chalcopyrite and attributed these effects to the band theory. Several chemical reactions have been proposed to describe chemical transformations of chalcopyrite and intermediates during the leaching processes. Sequences of the reactions have been characterized with electrochemical techniques and have revealed the formation of secondary sulfides (e.g., covellite, chalcocite, bornite, and other Fe-deficient sulfides) and S^0 in chemical and biological leaching experiments with chalcopyrite (Biegler & Horne, 1985; Biegler & Swift, 1979; Dutrizac & MacDonald, 1974; Elsherief, 2002; Hiroyoshi et al., 2004; Holliday & Richmond, 1990; Majuste et al., 2012; Munoz et al., 1979; Sohn & Wadsworth, 1980; Warren et al., 1982). Analyses of cathodic current (or reduction) and anodic current (or oxidation) peaks obtained with the cyclic voltammetry technique have greatly contributed to the understanding of chalcopyrite dissolution.

Many kinetic studies demonstrate that data on the chemical leaching and bioleaching of chalcopyrite in various solution compositions fit the shrinking core model. The rate limitation is a surface reaction, and the leaching over time becomes diffusion-controlled at ambient temperatures and sometimes chemical reaction-controlled at elevated temperatures (e.g., Hidalgo et al., 2019; Jordan et al., 2006; Kaplun et al., 2011; Koleini et al., 2010; Liao et al., 2020). Some chemical leaching results also indicate data fit in mixed kinetic models, changing with the time course (Ranjbar et al., 2020). Although there is only little consensus about the underlying reasons of the slow kinetics of copper extraction from chalcopyrite, poor leaching efficiency has been reported at high solution potentials, while at lower potentials (380–450 $mV_{Ag|AgCl|KClsat}$), the dissolution can reach near completion (Bevilaqua et al., 2014; Castro & Donati, 2016; Gu et al., 2013; Kametani & Aoki, 1985; Petersen & Dixon, 2006; Sandström et al., 2005; Santos et al., 2017; Third et al., 2000, 2002; Vilcáez et al., 2008; Yang et al., 2018; Zhao et al., 2015c). Studies with controlled solution potential have been carried out to better understand the role of factors that affect the bioleaching of chalcopyrite. The potential of the redox couples in the solution can be electrochemically controlled by the application of an external electrical potential with the use of electrodes or chemically using reducing agents (e.g., Na_2SO_3) or oxidizing agents (e.g., H_2O_2, O_2, $KMnO_4$). The solution potential is also influenced by the bulk microbial biomass (Li et al., 2013), although the prime redox couple in many cases is the iron shuttle, Fe^{3+}/Fe^{2+}.

Sandström et al. (2005) investigated the chemical and bacterial leaching at solution potentials maintained at 420 mV and 600 $mV_{Ag|AgCl|KClsat}$ at 65 °C, controlled with airflows and additions of $NaHSO_3$ and $KMnO_4$ solutions. Relatively high dissolution of chalcopyrite at 420 mV was observed, mainly in the chemical leaching. In the bioleaching process, a low amount of sulfur was formed because of its concurrent oxidation by bacteria at 600 mV. Because intermediate sulfur compounds were present in insignificant amounts, it was concluded that the formation of jarosite-type precipitates is key to passivation in the chalcopyrite bioleaching. Sandström et al. (2005) also noted that the dissolution rate increases with higher Cu^{2+} concentrations. Nazari and Asselin (2009) used computer simulations, based on the theory of percolation, to explain the morphology of secondary precipitates associated with the leaching of chalcopyrite in acidic ferric sulfate solution. A high copper extraction is obtained by maintaining the redox potential in the active sites of chalcopyrite within the optimal range, thus controlling the concentration and ratio of Fe^{3+} and Fe^{2+}. Without the reduction of Fe^{3+}, which slows the reaction in the active region of chalcopyrite, a Fe-deficient Cu-polysulfide passivation layer is formed, which inhibits the leaching of chalcopyrite.

Ahmadi et al. (2010) performed four types of experiments using a chalcopyrite concentrate sample in a bioreactor at 35 and 50 °C: chemical leaching, electrochemical leaching, bioleaching, and electrochemical bioleaching. In electrochemical bioleaching experiments the redox potential was controlled in the 400–450 $mV_{Ag/AgCl|KClsat}$ interval by applying an external electrical potential. The formation of a passive, Fe(III)-rich layer on chalcopyrite was limited and the highest yields of copper were obtained in these redox potential-controlled experiments (Ahmadi et al., 2010). Gericke et al. (2010) manipulated the redox potential in chalcopyrite bioreactor experiments by controlling the available oxygen concentration through aeration. About 90% copper dissolution from a chalcopyrite sample was obtained at potentials at 410–440 $mV_{Ag|AgCl}$ range, in contrast to approx. 40% extraction at 580 mV.

Velásquez-Yévenes et al. (2010) reported that the dissolution rate of chalcopyrite in 0.2 mol L^{-1} HCl with 0.5 g L^{-1} Cu^{2+} at 35 °C was strongly dependent on the solution redox potential in the range of 345–415 $mV_{Ag|AgCl|KClsat}$. The redox potential was controlled using three strategies: (i) electrochemical (a passage of an appropriate current between platinum electrodes allowing the electrode potential to be controlled to a defined value), (ii) chemical based on potentiometric titration with permanganate, and (iii) gaseous with mixtures of N_2 and O_2. The dissolution rate decreased substantially at potentials below 335 mV and above 415 mV. Chalcopyrite passivation occurring above this redox range was partially reversible if the redox potential was lowered to a more favorable value.

Gu et al. (2013) evaluated both electrochemical and bioleaching experiments using cyclic voltammetry with a chalcopyrite electrode containing *L. ferriphilum*. Chalcocite was identified during the bioleaching of chalcopyrite at low potentials. The reductive conditions enhanced the dissolution kinetics of chalcopyrite. Jarosite was formed but was not deemed a passivating component because of its loose and porous texture. In the cyclic voltammetry test, the potential sweep followed the common path from the open-circuit potential (OCP) or 381 $mV_{Ag\,|\,AgCl\,|\,KClsat}$ to 801 mV, then to −999 $mV_{Ag\,|\,AgCl\,|\,KClsat}$ and back to the OCP value. Among the cathode peaks obtained, the reductive formation of talnakhite ($Cu_9Fe_8S_{16}$) or bornite as well as chalcocite was proposed. When the potential sweep was reversed toward the positive direction, sulfur and non-stoichiometric Cu-polysulfides such as djurleite (ideally $Cu_{31}S_{16}$, the formula varies) and digenite (Cu_9S_5) were detected.

Bevilaqua et al. (2014) evaluated the effect of the solution potential and the ratio of Fe^{3+} and Fe^{2+} concentrations on the chemical leaching and bioleaching of two types of chalcopyrite concentrates. Initial redox potentials between 350 and 600 $mV_{Ag\,|\,AgCl\,|\,KClsat}$ were adjusted with different ratios of $[Fe^{3+}]/[Fe^{2+}]$. Chalcopyrite dissolution was hindered when Fe^{2+} was completely oxidized in these experiments. The leaching rate declined when the solution potential increased to 580 mV. Enhanced copper dissolution was observed at high Fe^{2+} concentrations that suppressed the redox potential to <370 mV.

Santos et al. (2017) reported 90% copper extraction in the chemical leaching at 200 mmol L^{-1} Fe^{2+} at redox potentials <420 $mV_{Ag\,|\,AgCl\,|\,KClsat}$ (Fig. 4). Relatively low copper dissolution of 17% was obtained at 610 $mV_{Ag\,|\,AgCl\,|\,KClsat}$ potentials in the presence of *A. ferrooxidans*. The bacteria catalyzed the oxidation and Fe^{2+}, therefore increasing the redox potential. Under both conditions, the formation of passivating species was observed, but the dissolution of chalcopyrite was not impeded. Thus, the maintenance of a low range of redox potential in the presence of Fe^{2+} favored the leaching of chalcopyrite (Fig. 5). Factors causing the passivation of chalcopyrite leaching were not verified by Santos et al. (2017). Li et al. (2017) used synchrotron scanning photoelectron microscopy to examine chalcopyrite surfaces and residues after partial leaching at a controlled redox potential (451 $mV_{Ag\,|\,AgCl\,|\,KClsat}$), which was maintained due to the concurrent oxidation of pyrite. Intermediate sulfur species (S^{2-}, S_2^{2-}, S_n^{2-}, S^0) were identified on the surface of chalcopyrite, but their heterogeneous distribution on mineral surfaces did not indicate passivation. Evidence to date demonstrates that the bioleaching of chalcopyrite is efficient at low redox potentials and the efficiency decreases with increasing redox potential. The dominant redox shuttle comprises Fe^{2+} and Fe^{3+}, and thus, the predominance of Fe^{2+} favors the bioleaching of chalcopyrite.

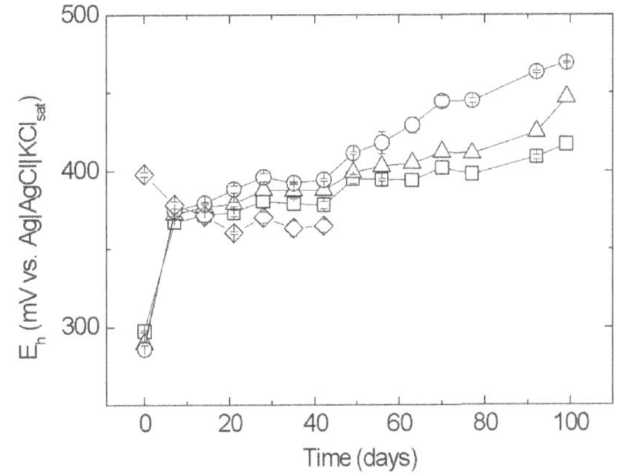

Fig. 4 Time course of copper dissolution and redox potential measurements on chalcopyrite leaching using different concentrations of ferrous ions: rhombus—0 mol L^{-1}; square—0.100 mol L^{-1}; triangle—0.200 mol L^{-1}; and circle—0.300 mol L^{-1} of ferrous ions (Santos et al., 2017)

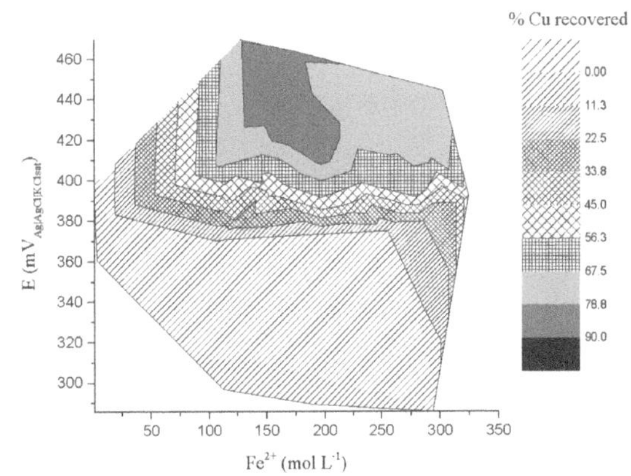

Fig. 5 Combined effect of ferrous iron concentration and redox potential on chalcopyrite dissolution under abiotic conditions (Santos et al., 2017)

Improved efficiencies would be expected with increasing Fe^{3+} concentrations, but this is not what happens at increasing redox potentials. Relatively fast rates of copper extraction have been reported for redox potentials up to a certain limit (380–480 mV versus Ag|AgCl|KCl (3 mol L^{-1}), and the presence of Cu^{2+} and Fe^{2+} also enhances the leaching of chalcopyrite (Bevilaqua et al., 2014; Biegler & Horne, 1985; Hiroyoshi et al., 1997, 2001, 2004, 2008; Sandström et al., 2005; Vilcáez et al., 2008; Nazari & Asselin, 2009; Viramontes-Gamboa et al., 2010; Velásquez-Yévenes et al., 2010; Gu et al., 2013; Zhao et al., 2015c).

In an attempt to explain the influence of solution potential and Fe^{2+}, Fe^{3+}, and Cu^{2+} on chalcopyrite leaching, Hiroyoshi et al. (2000) proposed a two-step model: chalcopyrite is initially reduced to chalcocite, followed by oxidation by Fe^{3+} and dissolved O_2 (Reactions 3–5).

$$CuFeS_2 + 3Cu^{2+} + 3Fe^{2+} \rightarrow 2Cu_2S + 4Fe^{3+}, \quad (3)$$

$$Cu_2S + 4Fe^{3+} \rightarrow 2Cu^{2+} + 4Fe^{2+} + S, \quad (4)$$

$$Cu_2S + 4H^+ + O_2 \rightarrow 2Cu^{2+} + S + 2H_2O. \quad (5)$$

This model stipulates that the redox potential remains within a range between the Nernst potentials of the chalcopyrite reduction and chalcocite oxidation (Reactions 3–5). The dissolution of chalcopyrite would be thermodynamically inhibited if the solution potential has a value greater than the Nernst potential of the chalcopyrite reduction Reaction (3) or less than the Nernst potential of the chalcocite oxidation Reactions (4 and 5). From a thermodynamic point of view, this model may be relevant approach to approximating limitations in the bioleaching of chalcopyrite. However, studies to date show that other intermediates besides chalcocite also participate in the chalcopyrite bioleaching process.

7 Concluding Remarks

Bioleaching of Cu-sulfide ores is practiced in heaps in many locations worldwide. This technology is particularly suitable for secondary Cu-sulfides such as chalcocite and covellite as well as Cu-oxides. The primary Cu-sulfide, chalcopyrite, is refractory in these bioprocesses and requires specific process control for enhancing the extent and kinetics of the bioleaching. In the 1970's and 1980's, it was generally recognized that bioleaching could not be applied to chalcopyrite concentrates because of the slow reaction kinetics. The underlying reasons and mechanisms have been elucidated in the past couple of decades, and the acid leaching reactions aided by acidophiles can be directed toward optimization by controlling the solution redox potential and the temperature regime. Chalcopyrite electrodes in acid leaching systems can be used to monitor the oxidation state of the mineral. Pyrite electrodes can also be useful if galvanic coupling is established. Redox potential, pH, chemical oxidants, and iron- and sulfur-oxidizing microorganisms are the main controlling factors of chalcopyrite surface chemistry. These parameters change during the time course of chemical leaching and bioleaching. Active acidophiles are best employed as consortia of Fe- and S-oxidizing prokaryotes (= bacteria and archaea) for the redox-temperature-pH-ferric-ferrous-iron conditions during optimal chalcopyrite bioleaching. The formation of secondary solid phases such as (Fe(III)-precipitates, S^0, and metastable Fe-deficient Cu-sulfides accounts for the diffusion control of the leaching kinetics. Chemical and microbiological leaching of chalcopyrite yields Fe^{2+}, S^0, sulfooxyanions, and non-stoichiometric Fe- and Cu-polysulfides. Instead of a strictly chemical leaching process, the kinetics are favorable for the bioleaching, preferentially with thermophiles especially. Both the chemical leaching and bioleaching are enhanced if chalcopyrite is initially reduced to bornite and other Cu-sulfides, followed by oxidation by Fe^{3+} and dissolved O_2. The pathways of Cu, Fe, and S may be different in the chemical leaching and bioleaching. Electrochemical techniques have given insight into the sequences of leaching reactions. Analysis of the thermodynamic and electrochemical properties of chalcopyrite has presented possibilities for external control of the chalcopyrite leaching process, for example, by the process temperature, redox shuttles, aeration, galvanic coupling, and voltammetry.

Glossary

C_{film}	Capacitance of biofilm or sulfur film partially covering the surface
CPE	Carbon paste electrodes
DFT	Density functional theory
E_C	Conduction band
EEC	Equivalent electrical circuits
E_F	Fermi level
EIS	Electrochemical impedance spectroscopy
ENA	Electrochemical noise analysis
E_{redox}	Redox potential
E_V	Valence band
ΔG_f°	Standard Gibbs free energy of formation
ΔH_f°	Standard enthalpy of formation
n	Exponent
OCP	Open-circuit potential
ρ_{pulp}	Pulp density
Q	Constant phase element
$Q_{\text{S-film}}$	Constant phase element of the film covering almost entirely the electrode surface
R_{act}	Charge transfer resistance of the active region of the electrode
R_{ct}	Charge transfer resistance
R_{film}	Resistance of biofilm or sulfur film partially covering the surface
R_n	Noise resistance
rpm	Revolutions per minute
Rs	Solution resistance
$R_{\text{S-film}}$	Resistance of the film covering almost entirely the electrode surface
SAT	Saturated
SHE	Standard hydrogen electrode
U	Hubbard-type correction parameter
W	Warburg element representing a semi-infinite linear diffusion
W_s	Warburg element, representing the diffusion through a finite layer thickness

XANES X-ray absorption near-edge spectroscopy
XPS X-ray photoelectron spectroscopy
Yo admittance
ZRA Zero Resistance Ammeter mode

References

Ahmadi, A., Schaffie, M., Manafi, Z., et al. (2010). Electrochemical bioleaching of high grade chalcopyrite flotation concentrates in a stirred bioreactor. *Hydrometallurgy, 104*, 99–105. https://doi.org/10.1016/j.hydromet.2010.05.001

Ahmadi, A., Ranjbar, M., & Schaffie, M. (2013). Effect of activated carbon addition on the conventional and electrochemical bioleaching of chalcopyrite concentrates. *Geomicrobiology Journal, 30*, 237–244. https://doi.org/10.1080/01490451.2012.665152

Akcil, A., Gahan, C. S., Erust, C., et al. (2013). Influence of chloride on the chemolithotrophic acidophiles in biohydrometallurgy: A review. In K. Pramanik & J. K. Patra (Eds.), *Industrial and environmental biotechnology* (pp. 45–69). Studium Press.

Arena, F. A., Suegama, P. H., Bevilaqua, D., et al. (2016). Simulating the main stages of chalcopyrite leaching and bioleaching in ferrous ions solution: An electrochemical impedance study with a modified carbon paste electrode. *Minerals Engineering, 92*, 229–241. https://doi.org/10.1016/j.mineng.2016.03.025

Arce, E. M., & González, I. (2002). A comparative study of electrochemical behavior of chalcopyrite, chalcocite and bornite in sulfuric acid solution. *International Journal of Mineral Processing, 67*, 17–28. https://doi.org/10.1016/S0301-7516(02)00003-0

Bevilaqua, D., Acciari, H. A., Benedetti, A. V., et al. (2006). Electrochemical noise analysis of bioleaching of bornite (Cu_5FeS_4) by *Acidithiobacillus ferrooxidans*. *Hydrometallurgy, 83*, 50–54. https://doi.org/10.1016/j.hydromet.2006.03.037

Bevilaqua, D., Acciari, H. A., Benedetti, A. V., et al. (2007). Electrochemical techniques used to study bacterial-metal sulfides interactions in acidic environments. In E. R. Donati & W. Sand (Eds.), *Microbial processing of metal sulfides* (pp. 59–76). Springer.

Bevilaqua, D., Diéz-Perez, I., Fugivara, C. S., et al. (2004). Oxidative dissolution of chalcopyrite by *Acidithiobacillus ferrooxidans* analyzed by electrochemical impedance spectroscopy and atomic force microscopy. *Bioelectrochemistry, 64*, 79–84. https://doi.org/10.1016/j.bioelechem.2004.01.006

Bevilaqua, D., Lahti-Tommila, H., Garcia Júnior, O., et al. (2014). Bacterial and chemical leaching of chalcopyrite concentrates as affected by the redox potential and ferric/ferrous iron ratio at 22 °C. *International Journal of Mineral Processing, 132*, 1–7. https://doi.org/10.1016/J.MINPRO.2014.08.008

Bevilaqua, D., Lahti, H., Suegama, P. H., et al. (2013). Effect of Na-chloride on the bioleaching of a chalcopyrite concentrate in shake flasks and stirred tank bioreactors. *Hydrometallurgy, 138*, 1–13. https://doi.org/10.1016/j.hydromet.2013.06.008

Bevilaqua, D., Suegama, P. H., Garcia Júnior, O., et al. (2011). Electrochemical studies of sulphide minerals in the presence and absence of *Acidithiobacillus ferrooxidans*. In L. G. Sobral, D. M. Oliveira, & C. E. G. Souza (Eds.), *Biohydrometallurgy processes: A practical approach* (pp. 143–167). CETEM/MCT.

Biegler, T., & Horne, M. D. (1985). The electrochemistry of surface oxidation of chalcopyrite. *Journal of the Electrochemical Society, 132*, 1363–1369. https://doi.org/10.1149/1.2114117

Biegler, T., & Swift, D. A. (1979). Anodic electrochemistry of chalcopyrite. *Journal of Applied Electrochemistry, 9*, 545–554. https://doi.org/10.1007/BF00610940

Bobadilla-Fazzini, R. A., & Poblete-Castro, I. (2021). Biofilm formation is crucial for efficient copper bioleaching from bornite under mesophilic conditions: unveiling the lifestyle and catalytic role of sulfur-oxidizing bacteria. *Frontiers in Microbiology, 12*, 761997. https://doi.org/10.3389/fmicb.2021.761997

Bott, A. W. (1998). Electrochemistry of semiconductors. *Current Separations, 27*, 87–91

Castro, C., & Donati, E. (2016). Effects of different energy sources on cell adhesion and bioleaching of a chalcopyrite concentrate by extremophilic archaeon *Acidianus copahuensis*. *Hydrometallurgy, 162*, 49–56. https://doi.org/10.1016/j.hydromet.2016.02.014

Crundwell, F. K. (2015). The semiconductor mechanism of dissolution and the pseudo-passivation of chalcopyrite. *Canadian Metallurgical Quarterly, 54*, 279–288. https://doi.org/10.1179/1879139515Y.0000000007

Crundwell, F. K. (1988). The influence of the electronic structure of solids on the anodic dissolution and leaching of semiconducting sulphide minerals. *Hydrometallurgy, 21*, 155–190. https://doi.org/10.1016/0304-386X(88)90003-5

Dopson, M., Holmes, D. S., Lazcano, M., et al. (2017). Multiple osmotic stress responses in *Acidihalobacter prosperus* result in tolerance to chloride ions. *Frontiers in Microbiology, 7*, 2132. https://doi.org/10.3389/fmicb.2016.02132

Dopson, M., & Okibe, N. (2023). Biomining microorganisms: Diversity and *modus operandi*. In D. B. Johnson, C. G. Bryan, M. Schlömann, & F. F. Roberto (Eds.), *Biomining technologies: Extracting and recovering metals from ores and waste* (pp. 89–110). Springer Nature. https://doi.org/10.1007/978-3-031-05382-5_5

Dutrizac, J. E. (1991). Ferric ion leaching of chalcopyrites from different localities. *Journal of Electronic Materials, 20*, 303–309. https://doi.org/10.1007/BF02816001

Dutrizac, J. E. (1978). The kinetics of dissolution of chalcopyrite in ferric ion media. *Metallurgical Transactions B, 9*, 431–439. https://doi.org/10.1007/BF02654418

Dutrizac, J. E. (1989). Elemental sulphur formation during the ferric sulphate leaching of chalcopyrite. *Canadian Metallurgical Quarterly, 28*, 337–344. https://doi.org/10.1179/cmq.1989.28.4.337

Dutrizac, J. E., & MacDonald, R. J. C. (1974). The kinetics of dissolution of covellite in acidified ferric sulphate solutions. *Canadian Metallurgical Quarterly, 13*, 423–433. https://doi.org/10.1179/cmq.1974.13.3.423

Elsherief, A. E. (2002). The influence of cathodic reduction, Fe^{2+} and Cu^{2+} ions on the electrochemical dissolution of chalcopyrite in acidic solution. *Minerals Engineering, 15*, 215–223. https://doi.org/10.1016/S0892-6875(01)00208-4

Fontana, M. G. (1987). *Corrosion engineering* (3rd ed.). McGraw-Hill.

García-Meza, J. V., Fernández, J. J., Lara, R. H., et al. (2013). Changes in biofilm structure during the colonization of chalcopyrite by *Acidithiobacillus thiooxidans*. *Applied Microbiology and Biotechnology, 97*, 6065–6075. https://doi.org/10.1007/s00253-012-4420-6

Gericke, M., Govender, Y., & Pinches, A. (2010). Tank bioleaching of low-grade chalcopyrite concentrates using redox control. *Hydrometallurgy, 104*, 414–419. https://doi.org/10.1016/j.hydromet.2010.02.024

Ghahremaninezhad, A., Asselin, E., & Dixon, D. G. (2010). Electrochemical evaluation of the surface of chalcopyrite during dissolution in sulfuric acid solution. *Electrochimica Acta, 55*, 5041–5056. https://doi.org/10.1016/j.electacta.2010.03.052

Ghahremaninezhad, A., Dixon, D. G., & Asselin, E. (2012). Kinetics of the ferric-ferrous couple on anodically passivated chalcopyrite ($CuFeS_2$) electrodes. *Hydrometallurgy, 125–126*, 42–49. https://doi.org/10.1016/j.hydromet.2012.05.004

Ghahremaninezhad, A., Dixon, D. G., & Asselin, E. (2013). Electrochemical and XPS analysis of chalcopyrite ($CuFeS_2$) dissolution in sulfuric acid solution. *Electrochimica Acta, 87*, 97–112. https://doi.org/10.1016/j.electacta.2012.07.119

Gu, G., Hu, K., Zhang, X., et al. (2013). The stepwise dissolution of chalcopyrite bioleached by *Leptospirillum ferriphilum*. *Electrochimica Acta, 103*, 50–57. https://doi.org/10.1016/j.electacta.2013.04.051

Hidalgo, T., Kuhar, L., Beinlich, A., et al. (2019). Kinetics and mineralogical analysis of copper dissolution from a bornite/chalcopyrite composite sample in ferric-chloride and methanesulfonic-acid solutions. *Hydrometallurgy, 188*, 140–156. https://doi.org/10.1016/j.hydromet.2019.06.009

Hiroyoshi, N., Hirota, M., Hirajima, T., et al. (1997). A case of ferrous sulfate addition enhancing chalcopyrite leaching. *Hydrometallurgy, 47*, 37–45. https://doi.org/10.1016/S0304-386X(97)00032-7

Hiroyoshi, N., Kitagawa, H., & Tsunekawa, M. (2008). Effect of solution composition on the optimum redox potential for chalcopyrite leaching in sulfuric acid solutions. *Hydrometallurgy, 91*, 144–149. https://doi.org/10.1016/j.hydromet.2007.12.005

Hiroyoshi, N., Kuroiwa, S., Miki, H., et al. (2004). Synergistic effect of cupric and ferrous ions on active-passive behavior in anodic dissolution of chalcopyrite in sulfuric acid solutions. *Hydrometallurgy, 74*, 103–116. https://doi.org/10.1016/j.hydromet.2004.01.003

Hiroyoshi, N., Miki, H., Hirajima, T., et al. (2001). Enhancement of chalcopyrite leaching by ferrous ions in acidic ferric sulfate solutions. *Hydrometallurgy, 60*, 185–197. https://doi.org/10.1016/S0304-386X(00)00155-9

Hiroyoshi, N., Miki, H., Hirajima, T., & Tsunekawa, M. (2000). Model for ferrous-promoted chalcopyrite leaching. *Hydrometallurgy, 57*, 31–38. https://doi.org/10.1016/S0304-386X(00)00089-X

Holliday, R. I., & Richmond, W. R. (1990). An electrochemical study of the oxidation of chalcopyrite in acidic solution. *Journal of Electroanalytical Chemistry, 288*, 83–98. https://doi.org/10.1016/0022-0728(90)80027-4

Horta, D. G., Acciari, H. A., Bevilaqua, D., et al. (2009a). The effect of chloride ions and *A. ferrooxidans* on the oxidative dissolution of the chalcopyrite evaluated by electrochemical noise analysis (ENA). *Advances in Materials Research, 71–73*, 397–400. https://doi.org/10.4028/WWW.SCIENTIFIC.NET/AMR.71-73.397

Horta, D. G., Bevilaqua, D., Acciari, H. A., et al. (2009b). Optimization of the use of carbon paste electrodes (CPE) for electrochemical study of the chalcopyrite. *Quimica Nova, 32*, 1734–1738. https://doi.org/10.1590/S0100-40422009000700010

Huynh, D., Giebner, F., Kaschabek, S. R., et al. (2019). Effect of sodium chloride on *Leptospirillum ferriphilum* DSM 14647^T and *Sulfobacillus thermosulfidooxidans* DSM 9293^T: Growth, iron oxidation activity and bioleaching of sulfidic metal ores. *Minerals Engineering, 138*, 52–59. https://doi.org/10.1016/j.mineng.2019.04.033

Johnson, D. B., & Quatrini, R. (2020). Acidophile microbiology in space and time. *Current Issues in Molecular Biology, 39*, 63–76. https://doi.org/10.21775/cimb.039.063

Jordan, H., Sanhueza, A., Gautier, V., et al. (2006). Electrochemical study of the catalytic influence of *Sulfolobus metallicus* in the bioleaching of chalcopyrite at 70 °C. *Hydrometallurgy, 83*, 55–62. https://doi.org/10.1016/j.hydromet.2006.03.038

Jung, H., Inaba, Y., & Banta, S. (2021). Genetic engineering of the acidophilic chemolithoautotroph *Acidithiobacillus ferrooxidans*. *Trends in Biotechnology, 40*, 677–692. https://doi.org/10.1016/j.tibtech.2021.10.004

Kaplun, K., Li, J., Kawashima, N., et al. (2011). Cu and Fe chalcopyrite leach activation energies and the effect of added Fe^{3+}. *Geochimica et Cosmochimica Acta, 75*, 5865–5878. https://doi.org/10.1016/j.gca.2011.07.003

Kametani, H., & Aoki, A. (1985). Effect of suspension potential on the oxidation rate of copper concentrate in a sulfuric acid solution. *Metallurgical Transactions B, 16*, 695–705. https://doi.org/10.1007/BF02667506

Khaleque, H. N., Kaksonen, A. H., Boxall, N. J., & et al. (2018). Chloride ion tolerance and pyrite bioleaching capabilities of pure and mixed halotolerant, acidophilic iron- and sulfur-oxidizing cultures. *Minerals Engineering, 120*, 87–93. https://doi.org/10.1016/j.mineng.2018.02.025

Khoshkhoo, M., Dopson, M., Shchukarev, A., et al. (2014). Chalcopyrite leaching and bioleaching: An X-ray photoelectron spectroscopic (XPS) investigation on the nature of hindered dissolution. *Hydrometallurgy, 149*, 220–227. https://doi.org/10.1016/j.hydromet.2014.08.012

Kinnunen, P.H.-M., Heimala, S., Riekkola-Vanhanen, M.-L., et al. (2006). Chalcopyrite concentrate leaching with biologically produced ferric sulphate. *Bioresource Technology, 97*, 1727–1734. https://doi.org/10.1016/j.biortech.2005.07.016

Klauber, C. (2008). A critical review of the surface chemistry of acidic ferric sulphate dissolution of chalcopyrite with regards to hindered dissolution. *International Journal of Mineral Processing, 86*, 1–17. https://doi.org/10.1016/j.minpro.2007.09.003

Klauber, C., Parker, A., Van Bronswijk, W., et al. (2001). Sulphur speciation of leached chalcopyrite surfaces as determined by X-ray photoelectron spectroscopy. *International Journal of Mineral Processing, 62*, 65–94. https://doi.org/10.1016/S0301-7516(00)00045-4

Koleini, S. M. J., Jafarian, M., Abdollahy, M., et al. (2010). Galvanic leaching of chalcopyrite in atmospheric pressure and sulfate media: Kinetic and surface studies. *Industrial and Engineering Chemistry Research, 49*, 5997–6002. https://doi.org/10.1021/ie100017u

Lara, R. H., García-Meza, J. V., González, I., et al. (2013). Influence of the surface speciation on biofilm attachment to chalcopyrite by *Acidithiobacillus thiooxidans*. *Applied Microbiology and Biotechnology, 97*, 2711–2724. https://doi.org/10.1007/s00253-012-4099-8

Lara, R. H., Vazquez-Arenas, J., Ramos-Sanchez, G., et al. (2015). Experimental and theoretical analysis accounting for differences of pyrite and chalcopyrite oxidative behaviors for prospective environmental and bioleaching applications. *Journal of Physical Chemistry C, 119*, 18364–18379. https://doi.org/10.1021/acs.jpcc.5b05149

Latorre, M., Cortés, M. P., Travisany, D., et al. (2016). The bioleaching potential of a bacterial consortium. *Bioresource Technology, 218*, 659–666. https://doi.org/10.1016/j.biortech.2016.07.012

Liao, R., Wang, X., Yang, B., et al. (2020). Catalytic effect of silver-bearing solid waste on chalcopyrite bioleaching: A kinetic study. *Journal of Central South University, 27*, 1395–1403. https://doi.org/10.1007/s11771-020-4375-1

Li, Y., Kawashima, N., Li, J., et al. (2013). A review of the structure, and fundamental mechanisms and kinetics of the leaching of chalcopyrite. *Advances in Colloid and Interface Science, 197–198*, 1–32. https://doi.org/10.1016/j.cis.2013.03.004

Li, Y., Qian, G., Brown, P. L., et al. (2017). Chalcopyrite dissolution: Scanning photoelectron microscopy examination of the evolution of sulfur species with and without added iron or pyrite. *Geochimica et Cosmochimica Acta, 212*, 33–47. https://doi.org/10.1016/j.gca.2017.05.016

Liu, H.-C., Xia, J.-I., Nie, Z.-Y., et al. (2017). Comparative study of S, Fe and Cu speciation transformation during chalcopyrite bioleaching by mixed mesophiles and mixed thermophiles. *Minerals Engineering, 106*, 22–32. https://doi.org/10.1016/j.mineng.2017.01.013

Liu, Q., & Li, H. (2011). Electrochemical behaviour of chalcopyrite ($CuFeS_2$) in $FeCl_3$ solution at room temperature under differential stress. *International Journal of Mineral Processing, 98*, 82–88. https://doi.org/10.1016/j.minpro.2010.10.010

Lv, X., Wang, J., & Zeng, X., et al. (2021). Cooperative extraction of metals from chalcopyrite by bio-oxidation and chemical oxidation. *Geochemistry, 81*, 125772. https://doi.org/10.1016/j.chemer.2021.125772

Mahmoud, A., Cézac, P., Hoadley, A. F. A., et al. (2017). A review of sulfide minerals microbially assisted leaching in stirred tank reactors. *International Biodeterioration and Biodegradation, 119*, 118–146. https://doi.org/10.1016/j.ibiod.2016.09.015

Majuste, D., Ciminelli, V. S. T., Osseo-Asare, K., et al. (2012). Electrochemical dissolution of chalcopyrite: Detection of bornite by synchrotron small angle X-ray diffraction and its correlation with the hindered dissolution process. *Hydrometallurgy, 111–112*, 114–123. https://doi.org/10.1016/j.hydromet.2011.11.003

Martins, F. L., & Leão, V. A. (2023). Chalcopyrite bioleaching in chloride media: A mini-review. *Hydrometallurgy, 216*, 105995. https://doi.org/10.1016/j.hydromet.2022.105995

Martins, F. L., Patto, G. B., & Leão, V. A. (2019). Chalcopyrite bioleaching in the presence of high chloride concentrations. *Journal of Chemical Technology and Biotechnology, 94*, 2333–2344. https://doi.org/10.1002/jctb.6028

Memming, R. (2015). *Semiconductor electrochemistry* (2nd ed.). Wiley-VCH Verlag.

Méndez, A., Álvarez, M. L., Fidalgo, J. M., et al. (2022). Biomass-derived activated carbon as catalyst in the leaching of metals from a copper sulfide concentrate. *Minerals Engineering, 183*, 107594. https://doi.org/10.1016/j.mineng.2022.107594

Mikhlin, Y., Nasluzov, V., Romanchenko, A., et al. (2017). Layered structure of the near-surface region of oxidized chalcopyrite ($CuFeS_2$): Hard X-ray photoelectron spectroscopy, X-ray absorption spectroscopy and DFT+*U* studies. *Physical Chemistry Chemical Physics, 19*, 2749–2759. https://doi.org/10.1039/c6cp07598c

Mikhlin, Y. L., Tomashevich, Y. V., Asanov, I. P., et al. (2004). Spectroscopic and electrochemical characterization of the surface layers of chalcopyrite ($CuFeS_2$) reacted in acidic solutions. *Applied Surface Science, 225*, 395–409. https://doi.org/10.1016/j.apsusc.2003.10.030

Moya-Beltrán, A., Simón Beard, S., Rojas-Villalobos, C., et al. (2021). Genomic evolution of the class Acidithiobacillia: Deep-branching Proteobacteria living in extreme acidic conditions. *ISME Journal, 15*, 3221–3238. https://doi.org/10.1038/s41396-021-00995-x

Munoz, P. B., Miller, J. D., & Wadsworth, M. E. (1979). Reaction mechanism for the acid ferric sulfate leaching of chalcopyrite. *Metallurgical Transactions B, B10*, 149–158. https://doi.org/10.1007/BF02652458

Nasluzov, V., Shor, A., Romanchenko, A., et al. (2019). DFT + *U* and low-temperature XPS studies of Fe-depleted chalcopyrite ($CuFeS_2$) surfaces: A focus on polysulfide species. *Journal of Physical Chemistry C, 123*, 21031–21041. https://doi.org/10.1021/acs.jpcc.9b06127

Natarajan, K. A., & Kumari, A. (2014). Role of applied potentials on bioleaching of chalcopyrite concentrate and growth of *Acidithiobacillus ferrooxidans*. *Minerals and Metallurgical Processing, 31*, 215–222.

Nava, D., González, I., Leinen, D., et al. (2008). Surface characterization by X-ray photoelectron spectroscopy and cyclic voltammetry of products formed during the potentiostatic reduction of chalcopyrite. *Electrochimica Acta, 53*, 4889–4899. https://doi.org/10.1016/j.electacta.2008.01.088

Nazari, G., & Asselin, E. (2009). Morphology of chalcopyrite leaching in acidic ferric sulfate media. *Hydrometallurgy, 96*, 183–188. https://doi.org/10.1016/j.hydromet.2008.09.004

Nicol, M. (2021). The initial stages of the dissolution of chalcopyrite in chloride solutions: Validity of mixed potential model and comparison with sulfate solutions. *Hydrometallurgy, 204*, 105721. https://doi.org/10.1016/j.hydromet.2021.105721

Nicol, M., Miki, H., & Zhang, S. (2017). The anodic behaviour of chalcopyrite in chloride solutions: Voltammetry. *Hydrometallurgy, 171*, 198–205. https://doi.org/10.1016/j.hydromet.2017.05.016

Nicol, M., & Zhang, S. (2017). The anodic behaviour of chalcopyrite in chloride solutions: Potentiostatic measurements. *Hydrometallurgy, 167*, 72–80. https://doi.org/10.1016/j.hydromet.2016.10.008

Nicol, M. J. (2017a). The use of impedance measurements in the electrochemistry of the dissolution of sulfide minerals. *Hydrometallurgy, 169*, 99–102. https://doi.org/10.1016/j.hydromet.2016.12.013

Nicol, M. J. (2017b). The anodic behaviour of chalcopyrite in chloride solutions: Overall features and comparison with sulfate solutions. *Hydrometallurgy, 169*, 321–329. https://doi.org/10.1016/j.hydromet.2017.02.009

Nicol, M. J. (2019). The electrochemistry of chalcopyrite in alkaline solutions. *Hydrometallurgy, 187*, 134–140. https://doi.org/10.1016/j.hydromet.2019.05.016

Nuñez, H., Moya-Beltrán, A., Covarrubias, P. C., et al. (2017). Molecular systematics of the genus *Acidithiobacillus*: Insights into the phylogenetic structure and diversification of the taxon. *Frontiers in Microbiology, 8*, 30. https://doi.org/10.3389/fmicb.2017.00030

Nyembwe, K. J., Fosso-Kankeu, E., Wanders, F., & et al. (2018). Mineralogical observation made during the kinetic dissolution study of chalcopyrite mineral in sulphate media under free pH at room temperature. In *10th International conference on advances in science, engineering, technology and healthcare* (pp. 148–152) Cape Town, South Africa. https://doi.org/10.17758/eares4.eap1118240

O'Connor, G. M., & Eksteen, J. J. (2020). A critical review of the passivation and semiconductor mechanisms of chalcopyrite leaching. *Minerals Engineering, 154*, 106401. https://doi.org/10.1016/j.mineng.2020.106401

Olvera, O. G., Quiroz, L., Dixon, D. G., et al. (2014). Electrochemical dissolution of fresh and passivated chalcopyrite electrodes. Effect of pyrite on the reduction of Fe^{3+} ions and transport processes within the passive film. *Electrochimica Acta, 127*, 7–19. https://doi.org/10.1016/j.electacta.2014.01.165

Osseo-Asare, K. (1992). Semiconductor electrochemistry and hydrometallurgical dissolution processes. *Hydrometallurgy, 29*, 61–90. https://doi.org/10.1016/0304-386X(92)90006-L

Parker, A., Klauber, C., Kougianos, A., et al. (2003). An X-ray photoelectron spectroscopy study of the mechanism of oxidative dissolution of chalcopyrite. *Hydrometallurgy, 71*, 265–276. https://doi.org/10.1016/S0304-386X(03)00165-8

Pathnak, P. A., Morrison, L., et al. (2017). Catalytic potential of selected metal ions for bioleaching, and potential techno-economic and environmental issues: A critical review. *Bioresource Technology, 229*, 211–221. https://doi.org/10.1016/j.biortech.2017.01.001

Petersen, J., & Dixon, D. G. (2006). Competitive bioleaching of pyrite and chalcopyrite. *Hydrometallurgy, 83*, 40–49. https://doi.org/10.1016/j.hydromet.2006.03.036

Peters, E. (1977). The electrochemistry of sulphide minerals. In J. O'M Bockris, D. A. J. Rand, & B. J. Welch (Eds.), *Trends in electrochemistry* (pp. 267–290) Plenum Publishing.

Plumb, J. J., Haddad, C. M., Gibson, J. A. E., et al. (2007). *Acidianus sulfidivorans* sp. nov., an extremely acidophilic, thermophilic archaeon isolated from a solfatara on Lihir Island, Papua New Guinea, and emendation of the genus description. *International Journal of Systematic and Evolutionary Microbiology, 57*, 1418–1423. https://doi.org/10.1099/ijs.0.64846-0

Quatrini, R., & Johnson, D. B. (2018). Microbiomes in extremely acidic environments: Functionalities and interactions that allow survival and growth of prokaryotes at low pH. *Current Opinion in Microbiology, 43*, 139–147. https://doi.org/10.1016/j.mib.2018.01.011

Ram, R., Coyle, V. E., Bond, A. M., et al. (2020). A scanning electrochemical microscopy (SECM) study of the interfacial solution chemistry at polarised chalcopyrite ($CuFeS_2$) and chalcocite (Cu_2S). *Electrochemistry Communications, 115*, 106730. https://doi.org/10.1016/j.elecom.2020.106730

Ranjbar, M., Hamghavandi, M. R., Fazaelipoor, M. H., et al. (2020). Development of a kinetic model of bacterial dissolution of copper concentrate. *Mining, Metallurgy and Exploration, 37*, 345–353. https://doi.org/10.1007/s42461-019-00114-7

Sadeghieh, S. M., Ahmadi, A., & Hosseini, M. R. (2020). Effect of water salinity on the bioleaching of copper, nickel and cobalt from the sulphidic tailing of Golgohar iron mine, Iran. *Hydrometallurgy, 198*, 105503. https://doi.org/10.1016/j.hydromet.2020.105503

Safar, C., Castro, C., & Donati, E. (2020). Importance of initial interfacial steps during chalcopyrite bioleaching by a thermoacidophilic archaeon. *Microorganisms, 8*, 1009. https://doi.org/10.3390/microorganisms8071009

Sandström, Å., Shchukarev, A., & Paul, J. (2005). XPS characterisation of chalcopyrite chemically and bio-leached at high and low redox potential. *Minerals Engineering, 18*, 505–515. https://doi.org/10.1016/j.mineng.2004.08.004

Santos, A. L. A., Arena, F. A., Benedetti, A. V., & et al. (2017). Effect of redox potential on chalcopyrite dissolution imposed by addition of ferrous ions. *Eclética Química, 42*, 40. https://doi.org/10.26850/1678-4618eqj.v42.1.2017.p40-50

Sasaki, K., Takatsugi, K., & Tuovinen, O. H. (2012). Spectroscopic analysis of the bioleaching of chalcopyrite by *Acidithiobacillus caldus*. *Hydrometallurgy, 127–128*, 116–120. https://doi.org/10.1016/j.hydromet.2012.07.013

Sedriks, A. J. (1996). *Corrosion of stainless steel* (2nd ed.). John Wiley & Sons.

Sohn, H.-J., & Wadsworth, M. E. (1980). Reduction of chalcopyrite with SO_2 in the presence of cupric ions. *Journal of Metals, 32*, 18–22. https://doi.org/10.1007/BF03354577

Tanne, C., & Shippers, A. (2021). Electrochemical investigation of microbially and galvanically leached chalcopyrite. *Hydrometallurgy, 202*, 105603–105611. https://doi.org/10.1016/j.hydromet.2021.105603

Third, K. A., Cord-Ruwisch, R., & Watling, H. R. (2000). Role of iron-oxidizing bacteria in stimulation or inhibition of chalcopyrite bioleaching. *Hydrometallurgy, 57*, 225–233. https://doi.org/10.1016/S0304-386X(00)00115-8

Third, K. A., Cord-Ruwisch, R., & Watling, H. R. (2002). Control of the redox potential by oxygen limitation improves bacterial leaching of chalcopyrite. *Biotechnology and Bioengineering, 78*, 433–441. https://doi.org/10.1002/bit.10184

Tian, Z., Li, H., Wei, Q., & et al. (2021). Effects of redox potential on chalcopyrite leaching: An overview. *Minerals Engineering, 172*, 107135. https://doi.org/10.1016/j.mineng.2021.107135

Toledo, A. G. R., Tayar, S. P., Arena, F. A., et al. (2022). New insights into oxidative-reductive leaching of chalcopyrite concentrate using a central composite factorial design. *Minerals Engineering, 180*, 107467. https://doi.org/10.1016/j.mineng.2022.107467

Uhlig, H. H. (1978). History of passivity, experiments and theories. In R. P. Frankenthal & J. Kruger (Eds.), *Passivity of metals* (pp. 1–28). The Electrochemical Society.

Vakylabad, A. B., Nazari, S., & Darezereshki, E. (2022). Bioleaching of copper from chalcopyrite ore at higher NaCl concentrations. *Minerals Engineering, 175*, 107281. https://doi.org/10.1016/j.mineng.2021.107281

Varotsis, C., Tselios, C., Yiannakkos, K. A., Andreou, C., et al. (2022). Application of double-pulse laser-induced breakdown spectroscopy (DP-LIBS), Fourier transform infrared micro-spectroscopy and Raman microscopy for the characterization of copper sulfides. *RSC Advances, 12*, 631–639. https://doi.org/10.1039/D1RA07189K

Velásquez-Yévenes, L., Nicol, M., & Miki, H. (2010). The dissolution of chalcopyrite in chloride solutions. *Hydrometallurgy, 103*, 108–113. https://doi.org/10.1016/j.hydromet.2010.03.001

Vilcáez, J., & Inoue, C. (2009). Mathematical modeling of thermophilic bioleaching of chalcopyrite. *Minerals Engineering, 22*, 951–960. https://doi.org/10.1016/j.mineng.2009.03.001

Vilcáez, J., Suto, K., & Inoue, C. (2008). Bioleaching of chalcopyrite with thermophiles: Temperature–pH–ORP dependence. *International Journal of Mineral Processing, 88*, 37–44. https://doi.org/10.1016/j.minpro.2008.06.002

Viramontes-Gamboa, G., Peña-Gomar, M. M., & Dixon, D. G. (2010). Electrochemical hysteresis and bistability in chalcopyrite passivation. *Hydrometallurgy, 105*, 140–147. https://doi.org/10.1016/j.hydromet.2010.08.012

Viramontes-Gamboa, G., Rivera-Vasquez, B. F., & Dixon, D. G. (2007). The active-passive behavior of chalcopyrite: Comparative study between electrochemical and leaching responses. *Journal of the Electrochemical Society, 154*, C299. https://doi.org/10.1149/1.2721782

Viramontes-Gamboa, G., Rivera-Vásquez, B. F., & Dixon, D. G. (2006). The active-to-passive transition of chalcopyrite. *ECS Transactions, 2*, 165–175. https://doi.org/10.1149/1.2196007

Wang, X., Ma, L., Wu, J., et al. (2020). Effective bioleaching of low-grade copper ores: Insights from microbial cross experiments. *Bioresource Technology, 308*, 123273. https://doi.org/10.1016/j.biortech.2020.123273

Wang, L., Yin, S., Deng, B., et al. (2022). Copper sulfides leaching assisted by acidic seawater-based media: Ionic strength and mechanism. *Minerals Engineering, 175*, 107286. https://doi.org/10.1016/j.mineng.2021.107286

Warren, G. W., Wadsworth, M. E., & El-Raghy, S. M. (1982). Passive and transpassive anodic behavior of chalcopyrite in acid solutions. *Metallurgical Transactions B, 13*, 571–579. https://doi.org/10.1007/BF02650014

Watling, H. R. (2013). Chalcopyrite hydrometallurgy at atmospheric pressure: 1. Review of acidic sulfate, sulfate–chloride and sulfate–nitrate process options. *Hydrometallurgy, 140*, 163–180. https://doi.org/10.1016/j.hydromet.2013.09.013

Weisener, C. G., Smart, R. S. C., & Gerson, A. R. (2003). Kinetics and mechanisms of the leaching of low Fe sphalerite. *Geochimica et Cosmochimica Acta, 67*, 823–830. https://doi.org/10.1016/S0016-7037(02)01276-0

Yang, B., Gan, M., Luo, W., et al. (2017). Synergistic catalytic effects of visible light and graphene on bioleaching of chalcopyrite. *RSC Advances, 7*, 49838–49848. https://doi.org/10.1039/C7RA10015A

Yang, B., Luo, W., Wang, X., et al. (2020). The use of biochar for controlling acid mine drainage through the inhibition of chalcopyrite biodissolution. *Science of the Total Environment, 737*, 139485. https://doi.org/10.1016/j.scitotenv.2020.139485

Yang, C., Qin, W., Zhao, H., et al. (2018). Mixed potential plays a key role in leaching of chalcopyrite: Experimental and theoretical analysis. *Industrial and Engineering Chemistry Research, 57*, 1733–1744. https://doi.org/10.1021/acs.iecr.7b02051

Yang, Y., Harmer, S., & Chen, M. (2014). Synchrotron X-ray photoelectron spectroscopic study of the chalcopyrite leached by moderate thermophiles and mesophiles. *Minerals Engineering, 69*, 185–195. https://doi.org/10.1016/j.mineng.2014.08.011

Yang, Y., Liu, W., Gao, X., et al. (2019). An XAS study of silver species evolution in silver-catalysed chalcopyrite bioleaching. *Hydrometallurgy, 186*, 252–259. https://doi.org/10.1016/j.hydromet.2019.04.023

Yu, P. H., Hansen, C. K., & Wadsworth, M. E. (1973). A kinetic study of the leaching of chalcopyrite at elevated temperatures. *Metallurgical Transactions, 4*, 2137–2144. https://doi.org/10.1007/BF02643279

Zammit, C. M., Mangold, S., Jonna, V. R., & et al. (2012). Bioleaching in brackish waters—Effect of chloride ions on the acidophile population and proteomes of model species. *Applied Microbiology and Biotechnology, 93*, 319–329. https://doi.org/10.1007/s00253-011-3731-3

Zeng, W., Liu, Z., Amanze, C., et al. (2023). In situ detection of Cu^{2+}, Fe^{3+} and Fe^{2+} ions at the microbe-mineral interface during bioleaching of chalcopyrite by moderate thermophiles. *Minerals Engineering, 191*, 107936. https://doi.org/10.1016/j.mineng.2022.107936

Zhang, X., Liu, X., Li, L., et al. (2019). Phylogeny, divergent evolution, and speciation of sulfur-oxidizing *Acidithiobacillus* populations. *BMC Genomics, 20*, 438. https://doi.org/10.1186/s12864-019-5827-6

Zhao, H., Huang, X., Wang, J., et al. (2017). Comparison of bioleaching and dissolution process of *p*-type and *n*-type chalcopyrite. *Minerals Engineering, 109*, 153–161. https://doi.org/10.1016/j.mineng.2017.03.013

Zhao, H., Wang, J., Gan, X., et al. (2015a). Effects of pyrite and bornite on bioleaching of two different types of chalcopyrite in the presence of *Leptospirillum ferriphilum*. *Bioresource Technology, 194*, 28–35. https://doi.org/10.1016/j.biortech.2015.07.003

Zhao, H., Wang, J., Qin, W., et al. (2015b). Electrochemical dissolution process of chalcopyrite in the presence of mesophilic microorganisms. *Minerals Engineering, 71*, 159–169. https://doi.org/10.1016/j.mineng.2014.10.025

Zhao, H., Wang, J., Qin, W., et al. (2015c). Surface species of chalcopyrite during bioleaching by moderately thermophilic bacteria. *Transactions of Nonferrous Metals Society of China, 25*, 2725–2733. https://doi.org/10.1016/S1003-6326(15)63897-3

Zhao, H., Wang, J., Yang, C., & et al. (2015d). Effect of redox potential on bioleaching of chalcopyrite by moderately thermophilic bacteria: An emphasis on solution compositions. *Hydrometallurgy, 151*, 141–150. https://doi.org/10.1016/j.hydromet.2014.11.009

Zhao, H., Zhanga, Y., Zhang, X., et al. (2019). The dissolution and passivation mechanism of chalcopyrite in bioleaching: An overview. *Minerals Engineering, 136*, 140–154. https://doi.org/10.1016/j.mineng.2019.03.014

Zhao, C., Yang, B., Liao, R., et al. (2022). Combined effect and mechanism of visible light and Ag^+ on chalcopyrite bioleaching. *Minerals Engineering, 175*, 107283. https://doi.org/10.1016/j.mineng.2021.107283

Zheng, X.-F., Nie, Z.-Y., Jiang, Q, & et al. (2021). The mechanism by which FeS_2 promotes the bioleaching of $CuFeS_2$: An electrochemical and DFT study. *Mineral Engineering, 173*, 107233. https://doi.org/10.1016/j.mineng.2021.107233

Bioleaching of Lateritic Nickel Ores

Hadi Abdollahi, Marzieh Hosseini Nasab and Ali Yadollahi

Abstract

Despite nickel-bearing sulfide deposits having a large share of the world's nickel extraction, lateritic ore deposits contain more than 70% of the world's nickel reserves. Considering the limitations of producing nickel from sulfide reserves, the use of oxide reserves (laterites) for the production of nickel will be of great importance in the future. In this chapter, the applications of nickel and cobalt in various industries were described. Nickel and cobalt are mainly used in alloys of other metals. In addition, the most effective methods for extracting nickel and cobalt from lateritic nickel ores were examined. Due to the need for high energy, pyrometallurgical methods, as well as acid leaching, which uses a high amount of acid, are rarely used today. Therefore, the bacterial and fungal leaching methods (bioleaching), which is another hydrometallurgical process, and their mechanisms were explained. Bioleaching is a new prospective method for extracting valuable elements from hard-to-treat ores. The benefits of bioleaching low-grade ores are numerous in comparison to traditional methods due to their simplicity, using unskilled labor, low capital and operating costs, low energy consumption, and also the lowest negative environmental effects. In this processing operation, metals are dissolved from low-grade deposits by using microorganisms and their metabolic products. In addition, the final concentrations of iron in PLS can be decreased by biological methods. The most effective factors in the bioleaching process such as pH, size of sample particles, type of microorganism species, type of substrate, amount of inoculation, type of produced metabolic acid, the pulp solid to liquid ratio, salinity, temperature, and leaching time were explained. Heterotrophic bacteria such as *Aspergillus, Penicillium, Pseudomonas,* and *Delftia* were also successful at dissolving laterites, in addition to autotrophic bacteria such as *At.ferrooxidans* and *At.thiooxidans*. The presence of O_2 is considered a key factor in increasing the bio-reduction dissolution of nickel and cobalt of iron-containing minerals. In addition, high temperature, low density, and pH gained a higher dissolution rate of nickel and cobalt. The main mechanisms for autotrophic acidophilic (iron-oxidizing) and iron-reducing (dissimilatory iron-reducing bacteria) were acidolysis and redoxolysis. In general, biological dissolution and chemical control, respectively, had a greater effect compared with chemical dissolution and diffusion control on the dissolution rate of nickel and cobalt from the laterites. It was found that optimizing factors that affect the bioleaching of nickel and cobalt from nickel-containing laterites greatly increased the dissolution rate, recovered nickel and cobalt, and reduced iron dissolution.

Keywords

Laterite · Nickel · Cobalt · Bioleaching

1 Application of Nickel in Various Industries and Characteristics of This Element

Nickel is a strategic metal with a wide range of industrial and metallurgical uses (Sadat et al., 2016). The use of nickel has now spread to a wide range of sectors, including steel alloying elements, space applications, and rechargeable batteries (Buyukakinci, 2008). Nickel is highly corrosion-resistant in contact with air, seawater, and organic and

H. Abdollahi (✉) · A. Yadollahi
School of Mining Engineering, College of Engineering, University of Tehran, Tehran, Iran
e-mail: h_abdollahi@ut.ac.ir

M. Hosseini Nasab
Department of Mining Engineering, College of Engineering, University of Sistan and Baluchestan, Zahedan, Iran

S. Panda et al. (eds.), *Biotechnological Innovations in the Mineral-Metal Industry*, Advances in Science, Technology & Innovation, https://doi.org/10.1007/978-3-031-43625-3_3

non-oxidizing acids; on the other hand, it is being eroded by solutions of aqueous ammonia. Nickel dissolves in dilute mineral acids; in contrast, it is insoluble in pure nitric acid. Among ionic states −1, 0, +2, +3, and +4, the ionic number +2 is the most important form of nickel ion. In aqueous solutions, this metal is stable in the +2 state (Buyukakinci, 2008). Nickel is most commonly used in alloys with other metals. This metal increases the strength, stiffness, and corrosion resistance of alloy metals. The other important application of nickel alloy is making stainless steel which uses more than 60% of the world's nickel to make it stronger and more resistant to corrosion. Nickel alloy production has the second rank in the world for using nickel. Because of their great thermal resistance and fracture toughness, nickel alloys have a wide variety of uses (Buyukakinci, 2008).

2 Types of Nickel Ores

In iron meteorites, nickel is typically found in the form of kamacite and taenite alloys. Nickel ores are divided into two categories (Sadat et al., 2016):

(1) Pentlandite [(Ni, Fe)$_9$S$_8$] is the major nickel mineral associated with this type of ore, which is linked to mafic and ultramafic rocks (magmatic sulfide deposits).
(2) Near-surface laterite deposits formed on olivine-rich host rocks after extensive weathering.

To date, most of the nickel ores are being extracted from the sulfide type minerals and sulfide deposits have been the main source of nickel for a long time as a result of the ease of nickel processing due to its sulfide-independent mineral (Dalvi et al., 2004; Sahu et al., 2011). Despite the limitations in nickel extraction from these sources, oxide ores (laterites) are becoming more and more prevalent (Dalvi et al., 2004; Pawlowska & Sadowski, 2017; Sahu et al., 2011; Valix et al., 2009). Over 70% of nickel reserves around the world come from laterite deposits (Dalvi et al., 2004; Pawlowska & Sadowski, 2017; Petrus et al., 2018; Sahu et al., 2011), recent advancements have weakened the supremacy of sulfide ores and allowed laterites to be used in new studies (Krstev et al., 2012). As a result, laterites will play a larger role in nickel and cobalt production in the future, accounting for 80% of nickel reserves and 90–95% of cobalt reserves (Valix et al., 2009). It's worth noting that laterite deposits account for just 40% of global nickel output at the moment (Petrus et al., 2018). The great majority of cobalt sources are found in laterite ores. A variety of catalytic processes, renewable batteries, super alloys, and chemical processes have put cobalt in high demand globally. Hence, laterite ore extraction is becoming more essential (Li et al., 2010). Pure nickel is not found in nature. Among the nickel ores, only pentlandite (sulfide [Ni, Fe)$_9$S$_8$], garnierite (hydrated silicate [Ni, Mg)$_6$Si$_4$O$_{10}$(OH)$_{10}$], and nickel limonite oxides ([(Fe, Ni)O(OH)] have economic importance (Buyukakinci, 2008; Kim et al., 2010).

3 Methods of Processing Nickel from Laterites

Nickel is extracted from laterite ores by pyrometallurgical and hydrometallurgical methods. Figure 1 depicts flowsheets for pyrometallurgical and hydrometallurgical nickel extraction from laterites.

In the following, each of the laterites processing methods will be briefly explained.

3.1 Pyrometallurgical Operations of Laterites

The main stages of pyrometallurgical processes consist of drying, calcination, reduction, and electric melting furnaces (Buyukakinci, 2008). This method is more suitable for saprolite (garnierite) deposits which contain lower amounts of iron and cobalt compared to limonite deposits. The Ni/Co ratio of feed of the melting section is usually around 40 (Fatahi et al., 2014). The minimum acceptable nickel content in this method with a cheap energy source is 1.7% and otherwise is 2.1% (Dalvi et al., 2004). Laterite ores have high humidity (more than 45%) in addition to the water in the form of hydroxides. Drying operations are carried out to lower the ore's moisture content to reduce problems in the following processes. During the melting stage, nickel is selectively reduced and high-grade metal is produced (Buyukakinci, 2008). Because of their high iron content, limonite ores are not suited for this procedure. Furthermore, laterites have a high humidity level that requires significant amounts of heat (around 800 to 900 °C) to dry. For this reason, the production costs will be increased per ton of nickel (Buyukakinci, 2008). As a result, hydrometallurgical processes are better for the treatment of high-grade iron limonite laterites, however, pyrometallurgical methods are better for magnesium-rich saprolite laterites processing (Kursunoglu & Kaya, 2016).

3.2 Hydrometallurgical Operations of Laterites

The hydrometallurgy method or the combination of hydrometallurgical and pyrometallurgical operations of the laterites depends on the chemical homogeneity and mineralogical distribution of the laterites. In laterites, nickel can be found in goethite, serpentine, smectite, manganese oxides

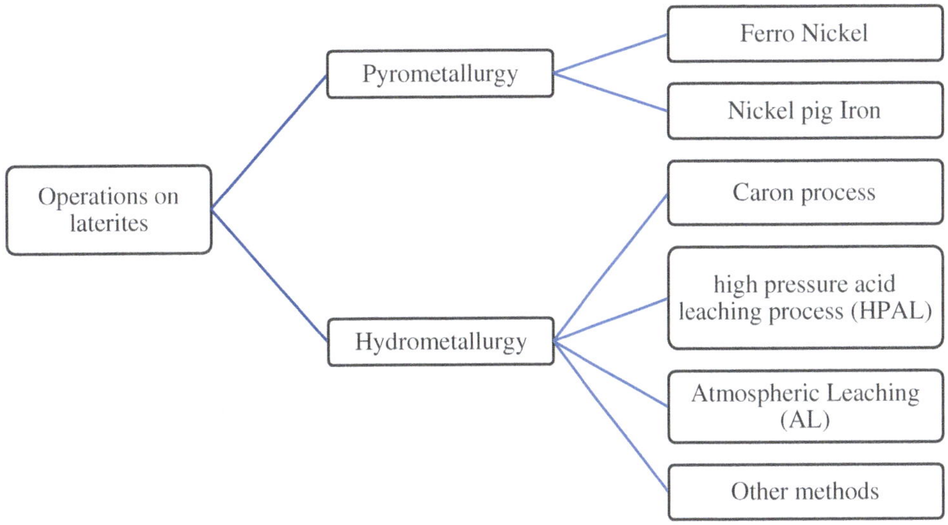

Fig. 1 Operations on laterites (Buyukakinci, 2008)

as well as cobalt (Buyukakinci, 2008). Hydrometallurgical processes are more commonly used for limonite or garnierite laterite ore. Although the nickel of saprolite laterites (highly weathered) is often richer than limonite ores, besides, their high magnesium content leads to high acid consumption (Dalvi et al., 2004).

Various hydrometallurgical methods have been studied on laterites. At the commercial level, only two methods are used to extract industrially nickel and cobalt which are the Caron process and the acid leaching process that uses sulfuric acid (Buyukakinci, 2008).

The Caron process includes reduction with high temperature and then ammonia leaching with ammonium carbonate (Buyukakinci, 2008). The acceptable range of nickel grade in these ores in the Caron method is 1–1.7% (Buyukakinci, 2008). The first step in the manufacturing process is drying the ore to reduce its moisture content. Then, the ore is reduced in the calcination plant using airless petroleum fuel at a temperature of about 700 °C. In the end, the extraction method with an organic solvent has been used to separate nickel and cobalt in an ionic solution. A concentrated ammonia carbonate solution is then used to wash the nickel from the solvent. The problem with the Caron process is its high-energy consumption which economically increases the cost of production. Therefore, the use of the Caron process has been limited due to the higher costs per ton of nickel production compared to the pressure acid leaching (Buyukakinci, 2008). Figure 2 shows the complete steps of the Caron process.

Pressure acid leaching has been applied to the Caron process to reduce high-energy consuming phases in direct ore leaching. Direct leaching eliminates the need for drying and reduction steps, which then leads to energy savings (Buyukakinci, 2008). The acceptable nickel content in this method is 1–5% (Dalvi et al., 2004). Also, the crushed ore is dissolved directly with sulfuric acid at high pressure and temperature, which leads to the dissolution of goethite. Then, the iron precipitates rapidly in the form of hematite. This precipitation reduces the ratio of iron to nickel in the solution. It has also been reported that when the acidity at the reaction temperature is less than 0.1 mol/L, nickel and magnesium can precipitate as a mixture of sulfated salts. High magnesium oxide increases acid consumption and consequently increases the costs of the operation process (Buyukakinci, 2008). One of the ways to extract nickel metal from this method is electrolytic extraction. However, the presence of by-products such as copper, cobalt, and zinc in nickel solutions from pressure acid leaching proves the importance of this method (Moskalyk & Alfantazi, 2002).

The cost of titanium autoclaves used in the procedure, as well as the corrosion of high-pressure valves, are the main disadvantages of pressure acid leaching (Buyukakinci, 2008). The high amount of required sulfuric acid, the difficulty of maintaining the permeability of the substrate under acidic leaching conditions, and the need for the proper time and water management are other disadvantages of this method (Kursunoglu & Kaya, 2016). Pressure acid leaching is currently the most acceptable operation for producing nickel from low-grade laterites in the absence of rival methods (Buyukakinci, 2008).

Laterite acid leaching at atmospheric pressure has been proposed as a substitute for pressure acid leaching. It is used for relatively low-grade nickel (less than 1.5% nickel), limonite, and saprolite leaching due to a low investment cost. In addition, the recovery of nickel and cobalt in this method is higher compared to heap leaching (Buyukakinci, 2008; Kursunoglu & Kaya, 2016). This process is performed at temperatures below 100 °C for a maximum of 12 h (Kursunoglu & Kaya, 2016). Direct leaching of laterites with organic and inorganic acids using agitation and

Fig. 2 Complete flowsheet of the Caron process (Nicol & Zainol, 2003)

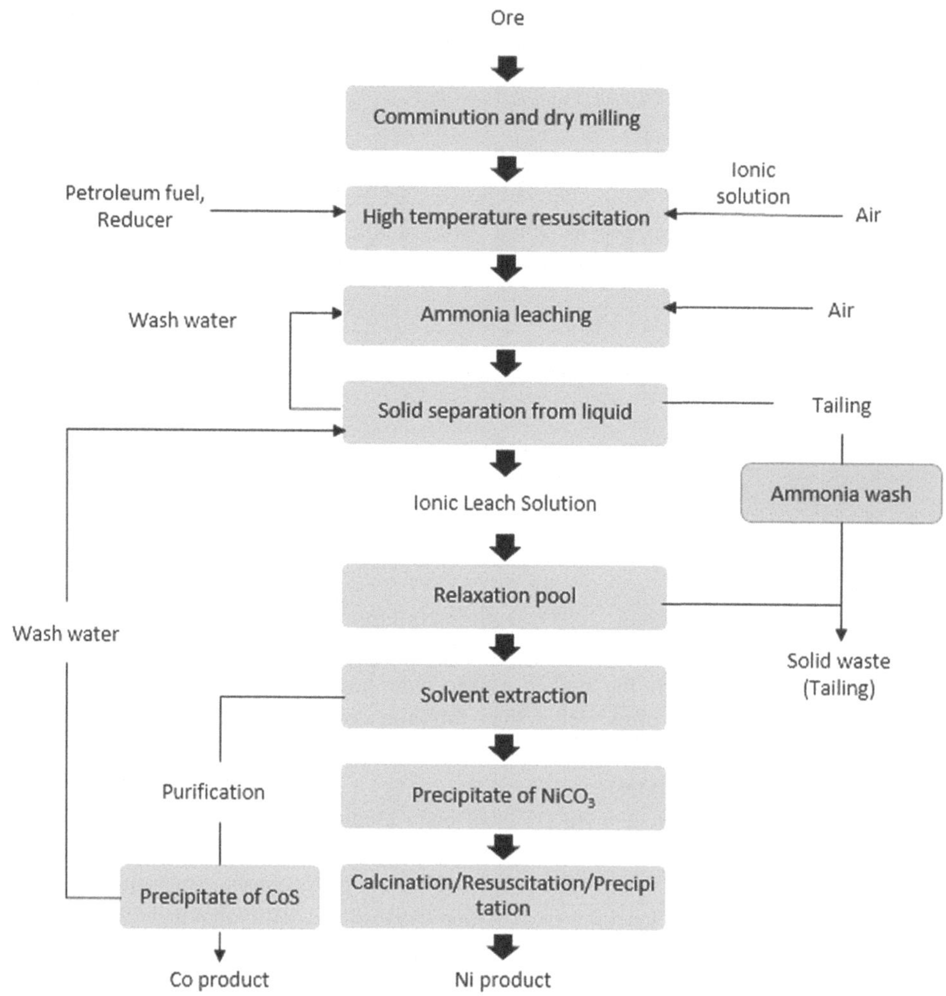

heap leaching are two of the proposed leaching methods (Buyukakinci, 2008). Leaching using different acids has been studied; nevertheless, for reasons especially economic, sulfuric acid is preferred. Nickel extraction from lateritic ores is influenced by the type of nickel-bearing minerals, leaching temperature, and sulfuric acid concentration. Studies on the leaching kinetics of laterites have shown that magnesium dissolved rapidly during leaching (Beukes et al., 2000). The magnesium in serpentine is the largest consumer of acid in the leaching process of serpentine-containing laterites (Whittington et al., 2003a). Therefore, nickel in limonite or saprolite compounds is easily soluble in sulfuric acid (Büyükakinci & Topkaya, 2009). However, the nickel in nontronite laterites dissolves incompletely. In these laterites, some nickel is extracted without destroying the structure of this type of laterites which is either in the form of nickel ion exchange from nontronite or due to the presence of nickel-rich minerals that are rapidly soluble in nontronite ores (Whittington et al., 2003b). The dissolution of nickel linked with iron has been described as a time-dependent process that frequently necessitates the total dissolution of iron (Büyükakinci & Topkaya, 2009). It is noteworthy that the solubility of lizardite, magnetite, hematite, and goethite in sulfuric acid, which are from iron minerals, decreases, respectively. Above all, the solubility of magnetite and hematite are almost equal (Liu et al., 2009). According to research, preheating laterites increases porosity and surface area, making it more appropriate for leaching, in addition to modifying the mineralogical composition. When raw ore and roasted ores are leached at different temperatures, the results reveal that increasing the roasting temperature to 300 °C gives optimal nickel recovery, whereas more heating is detrimental for nickel recovery (Li et al., 2009). Laterites leaching with sulfuric acid consume large amounts of acid. However, researches have shown that it is better to reduce the amount of calcite using a strong magnetic field despite the significant amount of calcite mineral in laterite ores which of course depends on the size of the feed particles (Agatzini-Leonardou et al., 2004). In addition, the presence of calcite harms the permeability of the ore in heap leaching due to gypsum formation (Agatzini-Leonardou & Zafiratos, 2004).

Heap leaching uses different methods such as diluting sulfuric acid solution at a limited temperature, extraction of nickel and cobalt from the leaching solution with solvent extraction and electrolysis, and purification of the leaching solution by chemical precipitation at atmospheric pressure (Buyukakinci, 2008). In HPAL and heap leaching, the final product is $NiSO_4$, which is used primarily for the production of batteries. Iron, a critical component that influences acid consumption, makes selective metal recovery from leaching solutions difficult. Due to the high concentration of residual acid, substantial volumes of neutralizing agents such as limestone are required during the iron removal from the atmospheric leaching process. In addition, the high initial Fe/Ni ratio in the leaching solution can cause severe loss of nickel and cobalt during iron removal wherefore increasing the cost of the product per ton of nickel production (Büyükakinci & Topkaya, 2009). The Vieille Montagne (V.M) process has been proposed for iron precipitation in the form of goethite from an aqueous solution of laterites ore leaching using air as the oxidizing agent. Therefore, this method does not require the reduction of ferric iron ions at the start, and the iron concentration in the final solution might be less than 1 g/L. The oxidation of ferric ions and the loss of nickel in solid waste are both influenced by the pH value. The oxidation rate of ferric ions increases with increasing the pH, in contrast, leads to the loss of more nickel in the solid waste. At pH between 2.5 and 3, the oxidation rate of ferric ion is about 53.8 mg/min. L, and nickel loss is 4.1%. Therefore, some nickel co-precipitates with iron in the precipitation process. The precipitated nickel cannot be recovered by leaching using weak acid (Chang et al., 2010).

In agitation leaching, first, the ore is crushed then milled, and at the end dissolved in a leaching tank. The temperature can be regulated by heating the tank, which will cause the metal to dissolve faster in the acidic solution, speeding up the process. Then, the ionized solution is collected to recover the metal (Buyukakinci, 2008). The results of the studies showed that the leaching time extension of the saprolite laterites causes the loss of nickel in the solid residues of waste, on the contrary, these laterites can be dissolved at low temperatures with a longer time and more free acid (Luo et al., 2009). In addition, the influence of sulfuric acid concentration on nickel dissolution rate has been studied. The amount of nickel dissolved increases by 60% with rising sulfuric acid concentration at 95 °C; The highest nickel dissolution (99.2%) was achieved at 95 °C in 120 min of leaching with a solution of 60% sulfuric acid concentration (Girgin et al., 2011). Another way to improve the atmospheric leaching of laterites using sulfuric acid is to blow SO_2 gas as a reduction agent. When Cu (II) is used as a catalyst, the rate of leaching increases as well. The particle size does not affect the leaching process in this case. However, nickel recovery increases from 74 to 90% which leads to increases of iron recovery from 47 to 73% in this case an optimal mode must be selected (Das & Lange, 2011). Another method includes three main steps of sulfating-roasting-leaching. With this method, 75% to 78% of nickel can be extracted and only 5% to 10% of iron is dissolved in the solution (Li et al., 2010).

Metal-rich leaching solution selectively precipitated to recover the metal using various methods. These methods include hydroxide precipitation, sulfide precipitation, or a combination of both methods (Büyükakinci & Topkaya, 2009). For example, it has been suggested that calcium hydroxide $[Ca(OH)_2]$ can be used to remove magnesium from the leached sulfate solution, which precipitates magnesium in the form of hydroxide $[Mg(OH)_2]$, and calcium added in the form of gypsum ($CaSO_4.2H_2O$). This method can remove rapidly 90 to 99% of magnesium from the solution. Temperature and precipitation amount are two of the most critical elements in this process (Karidakis et al., 2005). Under certain conditions, nickel metal can also precipitate into one of the oxides, hydroxide, and even pure metal forms. According to Fig. 3, nickel precipitates at pH between 6 and 12, and the Eh value determines the chemical composition of the precipitate (Beukes et al., 2000).

Atmospheric leaching versus pressure leaching has been challenged for several years and has a long list of strengths and weaknesses summarized in Table 1.

Among the three well-known hydrometallurgical processes for recovering nickel from lateritic ores with the least amount of energy expenditure, agitation leaching at atmospheric pressure is the least energy-intensive, followed by heap leaching, ferronickel, and high-pressure acid leaching, based on the available evidence and hypotheses. Furthermore, the Caron process consumes double as much

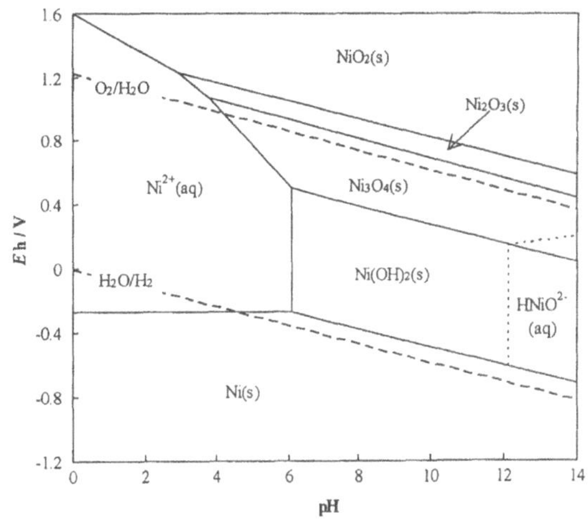

Fig. 3 Eh–pH diagram for Ni-H_2O at 25 °C temperature and 1 atm pressure (Beukes et al., 2000)

Table 1 Compared to pressure acidic leaching, the flaws and strengths of atmospheric leaching (tank and heap leaching) (McDonald & Whittington, 2008a)

Strengths	Weaknesses
Common, cheap, and ordinary technical equipment	Consume more acid to achieve acceptable nickel recovery
Lower maintenance costs	Consume more lime
Faster development	Not applicable for all types of stones
Ease of starting and stopping	Lack of economic justification for the present time
Less specialized worker for construction, operation, and maintenance	Slower extraction kinetics and longer retention time
Lower energy consumption	Produce a more contaminated solution with a lower Ni/Fe ratio
Easier process control	Extraction of less nickel and cobalt
Ease of water recycling in the process	The presence of jarosite in the tailings is a potential environmental problem
Applicable for low grades	
Suitable for remote and undeveloped places	
Fewer problems related to precipitation and corrosion	
Economical due to the availability of cheap sulfuric acid	
Use seawater to provide sodium to remove iron	

energy as other approaches (Kursunoglu & Kaya, 2016; Norgate & Jahanshahi, 2011).

The extraction of nickel-containing laterites by using organic acids is still in its early stages. The extraction of clay minerals is greatly improved by the use of citric and oxalic acids (Agatzini-Leonardou et al., 2004). The highest nickel recovery rate and magnesium selectivity are achieved by citric acid. Using the chelation mechanism, this acid generates a soluble ligand complex (Petrus et al., 2018). In addition, studies have shown that tank leaching is probably more effective than heap leaching. However, more research is still needed to improve metal recovery (Agatzini-Leonardou et al., 2004). In the alkaline leaching method to remove silica, silicon is first dissolved using concentrated sodium hydroxide. Then, magnesium is removed from the solid residue of the previous step using carbonation as magnesium carbonate. Finally, nickel is obtained as an oxide using ammonium carbonate (McDonald & Whittington, 2008b). Figure 4 shows a complete flowsheet of this process. This method has been used in the laboratory for a relatively high-grade silicate laterite sample. Table 2 shows the grades at different stages of this process.

Another proposed method for processing nickel-bearing laterites as shown in Fig. 5 includes milling, calcination, and magnetic separator. The magnetic property of laterite may change with calcination because the Curie temperature of magnetite is about 575 °C (Kim et al., 2010). Generally, increases in the temperature to 600 °C for limonite provide optimal recovery, while reduction of weathered saprolite at 800 °C is desirable. Forsterite is formed when weathered saprolite is reduced following calcination at temperatures above 700 °C. The formed phases at 800 °C in laterite minerals do not return to their original state upon cooling (Valix & Cheung, 2002).

There has been a development of a nickel extraction method for low-grade nickel-containing laterites with a grade of less than 2%. In this method, an electrochemical operation is used. In an electrolytic cell, nickel is electrochemically leached and then separated into two halves by a PVC separator. The cell current density varies in the range of 3.2–32 mA/cm^2. In the presence of SO_2 gas, nickel dissolves in iron-containing laterites through electrolytic reduction. In electrolysis, the reductive dissolution of iron is suspected to be a contributing factor to nickel extraction from laterites by electrochemical leaching. The experimental procedure is shown in Fig. 6 (Lee et al., 2005).

This method, unlike the hydrometallurgical method, is applicable at limited temperatures and a low cost. Also, it can be used for any type of magnesium-bearing or iron-bearing laterite. Nickel recovery in this method is between 50 and 75% and depends on the type of nickel and the amount of iron. The used electrodes and the cell design have a significant impact on extraction efficiency and energy consumption. On the other side, in this method, it is necessary to minimize the pollution caused by the release of SO_2 gas into the environment during electrolysis and exposure to dissolution wastes caused by nickel extraction (Lee et al., 2005).

Metal demand has increased as a result of industrialization and population development, while high-grade (sulfide) ores have declined, however, there are reserves of low-grade and low-profit for extraction in large volumes. It is very expensive to recover metals from low-grade reserves using conventional methods due to high-energy consumption and a high cost of capital. The high level of pollution caused by low-grade reserves is another major drawback of these reserves. Biotechnology is a more promising solution to these problems than pyrometallurgy or hydrometallurgy. In biotechnology, chemical reactions take place under gentle conditions and usually without the addition of toxic chemicals. Bioleaching techniques produce aqueous solutions, which are easier to process and regulate than gaseous wastes (Krstev et al., 2012).

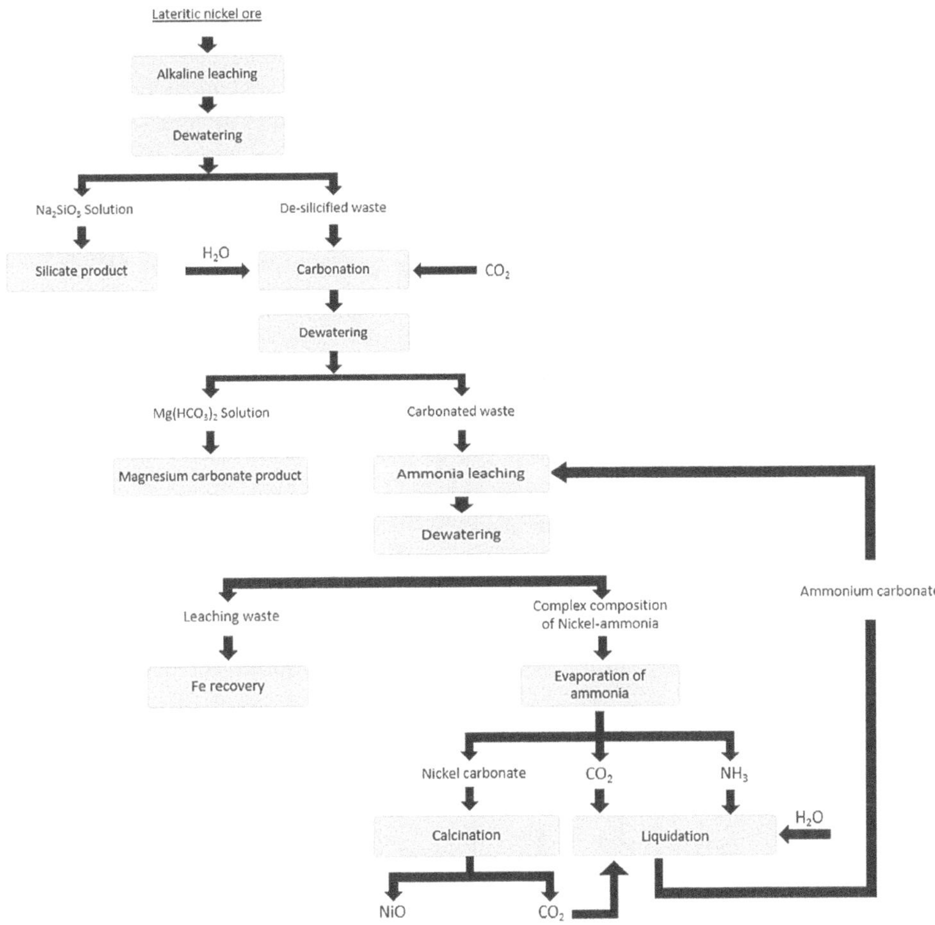

Fig. 4 Leaching flowsheet of a laterite sample using sodium hydroxide (McDonald & Whittington, 2008b)

Table 2 Grade in different stages of leaching of laterite samples using sodium hydroxide (McDonald & Whittington, 2008b)

	Grade (%)				
Sample	MgO	Fe_2O_3	SiO_2	NiO	Cr_2O_3
Feed	27.2	15.8	49.3	0.98	0.63
De-silicified waste	39.78	23.87	6.51	1.41	0.28
Carbonated waste	8.22	51.11	12.52	2.96	0.57
Residual waste	12.46	59.72	13.05	0.23	0.86

4 The Mechanism for Bacterial and Fungal Leaching (Bioleaching) of Laterites

Biological leaching has emerged as a promising new approach for recovering important elements from difficult-to-process ores in recent years which is the process of dissolving metals from low-grade deposits using microorganisms and their metabolic products (Li et al., 2010; Sahu et al., 2011). Microorganism-based processes make it feasible to recover metals from mineral deposits that are now inaccessible by conventional mining methods. During leaching processes, bacteria and fungi convert metal compounds into water-soluble forms by using biocatalysis (Krstev et al., 2012). At low metal concentrations, bacterial leaching is possible and takes little energy. Even when the extraction recovery is above 85–90%, this procedure is environmentally friendly (Krstev et al., 2012).

There are two mechanisms by which bacteria promote metal solubilization: the contact (direct) mechanism, which involves oxidation by microorganisms attaching to the metal substrate's surface and the non-contact (indirect) mechanism, which involves mineral oxidation by the oxidizing agent, which is usually ferric ion Fe (III) produced by bacteria in the solution (Borja et al., 2016; Santos et al., 2006). In the first mechanism, the bacteria adhere to the mineral surface, then produces compounds that break sulfur-metal and sulfur-sulfur bonds (Santos et al., 2006). Cysteine plays an important role in this mechanism, especially in the bio-oxidation of pyrite (Borja et al., 2016; Santos et al., 2006). On the other hand, bacteria oxidize Fe (II) to Fe (III) and consume electrons in the second operation. Fe(III) is a strong oxidant that can oxidize most sulfide minerals (Santos et al., 2006).

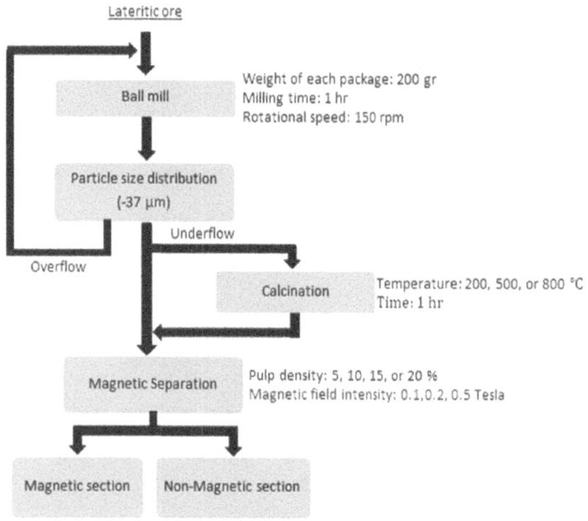

Fig. 5 Proposed flowsheet for nickel-bearing laterite processing (Kim et al., 2010)

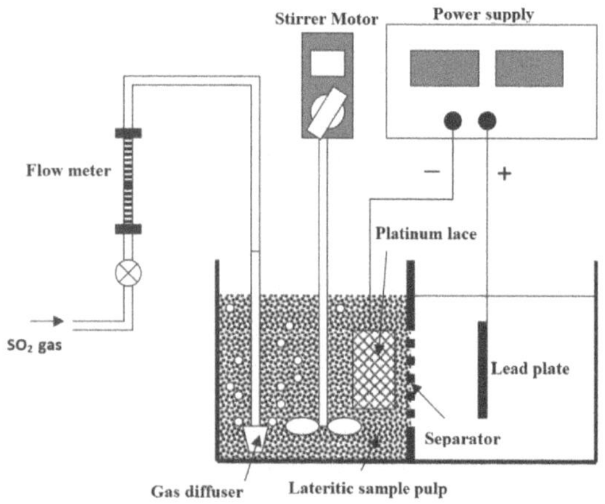

Fig. 6 Schematic of laboratory equipment for the extraction of nickel from nickel-containing laterites (Lee et al., 2005)

Biological oxidation of ferrous iron to ferric iron is shown in reaction (1): (Borja et al., 2016).

$$8Fe^{2+} + 2O_2 + 8H^+ \overset{\text{bacteria}}{\rightarrow} 8Fe^{3+} + 4H_2O \quad (1)$$

Fungal species, unlike chemolithotrophic species, require organic additions as a source of energy (Borja et al., 2016). Fungal leaching of metals is generally an indirect process. This process results in the production of organic acids, amino acids, and other metabolites (Simate et al., 2010). The following bioleaching processes are used by fungi to dissolve metals: (1) acidolysis (acid–base reaction), (2) redoxolysis (reduction processes), (3) chelation/complex formation, and (4) bioaccumulation (Dusengemungu et al., 2021; Simate et al., 2010). The first three steps are triggered by fungi's metabolites, whereas the fourth occurs when fungi take metal ions from the solution and disrupt the equilibrium between solids and dissolved metal, resulting in the metal's ongoing dissolution (Simate et al., 2010). The controlling process in fungal metal leaching is acidolysis. Metal detachment and metabolite formation are supported by the quick protonation of oxygen atoms covering the metal's complex surface and their interaction with water; protons stimulate solubilization processes without neutralizing them (Borja et al., 2016).

Therefore, the main processes that occur for nickel ions during the dissolution of laterite are listed below:

(A) Acidolysis (Proton attack) (Simate et al., 2010)

$$NiO + 2H^+ \rightarrow Ni^{2+} + H_2O \quad (2)$$

$$MCO_3 + 2H^+ \rightarrow M^{2+} + H_2O + CO_2 \quad (3)$$

That M can be Fe, Mg, Al, Mn, Ca, etc.

In Eq. (2), organic acids produce protons that ease the mineral's dissolution. In Eq. (3), the acid reacts to any carbonate in the ore, which increases the release of nickel. Particularly, this reaction necessitates the generation of an excessive amount of acid.

(B) Reduction (Simate et al., 2010)

$$MnO_2 + 2e^- + 4H^+ \rightarrow Mn^{2+} + 2H_2O \quad (4)$$

The reduction of soluble manganese in Eq. (4) can lead to an equilibrium between Mn^{4+} in the solid phase and Mn^{2+} in the soluble phase, which somewhat accelerates the dissolution of the mineral and also the release of nickel.

(C) Complexation/chelation (Simate et al., 2010)

$$Ni^{2+} + C_6H_8O_7 \rightarrow Ni(C_6H_5O_7)^- + 3H^+ \quad (5)$$

Equation (5) shows how metal ions are inhibited by organic acids increases the mineral's apparent solubility while the metal's activity is diminished (Simate et al., 2010).

The ability of microorganisms to create hydroxyl-carboxylic acids (citric, lactic, gluconic, pyruvic, and tartaric acids) and other metabolites eliminated from the culture media, as well as their resistance to heavy metals, determines their usage) (Simate et al., 2010). For example, citric acid is made by fermenting sucrose with the *Aspergillus niger* fungus, which generates a soluble ligand complex via the chelation mechanism. Metabolic citric acid is a type of citric acid produced by *Aspergillus niger's* excretory activity (MCA) (Petrus et al., 2018). The brucite layer of clay ores contains nickel from silicate ores, which forms weak bonds and dissolves quickly. Goethite and hematite have different adsorption capacities for nickel, based on surface

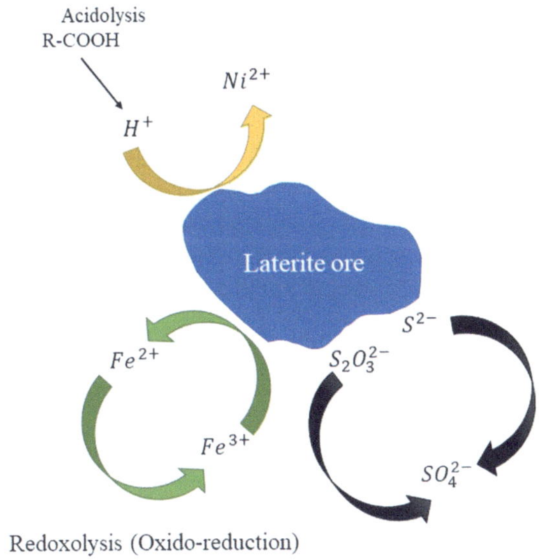

Fig. 7 Main mechanisms for bioleaching of laterites

area; for example, the lower the surface area, the greater the adsorption capacity) (Simate et al., 2010).

Figure 7 depicts the mechanisms of bacterial and fungal leaching. The mechanism for autotrophic acidophilic (iron-oxidizing) and iron-reducing (dissimilatory iron-reducing bacteria) bacterial leaching has been found.

4.1 Laterite's Bioleaching

The processes of nickel-based laterites have been impeded by the complicated mineralogy and limited applicability of existing technologies, hence, making nickel and cobalt extraction from commercial-grade laterites energy-intensive and costly. Therefore, alternative technologies that are technical, economic, and environmentally friendly are needed (Valix et al., 2009).

Metals have been recovered from laterite ores in recent years using organic acid metabolites generated by microorganisms. Nevertheless, several technical challenges must be overcome before industrialization, including a long leaching duration and low metal recovery. The organic acid is a good choice for extracting specific metals from laterite ores due to its high selectivity. Citric acid is the most effective organic acid for extracting nickel from serpentine-type laterite ores. On the contrary, it cannot dissolve nickel from limonite-type laterite ores because of its low reactivity with goethite where nickel is embedded in a goethite structure (FeO(OH)) (Li et al., 2010). Nickel-bearing laterite is not able to participate in the bacterial oxidation of primary chemolithotrophs because it does not contain ferrous iron and a significant amount of reduced sulfur (Krstev et al., 2012). Autotrophic and heterotrophic microorganisms can extract nickel from ore (Sahu et al., 2011). Industrial-scale techniques for the heterotrophic leaching of oxides, carbonates, and silicates lacked progress early on and are still awaiting development. Since low-grade nickel ores processing by conventional methods requires a lot of energy and high operation costs, for the recovery of nickel from laterite ores, the most environmentally friendly and economical technologies were used such as chemoorganotrophic bioleaching which is required to develop for use in industrial applications. More research is needed to improve the procedures for processing low-grade lateritic ores, as it appears that chemoorganotrophic bioleaching of oxide ores has a lot of potential (Chaerun et al., 2017).

The use of heterotrophic bacteria in the leaching of non-sulfide ores has recently attracted a lot of attention (Mubarok et al., 2013; Sahu et al., 2011). There has been evidence of heterotrophic bacteria leaching non-sulfide minerals successfully by *Pseudomonas* subtypes. Because non-sulfide ores lack an energy source for microorganisms when there is a carbon supply for energy and growth, heterotrophic bacteria and fungus can dissolve them. The metabolic products produced by organisms can contain organic acids and compounds that are formed from at least two hydrophilic reactions. Mineral surfaces are affected by secondary metabolites, which are made by organotrophic organisms that make energy from organic material. These metabolites generate exopolysaccharides, amino acids, and proteins, as well as many organic acids such as acetic acid, citric acid, oxalic acid, and ketoglutaric acid, which can be used to dissolve metals through various methods. The protons and anions of metal complexes present in organic acids make them useful for leaching. (Sahu et al., 2011).

Heterotrophic fungi and *Acidithiobacilli* species can react with lateritic oxides. The sulfuric acid made by acidophilic bacteria dissolves metals more readily than the organic acids made by fungi. Particularly, the environment in the in-situ and direct bioleaching method has a very important role in the success of this technology. In particular, *Acidithiobacillus thiooxidans* is a mesophilic acidophil that can tolerate severely acidic environments (pH = 1 or less). Its metabolite, predominantly sulfuric acid, can help in the dissolving of host minerals as well as pH regulation, which can help prevent metal extraction from lateritic ores) (Jang & Valix, 2017). The most effective organisms for dissolving laterites are *Aspergillus* and *Penicillium* species (Tang & Valix, 2004, 2006). Nickel and cobalt can be extracted from laterite ores using the species *Aspergillus foetidus* and *Aspergillus niger* exclusively from *Aspergillus* species (Mubarok et al., 2013).

5 Importance of Bioleaching Method Among Other Methods of Processing Nickel from Laterites

When compared to conventional processing technologies (high temperature/pressure), the biological processing method for extracting base metals from oxidized ores such as limonite offers significant advantages (in terms of environmental and carbon effects) (Smith et al., 2017). Table 3 summarizes the benefits and drawbacks of several nickel recovery techniques.

A variety of uses are available in the mineral industry for biological leaching of low-grade ores because of its relative simplicity, low operational costs, low investment costs, low input energy requirements, and absence of harmful environmental implications (Ahmadi et al., 2015; Li et al., 2014; Mohapatra et al., 2007; Simate & Ndlovu, 2008; Simate et al., 2010).

5.1 The Merits and Demerits of Bacterial and Fungal Applications

The microbes used in biomining have been designed to work under extreme environments, such as low solution pH, high temperature, and high heavy metal concentrations. Bioleaching, as an ecologically friendly method, will play a key role in minerals extraction more sustainable in the near future by selectively targeting exposed mineral grains (Borja et al., 2016).

Filamentous fungi are utilized in bioleaching because of their capacity to release organic acids that cooperate in the solubilization of metal ions from the solution phase while tolerating high concentrations of metal ions. The value of these organic acids is that they keep the pH low, which is necessary for a better bioleaching process, and they offer needed protons. In the fight to improve low-grade mineral bioleaching, fungi's great adaptability and tolerance to

Table 3 Advantages and disadvantages of various processes for nickel recovery (Simate et al., 2010)

	Advantages	Disadvantages
Bioleaching	• Low-temperature process • Done at atmospheric pressure • Low need for process control • Can be used for laterite ores with low-grade nickel • Operation ability at a low cost • Requires low investment costs	• Need for a long time • It has only been successful on a semi-industrial scale
High-pressure acid leaching	• Proven technology • Short Residence Time Distribution (RTD)	• Along with environmental concerns • Expensive administration • High-temperature process • Run at high pressures • Need more process control • Need for an oxygen plant to provide the needed oxygen • Only effective for certain types of laterite ores with certain grades (often limonite) • Need for high investment costs
Caron process	• Proven technology	• Along with environmental concerns • Expensive administration • High-temperature process • Only effective for certain types of laterite ores with certain grades (limonite or a mixture of limonite and saprolite) • Requires a variety of reactants • Need for high investment costs
Melting	• Proven technology	• Along with environmental concerns • Expensive administration • High-temperature process • Effective only for certain types of laterite ores with certain grades (often saprolite) • Need for high investment costs

environmental pressures such as metal contamination have been recognized as a benefit (Dusengemungu et al., 2021).

Long extraction times, the further process for produced by-products, and metal toxicity to biomining microorganisms are the key obstacles that bioleaching of mineral ores and concentrates encounters. New strategies should focus on generating new catalysts that can increase microbe-mineral interactions while also speeding up the kinetics. In laterite bioleaching, the catalyst should perform two key functions: (1) activation of the mineral surface for faster interactions with microorganisms; and (2) continual provision of nutrients or electrons to the bacteria. Those functions might help to speed up the kinetics (Borja et al., 2016).

Technological advancements should be made on both the process and equipment development fronts. However, further testing is required to make biohydrometallurgy suitable for mineral extraction.

6 Research Background (Leaching and Bioleaching) of Laterite Samples

Nickel bioleaching is a complicated process that is influenced by some factors including temperature, pH, bacterial type, sulfur content, and cell concentration (Chang et al., 2008). Many researchers studied the bioleaching of nickel-bearing laterite ores by changing these parameters and reached an acceptable recovery for a nickel. Some of the more new studies are presented in the following:

Gadd (2001) stated that organic acids generated by fungi have a reciprocal influence on the supply of hydrogen ions for the acidolysis of minerals and the creation of metal complexes, due to their chelating capacity (Gadd, 2001). Valix et al. (2001a) investigated the leaching of a variety of minerals, including saprolite, weathered saprolite, limonite, and nontronite. Their study showed that *Aspergillus* and *Penicillium* species were the most effective organisms and chemical dissolution was shown to be less successful than biological dissolution which increases the dissolution rate of nickel from saprolite or silica-rich phases. The accompaniment of nickel with goethite prevents the effective nickel recovery of limonite. On the other hand, cobalt dissolves easily from the limonite. Cobalt recovery is low from saprolite due to metal adsorption on the gangue rather than a lack of mineral solubility. The effective and selective dissolving of particular species of cobalt and nickel (not observed in the chemical dissolution process) leads to the conclusion that biological activity impacts the dissolution process itself in addition to bio-acid generation (Valix et al., 2001a). A study by Tang and Valix (2004) investigated nickel and cobalt solubility in limonite and weathered saprolites. The results showed that nickel and cobalt dissolution is influenced by the host minerals' structure and the ability of the acid to dehydroxylate these gangue minerals. The effect of electrosorption is minimized by adjusting the pulp pH below the pH_{IEP} (zero charge point) of the gangue in the ore, which can improve nickel recovery by up to 200%. This phenomenon has the least effect on the dissolution of cobalt. pH_{IEP} is 6.6 for limonite and 3.8 for weathered saprolite. Dissolution efficiency depends on the concentration of hydronium (H_3O^+) rather than the acid's type. Hydronium concentration is determined by the strength of the acids or how readily they dissolve in solution. Citric, malic, and lactic acids are listed in order of acid resistance. Weathered saprolites have the lowest solid to liquid ratio, which is linked to a larger amount of acid-neutralizing magnesium minerals. In weathered saprolites, the higher ability of nickel to be dissolved compared to cobalt is because in weathered saprolites, nickel is more reactive with the serpentine phase and cobalt is partly associated with goethite. Similarly, cobalt dissolves more readily in limonite than nickel because cobalt is connected with serpentine minerals in limonite and nickel with goethite (Tang & Valix, 2004). Swamy et al. (2005) showed that the bioleaching processes of lateritic nickel ores can be improved by ultrasound due to the increased growth of *Aspergillus niger* with ultrasound. Ultrasound speeds up the transfer of oxygen and nutrients to the cell (Swamy et al., 2005). Tang and Valix (2006) used fungal metabolic acids to dissolve low-grade limonite and nontronite ores. Cobalt had a maximum recovery of 90%, while nickel had a maximum recovery of 33% (Tang & Valix, 2006). Le et al. (2006) extract nickel and cobalt in New Caledonian weathered saprolite ores containing goethite, hematite, titaniferous pyrite, and nickel-ferrous pyrite, using *Aspergillus foetidus*. The extraction of nickel and cobalt was 28% and 31%, respectively (Le et al., 2006). Thangavelu et al. (2006) improved Le et al. (2006) research by adding NaCl to the bioleaching process while using the same fungus, resulting in a 45% improvement in nickel extraction (Thangavelu et al., 2006). Researchers Doshi and Mishra (2007) investigated nickel extraction from laterite ores containing nickel. Nickel was coupled with goethite in the laterite nickel ore, while cobalt was first found in the manganese mineral phase. The ore was soft and agglomerated with very fine porous particles with a very high surface area. Goethite, hydrated iron oxide (alpha-FeOOH), and dissolved residue including some jarosite were found in the analyzed laterite ore. The results of this study showed that nickel extraction decreases with increasing pulp density. With a pulp density of 2%, the maximum nickel extraction was achieved in 15 days and the percentage of nickel extracted was influenced by bacterial activity (Doshi & Mishra, 2007).

To dissolve nickel-bearing laterites, Simate and Ndlovu (2007) used a culture media containing a combination of chemolithotrophic bacteria (*Acidithiobacillus ferrooxidans,*

Acidithiobacillus caldus, and *Leptospirillum ferrooxidans*). Nickel was recovered from low-grade chromite ores using a culture media of a combination of acidophiles which *At. ferrooxidans* was predominating. It was found that a higher concentration of ferrous ions in the culture media resulted in a higher nickel recovery. Adaptation can improve an organism's structure to tolerate heavy metals. Heavy metal tolerance is highest in *Penicillium funiculosum* and *Aspergillus foetidus*. Generally, nickel and cobalt are not tolerated well by *penicillium simplicissimum*. Using the statistical design method of experiments, the effective factors in the bacterial dissolution of nickel-bearing laterites were examined in this work. Results showed that the amount of inoculation was not statistically significant, while the other factors were significant. A weak interaction between variables was observed in the study area. Findings show that recovery is highest at low pH, low pulp density, and big particles. Furthermore, nickel recovery was influenced by particles smaller than 38 microns and the sulfur substrate outperformed the pyrite substrate (Simate & Ndlovu, 2007). Mohapatra et al., (2009a) also examined the effect of laterite ore in various forms such as raw ore, calcined ore, and calcined ore and then soaked in dilute sulfuric acid, before bacterial dissolution. The results of this study showed that the highest percentage of nickel extraction with calcined ore at 600 °C for 31 days was obtained at a pH of 1.5, which was able to extract 77.23% of nickel. The minimum nickel extraction was in the raw ore state which extracted only 9.47% of nickel (Mohapatra et al., 2009a). Simate et al. (2009a) studied the efficacy of using elemental sulfur or pyrite in the bacterial breakdown of nickel-bearing laterites. Hybrid culture media (cultured in standard nutrient 9 K medium) including *At. ferrooxidans*, *A. caldus*, and *Leptospirillum ferrooxidans* were studied. The results revealed that in the presence of bacteria, the dissolution rate of nickel-bearing laterite was high at low pH and high ORP. In the presence of pyrite substrate, nickel recovery was marginally higher than in the presence of sulfur substrate. Consequently, bacterial influence on the dissolution of minerals that cannot participate in the bacterial oxidation process of primary chemolithotrophs appears to be determined by the interaction between the bacteria and the added substrate. In other words, in the early stages of leaching, bio-oxidation of the substrate is considered as a potential factor (Simate et al., 2009a). Nickel was extracted from the laterite ores of the Sukinda mines in Orissa in India with biological dissolution using *Aspergillus niger* by Mohapatra et al. (2009a). To match the experimental model reduction of the second-order system, the experiments were carried out using a central composite design method. The model is significant at a level of less than 0.01%, according to the F-value (for example 99.99% confidence level). Based on a pulp density of 8.75%, a sucrose content of 10.04 g/L, and a 33.8 °C temperature for 37.5 days, the maximum extraction predicted for nickel was 31.34% (Mohapatra et al., 2009b). Buyükakinci and Topkaya (2009) examined the leaching of lateritic nickel ores under atmospheric pressure using agitation leaching. It turned out that particle size did not affect agitation leaching at atmospheric pressure. Finally, nickel recovery was 93.1% and cobalt recovery was 75% (Büyükakinci & Topkaya, 2009). Valix et al. (2009) report that the high salinity of water and soil near laterite ores caused the use of fungi in in-situ bioleaching. To address this issue, *Aspergillus foetidus* adjusts to saline environments over time to make itself more resistant to salinity. In this study, the effect of this adaptation on fungal growth, metabolism, and leaching behavior of weathered saprolite ores with an average size of 53 microns was studied. This saprolite was rich in magnesium silicate. To increase *Aspergillus foetidus'* salinity tolerance, its gradual adaptation to salinity can be exploited. Thus, compared to conditions of 0 ppm salt, salinity slowed fungal metabolism, reduced bio-acid production, and reduced sucrose to bio-acid conversion efficiency. Nickel and cobalt had less metal dissolution in the first leaching stage and some metals (Co and Fe) experienced an adsorption delay (5–8 days), but high salinity (500 ppm NaCl) more effectively overcomes secondary reactions and metal loss, resulting in higher metal recovery. High salinity also stabilizes the selectivity of nickel over cobalt and iron during a long time. High salt concentrations appear to have two impacts in general. Where salinity is likely to be effective, precise salinity control requires optimization of metal recovery using bioleaching process (Valix et al., 2009). Simate et al. (2010) discussed new advances in fungi (chemoorganotrophic) and chemolithotrophic microorganisms. This study found that non-sulfide ores could be reacted with heterotrophic bacteria and fungi. The research on fungal leaching for nickel-bearing laterites is summarized in Table 4 (Simate et al., 2010). The most effective dissolving agent is citric acid, while oxalic acid has the least effect. One explanation for this rating may be the precipitation of nickel oxalate by oxalic acid, which has extremely low solubility (Simate et al., 2010).

Sahu et al. (2011) studied the dissolution kinetics of nickel-bearing laterite ores in a batch bioreactor using aqueous acid solutions containing three metabolic acids: citric acid, oxalic acid, and acetic acid. In nickel-bearing laterite ores, mineralogical studies have revealed that there is no discrete mineral phase from nickel. Iron is found in goethite, which is the host rock or principal phase in raw laterite ore which contains the majority of nickel. The laterite ore also contained hematite, nickel ferrite ($NiFe_2O_4$), garnierite, and quartz. Ashes-layer-diffusion control is a type of

Table 4 A research summary on fungal leaching of nickel-bearing laterites (Simate et al., 2010)

Ore/minerals	Type of fungus or species	Types of acids	Ni recovery (wt. percentage)	References
Calamine, garnierite	*Aspergillus niger*	Citric, oxalic	Up to 78	Castro et al. (2000)
Saprolite, limonite, nontronite	*Aspergillus, Penicillum*	Not specified	Up to 36	Valix et al. (2001)
Saprolite, weathered saprolite, limonite, nontronite	*Aspergillus, Penicillum*	Not specified	Up to 35	Valix et al. (2001b)
Weathered saprolite	*Aspergillus foetidus*	Not specified	28	Le et al. (2006)
Chromite slag	*Aspergillus*	Citric, oxalic, gluconic	Up to 34	Mohapatra et al. (2007)

heterogeneous reaction that includes the reaction of laterite ore particles (which are present in the goethite core) with metabolic acids such as citric acid, oxalic acid, and acetic acid. The shrinking core model was found to be compatible with laboratory evidence. Through the ash layer of the ore spherical particle, citric acid, oxalic acid, and acetic acid showed different diffusion coefficients. Effective diffusion coefficients of acids, besides activation energy profiles, can be useful for the development of basic prototypes to simulate, control, and increase the scale of biological leaching processes for nickel ores (Sahu et al., 2011). Krstev et al. (2012) reviewed heat-resistant ores or low-grade nickel oxides (laterites) in Macedonia by magnetic separation–flotation method, and production and melting of ferronickel. At the same time, this study pays attention to the possibility of dissolution by chemolithotrophic microorganisms (bioleaching). The results showed that the average grade obtained from the separation–flotation process in the form of nickel concentrate was 80–90% nickel and the recovery was 80–85%. The average recovery for nickel leaching was 87–90% while 93–95% for its bioleaching (Krstev et al., 2012). Aspergillus niger, along with acid mine drainage from a coal mine, was used as organic substrate (for carbon and as a source of fungi nutrients) by Mubarok et al. (2013) for direct bioleaching of nickel from Indonesian Pomalaa laterite ore. Goethite and quartz were the main minerals of limonite ores. The type of organic acids generated by the microbes was determined using high-performance liquid chromatography (HPLC). The optimal amount of acid mine drainage should be determined to increase nickel extraction and reduce the simultaneous dissolution of iron and magnesium. Saprolite ore dissolves better with fungal metabolite acids than limonite, according to the findings. The distribution of coarse particle size (+177 −250 microns) in both saprolite and limonite ores resulted in superior nickel extraction than the finer particle size of these ores. In the finer particle size distribution, the reduction in nickel extraction may be due to the adsorption of dissolved nickel ions into the gangue minerals and at the same time the higher dissolution of the metal impurities. The findings revealed that *Aspergillus niger* can grow on an acid mine drainage organic substrate derived from a coal mining location. The addition of elemental sulfur did not affect the nickel extraction rate (Mubarok et al., 2013).

Economical metals are also present in oxidized ores (such as nickel in laterite). On the other hand, because these ores cannot have oxidative dissolution, the ferric iron minerals present in these ores can be dissolved by reducing the iron and releasing the metals. Johnson et al. (2013) used the acidophilic bacterium *At. ferrooxidans* capacity to be permanently anaerobic which oxidation of sulfur is linked to ferric iron reduction in the goethite of nickel limonite ores. When anaerobic culture media were placed in bioreactors temperature and pH were adjusted, *At. ferrooxidans* accelerated the reductive dissolution of goethite and thus increase the dissolution rate of nickel in the ore. The quantity of ferrous iron in the solution was strongly correlated with the amount of nickel in the solution; on the other hand, goethite was found to be the source of the majority of the nickel. Furthermore, the proportion of sulfuric acid needed to keep the pH of the leach solution at 1.8 and the amount of dissolved iron (1.75:1) was close to the theoretical value (1.7:1) in Eq. (6), indicating that the predominant form of dissolved sulfate at pH 1.8 is HSO_4^- instead of SO_4^{2-} (Johnson et al., 2013).

$$24\mathrm{Fe}\cdot\mathrm{OH} + 4S^{\circ} + 40\mathrm{H}^+ \rightarrow 24\mathrm{Fe}^{2+} + 4\mathrm{SO}_4^{2-} + 32\mathrm{H}_2\mathrm{O} \quad (6)$$

In contrast, when iron dissolution proceeds entirely through acidic dissolution (according to Eq. 7), the ratio of proton consumption to the amount of dissolved iron is 3:1.

$$24\mathrm{FeO}\cdot\mathrm{OH} + 72\mathrm{H}^+ \rightarrow 24\mathrm{Fe}^{3+} + 48\mathrm{H}_2\mathrm{O} \quad (7)$$

In addition to the dissolution of iron and nickel, the concentration of dissolved cobalt and manganese also increases during the bioleaching of the laterite ore. Bio-processing of lateritic ores under anaerobic conditions contained higher amounts of dissolved manganese, nickel, iron, and cobalt within 20 days compared to aerobic conditions. Anaerobic conditions accelerate the reduction of ferric iron to ferrous iron by using elemental sulfur as a source of electrons provided by the *At.ferrooxidans* bacterium (Johnson et al.,

2013). Fatahi et al. (2014) examined the extraction of nickel from laterites using agitation leaching at atmospheric pressure. In this study, sulfuric acid and nitric acid were used for dissolution. As a result, 83% of recovery was achieved in the optimal conditions (Fatahi et al., 2014). MacCarthy et al. (2014) performed acid leaching at atmospheric pressure using 98% sulfuric acid on a laterite sample. The results showed that all pulps have non-Newtonian behavior and the maximum recovery was 87.78% for goethite-siliceous nickel-bearing laterites and 55.42% for goethite nickel-bearing laterites (MacCarthy et al., 2014). Önal and Topkaya (2014) replaced the pressure acid leaching with heap leaching by studying a sample of Çaldağ laterite ores. Variable parameters included acid to ore ratio, temperature, and time. Nickel had a maximum recovery of 94.1%, and cobalt had a maximum recovery of 94%. It was also found that nickel extraction is temperature sensitive and cobalt extraction is acid sensitive (Onal & Topkaya, 2014).

Microbial processing of lateritic reserves still needs to be improved so that this processing method can be applied on an industrial scale. Microorganisms play an essential role in the biological processing of minerals because they accelerated the extraction of metals from minerals. Therefore, the discovery of physiological and biomolecular mechanisms involved in this interaction of microbes and minerals is inevitable. Advances in nickel extraction from laterites were discussed by Behera and Mulaba-Bafubiandi (2015). Application of heterotrophic microbes (fungi of the *Aspergillus* and *Penicillium* genus and bacteria of the *Bacillus* and *Pseudomonas* species), chemolithotrophic microorganisms (mainly *Acidithiobacillus*, *Leptospirillum*, and *Sulphobacillus*), and iron-degrading and reducing bacteria (mainly from *Shewanella*, *Geobacter*, and *Geovibrio* species) was examined in this study. The use of chemolithotrophic microorganisms for the processing of oxidized minerals (laterites) using conventional bio-oxidation leaching is not used because oxidized minerals do not have the nutritional support (reduced sulfur compounds or divalent iron compounds) for such microbes. However, the microbial reduction process, which uses DIRB and chemolithotrophic bacteria, provides a clearer path to the development of a technical process that extracts nickel from laterites in an economically viable manner (Behera & Mulaba-Bafubiandi, 2015). Based on the results of Astuti et al.'s (2016) studies, it was determined that citric acid is more efficient than other organic and inorganic acids for saprolite leaching in Indonesia. They compared the efficiency of citric acid with that of other acids (lactic acid, oxalic acid, sulfuric acid, hydrochloric acid, nitric acid, as well as the combination of citric acid and sulfuric acid). Citric acid, as well as sulfuric acid, separately provided a recovery of 72%, while nitric acid and hydrochloric acid with a ratio of 1 + 1 provided a recovery of 43–45%. Furthermore, recovery with a 0.75 + 0.25 ratio using citric acid and sulfuric acid yielded 80% in one sample and 100% in the other. Sulfuric acid was also discovered to be superior to citric acid for leaching laterites containing goethite. Nickel was recovered from goethite as effectively as serpentine with inorganic acids such as sulfuric acid, nitric acid, and hydrochloric acid. Although citric acid is a weak acid, it is an excellent metal chelator and binder through the mechanism of acidolysis and complexolysis. With the combination of citric acid and sulfuric acid, nickel recovery and selectivity are both high, and acid consumption is lowered, making this system environmentally friendly (Astuti et al., 2016). The selective dissolution of nickel from limonite laterites was studied by Chang et al. (2016) under moderate pressure conditions. It was found that hematite precipitation lowers the iron concentration to less than 1 g/L when nickel is recovered at 95.2% under optimal conditions (Chang et al., 2016). Kursunoglu and Kaya (2016) studied leaching at atmospheric pressure on a sample of Caldag laterite ore. The nickel and cobalt recoveries were respectively 91.9% and 93.5%, while the iron recovery was 80.5%. The results showed that crushing finer than 0.150 mm did not affect nickel recovery, while the highest cobalt extraction was obtained with the smallest particle sizes. The nickel, asbolane, goethite/hematite, and clay phases were detected using XRD and SEM–EDX. Furthermore, the nickel extraction kinetics revealed a 51.6 kJ/mol activation energy. Thermal pretreatment, high mineral acid concentrations, and high-temperature leaching are methods used to recover significant quantities of nickel and cobalt (Kursunoglu & Kaya, 2016). Chaerun et al. (2016) evaluated nickel recovery from limonite and saprolite ores using bioleaching with agitation tank reactors with native bacteria and a combination of iron-oxidizing bacteria including *Alicyclobacillus ferrooxydans*, *Comamonas testosterone*, and *Pantoea septic* species, which was isolated from Indonesian ore. With a 10% pulp density, coarse limonite, and weathered saprolite ore were tested for their pulp density. The bioleaching process was done by measuring the content of iron, nickel, and the pH of the leaching solution. In addition, the FDA hydrolytic activity was used to calculate the overall bacterial enzyme activity. FDA hydrolytic activity was 17.2 μg fluorescein/mL for limonite and 16.9 μg fluorescein/mL for saprolite. At pH = 2.5–3.5, the amount of nickel and iron dissolved for limonite ore (30% Ni, 5.6% Fe) was higher than their amount in biocontrol (1% Ni, 0.1% Fe) but in the saprolite ore, a species of hybrid bacterium, was only able to bioleach 2.5% of nickel and 6% of iron (similar to limonite). In biological control, the pH of the environment remained relatively constant (about 6) (Chaerun et al., 2016).

Most researchers used the terms single-stage bioleaching (fungi and ores enter media simultaneously), two-stage bioleaching (fungus was added first, then ore after metabolite production began), and spent medium bioleaching to describe bioleaching with fungi (biometabolites were produced by fungi after a determinate time). The three different methods of fungi bioleaching have been described by Chaerun et al. (2017) as direct bioleaching (one-stage bioleaching), semi-direct bioleaching (two-stage bioleaching), and indirect bioleaching (spent medium) (Chaerun et al., 2017).

Fungal metabolic organic acids were used to indirectly bioleaching low-grade nickel-bearing laterite ores with *Aspergillus niger* due to the following reasons:

(1) Chemoorganotrophic microorganisms are better suited for the treatment of ores that are not sterile and cannot be sanitized, due to their ability to flourish in less selective media and higher tolerance than acidophilic chemolithotrophs.
(2) It seems that the growth of chemoorganotrophic microorganisms on the ore causes the degradation of produced acidulates/ligands.
(3) Organic acids, unlike inorganic acids such as sulfuric acid, are less corrosive.
(4) Nickel is extracted from laterite ores with a low nickel content using *Aspergillus niger* which is caused by the adsorption of dissolved nickel ions into the hyphae of the fungus (Chaerun et al., 2017).

In this study, the parameters include sources of organic carbon (molasses and starch) as substrate, sources of nitrogen and phosphorus ($(NH_4)_2SO_4$, KH_2PO_4), micronutrients ($MgSO_4$, $MnSO_4$), and aeration which improve the bio-production of fungal metabolic organic acids, were optimized. Chaerun et al. (2017) also used limonite and saprolite from an Indonesian mine to measure the types of organic acids produced and their effects on nickel leaching processes, as the types of organic acids produced by fungi differ with the sources of metabolized carbon (Chaerun et al., 2017). Aspergillus niger produces fungal metabolic organic acids which were used in the experiment to demonstrate greater selectiveness of nickel bioleaching compared to iron (S_{Fe}) and magnesium (S_{Mg}) at different temperatures (25, 65, and 95 °C). Hence, leaching with organic acid created by bio-production was strongly selective for magnesium and relatively non-selective for iron. *Aspergillus niger* produced the highest amount of organic acid with starch as a carbon source when it was incubated with KH_2PO_4, $(NH_4)_2SO_4$, and $MgSO_4$ under aeration conditions, which was then used for indirect bioleaching. At a pH of 1.4, starch produced more organic acids than molasses after 16 days, although the rate of synthesis was slow so it will need further studies for industrial uses. Saprolite and limonite ores both dissolved nickel at a maximum efficiency of 88.9% and 84.9%, respectively (Chaerun et al., 2017).

Jang and Valix (2017) studied the gradual adaptation of *A. thiooxidans* to dissolve heavy metals from Riverina-related nickel-bearing saprolitic laterites in Australia. For heavy metal adaption trials, the media was sterilized. The adaptation, drawn in terms of the adaptation index, depended on the metal concentration, acid production, and adaptation period. Simulation of cell growth and acid production proceeded using the effect of heavy metals on bacteria. The use of adapted bacteria increased nickel recovery by 20% and cobalt recovery by 7%. The results obtained in this study confirmed the strong structure and high stability of *A. thiooxidans* in the direct bioleaching of nickel-bearing ores. Acid production improves metal stability in solution and lowers metal adherence to the cell wall, as well as its toxic effects. Furthermore, as compared to unadapted bacteria, adapted *A. thiooxidans* demonstrated up to a tenfold improvement in cobalt leaching selectivity (Jang & Valix, 2017).

Several studies have discussed leaching processes using various organic acids, but only a few types of research have been done on the effect of surface area and particle size distribution on the leaching process. Pawlowska and Sadowski (2017) study focused on how microorganisms produced citric acid and added citric acid impact the specific surface area and particle size distribution of minerals chemically. The effect of chemical citric acid and filtered citric acid on nickel extraction following cultivation of *Aspergillus niger* was also compared in this article. Besides, laterite ore was acquired from the Szklary deposit in Poland. The microbes were cultivated for four days before being autoclaved and filtered to eliminate the dead biomass. It was observed that higher nickel recovery is the main factor of higher particle surface area. The highest recovery in nickel extraction was obtained for the 1 mol/dm^3 (67%) concentration of citric acid. After leaching, the specific surface area was 159 m^2/g. After leaching with citric acid, a particle size distribution study revealed an increase in average particle size, owing to a drop in the percentage of the smallest grains. Polish laterites also disintegrated according to the shrinking core-shrinking particle model (Pawlowska & Sadowski, 2017).

Marrero et al. (2017) demonstrated anaerobic reduction dissolution of iron, nickel, cobalt, and manganese from laterites in laboratory bioreactors using *At. ferrooxidans* and *A. Thiooxidans*. The chemical mobility of ferric ions in inoculated anaerobic bioreactors occurred at pH = 0.8. After 7 days, the pH needed to be increased from 0.8 to 1.8 to increase the dissolution rate with a biological decrease in ferric ions. Aerobic reactors by maintaining the pH at 0.8 (inoculated or not inoculated) resulted in more release of ferric ions in laterites than anaerobic reactors. The process of aerobic reduction dissolution was much more efficient than the anaerobic reduction solution for extracting total iron and ferrous iron, as well as for dissolving manganese, cobalt, and nickel. The presence of O_2 is considered a key factor in increasing the bio-reduction dissolution of iron-containing minerals (Marrero et al., 2017). Rasti and Rajabzadeh (2017) studied the geochemical and mineralogical characteristics of nickel-containing laterites of serpentine origin, the application of the lateralization process, and the classification of nickel-containing laterites related to the Fars region in Iran (Rasti & Rajabzadeh, 2017). For low-grade nickel ores in the Sarbisheh region, Javanshir et al. (2018) used atmospheric pressure leaching. The ore has a high content of iron and low content of nickel, with quartz, goethite, and dolomite being the major crystalline phases; the nickel-ferrous laterite ore was of the goethite type. The results of their study showed that goethite and hematite were the main minerals of the sample and nickel are scattered throughout the ore area. Magnetic separation was used to pre-concentrate the ore in this investigation, although it was unsuccessful. Hematite (the main mineralization phase) was formed when the laterite ore was heated, which was easier to dissolve, so calcination was an effective step to increase nickel extraction. TG/DTA analysis showed two endothermic peaks, dihydroxylation of goethite to hematite occurred at 280 °C and XRD analysis before and after calcination confirmed this conversion. Leaching with sulfuric acid obtained higher nickel values than leaching with hydrochloric acid (Javanshir et al., 2018). Li et al. (2018) examined the amount of nickel and cobalt extraction from limonitic laterites by changing the type and concentration of acid as well as the calcination temperature. Phosphoric acid and sulfuric acid were the selected acids for this selective leaching. The maximum recovery was 98.7% for nickel and 89.8% for cobalt, and 98.3% of the iron was converted to insoluble iron. After calcination, recovery was 95.9% for nickel and 91.4% for cobalt. In addition, 92.4% of iron was converted to insoluble iron. This study determined that the formation of iron phosphate with phosphoric acid was more effective than that using sulfuric acid and hydrochloric acid (Li et al., 2018). Petrus et al. (2018) used limonite-type laterite nickel ores from the Pomalaa region of southern Sulawesi, Indonesia. *Aspergillus niger* under optimal conditions was used to produce metabolic citric acid as the dissolving agent. The results showed that the maximum metal recovery increases with decreasing pulp density. The pulp density (solid to liquid ratio) of 5% obtained maximum recovery; the recovery rate at this pulp density was 1.63% for Ni, 0.47% for Al, 0.23% for Fe, and 1.09% for Mg. These recovery values were lower than those obtained from synthetic citric acid, which dissolved about 90% of nickel. This necessitates a change in the dissolution mechanism involving metabolic citric acid. This modification is done by adding H_2O_2 to decrease metals such as Fe and Mn, or by fine-tuning the metabolic citric acid synthesis pathway to produce higher acid concentrations. The shrinking core model of leaching is illustrated by the effect of particle size on leaching process. Three steps affect the reaction rate:

(1) Diffusion through a coating on the particle's surface,
(2) Diffusion through the particle's ash layer,
(3) Surface reaction on the particle's unreacted section.

The rate of the leaching reaction is determined by the slowest stage. In this research, the diffusion-controlled reaction rate in the leaching process. The maximum metal recovery occurred with a particle size of 125–149 μ and the recovery rate was 1.37% for Ni, 0.72% for Al, 0.1% for Fe, and 0.62% for Mg. Due to the broader radius of the ash layer generated for larger particles, which prevents the reactant from adsorbing to the reaction region and the product from desorbing to the liquid section, the recovery rate was lower in particle sizes larger than 200 microns than in 125–149 microns. The reaction rate increases with smaller particle sizes, which means a bigger surface area, but as the reaction rate increases and complex metal molecules form on the reaction area's surface, a space barrier is generated for the result to be desorbed (Petrus et al., 2018).

Giese et al. (2019) conducted bioleaching experiments using the heterotrophic species *Bacillus subtilis* on a Brazilian sample of nickel-containing laterite ores. The ore mainly contained lizardite (32.6% MgO) and 1% NiO (0.85% Ni). The results of SEM analysis showed that the primary nickel-bearing mineral was goethite. Discontinuous bioleaching experiments showed that 8.1% Ni (0.7 mg nickel per gram of ore) was dissolved using *Bacillus subtilis* within 7 days. In this study, diluted laterite with water was heated using a microwave as a pretreatment method. The biological extraction of nickel was raised from 8 to 26% using this pretreatment approach (2.3 mg nickel per gram of ore) (Giese et al., 2019). Yang et al. (2019) investigated the performance of direct or indirect bioleaching of cobalt and nickel from laterite and pyrite ores using *Aspergillus niger*. Imported laterite ores came from Kazakhstan, while pyrite ores came from Finland. Average cobalt recoveries from laterites and pyrite ores using direct bioleaching were obtained 65.9 ± 1.8% and 4.9 ± 2.7%, respectively, and average nickel recoveries were obtained 30.9 ± 0.6% and 10.9 ± 6.2%, respectively. Recovery of cobalt from indirect laterite bioleaching was significantly lower in the absence of fungal biomass compared to cobalt recovery in direct bioleaching. The most important bioleaching phase in laterites was cobalt-containing goethite, according to the X-ray powder diffraction (XRPD) study of mineral phases before and after bioleaching. Direct and indirect bioleaching yielded no significant changes in cobalt-containing synthesized goethite. As a result, Yang et al. proposed that bioleaching of laterites occurs in two stages:

(1) Cobalt-containing goethite was exposed to direct fungi-mineral interaction.
(2) Organic acids and metabolites produced by Aspergillus niger were used to dissolve cobalt-containing goethite.

The metal recovery process also revealed an inconsistent pattern of Co and Fe bioleaching from laterites (Yang et al., 2019).

Mulroy (2019) studied the geochemistry and mineralogical phases of nickel and cobalt oxides containing Mn (III/IV) and the microbiological interactions with these phases. The ore samples were inoculated with glucose to stimulate native microbial assemblies, while the asbolane mineral was inoculated with a pure culture medium of *Geobacter sulfurreducens* and acetate. Synchrotron-based XAS was used to determine the chemical properties of Co, Ni, and Mn crystals in both natural and synthetic samples before and after microbiological activity. The results showed that in fresh ore, Ni (II) compared to Co (III) is more diverse. Ni (II) is either incorporated in aluminum-bearing parts of lithiophorite or adsorbed on the edges of manganese oxide. The extent to which this occurs indicates oxidation/reduction conditions and microbial effects on the various laterite horizons. The microbial consortium reduces Mn (III / IV) and Co (III) to divalent forms. *G. sulfurreducens* has been shown to effectively degrade synthetic asbolane and release Co and Ni in solution. Then, the precipitation of Mn, Co, and Ni as carbonate phases represents a way to form products that are often used as intermediates in the various industrial process (Mulroy, 2019). Stankovic et al. (2020) reviewed the most important hydrometallurgical processes for the extraction of nickel from laterites. The results of this study evaluated high-pressure acid leaching (HPAL), Caron process, heap leaching (HL), atmospheric leaching (AL), direct nickel, Ferredox reductive bioleaching, and hydrochloric acid leaching (Neomet process). Using the direct nickel process (DNP), nickel ores are leached with nitric acid at 110 °C under atmospheric pressure in agitation tanks. The Direct Nickel Group of Australia, which processes Indonesian laterite ores, created the process. The duration of leaching is 2–4 h and the recovery of nickel and cobalt is more than 95%. DNP is a complicated technique that requires multiple pH adjustments throughout the purification stage. As a result of hydrolysis, which takes place at a pH of 2 to 3, ferrous iron is first removed. As a by-product, the hematite cake that collects on the filter is sold. To raise the pH and precipitate aluminum hydroxide, magnesia pulp (MgO) is used to treat the iron-free solution. To create products containing aluminum, aluminum hydroxide is filtered. The residual solution, which contains a small amount of aluminum, is prepared for the mixture hydroxide precipitation (MHP). In a two-stage circuit, the magnesia pulp is used to increase the pH once more, and the hydroxide mixture containing the majority of the nickel and cobalt recovered from the feed is precipitated. Eventually, nitrogen oxide gases are recovered during a series of adsorption steps to form nitric acid, which is returned to the leaching cycle (Stankovic et al., 2020). The Canadian Neomet company invented chloride leaching and the Neomet process. In tanks at atmospheric pressure and temperatures of 100–110 °C, hydrochloric acid is used to leach nickel and cobalt from limonite and saprolite ores. In this method, the recovery of nickel and cobalt is high (more than 90%). The removal of nickel from laterites is enabled by the biological reduction of oxidized iron and manganese minerals. Some anaerobic microorganisms include heterotrophic and autotrophic acidophilic bacteria. In hypoxic environments, these bacteria can use Fe^{3+} as an electron acceptor instead of oxygen in a respiratory chain that converts ferric iron (Fe^{3+}) to ferrous iron (Fe^{2+}), as shown in Eq. (8) (Stankovic et al., 2020).

$$24Fe^{3+} + C_6H_{12}O_6 + 6H_2O \rightarrow 24Fe^{2+} + 6CO_2 + 24H^+ \quad (8)$$

The reductive bioleaching process takes place at 30 °C and in a mildly acidic environment (pH = 1.7–2). As a result, when compared to conventional hydrometallurgical processes for extracting nickel and cobalt from laterites (excluding heap leaching), this process uses less energy and requires less expensive equipment, hence the Ferredox process's capital and operational costs may be lower. The consumption of sulfur and sulfuric acid has an impact on the economics of this process. Laterites are the part of oxidized ores so, unlike sulfide ores, they do not tend to catalyze the oxidative dissolution of ferric iron (Hallberg et al., 2011; Johnson, 2012). A strong bond between oxygen and ferric iron must be broken for the nickel in goethite or other host nickel minerals to be soluble. Several hydrometallurgical and pyrometallurgical methods have been used to dissolve the oxygen-iron bond (Mohapatra et al., 2007; Plessis et al., 2011; Simate & Ndlovu, 2008), but these methods require high energy or high costs of reactant and high capital costs for equipment. Hydrometallurgical processing methods result in the simultaneous dissolution of gangue minerals, raising the process complexity and increasing the cost of recovering precious metals and removing unwanted elements. A low-cost method for breaking the oxygen-iron bond in goethite can significantly reduce the economic costs of processing limonite containing nickel and cobalt. Oxygen-iron bond breakage is performed by the Ferredox process (Plessis et al., 2011). The Ferredox process with the reductive dissolution of goethite (the most important host mineral for nickel-bearing limonite) using elemental sulfur was identified as a reducing agent in environmental conditions (Johnson, 2012; Plessis et al., 2011). Reductive dissolution of nickel-bearing limonite led to the fundamental development of biomining. In short, this method, in contrast to the traditional method (oxidative dissolution, the use of ferric iron produced by bacteria), was used for the biological processing of metal ores (Johnson, 2012). The process, which was performed using pyrometallurgical methods at 800 °C, is performed at a temperature of 30 °C, which indicates the potential of microorganisms. It has been shown that indirect reductive dissolution of minerals, including asbolane and goethite, can be applied more broadly to the processing of oxidized ores (Johnson, 2012).

As pointed out before, the only commercial hydrometallurgical processes for the extraction of nickel and cobalt from laterites are HPAL and heap leaching. Hydrometallurgical plants have become more efficient with the production of cobalt, which is lost in pyrometallurgical processes. HPAL provides excellent recovery rates of nickel and cobalt and is quick to leach, however, it is complex and requires a large investment. HPAL is more expensive than pyrometallurgical activities in some instances, and it requires a higher operating cost, while heap leaching is more simple, with significantly lower operating costs and investment costs than HPAL and pyrometallurgy, however, the leaching rate is very slow. The possibility of obtaining cobalt and other valuable elements from laterite reserves makes hydrometallurgical processing a more appealing process, in contrast, pyrometallurgical processing extract 80% of nickel from laterite reserves (Stankovic et al., 2020). Santos et al. (2020) studied the biological leaching of three limonite-type laterite ores from Greek mines under reductive conditions using a consortium of acidophilic bacteria in stirred tank bioreactors at pH = 1.5 and 35 °C (using elemental sulfur as an electron source). *At. ferrooxidans*^{T}, *At. ferrooxidans*, species of *CF3*, *At. ferriphilus*^{T}, *At. ferridurans*^{T}, and *Sulfobacillus* (*Sb.*) *thermosulfidooxidans*^{T} are among the acidophilic bacteria used in this work. In this study, the target metal, which is cobalt, was successfully bioleached (40–50% within 30 days). Some other metals, such as iron, were recovered between 2 and 48%. In most cases, the concentration of dissolved cobalt was highly correlated with the concentration of dissolved manganese. This was under the finding of manganese (IV) minerals as a major source of cobalt. Extensive mineralogical analysis for these three limonite samples before and after biological processing revealed significant changes between these ores and showed that mineralogical and elemental changes could significantly affect the adaptability of the ore with reductive bioleaching (Santos et al., 2020). Esther et al. (2020) used Dissimilatory Iron Reducing Bacteria (DIRB) to remove iron impurities from various ores in bioremediation to eliminate the toxicity of heavy metals and toxic pollutants. This study aimed to investigate the effect of a consortium of anaerobic bacteria that decompose and reduce iron (III) on different phases of iron in lateritic nickel ores. Such changes are useful for better recovery of metals such as Ni and Co using bioleaching or acid leaching in later stages. 70% of nickel and 81% of cobalt are recovered using DIRB-treated acidic leaching of ore, while more than 54% of nickel and 57% of cobalt are recovered using acid leaching of untreated chromite overburden (COB). Therefore, COB pretreatment using DIRB can improve nickel recovery) (Esther et al., 2020).

Two of the most important challenges in implementing the acid leaching process at atmospheric pressure to extract nickel from laterites are the high acid consumption and the high final concentrations of iron in PLS. In a study conducted by de Alvarenga Oliveira et al. (2021) using pyrometallurgical and biohydrometallurgical operations, a new process for the extraction of nickel from laterites was proposed. In the laboratory, nickel limonite ores are reduced with hydrogen gas in a rotary furnace at 900 °C until all goethite is converted into metallic iron then reduced sample bioleached with 5% solid by mesophilic microbes cultured on Fe^{2+} (*At. ferrooxidans*) at 32 °C and pH<3. According to experiments conducted using 35 kg of H_2SO_4 per ton of reduced ore, researchers discovered that Fe, Ni, and Co will leach as Eh production from bacteria increases. Nickel dissolved at a rate of 92%, whereas cobalt at 35%. During these laboratory conditions, the concentration of iron in the aqueous phase produced was less than 5 mg/L due to the precipitation of Fe^{3+} as jarosite. A lower concentration of total iron in the aqueous phase of the leach was also observed (less than 200 mg/L), which was much lower than HPAL's reported values (Alvarenga Oliveira et al., 2021) (Fig. 8).

Fig. 8 Flowsheet and mass balance for the proposed procedure in nickel extraction (per ton of limonite ore) (Alvarenga Oliveira et al., 2021)

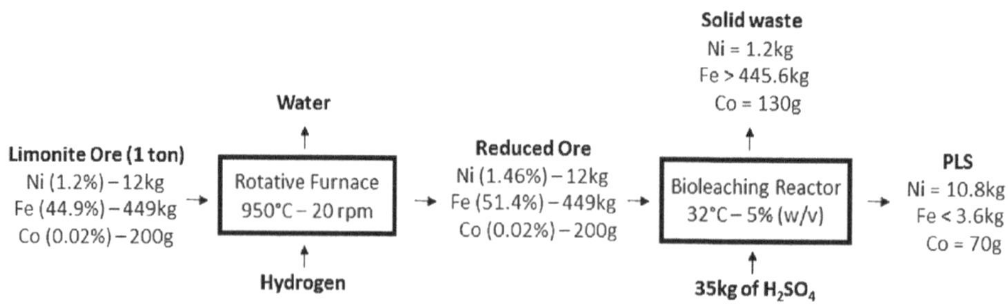

Acidophiles are a group of extremophiles whose optimum growth pH is 3 and below 3 (Gomes & Steiner, 2004; Rampelotto, 2013; Rothschild & Mancinelli, 2001). Organic acids inside acidophilic cells as respiratory chain separators are harmful, but the cell cleverly combines protons with organic acids inside the cell and converts organic acid to carbon dioxide (Nordstrom et al., 2000). Low pH strengthens and intensifies the population of acidophilic prokaryotes as well as increases the solubility of metals (Johnson & Hallberg, 2003; Nordstrom et al., 2000). Iron-reducing acidophiles use ferric iron as an electron acceptor (rather than oxygen) and need an electron donor (sulfur, hydrogen, or an organic molecule, depending on the species type) to reduce ferric iron (Hallberg et al., 2011; Johnson, 2012; Marrero et al., 2015). Due to the bacteria's ability to decrease the amount of soluble ferric iron, a balance between ferric iron in the mineral phase and ferric iron in the soluble phase is created, thus the mineral dissolution rate accelerated according to Eq. (9) (Hallberg et al., 2011):

$$\alpha - \mathrm{FeO.OH} \leftrightarrows \mathrm{Fe^{3+}} \rightarrow \mathrm{Fe^{2+}} \tag{9}$$

The reduction of goethite with sulfuric acid produced an alkaline (proton-consuming) reaction, as shown by Eq. (10) (Hallberg et al., 2011; Plessis et al., 2011):

$$6\alpha - \mathrm{FeO.OH} + \mathrm{S^0} + 10\mathrm{H^+} \rightarrow 6\mathrm{Fe^{2+}} + 8\mathrm{H_2O} + \mathrm{SO_4^{2-}} \tag{10}$$

For the reduction of ferric iron minerals in laterite ores, bacteria can release metals (particularly nickel) and ferric iron minerals into the solution. By doing the process in an acidic environment (for example, acidophilic microorganisms), it is possible to keep the pH of the solution low and thus the extracted nickel remains in the solution and its recovery is facilitated (Hallberg et al., 2011; Johnson, 2012).

The physical contact between bacteria and metals produced by their extracellular metabolites is the basis for metal dissolution by chemolithotrophic bacteria such as *At. Ferrooxidans* (Chang et al., 2008). In bioleaching, metabolites (mostly sulfuric acid) that suggest exponential growth play a crucial role. The pace of bioleaching is limited by acid generation. Metals dissolve when they come into physical contact with bacteria and when they produce acid, so acid production and bacteria can influence metal dissolution (Chang et al., 2008). Gram-negative chemolithotrophic, mesophilic, and voluntary anaerobic bacteria *At. ferrooxidans* have been the subject of extensive genetic, genomic, and phylogenetic studies and therefore have been the only model microorganisms to study the biochemical cycle and electron transfer during iron oxidation (Golyshina et al., 2000; Wu et al., 2008). This bacterium which oxidizes iron and sulfur (Hallberg et al., 2011; Mohapatra et al., 2007), uses energy sources that include Fe (II), H_2S, S^0, reduced inorganic sulfur compounds, and molecular hydrogen. Oxidation of sulfur produces sulfuric acid, which gives interesting properties of *At. ferrooxidans*. These bacteria grow under the temperature conditions of mesophiles (Wu et al., 2008). The use of chemolithotrophs in bioleaching, which uses sulfur and sulfide minerals such as pyrite as energy sources, eventually produces sulfuric acid during the oxidation process [Eqs. (11)–(14)] (Simate & Ndlovu, 2008):

$$\mathrm{FeS_2} + 6\mathrm{Fe^{3+}} + 3\mathrm{H_2O} \rightarrow \mathrm{S_2O_3^{2-}} + 7\mathrm{Fe^{2+}} + 6\mathrm{H^+} \tag{11}$$

$$\mathrm{S_2O_3^{2-}} + 8\mathrm{Fe^{3+}} + 5\mathrm{H_2O} \rightarrow 2\mathrm{SO_4^{2-}} + 8\mathrm{Fe^{2+}} + 10\mathrm{H^+} \tag{12}$$

$$8\mathrm{Fe^{2+}} + 2\mathrm{O_2} + 8\mathrm{H^+} \rightarrow^{\mathrm{bacteria}} 8\mathrm{Fe^{3+}} + 4\mathrm{H_2O} \tag{13}$$

$$0.125\mathrm{S_8} + 1.5\mathrm{O_2} + \mathrm{H_2O} \rightarrow^{\mathrm{bacteria}} \mathrm{SO_4^{2-}} + 2\mathrm{H^+} \tag{14}$$

Nickel-bearing laterite ores do not contain sulfur, so sulfur needs to be added to chemolithotrophs in the leaching of laterite ores (Simate & Ndlovu, 2008; Simate et al., 2010). For the growth of bacteria and to be able to effectively leach nickel-bearing laterites, the produced sulfuric acid maintains the desired pH level (Simate & Ndlovu, 2008). Biogenic sulfuric acid can be produced using chemolithotrophs such as *At. ferrooxidans* (Simate et al., 2010). The property of hydrogen ions to displace metal cations from the ore matrix is the reason that metal dissolution is enhanced (Eq. 15) (Simate & Ndlovu, 2008).

$$\mathrm{NiO} + 2\mathrm{H^+} \rightarrow \mathrm{Ni^{2+}} + \mathrm{H_2O} \tag{15}$$

The use of chemolithotrophic bacteria in nickel recovery from laterite ores has been relatively unknown until recently, owing to the ore's lack of sulfide concentration, which is required to make sulfuric acid (Simate & Ndlovu, 2008). There is uncertainty over whether ferric iron reduction is an indirect reaction from inorganic reduced sulfur compounds formed during sulfur oxidation or a direct reaction from sulfur oxidation enzymes (Marrero et al., 2015).

Simate et al. (2009) expressed that the use of a culture media with a combination of chemolithotrophs could improve the dissolution of nickel from nickel-containing laterites. This is due to the competition of using oxygen in iron-oxidizing bacteria which can lead to the reductive dissolution of ferric iron in nickel laterites and destabilize these ions (Simate et al., 2009b). As well as inhibiting sulfur dioxygenase, nickel ions make the conversion of elemental sulfur to sulfite faster, as well as making the conversion of sulfite to sulfate easier. Nickel surrounds the plasma membrane and prevents the entry of both enzymes, in other words, it prevents cell growth. While nickel interferes with the physiological functions of bacteria such as sulfur and iron oxidation, some bacteria can quickly adapt to high nickel concentrations (Watling, 2008). Simate and Ndlovu (2008) expressed that bacteria of the *caldus* genus extract nickel more efficiently than fungi by producing sulfuric acid through oxidizing elemental sulfur (Simate & Ndlovu, 2008). It has been shown that mesophiles can tolerate nickel in bioleaching environments better than thermophiles and moderate thermophiles. *At. ferrooxidans* are more tolerant of nickel than *A. thiooxidans* (Watling, 2008). Mohapatra et al. (2007) expressed in their study that the culture media contained *At. ferrooxidans* dissolve nickel more efficiently at temperatures of 30–37 °C than *Aspergillus* species and are unable to dissolve nickel at higher temperatures such as 45 °C. Optimal bacterial growth temperature of *At. ferrooxidans* is 30 °C. The results of their research showed that increases in ferrous-iron supplementation increase the leaching efficiency of *At. ferrooxidans*. Also, after 28 days of leaching, the maximum dissolution of nickel at 2% pulp density was equal to 40% and at 10% pulp density was equal to 24% (Mohapatra et al., 2007).

Although bacteria may not respond well to abrupt changes in metal ion concentrations, they can adapt to gradually increasing concentrations over time to increase their tolerance to metals. Natarajan and Iwasaki (1983) examined the growth of *At. ferrooxidans* in ferrous iron medium as nickel concentration increased gradually from 5 to 50 g/L of Ni^{2+} (Watling, 2008). Iron soluble in leaching leachate is often present in the form of iron, which makes the oxidation of bacterial ferrous ions a highly efficient process in bioleaching reactors. There is concern that the high concentration of iron in bioreactor leachates could affect the bio-oxidation of ferrous ions in acidophiles. Kawabe et al. (2003), Ojumu et al. (2006), and Molchanov et al. (2007) showed interest to study the effect of ferric ions on the oxidation of ferrous ions using *At. ferrooxidans* (Watling, 2008).

Acidiphilium was the first acidophilic heterotroph to be found to accelerate ferric iron reduction differently, and *Acidiphilium* SJH subsequently was found to be a strong reducing agent of iron and was able to use amorphous and crystalline ferric iron minerals as well as soluble ferric iron as an electron acceptor. *Acidiphilium* SJH:

(1) The ability to sustain extremely low pH levels is essential (a pH of at least 1.8 is necessary since the ferric iron reduction is an acid-producing reaction).
(2) Able to use a wide range of soluble electron sources with low molecular weight.
(3) Capable of reducing ferric iron in the presence of oxygen, removing the need for bioreactors to operate in an anaerobic environment.
(4) The inability of this species (as opposed to *At. ferrooxidans*) to oxidize ferrous iron, thus avoiding the fundamental problem of creating a cycle for iron (Johnson, 2012).

Coto et al. (2008) studied the recovery of nickel and cobalt from laterite wastes using organic and inorganic bio-acids. The results of their study showed that the recovery of metals in the two-stage experiments using biological sulfuric acid (79% of nickel and 55% of cobalt) was higher than that of biological citric acid, but the amount of recovery obtained in the two-stage experiments were lower than one-stage experiments after 15 days (100% of nickel and 80% of cobalt) (Coto et al., 2008). Hallberg et al. (2011) state that nickel and cobalt can be recovered from limonite ores by reducing goethite with *At. Ferrooxidans* (Hallberg et al., 2011; Plessis et al., 2011). After 14 days, 70% of the nickel had dissolved from the ore. In this study, initially, the pH of the culture media was raised to 1.8, and culture was performed at 30 °C with aeration for better growth of bacteria on sulfur. After the cultivation of the bacteria and reaching the number of bacteria to more than 5×10^8 cells per mL, the leaching process was performed under anaerobic conditions (Hallberg et al., 2011). Cabrera et al. (2005) used *At. ferrooxidans* to investigate the oxidation kinetics of ferrous ions to build a model that included the effect of metal ions (Watling, 2008). Although Pronk et al. (1991) were the first to reveal that *At. ferrooxidans* could reduce soluble ferric iron, it wasn't obvious until Hallberg et al. (2011) that this bacterium could also facilitate the reductive dissolution of ferric iron-containing rocks (Johnson, 2012). According to Johnson's (2012) findings, *At. ferrooxidans* facilitate the reductive dissolution of asbolane $((Ni,Co)_xMn(O,OH)_4{\cdot}nH_2O)$ in anaerobic bioreactors by producing ferrous iron. According to the study, the sulfuric acid utilized to maintain the appropriate acidity of the leach solution contains more than 90% of the sulfur used in the reductive solution for the processing of nickel-bearing laterites. The remaining used sulfur (10%) is as an electron donor to reduce ferric iron (Johnson, 2012). Simate and Ndlovu (2008) showed that using sulfur compared to pyrite as a substrate leads to higher nickel recovery due to higher acid production. It was also discovered that independent of the number of bacteria inoculated, the rate of acid production by bacteria via sulfur oxidation is larger than the rate of acid consumption (Simate & Ndlovu, 2008). *Desulfovibrios* are among the identified heterotrophs in the SRB (Kelly & Wood, 2000). SRBs are involved in the reduction of sulfates to hydrogen sulfide, which then binds the hydrogen sulfide produced to the metals to form an unstable solution precipitate. Then the metals in a stable state are removed from the solution (Luptakova & Kusnierova, 2005). A study by Johnson and McGinness (1991) indicated that many heterotrophic acidophiles are capable of reducing ferric iron (Toni & Bridge, 2000). Metal leaching by heterotrophic microorganisms is typically a side effect of the microbes' production of organic acids, amino acids, and other metabolites (Coto et al., 2008). Heterotrophic bacteria' leaching efficiency is determined by the formation of organic metabolites, which are expelled in the culture medium and cause a pH drop (Mohapatra et al., 2007). In the presence of reducing agents (e.g., Fe^{2+}), electron transfer at the surface of the oxidized substance forms highly reactive oxidized forms of Fe (II) (McDonald & Whittington, 2008a). The *Delftia* genus is of particular importance in biotechnology (Morel et al., 2016; Ubalde et al., 2012). One of the nickel tolerant species is this one (Zou et al., 2015). The phenotypic similarities between *Delftia acidovorans* and the *Pseudomonadaceae* family are shown (Barrionuevo & Vullo, 2012).

Newsome et al. (2020) noted that so far, the knowledge of how the biogeochemical cycle affects the behavior of cobalt has not been able to develop new technologies for cobalt recovery from ores. Metal reduction is stimulated by the addition of inexpensive organic substrates with simple access (acetate or glucose). At the end of the experiment, the amount of cobalt that could be easily recovered (soluble or extracted with acetic acid) increased from less than 1% to more than 64%, as well as manganese, with only a small fraction of iron which is transferred to a phase that is easily recoverable. The addition of organic substrates encouraged the growth of native prokaryotes, which are virtually exclusively associated with recognized manganese (IV) / iron (III) reductants, especially Clostridiale, according to microbial community sequencing. It was also discovered that *Penicillium* fungi can produce useful organic acids for the leaching of nickel and cobalt from laterites. The results demonstrated that the biogeochemical cycle of manganese using microorganisms is likely to affect the environmental behavior of cobalt in laterites. Interestingly, four laterites from different continents (Acoje, Alda, Piau, and Shevchenko) showed comparable behavior. A novel biological processing technique involves treating the laterites with an organic substrate to reduce the metal, then rinsing with acetic vinegar to remove it. Organic substrates are not only eco-friendly and may be generated from waste carbon substrates, but they also precipitate a little quantity of iron oxide, resulting in less waste production (Newsome et al., 2020). Hosseini Nasab et al. (2021) researched the Sarbisheh (Iran) laterite sample. The supernatant of two fungi, *Aspergillus niger* and *Penicillium bilaji*, and two bacteria, *Pseudomonas putida,* and *Pseudomonas koreensis*, were used to explore indirect bioleaching of nickel and cobalt. The most prominent carboxylic acids in the supernatant of these bacteria and fungi were gluconic acid, citric acid, and oxalic acid, according to HPLC analyses. Gluconic acid (10.8–14.4 g/l) was the most important metabolic acid in *Aspergillus niger* supernatant and *Pseudomonas* species, while citric acid (6.2 g/l) was the most important metabolic acid in *Penicillium bilaji*. The maximum nickel and cobalt recovery from *Pseudomonas putida* supernatant was 90.6% and 71.98%, respectively, after 3 h at 90 °C. The resulting activation energies also indicated a greater effect of chemical control compared to diffusion control on the dissolution rate of nickel and cobalt from the studied laterite sample (HosseiniNasab et al., 2020a). In another study, the two-stage bioleaching kinetics of nickel and cobalt from iron-rich laterite ores were investigated using the metabolic products of the supernatant *Salinivibrio kushneri* as a halophilic bacterium. In this study, recovery of nickel and cobalt was reported 58.4% and 60.6% after 3 h of bioleaching at 90 °C, respectively (HosseiniNasab et al., 2020b). In another study on the same sample, Hosseini Nasab et al. (2021) studied the bioleaching of nickel and cobalt from an iron-rich lateritic sample using the *Delftia acidovorans* (heterotroph) and *At. ferrooxidans* (autotrophs) acidophilic bacteria. The findings revealed that acidophilus bacteria metabolic products play a significant role in the bioleaching of this sample. Highest nickel recovery using *Delftia acidovorans* and *At. ferrooxidans* supernatants were 80.18% and 83.65%, respectively, and cobalt was 83.94% and 86.93%, respectively. The dissolution rate of iron in these conditions was 54.41% and 64.34% for the two studied bacteria, respectively. The extraction of nickel and cobalt in the indirect bioleaching method using *Delftia acidovorans* supernatant was 29.84% and 23.75% higher than the direct bioleaching method (Hosseini Nasab et al., 2021).

Major investigations relating to the nickel laterite leaching by microorganisms are summarized in Table 5.

Figure 9 summarizes the bioleaching process of the laterites as schematically by iron-oxidizing and iron-reducing bacteria, acidophilic bacteria, and fungi.

7 Conclusions

Bioleaching is mostly used in the rehabilitation of the environment and the recovery of valuable metals. Researchers are interested in metal recovery from mineral ores, metal concentrates, and mine tailings. According to the research on the biological dissolution (bacterial and fungal leaching) of laterites, it can be concluded that the most parameters affecting the bioleaching process of nickel and cobalt from the laterites include the effect of ultrasonic waves, the salinity of the culture media, and growth medium, density, pH, particle size, and species type of microorganism. The optimization of these parameters will substantially aid in increasing nickel and cobalt dissolving rates and recovery from nickel-bearing laterites. Among the studied parameters, the change of species type has attracted more attention compared to other parameters. Heterotrophic fungi and bacteria can react with laterite. New studies have focused on indirect bioleaching by metabolites produced

Table 5 Summary of results of previous studies on bioleaching of lateritic ore

Studied subject	Result	References
Fungal bioleaching	Chemical dissolution was shown to be less successful than biological dissolution	Valix et al. (2001)
Fungi metabolic acids leach low-grade nickel ores	Cobalt is more soluble in limonite than nickel	Tang and Valix (2004)
Chemolithotrophic bacteria are used in bioleaching (*Acidithiobacillus ferrooxidans*)	Nickel extraction is reduced by increasing pulp density	Doshi and Mishra (2007)
Statistical design of experiments for bacterial leaching	The maximum recovery was found when the pH and the pulp density were low and the particles were large	Simate and Ndlovu (2007)
Nickel and cobalt biological leaching	The highest Ni and Co leaching efficiencies were achieved using a consortium primarily composed of *At. ferrooxidans*	Mohapatra et al. (2009b)
Bioleaching with halotolerant Aspergillus foetidus	The presence of salt slows down microbial metabolism, produces less bio-acid, and reduces the conversion of sucrose to bio-acid	Valix et al. (2009)
A new technique for bio-processing nickel laterites	Bioleaching of Ni laterites by *At. ferrooxidans*. Leaching was carried out under anaerobic circumstances. Over 70% of the nickel in the ore had been dissolved	Hallberg et al. (2011)
Bioleaching with *Aspergillus niger* filtrated culture	An oxalic acid concentration of 87% was found in the filtrate of a 21-day culture at a pulp density of 2% at 80 °C, while 67% of Ni and 65% of Co were recovered from roasted ore samples	Biswas et al. (2013a)
Aspergillus niger was used to leach chromite-bearing laterite overburden	Bioleaching recovered 65% Ni and 59% Co from the roasted ore, while oxalic acid brought back 26.87% Ni and 31.3% Co	Biswas et al. (2013b)
Biological leaching by mixotrophic bacteria	Nickel and iron were present in greater quantities in limonite ore (30% Ni and 5.6% Fe) than the biocontrol (1% Ni and 0.1%)	Chaerun et al. (2016)
Direct bioleaching uses the fungi's metabolic organic acids generated by Aspergillus niger	The rate of synthesis was slower for starch at pH = 1.4 than molasses after 16 days, despite producing more organic acids	Chaerun et al. (2017)
To overcome the bacteriostatic effects of heavy metals through direct bioleaching	The use of modified bacteria resulted in a 20% increase in nickel recovery and a 7% increase in cobalt recovery (*Acidithiobacillus thiooxidans*)	Jang and Valix (2017)
A lateritic ore is bioleached by mesophilic acidophile cultures and mixed cultures	A mixed culture of mesophilic acidophiles recovered 97% of Ni and 95% of Co	Ciftci and Atik (2017)
Chemical and bioleaching effects on laterite surface area	The maximum Ni recovery of 67% was achieved by adjusting citric acid content, surface area, and particle size	Pawlowska and Sadowski (2017)
Laterite is chemically and biologically leached by *Aspergillus niger*	For bio-catalytic processes containing 25% sucrose and 1% solids, nickel, cobalt, and iron recovered at 95, 3%, 74.3%, and 50%, respectively	Ciftci et al. (2018)
Redox cycling of manganese and cobalt in laterites	A new biological processing technique suggests treating laterites with an organic substrate to reduce the metal, followed by rinsing them with vinegar to remove cobalt	Newsome et al. (2020)
Using the metabolic carboxylic acid produced by *Aspergillus niger, Penicillium bilaji, Pseudomonas putida*, and *Pseudomonas koreensis*, this study examined the indirect bioleaching of Co and Ni from laterite ore with iron	Gluconic acid was the most carboxylic acid in the supernatant of these bacteria and fungi. After 3 h at 90 °C, *Pseudomonas putida* supernatant had the maximum recovery of Ni and Co. Chemical control was more successful on the dissolution rate of Ni and Co from the examined laterite sample than diffusion control	Hosseini Nasab et al. (2021)
Salinivibrio kushneri supernatant metabolites are used in a two-step bioleaching process to remove nickel and copper from iron-rich laterite	After 3 h of bioleaching at 90 °C, the supernatant *Salinivibrio kushneri*, a halophilic bacterium, recovered 58.4% Ni and 60.6% Co from the lateritic sample	Hosseini Nasab et al. (2021)
Direct and indirect bioleaching of Co and Ni from iron-rich laterite ore	The acidophilic bacteria (*Delftia acidovorans* and *Acidithiobacillus ferrooxidans*) were used to explore the bioleaching of nickel and cobalt from an iron-rich lateritic material. The bioleaching of this sample was aided by the metabolic products of acidophilus bacteria	Hosseini Nasab et al. (2021)

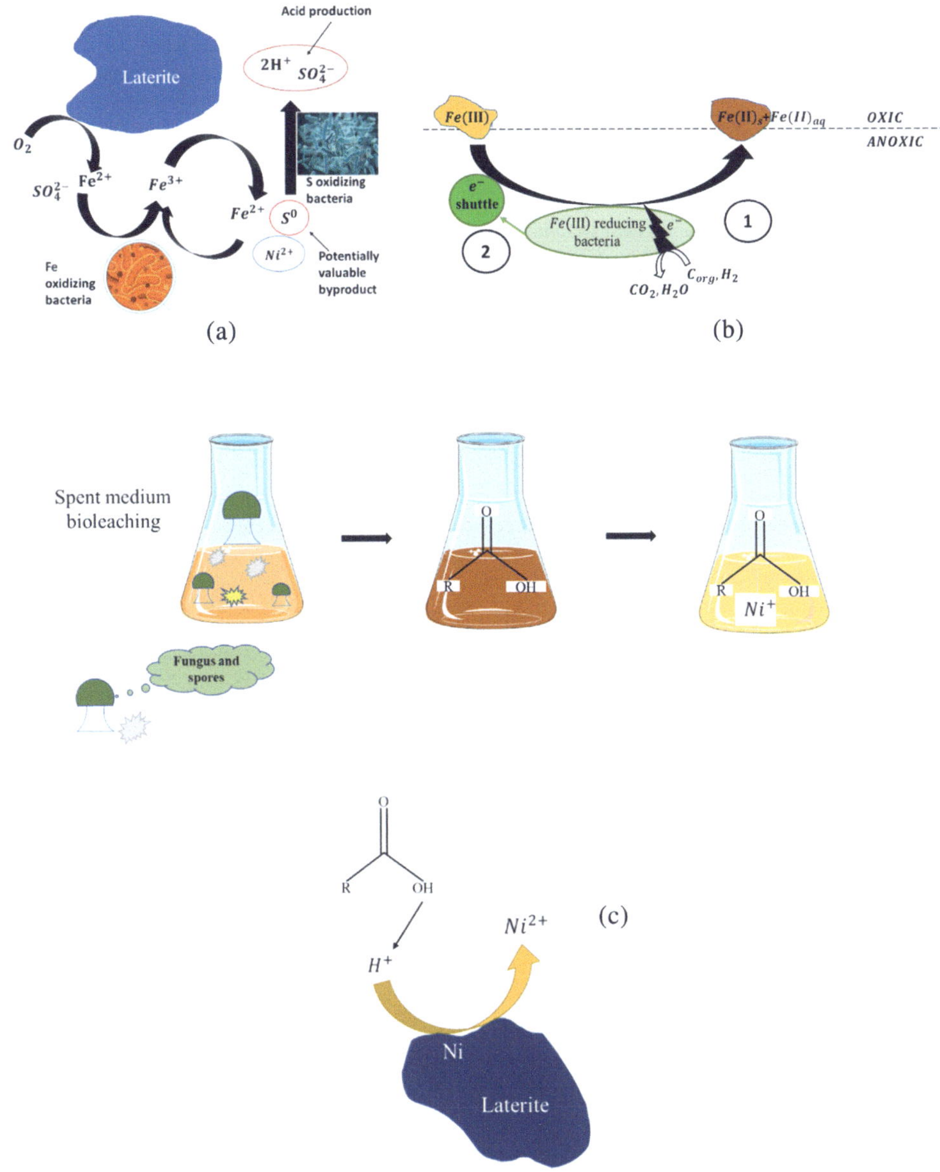

Fig. 9 The mode of action for **a** iron-oxidizing and acidophile bacteria, **b** iron-reducing bacteria, and **c** fungus, in the bioleaching process of the laterites

from microorganisms then removing impurities such as iron from bioleaching solution to increase the recovery of nickel and cobalt from laterites. *Aspergillus* and *Penicillium* species are the most effective fungi for dissolving laterites. Metabolites of the *Delftia* genus, which is one of the nickel tolerant heterotrophic species, well recovered Ni and Co from studied laterite sample. In addition, *At. ferrooxidans and At. thiooxidans* are acidophilic autotrophic bacteria that mainly were used in different research. In general, biological dissolution was more effective compared with chemical dissolution by mineral and organic acids. Iron dissolution simultaneously with nickel and cobalt dissolution in laterites is a serious problem that needs the complicated process of iron precipitation. Finally, bioleaching, due to many advantages in comparison to traditional methods was introduced as a new and promising technology for recovering valuable elements from laterites. However, still there is a weakness in the industrial development of this technology.

References

Agatzini-Leonardou, S., & Zafiratos, I. G. (2004). Beneficiation of a Greek serpentinic nickeliferous ore Part II. Sulphuric acid heap and agitation leaching. *Hydrometallurgy, 74*(3–4), 267–275.

Agatzini-Leonardou, S., Zafiratos, I. G., & Spathis, D. (2004). Beneficiation of a Greek serpentinic nickeliferous ore: Part I. Mineral processing. *Hydrometallurgy, 74*(3–4), 259–265.

Ahmadi, A., Khezri, M., Abdollahzadeh, A. A., & Askari, M. (2015). Bioleaching of copper, nickel and cobalt from the low grade sulfidic tailing of Golgohar Iron Mine, Iran. *Hydrometallurgy, 154*, 1–8.

Astuti, W., Hirajima, T., Sasaki, K., & Okibe, N. (2016). Comparison of effectiveness of citric acid and other acids in leaching of low-grade Indonesian saprolitic ores. *Minerals Engineering, 85*, 1–16.

Barrionuevo, M., & Vullo, D. L. (2012). Bacterial swimming, swarming and chemotactic response to heavy metal presence: Which could be the influence on wastewater biotreatment efficiency? *World Journal of Microbiology & Biotechnology, 28*, 2813–2825.

Behera, S. K., & Mulaba-Bafubiandi, A. F. (2015). Advances in microbial leaching processes for nickel extraction from lateritic minerals-A review. *Korean Journal of Chemical Engineering, 32*(8), 1447–1454.

Beukes, J., Giesekke, E., & Elliott, W. (2000). Nickel retention by goethite and hematite. *Minerals Engineering, 13*(14–15), 1573–1579.

Biswas, S., Dey, R., Mukherjee, S., Banerjee, P. C. (2013a). Bioleaching of nickel and cobalt from lateritic chromite overburden using the culture filtrate of *aspergillus niger*. *Applied biochemistry and biotechnology, 170*(7), 1547–1559.

Biswas, S., Samanta, S., Dey, R., Mukherjee, S., Banerjee, P. C. (2013b). Microbial leaching of chromite overburden from Sukinda mines, Orissa, India using *aspergillus niger*. *International Journal of Minerals, Metallurgy, and Materials, 20*(8), 705–712.

Borja, D., Nguyen, K. A., Silva, R. A., Park, J. H., Gupta, V., Han, Y., Lee, Y., & Kim, H. (2016). Experiences and future challenges of bioleaching research in South Korea. *Minerals, 6*(4), 128.

Buyukakinci, E. (2008). *Extraction of nickel from lateritic ores*. Yüksek Lisans Tezi, Orta Doğu Teknik Üniversitesi.

Büyükakinci, E., & Topkaya, Y. (2009). Extraction of nickel from lateritic ores at atmospheric pressure with agitation leaching. *Hydrometallurgy, 97*(1–2), 33–38.

Cabrera, G., Gomez, J. M., Cantero, D. (2005). Kinetic study of ferrous sulphate oxidation of *Acidithiobacillus ferrooxidans* in the presence of heavy metal ions. *Enzyme and Microbial Technology, 36*, 301–306.

Castro, I. M., Fieto, J. L. R., Vieira, R. X., Trópia, M. J. M., Campos, L. M. M., Paniago, E. B., Brandão, R. L. (2000). Bioleaching of zinc and nickel from silicates using *Aspergillus niger* cultures. *Hydrometallurgy, 57*, 39–49.

Chaerun, S. K., Alting, S. A., Mubarok, M. Z., & Sanwani, E. (2016). Bacterial bioleaching of low grade nickel limonite and saprolite ores by mixotrophic bacteria. In *E3S web of conferences*. EDP Sciences.

Chaerun, S. K., Minwal, W. P., & Mubarok, M. Z. (2017). Indirect bioleaching of low-grade nickel limonite and saprolite ores using fungal metabolic organic acids generated by *Aspergillus niger*. *Hydrometallurgy, 174*, 29–37.

Chang, J. H., Hocheng, H., Chang, H. Y., & Shih, A. (2008). Metal removal rate of *Thiobacillus thiooxidans* without pre-secreted metabolite. *Journal of Materials Processing Technology, 201*, 560–564.

Chang, Y., Zhai, X., Li, B., & Fu, Y. (2010). Removal of iron from acidic leach liquor of lateritic nickel ore by goethite precipitate. *Hydrometallurgy, 101*(1–2), 84–87.

Chang, Y., Zhao, K., & Pešić, B. (2016). Selective leaching of nickel from prereduced limonitic laterite under moderate HPAL conditions-Part I: Dissolution. *Journal of Mining and Metallurgy B: Metallurgy, 52*(2), 127–134.

Ciftci, H., Atik, S. (2017). Microbial leaching of metals from a lateritic nickel ore by pureand mixed cultures of mesophilic acidophiles. *Metallurgical Research & Technology, 114*(5), 508.

Ciftci, H., Atik, S., Gurbuz, F. (2018). Biocatalytic and chemical leaching of a low-grade nickel laterite ore. *Metallurgical Research & Technology, 115*(3), 305.

Coto, O., Galizia, F., Hernández, I., Marrero, J., & Donati, E. (2008). Cobalt and nickel recoveries from laterite tailings by organic and inorganic bio-acids. *Hydrometallurgy, 94*, 18–22.

Dalvi, A. D., Bacon, W. G., & Osborne, R. C. (2004). The past and the future of nickel laterites. In *PDAC 2004 International Convention, Trade Show & Investors Exchange*. The prospectors and Developers Association of Canada.

Das, G., & De Lange, J. (2011). Reductive atmospheric acid leaching of West Australian smectitic nickel laterite in the presence of sulphur dioxide and copper (II). *Hydrometallurgy, 105*(3–4), 264–269.

de Alvarenga Oliveira, V., Rodrigues, M. L. M., & Leao, V. A. (2021). Reduction roasting and bioleaching of a limonite ore. *Hydrometallurgy, 200*, 105554.

Doshi, J., & Mishra, S. D. (2007). *Bioleaching of lateritic nickel ore using chemolithotrophic micro organisms (Acidithiobacillus ferrooxidans)*. A thesis submitted in partial fulfillment of the requirements for the degree of bachelor of technology in chemical engineering, National Institute of Technology, Rourkela.

du Plessis, C. A., Slabbert, W., Hallberg, K. B., & Johnson, D. B. (2011). Ferredox: A biohydrometallurgical processing concept for limonitic nickel laterites. *Hydrometallurgy, 109*, 221–229.

Dusengemungu, L., Kasali, G., Gwanama, C., & Mubemba, B. (2021). Overview of fungal bioleaching of metals. *Environmental Advances, 5*, 100083.

Esther, J., Pattanaik, A., Pradhan, N., & Sukla, L. B. (2020). Applications of dissimilatory iron reducing bacteria (DIRB) for recovery of Ni and Co from low-grade lateritic nickel ore. *Materials Today: Proceedings, 30*, 351–354.

Fatahi, M., Noaparast, M., & Shafaei, S. Z. (2014). Nickel extraction from low grade laterite by agitation leaching at atmospheric pressure. *International Journal of Mining Science and Technology, 24*(4), 543–548.

Gadd, G. M. (2001). Microbial metal transformations. *The Journal of Microbiology, 39*(2), 83–88.

Giese, E. C., Carpen, H. L., Bertolino, L. C., & Schneider, C. L. (2019). Characterization and bioleaching of nickel laterite ore using Bacillus subtilis strain. *Biotechnology Progress, 35*(6), e2860.

Girgin, I., Obut, A., & Üçyildiz, A. (2011). Dissolution behaviour of a Turkish lateritic nickel ore. *Minerals Engineering, 24*(7), 603–609.

Golyshina, O. V., Pivovarova, T. A., Karavaiko, G. I., Kondratéva, T. F., Moore, E. R., Abraham, W. R., Lünsdorf, H., Timmis, K. N., Yakimov, M. M., & Golyshin, P. N. (2000). *Ferroplasma acidiphilum* gen. nov., sp. nov., an acidophilic, autotrophic, ferrous-iron-oxidizing, cell-wall-lacking, mesophilic member of the Ferroplasmaceae fam. nov., comprising a distinct lineage of the Archaea. *International Journal of Systematic and Evolutionary Microbiology, 50*(3), 997–1006.

Gomes, J., & Steiner, W. (2004). The biocatalytic potential of extremophiles and extremozymes. *Food Technol. Biotechnol., 42*(4), 223–235.

Hallberg, K. B., Grail, B. M., du Plessis, C. A., & Johnson, D. B. (2011). Reductive dissolution of ferric iron minerals: A new approach for bio-processing nickel laterites. *Minerals Engineering, 24*, 620–624.

Hosseini Nasab, M., Noaparast, M., & Abdollahi, H. (2021). Direct and indirect bioleaching of Co and Ni from iron rich laterite ore using *Delftia acidovorans* and *Acidithiobacillus ferrooxidans*. *Journal of Mining and Environment, 12*(2), 471–489.

Jang, H. C., & Valix, M. (2017). Overcoming the bacteriostatic effects of heavy metals on *Acidithiobacillus thiooxidans* for direct bioleaching of saprolitic Ni laterite ores. *Hydrometallurgy, 168*, 21–25.

Javanshir, S., Mofrad, Z. H., & Azargoon, A. (2018). Atmospheric pressure leaching of nickel from a low-grade nickel-bearing ore. *Physicochemical Problems of Mineral Processing, 54*(3), 890–900.

Johnson, D. B. (2012). Reductive dissolution of minerals and selective recovery of metals using acidophilic iron- and sulfate-reducing acidophiles. *Hydrometallurgy, 127–128*, 172–177.

Johnson, D. B., Grail, B. M., & Hallberg, K. B. (2013). A new direction for biomining: Extraction of metals by reductive dissolution of oxidized ores. *Minerals, 3*, 49–58.

Johnson, D. B., & Hallberg, K. B. (2003). The microbiology of acidic mine waters. *Research in Microbiology, 154*(7), 466–473.

Johnson, D. B., McGinness, S. (1991). Ferric iron reduction by acidophilic heterotrophic bacteria. *Applied and Environmental Microbiology, 57*, 207–211.

Karidakis, T., Agatzini-Leonardou, S., & Neou-Syngouna, P. (2005). Removal of magnesium from nickel laterite leach liquors by chemical precipitation using calcium hydroxide and the potential use of the precipitate as a filler material. *Hydrometallurgy, 76*(1–2), 105–114.

Kawabe, Y., Inoue, C., Suto, K., Chida, T. (2003). Inhibitory effect of high concentrations of ferric ions on the activity of *Acidithiobacillus ferrooxidans*. *Journal of Bioscience and Bioengineering, 96*, 375–379.

Kelly, D. P., & Wood, A. P. (2000). Reclassification of some species of *Thiobacillus* to the newly designated genera *Acidithiobacillus* gen. nov., *Halothiobacillus* gen. nov. and *Thermithiobacillus* gen. nov. *International Journal of Systematic and Evolutionary Microbiology, 50*(2), 511–516.

Kim, J., Dodbiba, G., Tanno, H., Okaya, K., Matsuo, S., & Fujita, T. (2010). Calcination of low-grade laterite for concentration of Ni by magnetic separation. *Minerals Engineering, 23*(4), 282–288.

Krstev, B., Krstev. A., Golomeova, M., & Golomeov, B. (2012). The recent trends and perspectives of leaching or bioleaching from nickel oxided ores. In *Scientific Proceedings IX International Congress "Machines, Technologies, Materials"* (pp. 107–109).

Kursunoglu, S., & Kaya, M. (2016). Atmospheric pressure acid leaching of Caldag lateritic nickel ore. *International Journal of Mineral Processing, 150*, 1–8.

Le, L., Tang, J., Ryan, D., & Valix, M. (2006). Bioleaching nickel laterite ores using multi-metal tolerant *Aspergillus foetidus* organism. *Minerals Engineering, 19*(12), 1259–1265.

Lee, H. Y., Kim, S. G., & Oh, J. K. (2005). Electrochemical leaching of nickel from low-grade laterites. *Hydrometallurgy, 77*(3–4), 263–268.

Li, D., Kyung-ho, P., Wu, Z., & Xue-yi, G. (2010). Response surface design for nickel recovery from laterite by sulfation-roasting-leaching process. *Transactions of Nonferrous Metals Society of China, 20*, 92–96.

Li, G. H., Rao, M. J., Qian, L., Peng, Z. W., & Jiang, T. (2010). Extraction of cobalt from laterite ores by citric acid in presence of ammonium bifluoride. *Transactions of Nonferrous Metals Society of China, 20*(8), 1517–1520.

Li, G., Zhou, Q., Zhu, Z., & Jiang, T. (2018). Selective leaching of nickel and cobalt from limonitic laterite using phosphoric acid: An alternative for value-added processing of laterite. *Journal of Cleaner Production, 189*, 620–626.

Li, J., Li, X., Hu, Q., Wang, Z., Zhou, Y., Zheng, J., Liu, W., & Li, L. (2009). Effect of pre-roasting on leaching of laterite. *Hydrometallurgy, 99*(1–2), 84–88.

Li, S., Zhong, H., Hu, Y., Zhao, J., He, Z., & Gu, G. (2014). Bioleaching of a low-grade nickel–copper sulfide by mixture of four thermophiles. *Bioresource Technology, 153*, 300–306.

Liu, K., Chen, Q., & Hu, H. (2009). Comparative leaching of minerals by sulphuric acid in a Chinese ferruginous nickel laterite ore. *Hydrometallurgy, 98*(3–4), 281–286.

Luo, W., Feng, Q., Ou, L., Zhang, G., & Lu, Y. (2009). Fast dissolution of nickel from a lizardite-rich saprolitic laterite by sulphuric acid at atmospheric pressure. *Hydrometallurgy, 96*(1–2), 171–175.

Luptakova, A., & Kusnierova, M. (2005). Bioremediation of acid mine drainage contaminated by SRB. *Hydrometallurgy, 77*(1–2), 97–102.

MacCarthy, J., Nosrati, A., Skinner, W., & Addai-Mensah, J. (2014). Atmospheric acid leaching of nickel laterite: Effect of temperature, particle size and mineralogy. In *Chemeca, Processing excellence; Powering our future* (p. 1273). Western Australia.

Marrero, J., Coto, O., Goldmann, S., Graupner, T., & Schippers, A. (2015). Recovery of Nickel and cobalt from laterite tailings by reductive dissolution under aerobic conditions using *Acidithiobacillus* species. *Environmental Science and Technology, 49*, 6674–6682.

Marrero, J., Coto, O., & Schippers, A. (2017). Anaerobic and aerobic reductive dissolutions of iron-rich nickel laterite overburden by *Acidithiobacillus*. *Hydrometallurgy, 168*, 49–55.

McDonald, R. G., & Whittington, B. I. (2008a). Atmospheric acid leaching of nickel laterites review: Part I. Sulphuric acid technologies. *Hydrometallurgy, 91*(1–4), 35–55.

McDonald, R., & Whittington, B. (2008b). Atmospheric acid leaching of nickel laterites review. Part II. Chloride and bio-technologies. *Hydrometallurgy, 91*(1–4), 56–69.

Mohapatra, S., Bohidar, S., Pradhan, N., Kar, R. N., & Sukla, L. B. (2007). Microbial extraction of nickel from Sukinda chromite overburden by *Acidithiobacillus ferrooxidans* and *Aspergillus* strains. *Hydrometallurgy, 85*, 1–8.

Mohapatra, S., Sengupta, C., Nayak, B. D., Sukla, L. B., & Mishra, B. K. (2009a). Biological leaching of nickel and cobalt from lateritic nickel ore of Sukinda mines. *Korean Journal of Chemical Engineering, 26*(1), 108–114.

Mohapatra, S., Pradhan, N., Mohanty, S., & Sukla, L. B. (2009b). Recovery of nickel from lateritic nickel ore using *Aspergillus niger* and optimization of parameters. *Minerals Engineering, 22*(3), 311–313.

Molchanov, S., Gendel, Y., Lahav, O. (2007). Improved experimental and computational methodology for determining the kinetic equation and the extant kinetic constants of Fe(II) oxidation by *Acidithiobacillus ferrooxidans*. *Applied and Environmental Microbiology, 73*, 1742–1752.

Morel, M. A., Iriarte, A., Jara, E., Musto, H., & Sowinski, S. C. (2016). Revealing the biotechnological potential of *Delftia* sp. JD2 by a genomic approach. *AIMS Bioengineering, 3*(2), 156–175.

Moskalyk, R., & Alfantazi, A. (2002). Nickel laterite processing and electrowinning practice. *Minerals Engineering, 15*(8), 593–605.

Mubarok, M. Z., Kusuma, H., Minwal, W. P., & Chaerun, S. K. (2013). Effects of several parameters on nickel extraction from laterite ore by direct bioelaching using *Aspergillus niger* and acid rock drainage from coal mine as an organic substrate. In *Advanced materials research*. Trans Tech Publ.

Mulroy, D. (2019). *The microbiology of lateritic Co-Ni-bearing manganese oxides*.

Nasab, M. H., Noaparast, M., Abdollahi, H., & Amoozegar, M. A. (2020a). Indirect bioleaching of Co and Ni from iron rich laterite ore, using metabolic carboxylic acids generated by *P. putida*, *P. koreensis*, *P. bilaji* and *A. niger*. *Hydrometallurgy, 193*, 105309.

Nasab, M. H., Noaparast, M., Abdollahi, H., & Amoozegar, M. A. (2020b). Kinetics of two-step bioleaching of Ni and Co from iron rich-laterite using supernatant metabolites produced by *Salinivibrio kushneri* as halophilic bacterium. *Hydrometallurgy, 195*, 105387.

Natarajan, K. A., Iwasaki, I. (1983). Role of galvanic interactions in the bioleaching of Duluth gabbro copper-nickel sulfides. *Separation Science and Technology, 18*, 1095–1111.

Newsome, L., Arguedas, A. S., Coker, V. S., Boothman, C., & Lloyd, J. R. (2020). Manganese and cobalt redox cycling in laterites; Biogeochemical and bioprocessing implications. *Chemical Geology, 531*, 119330.

Nicol, M. J., & Zainol, Z. (2003). The development of a resin-in-pulp process for the recovery of nickel and cobalt from laterite leach slurries. *International Journal of Mineral Processing, 72*(1–4), 407–415.

Nordstrom, D. K., Alpers, C. N., Ptacek, C. J., & Blowes, D. W. (2000). Negative pH and extremely acidic mine waters from Iron Mountain, California. *Environmental Science and Technology, 34*(2), 254–258.

Norgate, T., & Jahanshahi, S. (2011). Assessing the energy and greenhouse gas footprints of nickel laterite processing. *Minerals Engineering, 24*(7), 698–707.

Ojumu, T. V., Petersen, J., Searby, G. E., Hansford, G. S. (2006). A review of rate equations proposed for the microbial ferrous-iron oxidation with a view to application to heap bioleaching. *Hydrometallurgy, 83*, 21–28.

Onal, M. A. R., & Topkaya, Y. A. (2014). Pressure acid leaching of Caldag lateritic nickel ore: An alternative to heap leaching. *Hydrometallurgy, 1*(42), 98–107.

Pawlowska, A., & Sadowski, Z. (2017). Influence of chemical and biogenic leaching on surface area and particle size of laterite ore. *Physicochemical Problems of Mineral Processing, 53*(2), 869–877.

Petrus, H., Wanta, K. C., Setiawan, H., Perdana, I., & Astuti, W. (2018). Effect of pulp density and particle size on indirect bioleaching of Pomalaa nickel laterite using metabolic citric acid. In *IOP Conference Series: Materials Science and Engineering*. IOP Publishing.

Pronk, J. T., Liem, K., Bos, P., Kuenen, J. G. (1991). Energy transduction by anaerobic ferric iron reduction in *Thiobacillus ferrooxidan*. *Applied and Environmental Microbiology, 57*, 2063–2068.

Rampelotto, P. (2013). Extremophiles and extreme environments. *Life, 3*(3), 482–485.

Rasti, S., & Rajabzadeh, M. A. (2017). Mineralogical and geochemical characteristics of serpentinite-derived Ni-bearing laterites from Fars Province, Iran: Implications for the Lateritization process and classification of Ni-laterites. *International Journal of Geological and Environmental Engineering, 11*(7), 603–608.

Rothschild, L. J., & Mancinelli, R. L. (2001). Life in extreme environments. *Nature, 409*(6823), 1092–1101.

Sadat, A. S., Ahmadi, A., & Zilouei, H. (2016). Separation of Cu from dilute Cu–Ni–Co bearing bioleach solutions using solvent extraction with Chemorex CP-150. *Separation Science and Technology, 51*(18), 2903–2912.

Sahu, S., Kavuri, N., & Kundu, M. (2011). Dissolution kinetics of nickel laterite ore using different secondary metabolic acids. *Brazilian Journal of Chemical Engineering, 28*(2), 251–258.

Santos, A. L., Dybowska, A., Schofield, P. F., Herrington, R. J., & Johnson, D. B. (2020). Sulfur-enhanced reductive bioprocessing of cobalt-bearing materials for base metals recovery. *Hydrometallurgy, 195*, 105396.

Santos, L. R. G., Barbosa, A. F., Souza, A. D., & Leão, V. A. (2006). Bioleaching of a complex nickel–iron concentrate by mesophile bacteria. *Minerals Engineering, 19*(12), 1251–1258.

Simate, G. S., & Ndlovu, S. (2007). Characterisation of factors in the bacterial leaching of nickel laterites using statistical design of experiments. In *Advanced Materials Research*. Trans Tech Publ.

Simate, G., Ndlovu, S., & Gericke, M. (2009a). The effect of elemental sulphur and pyrite on the leaching of nickel laterites using chemolithotrophic bacteria. In *Hydrometallurgy Conference*.

Simate, G. S., & Ndlovu, S. (2008). Bacterial leaching of nickel laterites using chemolithotrophic microorganisms: Identifying influential factors using statistical design of experiments. *International Journal of Mineral Processing, 88*, 31–36.

Simate, G. S., Ndlovu, S., & Gericke, M. (2009b). Bacterial leaching of nickel laterites using chemolithotrophic microorganisms: Process optimization using response surface methodology and central composite rotatable design. *Hydrometallurgy, 98*, 241–246.

Simate, G. S., Ndlovu, S., & Walubita, L. F. (2010). The fungal and chemolithotrophic leaching of nickel laterites—Challenges and opportunities. *Hydrometallurgy, 103*(1–4), 150–157.

Smith, S. L., Grail, B. M., & Johnson, D. B. (2017). Reductive bioprocessing of cobalt-bearing limonitic laterites. *Minerals Engineering, 106*, 86–90.

Stankovic, S., Stopic, S., Sokic, M., Markovic, B., & Friedrich, B. (2020). Review of the past, present and future of the hydrometallurgical production of nickel and cobalt from lateritic ores. *Metallurgical and Materials Engineering, 26*(2), 199–208.

Swamy, K., Narayana, K., & Misra, V. N. (2005). Bioleaching with ultrasound. *Ultrasonics Sonochemistry, 12*(4), 301–306.

Tang, J., & Valix, M. (2004). Leaching of low-grade nickel ores by fungi metabolic acids. In *Proceedings of Separations Technology VI: New Perspectives on Very Large-Scale Operations* (pp. 1–16).

Tang, J., & Valix, M. (2006). Leaching of low grade limonite and nontronite ores by fungi metabolic acids. *Minerals Engineering, 19*(12), 1274–1279.

Thangavelu, V., Tang, J., Ryan, D., & Valix, M. (2006). Effect of saline stress on fungi metabolism and biological leaching of weathered saprolite ores. *Minerals Engineering, 19*(12), 1266–1273.

Toni, D. B. J., & Bridge, A. M. (2000). Reductive dissolution of ferric iron minerals by *Acidiphilium* SJH. *Geomicrobiology Journal, 17*(3), 193–206.

Ubalde, M. C., Brana, V., Sueiro, F., Morel, M. A., Martinez-Rosales, C., Marquez, C., & Castro-Sowinski, S. (2012). The versatility of *Delftia* sp. isolates as tools for bioremediation and biofertilization technologies. *Current Microbiology, 64*, 597–603.

Valix, M., & Cheung, W. (2002). Study of phase transformation of laterite ores at high temperature. *Minerals Engineering, 15*(8), 607–612.

Valix, M., Thangavelu, V., Ryan, D., & Tang, J. (2009). Using halotolerant *Aspergillus foetidus* in bioleaching nickel laterite ore. *International Journal of Environment and Waste Management, 3*(3–4), 253–264.

Valix, M., Usai, F., & Malik, R. (2001a). Fungal bio-leaching of low grade laterite ores. *Minerals Engineering, 14*(2), 197–203.

Valix, M., Tang, J. Y., Malik, R. (2001b). The electro-sorption properties of nickel on laterite gangue leached with an organic chelating acid. *Minerals Engineering, 14*(2), 2005–2215.

Watling, H. R. (2008). The bioleaching of nickel-copper sulfides. *Hydrometallurgy, 91*, 70–88.

Whittington, B., McDonald, R. G., Johnson, J. A., & Muir, D. M. (2003a). Pressure acid leaching of arid-region nickel laterite ore: Part I: Effect of water quality. *Hydrometallurgy, 70*(1–3), 31–46.

Whittington, B., McDonald, R. G., Johnson, J. A., & Muir, D. M. (2003b). Pressure acid leaching of arid-region nickel laterite ore: Part II. Effect of ore type. *Hydrometallurgy, 70*(1–3), 47–62.

Wu, A., Zhang, Y., Zheng, C., Dai, Y., Liu, Y., Zeng, J., Gu, G., & Liu, J. (2008). Purification and enzymatic characteristics of cysteine desulfurase, IscS, in *Acidithiobacillus ferrooxidans* ATCC 23270. *Transactions of the Nonferrous Metals Society of China, 18*, 1450–1457.

Yang, Y., Ferrier, J., Csetenyi, L., & Gadd, G. M. (2019). Direct and indirect bioleaching of cobalt from low grade laterite and pyritic ores by *Aspergillus niger*. *Geomicrobiology Journal, 36*(10), 940–949.

Zou, G., Papirio, S., van Hullebusch, E. D., & Puhakka, J. A. (2015). Fluidized-bed denitrification of mining water tolerates high nickel concentrations. *Bioresource Technology, 179*, 284–290.

Microbial Leaching Strategies for Extraction of Rare Earth Elements from Primary and Secondary Resources

Alessandra Cesaro

Abstract

Rare Earth elements (REEs) are a group of 17 elements, including 15 lanthanides, coupled with chemically similar yttrium and scandium. Due to their unique physical and electrochemical characteristics, they are applied in a wide variety of sectors of the global economy, standing as elements of strategic importance. The rapid increase in the demand, together with the limitations of their availability, have addressed the study of alternative, secondary sources of REEs as well as the development of eco-friendly processes to pursue their sustainable recovery. The need to set environmentally sound treatments has thus driven the scientific and technical interest towards bio-metallurgy, as a promising alternative to conventional methods. This chapter focuses on the application of biotechnological strategies to leach REEs from both primary and secondary sources. Biological-mediated leaching processes are discussed to point out the main mechanisms driving the extraction of REEs as well as the factors influencing their yields. The current state-of-the-art of REE bioleaching processes is figured out in order to highlight the potential for scale up as well as to address future research perspectives.

Keywords

Biological mining strategies · Critical elements · Extraction · Microorganisms · Recovery

A. Cesaro (✉)
Department of Civil, Architectural and Environmental Engineering, University of Napoli Federico II, via Claudio 21, 80125 Napoli, Italy
e-mail: alessandra.cesaro@unina.it

1 Rare Earth Elements: Main Properties, Sources, and Applications

Rare Earth elements (REEs) are a set of seventeen elements, including fifteen lanthanides, coupled with chemically similar yttrium and scandium. Based on their atomic weight, they are usually classified as light rare Earths (LREEs), from lanthanum to europium and heavy rare Earths (HREEs), including the remaining lanthanides as well as yttrium (Table 1). Scandium is not included in this classification because its ionic radius is much smaller than that of the other REEs (Kumar Jha et al., 2016).

Although they are referred to as "rare", the Earth's crust is relatively rich in these elements: light REEs are as abundant as copper and the less abundant heavy REEs are much more abundant than precious metals like gold (Wall, 2021). Nonetheless, due to their chemical properties, they tend to be dispersed, not concentrated in mineral ores, where they may also occur together with radioactive elements, such as uranium and thorium. This condition, along with the chemical similarities among different elements, mainly account for their difficult separation.

Their unique physical and chemical properties make them fundamental in a wide variety of industrial applications (Wall, 2021), since they can contribute to several technological advantages, such as low power consumption, high performance, last long and thermal stability (Cao et al., 2021). Balaram (2019) defined REEs as the "vitamins of modern industry", because although used in small amounts, they play a key role in defining the performances of the goods they compose. Such pivotal role resulted in a violent increase of the REE market that grew by a factor of 20.5 from 1997 to 2007 (Charalampides et al., 2015).

China has been the leader in the production of REEs for decades. Chinese scientists discovered rare Earth resources in 1927 and started the production of REE concentrates in 1957. In 1998, Chinese resource of REEs was reported to be 92 million metric tonnes; in 2009 the Chinese Ministry

S. Panda et al. (eds.), *Biotechnological Innovations in the Mineral-Metal Industry*,
Advances in Science, Technology & Innovation, https://doi.org/10.1007/978-3-031-43625-3_4

Table 1 Classification into light REEs and heavy REEs (Kumar Jha et al., 2016; van Gosen et al., 2017)

	Rare Earth element[a]	Atomic weight	Ionic radius [nm]
Light rare Earth elements (LREEs)	Lanthanum (La)	138.91	1.061
	Cerium (Ce)	140.12	1.034
	Praseodymium (Pr)	140.91	1.013
	Neodymium (Nd)	144.24	0.995
	Promethium (Pm)	144.91	0.979
	Samarium (Sm)	150.36	0.964
	Europium (Eu)	1515.96	0.95
	Gadolinium (Gd)	157.25	0.938
Heavy rare Earth elements (HREEs)	Terbium (Tb)	158.92	0.923
	Dysprosium (Dy)	162.50	0.908
	Holmium (Ho)	164.93	0.894
	Erbium (Er)	167.26	0.881
	Thulium (Tm)	168.93	0.869
	Ytterbium (Yb)	173.04	0.858
	Lutetium (Lu)	174.97	0.848
	Yttrium (Y)	88.91	0.88

[a] excluding Scandium

of Land and Resources stated that the country had 18.6 million metric tonnes of rare Earth oxide reserves (Tse, 2011). In the last decade of the twentieth century, REE production in China increased over 450% and from the late 90s to the first decade of 2000, China has supplied over 80% of the global demand for rare Earths (Tse, 2011).

When in 2010 China announced the reduction of its exports of REEs, both the USA and the European Union claimed the criticality of these elements and the search for deposits outside of China as well as that for alternative sources increased rapidly (van Gosen et al., 2017). In this context, several efforts were carried out to promote the recycling of REEs from waste materials to reduce the dependence from foreign export, while promoting the sustainable handling of some residual streams.

1.1 Primary Sources

In nature, REEs occur as part of hosting minerals, that can be differentiated based on their REE content. These minerals may be either complex, if composed of all the lanthanides, or selective, when only some of them are present (Jordens et al., 2013). The most frequently extracted are bastnasite, xenotime and monazite, even though over 200 additional minerals are known to contain REEs.

Bastnasite is a fluorocarbonate mineral, which is considered the primary source of LREEs, with around 70% of rare Earth oxides (Dushyantha et al., 2020), mainly as Ce, La, Pr and Nd: for this reason, it may be considered a selective mineral. The bastnasite deposits distributed in China and the USA represent the largest portion of the global economic resources of REEs. At these deposits, the separation procedure for bastnasite ores includes a flotation process using a fatty-acid or hydroxamate-based collector system (Jordens et al., 2013).

Monazite is a phosphate mineral that represents the second largest source of REEs, with deposits located in Australia, Brazil, China, India, Malaysia, South Africa, Sri Lanka, Thailand and USA (Dushyantha et al., 2020). As bastnasite, monazite holds a content in REE oxides of approximately 70%, including mainly Ce, La, Pr and Nd; differently from bastnasite, monazite also contains thorium and uranium (Jordens et al., 2013). The extraction of REEs from monazite relies on different gravity separation steps, magnetic, electrostatic, and eventual flotation steps (Jordens et al., 2013). A similar processing is used for xenotime, an yttrium phosphate mineral, very close to monazite. It has a rare Earth oxide content lower than 70%, but it is a major source of HREEs (Jordens et al., 2013).

It is worth highlighting that ion-adsorbed clays are an additional, relevant source of these valuable elements, with a 60% content of HREEs (Kumar Jha et al., 2016). Differently from the mentioned minerals, these clays require little to no physical beneficiation, but can be directly processed (Jordens et al., 2013) using appropriate methods.

1.2 Secondary Sources

The chance to obtain REEs from natural deposits is tightly related to their local availability: when this is not the case, it is necessary to rely on import, which can turn to be not always economically favourable. This condition, together with the need to reduce the depletion of natural resources, while promoting the valorization of residual streams in accordance with the circular economy principles, has driven the search for alternative sources of REEs (Cesaro et al., 2018).

As REEs are largely used in electronic devices, waste electrical and electronic equipment (WEEE) is considered a relevant urban mine. Cathode ray tube (CRT), fluorescent lamps, magnets, accumulators, electrodes, semi-conductors, capacitors and electric contacts are among the main secondary sources of REEs (Menad & von Houwelingen, 2011; Kuma Jha et al., 2016). However, phosphor-containing products, permanent magnets, and batteries (Table 2) have been the focus of REE recycling from end-of-life devices (Tunsu et al., 2015).

As REEs are used in small amounts in the products, their concentrations in waste streams are generally low, with order of magnitude ranging from 10^{-1} and 10^{2} mg/kg of WEEE (Cucchiella et al., 2015).

Some kinds of industrial and mining wastes are other potential sources of REEs. Among the more common industrial wastes, mineral processing residues and fly ashes should be mentioned; mine tailings and acid mine drainage sources are the main mining wastes considered for REE extraction (Binnemans et al., 2015; Costis et al., 2021).

The use of waste materials in bioleaching processes for REE extraction requires its proper pretreatment, depending on the kind of waste and its original characteristics. From solid waste, it is desirable to obtain a powdered material, to be eventually sieved in order to achieve a specific particle size range. Occasionally, drying can be necessarily prior to bioleaching (Baniasadi et al., 2019).

2 Bioleaching Processes

The extraction of REEs from the source material is the first step to use them for the application of interest. Over time, several methods, including both hydro-metallurgical and pyro-metallurgical processes have been proposed and used. However, high operational costs, poor environmental sustainability and low recovery rates have been identified among the main drawbacks of these processes (Binnemans et al., 2013), increasing the interest towards a biological approach.

The biological extraction of valuable elements from the corresponding source materials is known as "bioleaching". It relies on the capability of selected microorganisms in solubilizing specific elements through bacterially assisted reactions. After the first applications in the 1950s for the extraction of copper, in the mid-1980s it has been extended to other metals such as gold (Brierley & Brierley, 2001). Applications to common and precious metals are common, but some studies pointed out the effectiveness of bioleaching towards different rare Earths.

Both autotrophic and heterotrophic microbial species have been considered for REE leaching from both primary (Brisson et al., 2016; Zhang et al., 2018) and secondary sources (Dev et al., 2020).

The microbial species, the source materials as well as a variety of abiotic factors can affect the yields of bioleaching, as detailed in the following paragraphs.

2.1 Main Mechanisms and Microorganisms Involved in REE Bioleaching Processes

The biological leaching of REEs has been largely debated and scientific literature proposes different approaches to the classification of the reactions and the categorization of the microbial species participating in these mechanisms. However, it is commonly reported that the bioleaching relies on three main processes, namely acidolysis, redoxolysis and complexolysis (Fig. 1).

These are mediated by different microbial species (Table 3) and usually result in the simultaneous mobilization of different REEs from the solid matrix to the aqueous phase. Then, the selective recovery of REEs by downstream processes is needed.

In acidolysis, the reactions are mediated mainly by sulphur-oxidizing or phosphate-solubilizing microorganisms and involve the generation of acids, which cover the surface of the source material and promote REE dissolution. Either the chemical reaction at the surface of the source material or the diffusion of the reagents through the liquid film layer/solid product layer control the leaching rate, in accordance with the shrinking core kinetics (Sethurajan et al, 2017).

Sulphur-oxidizing bacteria, such as *A. ferrooxidans*, *A. thiooxidans* and *S. acidophilus* are acidophiles chemolithotrophs that oxidize sulphide to sulphuric acid to determine REE dissolution as shown in Eq. (1) (Dev et al., 2020):

$$4(\mathrm{REE})\mathrm{S}_{(\mathrm{S})} + 2\mathrm{H_2O} + 7\mathrm{O_2} \rightarrow 4\mathrm{REE}^{+}_{(\mathrm{aq})} + 4\mathrm{H}^{+} + 4\mathrm{SO}_4^{2-}. \quad (1)$$

Table 2 Rare Earth elements (REEs) in some end-of-life products

Category of products	Kind of products	REEs
Phosphor-containing products	Fluorescent lamps	Eu, Y, Tb, La, Ce, Gd
	LEDs	Ce, Y, Gd, Eu
	Plasma display panels	Eu, Y, Gd
	CRT screens	Y, Eu, Tb, Ce, Nd, Sm
Permanent magnets	Speakers, headphones, electric motors, wind turbine generators, …	Nd, Dy, Pr, Gd, Tb, Sm
Batteries	NiMH batteries	Ce, La, Nd, Pr, Y

Adapted from Tunsu et al. (2015)

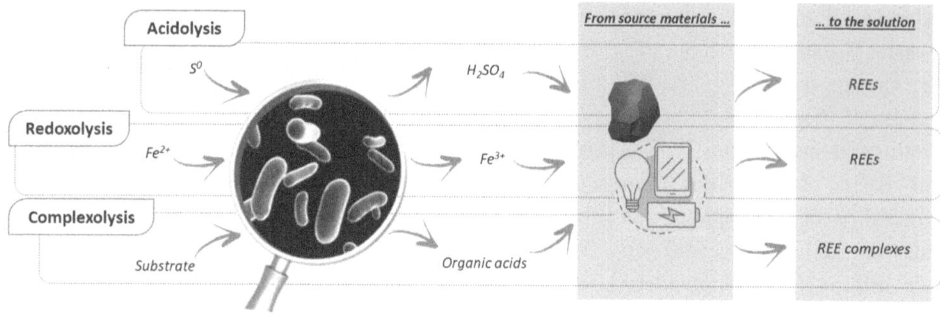

Fig. 1 Schematic examples of the main mechanisms for REE bioleaching

Table 3 Microorganisms involved in REE bioleaching process

Microorganism	Mechanism	Process condition	Leached REEs
Acidithiobacillus ferrooxidans, Acidithiobacillus thiooxidans	Redoxolysis, Acidolysis	Batch, $T = 30$ °C, pH = 1.8 Source material: monazite	Ce: 2–9% La: 1–5%
Acidithiobacillus thiooxidans	Acidolysis	Batch, $T = 30$ °C, pH = 1.2–1.5 Source material: WEEE	Ce, Eu, Nd: >99% La, Y: 80%
Acidithiobacillus thiooxidans, Acidithiobacillus ferrooxidans, Leptospirillum ferrooxidans	Acidolysis/Redoxolysis	$T = 25$ °C, pH = 1.8–3.2 Source material: Magnets	Pr: 100% Nd: 86.4%
Acidithiobacillus thiooxidans, Acidithiobacillus ferrooxidans, Acidithiobacillus caldus, Sulfobacillus sp.	Acidolysis	Batch, $T = 45$ °C, pH = 4 Source material: ash slag	Sc: 52% Y: 52.6% La: 59.5%
Acidithiobacillus ferrooxidans	Redoxolysis	Batch, $T = 25$ °C, pH = 2 Source material: gibbsite ore	REEs: 67.6%
Candida bombicola	Complexolysis/Acidolysis	Batch, $T = 28$ °C, pH = 3.3–3.5 Source material: coal fly ash	Yb: 67.7% Er: 64.6% Sc: 63%
Gluconobacter oxydans	Complexolysis/Acidolysis	Batch, T = 30 °C, pH = 3.3 Source material: fluid catalytic cracking catalyst, phosphor powder	Tb, Eu, Ce, La, Y: 49%
Acidithiobacillus ficuum *P. aeruginosa*	Complexolysis/Acidolysis	Batch, $T = 30$ °C, pH = 3–6 Source material: monazite	REEs: 75.4%
Acetobacter aceti	Complexolysis/Acidolysis	Batch, $T = 30$ °C, pH = 3 Source material: monazite	Ce, La: 0,0.1%
Gluconobacter oxydans NRRL B85	Complexolysis	Batch, $T = 25$ °C, pH = 2.1 Source material: synthetic phosphogypsum	Y: 91.2% Ce: 36.7% Nd: 42.8% Sm: 73.2% Eu: 50% Yb: 83.7%
Penicillium tricolor	Complexolysis	Batch, $T = 30$ °C, pH = 9–10.4 Source material: red mud	Y: 78% Sc: 75% Lu: 67% Yb: 66% Tm: 65% Er: 63%
Aspergillus niger, Aspergillus terreus, Paeciliomyces sp.	Complexolysis	Batch, $T = 25$–28 °C, pH = 2–2.8 Source material: monazite	Not mentioned
Acidithiobacillus ficuum	Complexolysis	Batch, $T = 28$ °C, pH = 3 Source material: Th-U concentration	La: 20% Ce: 33% Y: 2.5%

Adapted from Dev et al., (2020)

highlighted by Isildar et al. (2019), most chemolithoautotrophs exhibit high tolerance towards heavy metals and for this reason, they can be used to treat also complex source materials, like WEEE, by adding an external reduced sulphur.

Phosphate-solubilizing microorganisms (PSMs), such as *Pseudomonas*, *Enterobacter*, *Klebsiella*, *Bacillus*, *Rhizobium*, *Flavobacterium*, *Micrococcus*, *Erwinia*, *Acidithiobacillus*, *Serratia*, *Acetobacter* and *Streptomyces* (Fathollahzadeh et al., 2018) either release protons or secret organic acids (citric, gluconic, malonic, etc.). In both cases, phosphate liberation is achieved and REEs are solubilized, in accordance with Eq. (2) (Dev et al., 2020):

$$(\mathrm{REE})\mathrm{PO}_{4(\mathrm{S})} \rightarrow \mathrm{REE}^{3+}_{(\mathrm{aq})} + \mathrm{PO}_4^{3-}. \quad (2)$$

PSMs also release phosphatases, which are enzymes contributing to REE solubilization (Dev et al., 2020).

Another mechanism for REE bioleaching is redoxolysis, which produces the solubilization of the target elements through oxidation–reduction reactions: the electron transfers promote the microbial growth and lead to the dissolution of the target elements (Opare et al., 2021). Acidophiles are able to carry out redoxolysis reactions by reducing ferric ions enzymatically under anaerobic conditions and using hydrogen or sulphur as the electron donor (Baniasadi et al., 2019). Nevertheless, microorganisms can also catalyze the oxidation of ferrous to ferric ions [Eq. (3)], which determines the subsequent oxidative dissolution of solid phase REEs [Eq. (4)], with the release of ferrous ions (Dev et al., 2020) that are used back according to Eq. (3):

$$4\mathrm{Fe}^{2+} + \mathrm{O}_2 + 4\mathrm{H}^+ \rightarrow 4\mathrm{Fe}^{3+} + 2\mathrm{H}_2\mathrm{O}, \quad (3)$$

$$(\mathrm{REE})\mathrm{FeS}_{2(\mathrm{S})} + \mathrm{Fe}^{3+} \rightarrow 4\mathrm{Fe}^{2+} + \mathrm{REE}^{+}_{(\mathrm{aq})} + 2\mathrm{S}^0. \quad (4)$$

Among the microorganisms that are reported to participate in redoxolysis reactions, Dev et al. (2020) mentioned *L. ferrooxidans*, *Acidithiobacillus thiooxidans* and *Acidithiobacillus ferrooxidans*. The latter, for instance, is by far one of the most studied on both primary sources (Dev et al., 2020) and waste materials (Auerbach et al., 2019a; Marra et al., 2018). It was the first bacterium discovered to be capable of oxidizing minerals and it was for long considered the most important microorganism in biomining processes operated at temperature as high as 40 °C (Rawlings, 2002). Being a facultative anaerobe, *At. ferrooxidans* can catalyze the dissimilatory oxidation of iron, sulphur and hydrogen as well as the reduction of iron and sulphur (Quatrini & Johnson, 2019).

It seems interesting to highlight that acidophilic chemolithotroph bioleaching can occur through contact and non-contact mechanisms, depending on the extraction process mode. Tao and Dongwei (2014) clarified these mechanisms for chalcopyrite and pyrite, stating that in the contact (or direct) bioleaching the electrons are directly transferred from the REE sulphide to the cell attached to the mineral surface; in the non-contact (or indirect) bioleaching the electrons are transmitted to the ferric ions, acting as the oxidizing agent of the sulphide ores. In recent time, it was claimed how close these mechanisms are, since in the direct bioleaching the dissolution occurs through ferric ions as in the indirect one (Srichandan et al., 2019). In both contact and non-contact mechanism, the mobilization of the target elements by chemolithotrophs proceeds via the thiosulphate and the polysulphide pathways. In the former case, the element sulphides oxidize through electron extraction by ferric ions; these reduce to ferrous ones and reoxidize by the iron-oxidizing microbes to start another cycle. In the polysulphide pathway, acid soluble element sulphide dissolve by electron extraction via the combined action of Fe^{3+} and H^+: the sulphide bond, thus, breaks and the sulphur releases as H_2S which undergoes diverse oxidation reactions to elemental sulphur, as detailed by Srichandan et al. (2019).

Lastly, complexolysis occurs through the generation of microbial metabolites, namely organic acids and siderophores (Dev et al., 2020), which can bind to the target element forming stable complexes.

This mechanism is significant for the recovery of precious metals by cyanogenic bacteria, such as *Pseudomonas putida* that was found to be effective in gold solubilization from printed circuit boards (Isildar et al., 2016). Similarly, siderophore forming bacteria, like *Pseudomonas aeruginosa* were observed to induce the extraction of copper from metallurgical waste (Sethurajan et al., 2018). Dev et al. (2020) pointed out that *Acetobactor sp.* and *Gluconobacter oxydans* NRRL B85 have been successfully involved in the complexolysis of different REEs by the production of a variety of organic acids, including succinic, lactic, oxalic, acetic and gluconic.

Among siderophores, *Actinobacteria* and *Streptomyces* are those studied for the extraction of REEs, usually from minerals.

In the last decades, the fungal bioleaching has raised increasing interest, due to the capability of fungi to grow in a wide pH range and their great resistance to metal toxicity (Gravilescu, 2022). As it may involve diverse microorganisms, such as *Aspergillus niger*, *Penicillium simplicissimum*, *Marasmius oreades*, *Clitocybe* sp. and *Polysporus* sp. (Gravilescu, 2022), the fungal bioleaching can occur via complexolysis, acidolysis and redoxolysis or a combination of these three mechanisms (Desmarais et al., 2020). Although acidolysis is considered the fastest leaching mechanism, complexes and chelates can be formed with salts of organic acids, facilitating the mobilization of target elements according to the following, possible equations (Gravilescu, 2022):

$$Me^{2+} + C_6H_8O_7 \rightarrow Me(C_6H_8O_7)$$

$$Ni^{2+} + H_3(Cit) \rightarrow NiCit^- + 3H^+$$

$$Fe^{3+} + 0.5HO_2C.CO_2H \rightarrow Fe(O_2C.CO_2)_3^{3-} + 6H^+$$

Aspergillus niger, for instance, can produce siderophores able to bioleach approximately 50% of La and Ce from phosphorite at pH 7 and 30 °C. The siderophore generation can be enhanced by providing glucose, glycerol and NH_4Mo^{2+} as substrates (Dev et al., 2020), which in turn, resulted in the improved bioleaching of REEs. An *Aspergillus niger* strain, along with *Aspergillus terreus* ML3-1 and *Paecilomyces spp.* were applied for solubilizing REE from monazite. This mineral as well as a thorium-uranium concentrate were successfully mined for REE dissolution also by using *Aspergillus ficuum*: in this case, it was possible to reach approximately 75% and 63% REE extraction from the monazite and the concentrate, respectively (Isildar et al., 2019).

The fungal action can also rely on acydolysis: in this case, organic acids and carbonic acid from CO_2 respiration are used (Sethurajan et al., 2018).

Some attempts have also addressed the fungal bioleaching of waste for REE extraction. Two strains of *Aspergillus niger* were isolated from both pistachio husk and grape skin and applied to red mud: under the best operating conditions, these strains allowed 38% and 29% Sc extraction, respectively (Pedram et al., 2020). Hosseinzadeh et al. (2021) used the same fungal specie to extract REEs from spent automobile catalyst, under different operating conditions. At an initial pH of 5, with an initial glucose concentration of 100 g/L, inoculum percentage of 2.5% and a solid/liquid ratio of 1% (w/v), extraction efficiencies of 25.9% and 23.9% were obtained for Ce after 240 h and for La after 48 h, respectively.

2.2 Factors Influencing the Bioleaching of REEs

Several biotic as well as abiotic factors affect bioleaching process yields (Priya & Hait, 2017).

Biotic factors include the kind of microorganism, the inoculum size and maximum tolerance and adaptability to other possible constituents of the source material. The latter aspect is particularly important when REEs are supposed to be extracted from waste, as in the case of WEEE, which contains a wide variety of constituting materials that may be toxic to microorganisms, and these are usually present in much more relevant concentrations than REEs.

Among the abiotic factors, pH, temperature, particle size of the source material, pulp density, leaching time, aeration, stirring rate as well as the composition of leaching media are among the most important ones.

Both pH and temperature display their effect on the microbial growth as well as on the chemistry of the target element solubilization (Priya & Hait, 2017). The microorganisms involved in REE bioleaching are generally acidophilic and their optimum growth pH lays in the range 0.8–3.5, depending on the specific strain (Dev et al., 2020). Filamentous fungi can grow under various alkaline and acidic pH conditions (Dusengemungu et al., 2021). Hosseinzadeh et al. (2021) studied REE bioleaching using *Aspergillus niger* and investigated the effect of different parameters, including initial pH to find that, under the optimal conditions, an initial pH of 5 determined the extraction of 26% Ce, 43% Al for 240 h, and 24% La for 48 h.

As for the temperature, bioleaching processes are generally carried out by either mesophilic or thermophilic microbes, acting between 25–30 °C and 40–45 °C, respectively (Ilyas et al., 2007). However, mixed cultures of *A. ferrooxidans* and *A. thiooxidans* were found to be effective even at room temperature (Isildar et al., 2016; Marra et al., 2018). Fungi can grow over a wide range of temperature, depending on the strain; *Aspergillus*, *Fusarium* and *Hypocrea* could grow at 30 °C, whereas *Penicillium* was reported to improve its recovery activity of uranium when temperature rose from 20 to 50 °C (Dusengemungu et al., 2021).

The particle size of the source materials determines the specific surface area that can enter in contact with the microorganisms. Efficient leaching has been achieved using particle size ranging between 40 and 200 μm (Ilyas et al., 2007; Priya & Hait, 2017). In this regard, Auerbach et al. (2019b) analyzed the effects of two different particle size ranges in the bioleaching of REEs from incineration sludge. Two chemolithoautotrophic acidophilic iron-oxidizing bacteria and a chemolithoautotrophic acidophilic sulphur-oxidizing bacterium in co-culture with *A. ferrooxidans* were used for bioleaching experiments. Interestingly, these authors found that any significant increase of the element concentrations with the finer particles compared with the coarser occurred, likely due to the formation of a passivating sulphur layer on the sample surface.

Microorganism action is also influenced by the pulp density via different routes. This parameter can determine the extent of shear forces, limit O_2 and CO_2 fluxes, increase the load of toxic metals as well as affect the buffer capacity of the system and, in turn, the pH (Dev et al., 2020). The issue of the toxicity is basically related to the presence of heavy metals associated to minerals or present in the waste source materials: these may accumulate into the microbial cells, inhibit specific enzymes, and differently interfere with the metabolism of the microorganisms (Dev et al., 2020).

When WEEE is the REE source material, other materials like plastics, ceramic and glass may also contribute to the toxicity potential. Generally, the increase in pulp density results in the decrease of REE extraction by bioleaching. The extent of the effect of pulp density on the bioleaching yields is also depending on the microbial/fungal strain adopted, so that specific tests are needed to select the best operating conditions.

Marra et al. (2018) performed bioleaching tests with *A. thiooxidans* to extract REEs from WEEE shredding dust at pulp densities of 0.5, 1 and 2% w/v; results showed that the best mobilization yields were achieved at the pulp density values lower than 2% w/v. Qu et al. (2019) studied REE bioleaching from red mud using indigenous *Acetobacter* sp. at pulp densities of 2, 5 and 10% and observed leaching ratios of Lu, Y and Sc of 53%, 61% and 52% at the lower value of 2%. Red mud was also bioleached by a filamentous, acid-producing fungi, known as *Penicillium tricolor* RM-10; a total concentration of 2% (w/v) red mud gave the maximum REE leaching ratios under one-step bioleaching process, but the highest extraction yields were achieved under a two-step bioleaching process at 10% (w/v) pulp density (Qu and Lian, 2013).

However, it is possible to reduce such negative effects by adapting the microorganisms to increasing pulp density, up to an optimum threshold depending on the target element, the source material as well as the involved microorganisms (Jowkar et al., 2018; Qu & Lian, 2013). For instance, Muravyov et al. (2015) with 10% pulp density reported leaching yields of 52%, 52.6% and 59.5% for Sc, Y and La, respectively, after 10 days at 45 °C. More recently, Hanabe Muddana et al. (2021) adapted two different types of *A. ferrooxidans* at 20% pulp densities. These authors reported that the adapted strains were much more efficient in La leaching from spent catalysts than unadapted ones, because the exposure to higher load of spent catalyst reduced or nullified its toxicity to *A. ferrooxidans*.

It is worth noting that the optimization of the bioleaching from potentially toxic waste source materials may also rely on the adoption of specific methods to bring the source materials itself in contact with the bioleaching agents. Beyond adding simultaneously the microorganisms and the source materials (one-step process), it is possible to supply the material when the biological production of the leachants has already started in a two-step process (Srichandan et al., 2019). In the latter case, the medium is first inoculated and then pre-cultured without the waste source materials: this two-step method results in a better bacterial growth and appropriate Fe^{2+} oxidation rate, which account for higher mobilization efficiencies than the one-step process (Baniasadi et al., 2019). Additionally, bioleaching can be performed as a spent medium-step process, in which the leachants are biologically generated, separated from the microorganisms and used for the solubilization of the target elements (Srichandan et al., 2019). In this case, the non-direct contact of the biomass with the waste source material allows the biomass recycling as well as the optimization of metabolite production. This, in turn, may enable the increase of the source material load to the process (Baniasadi et al., 2019).

Compared with chemical processes, bioleaching is recognized to require longer operation time: an average leaching time of 7 days was reported by several studies. Moreover, in bioleaching experiments aeration is fundamental: when provided by agitation, this should not be excessive to avoid abrasion (Priya & Hait, 2017). In acidophilic chemolithotrophic bioleaching of REEs, oxygen serves as electron acceptor for the microbial metabolism. For instance, Liang et al. (2016) reported an increase of Fe^{2+} oxidation to Fe^{3+} by *A. ferrooxidans*, when the aeration rate was enhanced to increase the dissolved oxygen level from 5 to 25%. The higher dissolved oxygen level resulted, in turn, in improved dissolution of the REE waste source material. Conversely, Nancucheo et al. (2019) reported 9% and 5% bioleaching of Ce and La, respectively, by *A. thiooxidans* at pH 1.8 and 30 °C: such low yields were attributed to the anoxic conditions that interrupted Fe^{2+} reoxidation to Fe^{3+} cycle during the dissolution of the source material, namely the monazite. More recently, Zhang et al. (2020) proposed an aerobic/anaerobic two-stage bioleaching approach to extract Al and REEs from red mud. These authors observed that during the aerobic stage, under the optimal operating conditions, extraction rates of 82.4%, 86.8%, 85.3% and 78.6% were achieved for Ce, Gd, Y and Sc, respectively after 22 days. This outcome further suggests the long leaching periods required to pursue significant biological mobilization of target elements.

The rate of bioleaching is also affected by the composition of the leaching media in terms of nutrients supporting the microbial growth. The concentration of Fe^{2+} as well as S^0 influences the activity of autotrophic microorganisms (Isildar et al., 2019; Zhu et al., 2011) due to its relevant role in bioleaching processes, whereas the concentration of organic compounds regulates the growth of heterotrophs (Isildar et al., 2019). It is worth noting that the supply of external elements, such as iron and sulphur, promoting bioleaching is usually recommended, because their content in the source material is often not enough to properly promote the process (Srichandan et al., 2019).

3 Bioleaching Applications

In the last decades, the bioleaching of REEs has gained great attention. The possibility to extract valuable elements adopting a biological approach can indeed result in lower environmental burdens than conventional hydro-metallurgical leaching. Moreover, the biological generation of extractants may decrease the operating costs for chemical supply and this may be particularly interesting if the source material is a waste stream, such as WEEE: in this case, the low REE grade would result in high recovery, thus enhancing the potential economic benefits.

3.1 State-of-the-Art

The application of the bioleaching has focused on primary sources more than on secondary sources (Dev et al., 2020) and some patents have been developed. In this regard, Barmettler et al. (2016) provided a short overview, pointing out that most of the patents refer to ashes and slags as well as mining waste, whereas other REE containing solids are not considered, apart from luminescent powder originating from fluorescent lamps. At industrial scale, biotechnological processes have been applied for wastewater treatment, remediation of metal-contaminated sites and processing of sulphide ores, but in most cases the interest is not in the recovery of REEs but in the fulfilment of legislative requirements (Gravilescu, 2022).

Nevertheless, mainly due to the interest in REE extraction, several research efforts have been provided and bioleaching mechanisms have been studied under diverse leaching modes. Bioleaching can indeed be performed in shake flasks, batch or continuous reactors, columns and heaps. The former three configurations, namely shake flasks, batch and continuous reactors, are used at smaller scale, heaps at large scale and columns to perform experiments at an intermediate scale, often to study conditions to be then applied for heap processes (Srichandan et al., 2020).

Reed et al. (2016) performed bioleaching studies using shake flasks to evaluate microbial cultures for organic acid production and their potential utility for leaching REEs from fluorescent lamp wastes and spent petrochemical FCC catalyst. Similarly, shake flasks were adopted to better understand the role of *Enterobacter aerogens* in REE leaching from monazite (Fathollahzadeh et al., 2018) as well as to study the characteristics of siderophores produced by *Aspergillus niger* to mobilize REEs from phosphorites (Osman et al., 2019).

Additionally, batch tests have been largely used to examine bioleaching kinetics of REEs from diverse source materials, using different microbial and fungal strains (Hanabe Muddana & Baral 2021; Hopfe et al., 2017; Hosseinzadeh et al., 2021; Qu & Lian, 2013; Zhang et al., 2018).

Interesting outcomes can also be drawn by comparing batch and continuous tests, as reported by Qu et al. (2015). These authors studied the leaching of REEs from red mud using *Aspergillus niger* under both configurations and found that in batch leaching tests, the best performances were obtained at 2% pulp density. However, when moving to the continuous mode, the system could reach the steady state at higher pulp density values, due to the beneficial effect for the microorganisms of prolonged residence time (Qu et al., 2015).

Both shake flasks and batch leaching experiments are easy to carry out and can be used to reduce the process cost and the likely process complication, while gaining a deep understanding of the main operating aspects. Additional information and basic data for preliminary feasibility tests may be obtained from continuous trials. Successful outcomes are scaled up to column bioreactors, to be then extrapolated to heap processes. The column may be operated under different modes, depending on the bioleaching method to be adopted. Therefore, the column may contain the source material of the target elements, and in this case the bioleaching solution is added and the one rich in the elements of interest is drained once the process has been completed. Conversely, the pregnant solution may be drained at the bottom of the column and recirculated to the top to carry out multiple bioleaching cycles. Lastly, the column may be fed with the leaching solution of biological origin, produced in an external reactor (Srichandan et al., 2020). According to the specific bioleaching process to be studied, the column set-up varies.

Column bioleaching has been recently proposed to mobilize REEs from coal-based resources (Sarswat et al., 2020) as well as from a mineralized black shale, which was found to be suitable for heap leaching for the recovery of multiple elements (Watling et al., 2017). Heap as well as in situ bioleaching were developed to overcome the main limits of other metallurgical techniques to extract metals from low-grade ores and it has been widely used in the copper industry (Watling et al., 2017). Currently, any specific application for REE mobilization has been reported.

3.2 Future Research Perspective

The majority of literature studies dealing with bioleaching applications has been carried out at laboratory scale (Panda et al., 2021) and further research is needed to overcome the main limits for the process scale up, especially when considering the possible use of complex waste streams such as WEEE.

The main bottlenecks include: (i) the relatively slow kinetics; (ii) the need to work with low pulp density and (iii) the possible inhibition of the microbial activity (Yaashikaa et al., 2022).

Room for improvements still characterize the area of the basic research on REE bioleaching and future research efforts should be directed towards the optimization of both the process and its multiple operating parameters. The gradual scale up of successful lab-scale processes is fundamental to explore their economic feasibility. Since this aspect is largely affected from the fluctuation of the prices of REEs on the global market, it is crucial to limit the operating costs.

With regard to chemical leaching, which is reportedly characterized by high energy demand and large wastewater generation, bioleaching is considered cleaner, less costly and flexible enough to be applied to a wide variety of source materials (Barmettler et al., 2016). This condition makes it amenable to treat waste source materials, thus addressing cost reduction.

The use of different kinds of waste not only as the REE source materials, but also as substrate for microbial growth, may play a role as showed by Jin et al. (2019). These authors proposed the use of agricultural waste as substrate for *Gluconobacter oxydans*, to bioleach REEs from different kinds of industrial waste. They found that such approach was more economically competitive than the use of refined glucose as carbon source, with the total costs up to 22% lower (Jin et al., 2019). In the same regard, fungi have reportedly high metal recovery ability with cheaper substrates and nutrients than bacteria, showing better economic feasibility (Yaashikaa et al., 2022). Moreover, fungal bioleaching by heap processes was hypothesized to further reduce the costs by enhancing the economic competitiveness of the bioleaching itself (Dusengemungu et al., 2021). The selection of the most suitable microorganisms to extract REEs from a source material is another fundamental aspect to be considered in future studies and, in this regard, it is worth highlighting that great attention is being raised by genetically modified species. The study of metabolically engineered strains, as already experienced for precious metals (Tay et al., 2013), could be carried out to enhance the mobilization of REEs and address the issue of their selective extraction for recovery.

An additional interesting area of future research has been recently identified in the combination of the omics approach, relying on genetics, genomics, metabolomics and proteomic, with bioinformatic to clarify and specify the microorganism contribution within bioleaching processes (Baniasidi et al., 2019) as well as to predict and figure out possible metabolic pathways, as proposed for metals by Abdollahi et al. (2019).

It is worth highlighting that all these aspects requiring further study will be more effective and successful if carried out following an interdisciplinary approach and framed in a holistic perspective, to develop reliable processes and support their implementation at industrial scale.

4 Conclusive Remarks

The application of biotechnological strategies to leach REEs from both primary and secondary sources stands as an opportunity to provide a cleaner and efficient alternative to conventional processes, largely based on aggressive chemical leaching. Although the main mechanisms for REE extraction have been identified and the effects of the main factors influencing the mobilization of REEs from the source materials have been pointed out, further efforts are needed to provide the optimization of this strategy and address the reduction of its operational costs. To this end, lab-scale tests are necessary to identify the operating conditions to be scaled up and tested in a relevant environment.

Although the bioleaching has already some large-scale applications for common and precious metals recovery, further research should focus on REEs. To this end, an interdisciplinary approach, covering the fields of microbiology, chemistry and metallurgy among others, would be highly beneficial. Great support could be also gained by omics approach, which is claimed as one of the cutting-edge research areas in this field. Scientific studies would be even more effective if carried out in close cooperation with the industrial sector as well as with the support of policymakers, in order to frame research in the holistic perspective to promote the use of *green technologies*.

Acknowledgements Alessandra Cesaro would like to thank the former Italian Ministry of Education, University and Research (MIUR) who provided financial support for her position as Assistant Professor in the frame of the research project entitled "Dipartimenti di Eccellenza" per Ingegneria Civile Edile e Ambientale -CUPE65D18000820006.

References

Abdollahi, H., Noaparast, M., Shafaei, S. Z., et al. (2019). Prediction and optimization studies for bioleaching of molybdenite concentrate using artificial neural networks and genetic algorithm. *Minerals Engineering, 130*, 24–35. https://doi.org/10.1016/j.mineng.2018.10.008

Auerbach, R., Bokelmann, K., Stauber, R., et al. (2019a). Critical raw materials—Advanced recycling technologies and processes: Recycling of rare earth metals out of end of life magnets by bioleaching with various bacteria as an example of an intelligent recycling strategy. *Mineral Engineering, 134*, 104–117.

Auerbach, R., Ratering, S., Bokelmann, K., et al. (2019b). Bioleaching of valuable and hazardous metals from dry discharged incineration slag. An approach for metal recycling and pollutant elimination. *Journal of Environmental Management, 232*, 428–437. https://doi.org/10.1016/j.jenvman.2018.11.028

Balaram, V. (2019). Rare earth elements: A review of applications, occurrence, exploration, analysis, recycling, and environmental impact. *Geoscience Frontiers, 10*(4), 1285–1303.

Baniasadi, M., Vakilchap, F., Bahaloo-Horeh, N., et al. (2019). Advances in bioleaching as a sustainable method for metal recovery from e-waste: A review. *Journal of Industrial and Engineering Chemistry, 76*, 75–90. https://doi.org/10.1016/j.jiec.2019.03.047

Barmettler, F., Castelberg, C., Fabbri, C., et al. (2016). Microbial mobilization of rare earth elements (REE) from mineral solids—a mini review. *AIMS Microbiology, 2*(2), 190–204. https://doi.org/10.3934/microbiol.2016.2.190

Binnemans, K., Jones, P. T., Blanpain, B., et al. (2013). Recycling of rare earths: A critical review. *Journal of Cleaner Production, 51*, 1–22. https://doi.org/10.1016/j.clepro.2012.12.037

Binnemans, K., Jones, P. T., Blanpain, B., et al. (2015). Towards zero-waste valorization of rare-earth-containing industrial process residues: A critical review. *Journal of Cleaner Production, 99*, 17–38. https://doi.org/10.1016/j.clepro.2015.02.089

Brierley, J., & Brierley, C. (2001). Present and future commercial applications of biohydrometallurgy. *Hydrometallurgy, 59*, 233–239. https://doi.org/10.1016/S0304-386X(00)00162-6

Brisson, V. L., Zhuang, W. Q., & Alvarez-Cohen, L. (2016). Bioleaching of rare earth elements from monazite sand. *Biotechnology and Bioengineering, 113*, 339–348. https://doi.org/10.1002/bit.25823

Cao, Y., Shao, P., Chen, Y., et al. (2021). A critical review of the recovery of rare earth elements from wastewater by algae for resources recycling technologies. *Resources, Conservation and Recycling, 169*, 105519.

Cesaro, A., Marra, A., Kuchta, K., et al. (2018). WEEE management in a circular economy perspective: An overview. *Global Nest Journal, 20*(4), 743–750.

Charalampides, G., Vatalis, K. I., Apostoplos, B., et al. (2015). Rare earth elements: Industrial applications and economic dependecy of Europe. *Procedia Economics and Finance, 24*, 126–135.

Costis, S., Mueller, K. K., Coudert, L., et al. (2021). Recovery potential of rare earth elements from mining and industrial residues: A review and cases studies. *Journal of Geochemical Exploration, 221*, 106699. https://doi.org/10.1016/j.gexplo.2020.106699

Cucchiella, F., D'Adamo, I., Koch, L., et al. (2015). Recycling of WEEEs: An economic assessment of present and future e-waste streams. *Renewable and Sustainable Energy Reviews, 51*, 263–272. https://doi.org/10.1016/j.rser.2015.06.010

Desmarais, M., Pirade, F., Zhang, J., et al. (2020). Biohydrometallurgical processes for the recovery of precious and base metals from waste electrical and electronic equipments: Current trends and perspectives. *Bioresource Technology Reports, 11*, 100526. https://doi.org/10.1016/j.biteb.2020.100526

Dev, S., Sachan, A., Dehghani, F., et al. (2020). Mechanisms of biological recovery of rare-earth elements from industrial and electronic wastes: A review. *Chemical Engineering Journal, 397*, 124596. https://doi.org/10.1016/j.cej.2020.12596

Dusengemungu, L., Kasali, G., Gwanama, C., et al. (2021). Overview of fungal bioleaching of metals. *Environmental Advances, 5*, 100083. https://doi.org/10.1016/j.envadv.2021.100083

Dushyantha, N., Batapola, N., Ilankoon, I. M. S. K., et al. (2020). The story of rare earth elements (REEs): Occurrences, global distribution, genesis, geology, mineralogy and global production. *Ore Geology Reviews, 122*, 103521. https://doi.org/10.1016/j.oregeorev.2020.103521

Fathollahzadeh, H., Becker, T., Eksteen, J. J., et al. (2018). Microbial contact enhances bioleaching of rare earth elements. *Bioresource Technology Reports, 3*, 102–108. https://doi.org/10.1016/j.biteb.2018.07.004

Gravilescu, M. (2022). Microbial recovery of critical metals from secondary sources. *Bioresource Technology, 344*, 126208. https://doi.org/10.1016/j.biortech.2021.126208

Hanabe Muddana, M., & Baral, S. S. (2021). Bioleaching of rare earth elements from spent fluid catalytic cracking catalyst using Acidothiobacillus ferrooxidans. *Journal of Environmental Chemical Engineering, 9*, 104848. https://doi.org/10.1016/j.jece.2020.104848

Hopfe, S., Flemming, K., Lehmann, F., et al. (2017). Leaching of rare earth elements from fluorescent powder using the tea fungus Kombucha. *Waste Management, 62*, 211–221. https://doi.org/10.1016/j.wasman.2017.02.005

Hosseinzadeh, F., Omid Rastegar, S., & Ashengroph, M. (2021). Bioleaching of rare earth elements from spent catalyst as pretreatment method to improve Pt and Pd recovery: Process optimization and kinetic study. *Process Biochemistry, 105*, 1–7. https://doi.org/10.1016/j.procbio.2021.03.022

Ilyas, S., Anwar, M. A., Niazi, S. B., et al. (2007). Bioleaching of metals from electronic scrap by moderately thermophilic acidophilic bacteria. *Hydrometallurgy, 88*, 180–188. https://doi.org/10.1016/j.hydromet.2007.04.007

Isildar, A., van de Vossnberg, J., Rene, E. R., van Hullebusch, E. D., & Lens, P. N. L. (2016). Two-step bioleaching of copper and gold from discarded printed circuit boards (PCB). *Waste Management, 57*, 149–157. https://doi.org/10.1016/j.wasman.2015.11.033

Isildar, A., van Hullebusch, E. D., Lenz, M., et al. (2019). Biotechnological strategies for the recovery of valuable and critical raw materials from waste electrical and electronic equipment (WEEE)—a review. *Journal of Hazardous Materials, 362*, 467–481. https://doi.org/10.1016/j.jhazmat.2018.08.050

Jin, H., Reed, D. W., Thompson, V. S., et al. (2019). Sustainable bioleaching of rare earth elements from industrial waste materials using agricultural wastes. *ACS Sustainable Chemistry Amd Engineering, 7*, 15311–15319. https://doi.org/10.1021/acssuschemeng.9b02584

Jordens, A., Cheng, Y. P., & Waters, K. E. (2013). A review of the beneficiation of rare earth element bearing minerals. *Minerals Engineering, 41*, 97–114. https://doi.org/10.1016/j.mineng.2012.10.017

Jowkar, M. J., Bahaloo-Horeh, N., Mousavi, S. M., et al. (2018). Bioleaching of indium from discarded liquid crystal displays. *Journal of Cleaner Production, 180*, 417–429. https://doi.org/10.1016/j.jclepro.2018.01.136

Kumar Jha, M., Kumari, A., Panda, R., et al. (2016). Review on hydrometallurgical recovery of rare earth metals. *Hydrometallurgy, 165*, 2–26.

Liang, G., Li, P., Liu, W., et al. (2016). Enhanced bioleaching efficiency of copper from waste printed circuit boards (PCBs) by dissolved oxygen-shifted strategy in Acidithiobacillus ferrooxidans. *Journal of Material Cycles and Waste Management, 18*, 742–751.

Marra, A., Cesaro, A., Rene, E. R., et al. (2018). Bioleaching of metals from WEEE shredding dust. *Journal of Environmental Management, 210*, 180–190.

Menad, N., & van Houwelingen, J. A. (2011). Identification and recovery of rare earth metals in electric and electronic scrap. In Cossu et al. (Eds.), *Proceedings Sardinia—13th International Waste Management and Landfill Symposium*. CISA Publisher.

Muravyov, M. I., Bulaev, A. G., Melamud, V. S., et al. (2015). Leaching of rare earth elements from coal ashes using acidophilic chemolithotrophic microbial communities. *Microbiology, 84*, 194–201. https://doi.org/10.1134/S0026261715010087

Nancucheo, I., Oliveira, G., Lopes, M., et al. (2019). Bioreductive dissolution as a pretreatment for recalcitrant rare-earth phosphate minerals associated with lateritic ores. *Minerals, 9*, 136. https://doi.org/10.3390/min9030136

Opare, E., Struhs, E., & Mirkouei, A. (2021). A comparative state-of-technology review and future directions for rare earth element separation. *Renewable and Sustainable Energy Reviews, 143*, 110917. https://doi.org/10.1016/j.rser.2021.110917

Osman, Y., Gebreil, A., Mowafy, A. M., et al. (2019). Characterization of *Aspergillus niger* siderophore that mediates bioleaching of rare earth elements from phosphorites. *World Journal of Microbiology and Biotechnology, 35*, 93. https://doi.org/10.1007/s11274-019-2666-1

Panda, S., Costa, R. B., Shah, S. S., Mishra, S., Bevilaqua, D., & Akcil, A. (2021). Biotechnological trends and market impact on the recovery of rare earth elements from bauxite residue (red mud)—A review. *Resources Conservation and Recycling, 171*, 105645. https://doi.org/10.1016/j.resconrec.2021.105645

Pedram, H., Hosseini, M. R., & Bahrami, A. (2020). Utilization of *A. niger* strains isolated from pistachio husk and grape skin in the bioleaching of valuable elements from red mud. *Hydometallurgy, 198*, 105495. https://doi.org/10.1016/j.hydromet.2020.105495.

Priya, A., & Hait, S. (2017). Compaartive assessment of metallurgical recovery of metals from electronic waste with special emphasis on bioleaching. *Environmental Science and Pollution Research, 24*, 6989–7008. https://doi.org/10.1007/s11326-016-8313-6

Qu, Y., Li, H., Tian, W., et al. (2015). Leaching of valuable metals from red mud via batch and continuous proceses by using fungi. *Minerals Engineering, 81*, 1–4. https://doi.org/10.1016/j.mineng.2015.07.022

Qu, Y., Li, H., Wang, X., et al. (2019). Bioleaching of major, rare earth, and radioactive elements from red mud by using indigenous chemoheterotrophic Bacterium *Acetobacter* sp. *Minerals, 9*, 67. https://doi.org/10.3390/min9020067

Qu, Y., & Lian, B. (2013). Bioleaching of rare earth and radioactive elements from red mud using Penicillium tricolor RM-10. *Bioresource Technology, 136*, 16–23.

Quatrini, R., & Johnson, D. B. (2019). Acidithiobacillus ferrooxidans. *Trends in Microbiology, 27*(3), 282–283. https://doi.org/10.1016/j.tim.2018.11.009

Rawlings, D. E. (2002). Heavy metals mining using microbes. *Annual Review of Microbiology, 56*, 65–91. https://doi.org/10.1146/annurev.micro.56.012302.161052

Reed, D. W., Fujita, Y., Daubaras, D. L., et al. (2016). Bioleaching of rare earth elements form waste phosphors and cracking catalysts. *Hydrometallurgy, 166*, 34–40. https://doi.org/10.1016/j.hydromet.2016.08.006

Sarswat, P. K., Leacke, M., Allen, L., et al. (2020). Efficient recovery of rare earth elements form coal based resources: A ioleaching approach. *Materials Today Chemistry, 16*, 100246. https://doi.org/10.1016/j.mtchem.2020.100246

Sethurajan, M., Lens, P. N. L., Rene, E. R., et al. (2017). Bioleaching and selective biorecovery of zinc from zinc metallurgical leach residues from the Tres Marias zinc plant (Minas Gerais, Brazil). *Journal of Chemical Technology and Biotechnology, 3*(92), 512–521. https://doi.org/10.1002/jctb.5026

Sethurajan, M., van Hullebusch, E. D., & Nancharaiah, Y. V. (2018). Biotechnology in the management and resource recovery from metal bearing solid waste: Recent advances. *Journal of Environmental Management, 211*, 138–153. https://doi.org/10.1016/j.envman.2018.01.035

Srichandan, H., Kumar Mohapatra, R., Kumar Parhi, P., et al. (2019). Bioelaching approach for extraction of metal from secondary solid wastes: A critical review. *Hydrometallurgy, 189*, 105122. https://doi.org/10.1016/j.hydromet.2019.105.122

Srichandan, H., Kumar Mohapatra, R., Kumar Singh, P., et al. (2020). Column bioleaching applications, process development, mechanism, parametric effect and modelling: A review. *Journal of Industrial and Engineering Chemistry, 90*, 1–16. https://doi.org/10.1016/j.jiec.2020.07.012

Tao, H., & Dongwei, L. (2014). Presentation on mechanisms and applications of chalcopyrite andpyrite bioleaching in biohydrometallurgy–a presentation. *Biotechnology Reports, 4*, 107–119. https://doi.org/10.1016/j.btre.2014.09.003

Tay, S. B., Natarajan, G., Rahim, M. N., Tan, H. T., et al. (2013). Enhancing gold recovery from electronic waste via lixiviant metabolic engineering in Chromobacterium violaceum. *Scientific Reports, 3*(1), 2236. https://doi.org/10.1038/srep02236

Tse, P. K. (2011). *China's rare-earth industry*. U.S. Geological Survey Open-File Report 2011–1042 (11 p.). Available only online.

Tunsu, C., Petranikova, M., Gergoric, M., et al. (2015). Reclaiming rare earth elements from end-of-life products: A review of the perspectives for urban mining using hydrometallurgical unit operations. *Hydrometallurgy, 156*, 239–258. https://doi.org/10.1016/j.hydromet.2015.06.007

van Gosen, B. S., Verplanck, P. L., Seal, R. R. et al. (2017). Rare earth elements In K. J. Schultz, J. H. DeYoung, R. R. Seal, & D. C. Bradley (Eds.), *Critical mineral resources of the united states-economic and environmental geology and prospects for future supply, U.S. Geological Survey Professional Paper 1802* (pp. O1–O31). https://doi.org/10.3133/pp1802O.

Wall, F. (2021). Rare earth elements. In *Encyclopedia of geology* (2nd edn, pp. 680–693). https://doi.org/10.1016/B978-0-08-102908-4.00101-6

Watling, H. R., Collinson, D. M., Watling, R. J., et al. (2017). Simulated heap leaching and recovery of multiple elements from a mineralised black shale. *Hydrometallutgy, 167*, 48–57. https://doi.org/10.1016/j.hydromet.2016.10.015

Yaashikaa, P. R., Priyanka, B., Senthil Kumar, P., et al. (2022). A review on recent advancements in recovery of valuable and toxic metals from e-waste using bioleaching approach. *Chemosphere, 287*, 132230. https://doi.org/10.1016/j.chemosphere.2021.132230

Zhang, D. R., Chen, H. R., Nie, Z. Y., Xia, J. L., Li, E. P., Fan, X. L., & Zheng, L. (2020). Extraction of Al and rare earths (Ce Gd Sc Y) from red mud by aerobic and anaerobic bi-stage bioleaching. *Chemical Engineering Journal, 401*, 125914. https://doi.org/10.1016/j.cej.2020.125914

Zhang, L., Dong, H., Liu, Y., et al. (2018). Bioleaching of rare earth elements from bastnaesite-bearing rock by actinobacteria. *Chemical Geology, 483*, 544–557. https://doi.org/10.1016/j.chemgeo.2018.03.023

Zhu, N., Xiang, Y., Zhang, T., et al. (2011). Bioleaching of metal concentrates of waste printed circuit boards by mixed culture of acidophilic bacteria. *Journal of Hazardous Materials, 192*, 614–619. https://doi.org/10.1016/j.jhazmat.2011.05.062

Biotechnological Applications in Spent Lithium-Ion Battery Processing

Hadi Abdollahi, Roozbeh Saneie, Ahmad Rahmanian, Ehsan Ebrahimi, Amirhossein Mohammadzadeh and Ghazaleh Shakiba

Abstract

Lithium-ion batteries (LIBs) have emerged as the leading energy source for a diverse array of electronic devices, owing to their numerous benefits. Recycling LIBs is of significant importance since the ever-growing demand for them will shortly lead to massive disposal of spent LIBs and cause critical environmental problems. Spent LIBs can be considered as an excellent secondary source for various valuable metals since they approximately contain Mn (5–11%), Co (5%–25%), Ni (5%–10%), Li (5%–7%), Al (15–20%), Cu (5–7%), and graphite. Currently, LIBs' recycling is mainly through the pyrometallurgical or high-temperature hydrometallurgical approaches, which have been substantiated to be effective for metal extraction. However, these processes come with some disadvantages such as inefficient energy consumption, high operational cost, toxic gas emission and production of secondary hazardous waste. The application of bio-hydrometallurgical methods has demonstrated remarkable efficacy in extracting metals from ores, flotation concentrates, tailings, and diverse waste materials. As an eco-friendly, low-cost, and energy-efficient method, bioleaching can be a superior replacement for conventional LIB recycling processes. In this chapter, new trends of LIB bioleaching, various factors influencing the LIB bioleaching, mechanisms, microorganisms, and regeneration of black mass have been thoroughly discussed. Moreover, the dominant challenges for industrial application of LIB bioleaching and several approaches for upscaling were summarized.

H. Abdollahi (✉) · R. Saneie · A. Rahmanian · E. Ebrahimi · A. Mohammadzadeh · G. Shakiba
School of Mining Engineering, College of Engineering, University of Tehran, P.O. Box 1439957131, Tehran, Iran
e-mail: h_abdollahi@ut.ac.ir

Keywords

Bio-hydrometallurgy · Bioleaching · Lithium-ion battery · LIB · Recycling

1 Introduction

With the increasing demand for rechargeable lithium-ion batteries (LIBs), the enhanced production of these batteries will soon translate into enormous amounts of LIBs waste. The use of LIBs spans across a wide range of industries, including mobile communication devices like smartphones and tablets, various portable electronic gadgets, the rapidly expanding electric vehicle (EV) market, as well as the extensive landscape of computing devices and storage equipment. The demand for LIBs is continuing to surge, driven by a persistent growth route (Chen et al., 2019). LIBs possess a captivating array of qualities, including elevated voltage levels, remarkable longevity, minimal self-discharge, impressive energy density, a wide operational temperature range, compact dimensions, and lightweight construction. These exceptional electrochemical characteristics render them a superior choice over alternative battery technologies (Fergus, 2010). Extensive applications of LIBs are causing a surge in demand for the materials used in their production. These batteries require several key metals, including cobalt, lithium, nickel, manganese, copper, aluminum, and other elements, in order to be manufactured. The demand for these materials is expected to continue to grow as the use of LIBs becomes more prevalent in various industries, including the automotive and electronics sectors. As a result, there is an increasing focus on developing sustainable and responsible mining practices to ensure a stable supply chain for these critical materials. Some of these valuable metals are generally used for the formation of the LIBs cathode. For instance, cobalt-lithium oxide is used as an active cathode in LIBs because of its high-energy capacity per unit volume makes it perfect for portable electronics.

S. Panda et al. (eds.), *Biotechnological Innovations in the Mineral-Metal Industry*, Advances in Science, Technology & Innovation, https://doi.org/10.1007/978-3-031-43625-3_5

The cobalt supply risk is high; hence, aluminum, manganese, or nickel is usually used in the cathode to reduce cobalt consumption (Habib et al., 2016).

Furthermore, lithium, nickel, cobalt, and aluminum oxide provides a sufficient energy density for the batteries, which prepares them to be utilized in electric vehicles (Olivetti et al., 2017). China's market is poised to experience a notable surge of more than 13% by 2025, driven by its robust economic growth and the persistent advancements and expansions in the realm of vehicle manufacturing. Notably, according to real-time intelligence statistics provided by GSMA, a staggering 5.17 billion individuals currently possess a mobile phone device projected to reach an impressive 7.33 billion by the year 2023 (Ordoñez et al., 2016). Furthermore, by 2030, almost one-fifth of all vehicles on American roads will be electric, and many of these will be powered by LIBs. This shift is being driven in part by increasing concerns over climate change and a growing desire to reduce carbon emissions. As a result, there is a growing need for sustainable and responsible mining practices to ensure a stable supply of the materials required for the production of LIBs (Vikström et al., 2013; Wang et al., 2014a). Furthermore, due to the extensive application of LIBs and decreased lithium and other transitional metals resources, the recycling of LIBs presents a number of benefits, both from an environmental and strategic perspective. The proper disposal of spent LIBs can help to prevent pollution and reduce the amount of waste sent to landfills. Additionally, recycling these batteries can help to recover valuable materials, including lithium, cobalt, nickel, and other metals, which can then be used to produce new batteries. This process not only helps to conserve strategic materials, but also reduces the need for new mining operations. As a result, there is a growing focus on developing effective and sustainable recycling strategies to ensure that the valuable resources contained within spent LIBs are recovered and reused (Xu et al., 2008). Considering that the primary resources of lithium are gradually depleted during long-term exploitation (Jafari et al., 2019), the supply risk related to these metals and the reduction of their mineral resources will be a serious concern (Bardi et al., 2016). Thus, secondary resources such as spent LIBs have the great capacity to be a suitable replacement to the primary resources (Ambrose & Kendall, 2020). Furthermore, dangerous substances encompassing elements like lithium, nickel, cobalt, and manganese, along with harmful chemicals like electrolytes and binders, are utilized in the production of LIBs. These materials can result in significant harm to the environment and cause pollution, posing a threat to both ecosystems and human health (Kim et al., 2006; Shin et al., 2005). Hence, based on economic and environmental concerns, today's industrial recycling approaches are moving toward environmentally friendly methods, especially biotechnological methods in the recycling of electronic waste.

1.1 Structure of Lithium-Ion Batteries

Disparate conventional batteries, LIBs operate using a different mechanism to generate energy, not relying on a reduction–oxidation reaction. Instead, these batteries utilize the movement of lithium ions between the cathode and anode, which in turn forces electrons to travel along with them. This advanced type of battery includes various components such as the cathode, anode, electrolyte, shell, separator, and other parts. A visual representation of the structure of depleted LIBs can be observed in Fig. 1 (Tarascon & Armand, 2001).

1.1.1 Anode

Carbon (graphite) and lithium alloyed metals are the two most common anode materials used in LIBs (Mekonnen et al., 2016). Carbon-based anodes have emerged as the predominant choice for the advancement of LIBs owing to the cost-effectiveness of graphite production and its commendable electrochemical properties. By employing a layered graphite, the storage of Li-ions between carbon atoms (known as intercalation) occurs during the charging phase, while their controlled release transpires during discharging. However, the formation of dendrites during this process presents a challenge, as it leads to short circuits and instability in LIBs. Hence, the active anode materials in LIBs predominantly consist of carbon-based components, encompassing graphite, carbon black, carbon fiber, pyrolysis, petroleum coke, mesophase carbon microsphere bituminous, glass carbon, and more. These active materials are subsequently coated onto copper foil collectors, employing a layer of polyvinylidene fluoride (PVDF) binder for optimal performance (Chen & Xue, 2014; Zeng et al., 2014). In addition to graphite, LIBs utilize lithium alloy anodes such as lithium aluminum (Li-Al) and $LiTiO_2$, which are essential materials for the anode (Mekonnen et al., 2016). $LiTiO_2$, in particular, stands out for its exceptional electrochemical cycling performance. One of its notable attributes is its ability to undergo lithiation and dilithiation processes without experiencing any significant changes in volume. This unique characteristic contributes to its reliability and longevity, making it an excellent choice as an anode material for LIBs (Liu et al., 2019; Subhan et al., 2019). The use of graphite intercalation alloy can help to protect the lithium in LIBs during the charging and discharging process. This is because the metals present in the alloy act as a shield, preventing the lithium ions from reacting with the electrolyte and causing potential safety hazards. The alloy

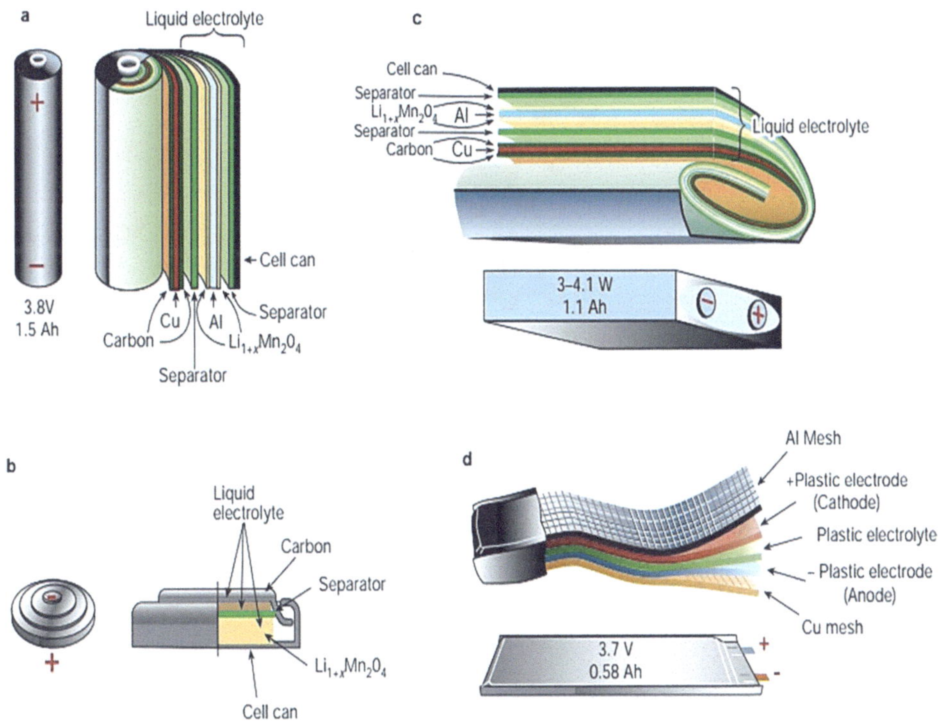

Fig. 1 Structure of spent LIB. **a** Cylindrical; **b** coin; **c** prismatic; and **d** thin and flat (Tarascon & Armand, 2001)

works by allowing the lithium ions to intercalate, or insert themselves between the layers of graphite, which helps to stabilize the battery's performance. This technology has been instrumental in improving the safety and reliability of LIBs, making them a popular choice for a variety of applications, including electric vehicles and portable electronics (Lavoie et al., 2017).

1.1.2 Cathode

Table 1 gives the compositions of LIBs cathodes (Methekar & Anwani, 2019). Figure 2 provides an illustration of five distinct cathode materials employed in LIBs, each with its unique atomic arrangement or crystal structure. These active materials encompass lithium cobalt oxide ($LiCoO_2$, also known as LCO), lithium nickel cobalt manganese oxide (LiNi1−x−yCoxMnyO2, referred to as NMC), lithium manganese oxide ($LiMn_2O_4$, identified as LMO), lithium iron phosphate ($LiFePO_4$, commonly referred to as LFP), and lithium nickel cobalt aluminum oxide ($LiNiCoAlO_2$, known as NCA). They can be further categorized into layered structures like LCO, NMC, NCA, and spinel LMO, along with olivine LFP, based on their distinctive compositions and arrangements. Table 2 compares numerous cathode materials' key characteristics and applications (He et al., 2015; Kwon et al., 2018).

1.1.3 Electrolyte

In a battery electrode, the transfer of ions from the cathode to the anode is facilitated by the presence of electrolytes which serve as a medium for this process. Through this transfer of ions, the chemical energy stored in the battery is converted into electrical energy. Typically, the electrolyte comprises an organic liquid containing soluble substances. LIBs are designed to utilize various types of electrolytes. These four distinct types of electrolytes commonly used in LIBs are: liquid electrolytes, colloidal electrolytes, polymer electrolytes, and ceramic electrolytes. Each type of electrolyte has its own unique properties and characteristics, which determine its effectiveness in facilitating the transfer of ions within the battery. $LiPF_6$, $LiBF_4$,

Table 1 Composition of a typical LIB active cathode material (Methekar & Anwani, 2019)

Metal	Weight %	Metal	Weight %	Metal	Weight %	Metal	Weight %	Metal	Weight %
Li	6.3	Cu	0.6	K	0.02	Si	0.09	Fe	0.11
Co	45.1	Al	0.67	Ca	0.03	Cl	0.02	Zr	0.01
Mn	11.8	P	0.32	Na	0.23	Ti	0.02	Nb	0.01
Ni	0.30	S	0.16	Mg	0.41	Cr	0.09	O_2	Rest

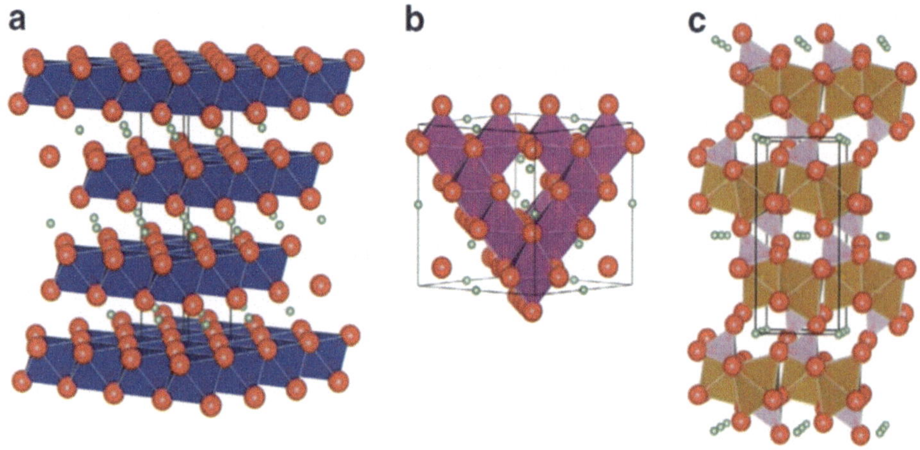

Fig. 2 Crystal structures of various types of cathode materials used in LIBs: **a** Layered structures such as LCO, NMC, and NCA; **b** spinel structures like LMO; and **c** olivine structures like LFP (Deng et al., 2016; Kwon et al., 2018)

Table 2 Cathode component of LIBs and each application (Kwon et al., 2018)

Type	LCO	NMC	LMO	LFP	NCA
Structure	Layered	Layered	Spinel	Olivine	Layered
Voltages (V)	3.0–4.2	3.0–4.2	3.0–4.2	2.5–3.65	3.0–4.2
Energy density (Wh/Kg)	150–200	150–220	100–150	90–120	200–260
Thermal runaway (°C)	150	210	250	270	150
Cycle life	Good	Medium	Poor	Good	Good
Stability	Good	Good	Good	Very good	Poor
Example of application	Portable electronics	E-bikes electric vehicles	Power tools electrical powertrains	High load currents and endurance	Industrial electric powertrain

$LiCF_3SO_3$, or Li $(SO_2CF_3)_2$ are some of the available electrolyte salts. However, $LiPF_6$ is the most widely used. Given that the lithium-ion cell voltage (~3.6 V) is higher than the standard water electrolysis potential (1.23 V at 25 °C), the presence of a non-aqueous solvent is essential; therefore, solvents with a high dielectric constant are needed. For electrolytes to conduct ions, they must also contain lithium salts (Kwon et al., 2018; Zeng et al., 2014; Zheng et al., 2017). Propylene carbonate (PC), ethylene carbonate (EC), and dimethyl sulfoxide (DMSO) are frequently used solvents that have the ability to dissolve a variety of lithium salts. However, these solvents have a high viscosity that can impede ion transfer and decrease conductivity. Therefore, actual electrolyte fluids consist of novel compounds and feature low viscosity solvents. Up until now, a variety of distinct polymer categories have been developed, comprising polyethylene oxide (PEO), polypropylene oxide (PPO), polyacrylonitrile (PAN), polymethyl methacrylate (PMMA), polyvinyl chloride (PVC), polyvinylidene fluoride (PVDF), and polyvinylidene fluoride-hexafluoropropylene (PVDF-HFP) (Hamidah et al., 2015; Zheng et al., 2017).

1.1.4 Separator

A crucial component in LIBs, the separator effectively safeguards against short circuits arising from direct contact between the anode and cathode. By creating a deliberate space between these electrodes, it acts as a protective barrier. This separator is typically composed of a finely porous material, commonly polyethylene (PE) or polypropylene (PP), which ensures regularity in its construction. Additionally, the separator serves as a safety mechanism: in the event of excessive heat generation, the porous strip melts, irreversibly maintaining a physical separation between the electrodes. However, electrical appliances will typically remove the cell from the circuit before such a problem occurs (Xu et al., 2015).

1.2 Environmental Risk of Spent LIBs and Importance of Recycling

The life cycle of LIBs encompasses various stages, starting from the product life cycle involving activities like selling, storage, use, reuse, gifting, and export. It extends further to the product's end-of-life phase, as evaluated through lifecycle analysis (LCA) (Liang et al., 2017), and material flow analysis (MFA) (Sommer et al., 2015) (recycling, landfilling, and incineration) (Tanskanen, 2013). As discussed in the previous section, LIBs are complex devices comprised of several components that, if not properly managed and recycled, can lead to environmental pollution, including contamination of soil and water (Zeng et al., 2015). LIBs contain hazardous materials that include metals, and toxic chemical materials (such as electrolytes and binders) that cause severe environmental damage and pollution, threatening ecosystems and human health. Lithium hydroxide and hydrogen gas are generated when lithium in the anode reacts with water. However, the reaction is not as intense as that of lithium metal. Excessive Li pollution into water and soil can harm animals and plants. Overcharged batteries will form a lithium coating on the surface of their anode. The most probable cause of discarding spent batteries is due to their inadequate performance, which means the inclusion of lithium metal in recycling procedures cannot be ignored (Shin et al., 2005). In addition to Li, lower amounts of Co in LIBs are helpful to humans because it increases the development of red blood cells. Cobalt is toxic and carcinogenic at high concentrations, and causes vomiting, nausea, eyesight, and cardiac difficulties (Kim et al., 2006). As anode material of LIBs, graphite carbon materials can react with strong oxidants and generate CO, CO_2, or other gases. The electrolytes in LIBs, for instance, $LiPF_6$, $LiBF_4$, or $LiClO_4$ can react with water and produce HF and PF_5 gases which are harmful to the atmosphere (Zeng et al., 2015). The utilization of PVDF in electrode production mandates the employment of harmful solvents like N-methyl-pyrrolidone (Versaci et al., 2017). With the escalating environmental concerns surrounding the disposal of LIBs, there is a growing demand for critical metals. In light of this, it becomes imperative to explore economically viable and eco-friendly recycling techniques that facilitate the recovery of these valuable metals from used Li-ion batteries. By adopting selective leaching methods and implementing efficient recovery processes, we can significantly reduce environmental impact and promote the sustainable utilization of secondary resources (Zheng et al., 2018).

1.3 Bioleaching of Spent LIBs

Bioleaching, a bio-based process, offers notable advantages characterized by its exceptional efficiency, safety, applicability at atmospheric pressure and room temperature, and reduced energy consumption (Vakilchap et al., 2016). Bioleaching is a cost-effective technique for treating disposed waste that employs lixiviants (leaching agents) generated biologically by microorganisms. It is highly beneficial with few industrial requirements (Bosecker, 1997; Islam et al., 2020). Alternative methods for metal recovery from primary and secondary sources, such as hydrometallurgy and pyrometallurgy have been effectively employed alongside bioleaching (Srichandan et al., 2019). Although both hydrometallurgy and pyrometallurgy are efficient procedures, they have significant drawbacks (Asghari et al., 2013). Hydrometallurgy involves the use of highly concentrated acids and bases to dissolve the metals. However, this process can generate significant amounts of acidic waste, which can be costly and challenging to manage. Downstream processing costs can be high due to the need for specialized equipment and techniques to safely handle and dispose of the waste generated (Asghari et al., 2013; Bharadwaj & Ting, 2013; Srichandan et al., 2014). Conversely, pyrometallurgy involves operating at high temperatures (1500–1700 °C), making it an energy-inefficient process. Additionally, it is linked to the emission of hazardous gases such as SO_2 (Bharadwaj & Ting, 2013; Srichandan et al., 2014). Hence, both approaches fail in energy, the environment and the economy.

On the other hand, bioleaching avoids the need for intense acid/base, is energy efficient, and produces no hazardous gases (Asghari et al., 2013). During the bioleaching process, bio-oxidation transforms insoluble compositions into water-soluble compounds, and the microbe generates energy by rupturing ores or wastes (Rohwerder et al., 2003). During the process of bioleaching, a range of microorganisms, including fungi and bacteria, can play an integral role in facilitating the extraction of valuable metals and minerals. These microorganisms are capable of secreting either inorganic or organic acids, which can aid in the breakdown of the ores and promote the solubilization of metals. Furthermore, microorganisms have the potential to augment enzymatic oxidation–reduction, proton-promoted mechanisms, and enhance the formation of ligands and complexes. These mechanisms play an important role in enhancing the efficacy of the bioleaching process and augmenting the overall recovery of extracted minerals and metals. With their distinct capabilities, microorganisms are rapidly emerging as an indispensable asset in the domain of bioleaching, propelling the advancement of sustainable and environmentally friendly techniques for mineral and metal extraction (Vakilchap et al., 2016; Xiang et al., 2010). Microorganisms that have the ability to bioleach metals can be divided into three categories (Abhilash & Natarajan, 2015): the first category includes autotrophic bacteria that utilize both acidolysis and redoxolysis mechanisms to break down ores and facilitate the extraction of metals. The second category comprises heterotrophic microorganisms such as fungi, which utilize acidolysis and complexolysis mechanisms to produce organic acids and facilitate the solubilization of metals. Lastly, the third category includes cyanogenic bacteria that utilize complexolysis mechanisms to facilitate the extraction of metals from ores. Through the utilization of these distinctive mechanisms, microorganisms play a fundamental part in facilitating the bioleaching process, allowing for the efficient and environmentally sustainable extraction of precious minerals and metals. The leaching process is executed by various microorganisms, including bacteria, fungi, and yeast. Acidophilic sulfur-oxidizing bacteria and iron-oxidizing bacteria are among the most extensively employed microorganisms, standing out as the primary choices in bioleaching techniques (Xin et al., 2009). *Acidithiobacillus ferrooxidans* and *Acidithiobacillus thiooxidans* are widely recognized autotrophic microorganisms renowned for their catalytic role in the oxidation process that converts ferrous iron ions to ferric iron ions (Johnson, 2018). Fungi can be utilized in the bioleaching process, like *Aspergillus niger* and *Penicillium simplicissimum* (Vakilchap et al., 2016; Wang et al., 2015). The process of bioleaching can be employed to separate the metallic constituents of batteries into various fractions, which can then be utilized to create novel batteries. Bioleaching offers benefits such as energy efficiency, cost-effectiveness, and a reduction in the presence of dangerous battery waste materials (Johnson, 2014; Vanitha & Balasubramanian, 2013). In comparison with conventional recycling approaches, bioleaching of LIBs presents a significant advantage as it creates a weak acid waste and releases minimal amounts of hazardous gases, thereby eliminating the necessity for supplementary treatment and reducing treatment expenses (Yu et al., 2020).

2 Preparation and Pretreatment of Waste LIBs

Pretreatment plays an important role in LIB recycling to increase recovery and reduce energy consumption. Therefore, it would be advantageous to systematically categorize and analyze the diverse range of novel approaches employed in the pretreatment process, along with their specific scopes and sequences. The main goal for LIB pretreatment is to separate active materials from metallic foils and other components. The listed sequences can be categorized into several groups, which include discharge, disassembly, comminution, classification and segregation, dissolution, and thermal processing (Golmohammadzadeh et al., 2018; Kim et al., 2021; Lai et al., 2021; Makuza et al., 2021; Roy et al., 2021a; Windisch-Kern et al., 2022).

2.1 Discharging and Dismantling (Disassembly)

When LIBs reach the end of their useful life, they lose a small amount of power; consequently, batteries burst through the recycling process. Batteries must be totally discharged prior to dismantling. The most popular way of deactivating the LIBs is to immerse them in a NaCl solution. Submerging metallic lithium in liquid nitrogen reduces its reactivity (Roy et al., 2021a).

The typical structure of an LIB's system in vehicles consists of battery packs and a battery management system. Within the market, there exists a vast array of power batteries with varying physical structures, battery types, and material systems, making it challenging to autonomously disassemble these systems. At present, the majority of battery disassembly is carried out manually or in a semi-automated manner, such as with an automatic disassembly screw. However, this approach is time-consuming and labor-intensive, which may lead to a decrease in battery performance and a reduction in the overall efficiency of battery material recycling. As such, there is a growing need for advanced and automated disassembly processes that can effectively and efficiently disassemble LIBs while minimizing the risk of battery damage and material waste (Lai et al., 2021).

2.2 Comminution and Mechanical Treatment

Before any further processing can take place, it is imperative to execute a mechanical pretreatment stage before any subsequent processing can occur. The primary aim of this stage is to segregate the Fe, Cu, and Al alloy fractions with significant value from the fines, commonly known as "black matter." The fine material is mainly composed of anode and cathode active materials of the LIB and contains crucial chemicals such as lithium, cobalt, nickel, and manganese that can be salvaged via downstream processes. Contemporary mechanical pretreatment methods typically employ advanced technologies such as rotary shears in double shaft shredders at low rpm or impact crushing at high rpm in hammer mills. These sophisticated techniques facilitate the effective and efficient separation of valuable materials from the black matter, leading to a maximization of the overall yield of precious metals and minerals throughout the recycling process (Windisch-Kern et al., 2022).

2.3 Classification and Separation

To separate and concentrate the constituent elements of depleted LIBs, several mechanical methods are utilized, including fragmentation, sifting, magnetic segregation, ultra-fine grinding, and sorting. By leveraging these mechanical separation techniques, it becomes possible to extract the crust and steel cases from spent LIBs, capitalizing on alterations in physical properties like density, magnetic behavior, and conductivity. However, the inability to completely separate all components of wasted LIBs precludes the effective use of mechanical techniques (Golmohammadzadeh et al., 2018).

2.4 Dissolution

During the classification and separation stages of LIBs recycling, certain active elements are liberated from the current collectors, while others are still held together by binders. In such cases, the binders or aluminum foils are commonly dissolved using suitable solvents. Acetone, dichloromethane, carbon tetrachloride, and N-methylpyrrolidone are among the various solvents utilized to separate the active cathode components from the aluminum foil (Kim et al., 2021).

2.5 Thermal Treatment (Heat Treatment, Pyrolysis)

Thermal pretreatment enables the regulated and safe deactivation and breakdown of the LIB's combustible organic component. Additionally, heat treatments can be employed to drain the LIB. Thermal treatment, depending on the specific temperature applied, can serve multiple purposes. It proves to be an effective approach for eliminating organic binder material. The application of elevated temperatures during thermal treatment leads to the decomposition of the binder, thereby breaking down the bonding force between the binder and active cathode material. Consequently, the cathode material can be readily separated through screening and other methods. Thermal pretreatment finds extensive applications in both laboratory and industrial settings, primarily aimed at removing carbon and organic constituents. Vacuum pyrolysis is one of the best thermal pretreatments with several advantages (Makuza et al., 2021).

3 Bioleaching of Waste LIBs with Autotrophic Bacteria

By oxidizing iron and sulfur, autotrophic bacteria can improve the dissolution of precious metals in LIBs. Most bacteria used are *Acidithiobacillus ferrooxidans, Acidithiobacillus thiooxidans,* and *Leptospirillum ferrooxidans*, or a combination of them. Higher levels of toxins in solution generally hamper bacterial activity in high pulp densities. The goal of most researchers in the bacterial dissolution process of LIBs has been to increase recovery and greater efficiency in higher pulp densities (Heydarian et al., 2018; Niu et al., 2014; Roy et al., 2021b, 2021c).

3.1 Mechanisms

Investigating the mechanism of battery dissolution aids significantly in properly identifying the processes and establishing the conditions for further efficiency. Many researchers have investigated how bacteria dissolve batteries materials (Wu et al., 2019; Xin et al., 2009; Zeng et al., 2013a). Lithium cobalt oxide and lithium nickel oxide are both included in LIBs. These compounds dissolve in the following manner (Heydarian et al., 2018):

$$4\,LiCoO_2(s) + 6\,H_2SO_4(aq) \rightarrow 2\,Li_2SO_4(aq) + 4\,CoSO_4(aq) + 6\,H_2O(aq) + O_2(g). \quad (1)$$

In the presence of ferrous sulfate ions:

$$2\,FeSO_4 + 2\,LiCoO_2 + 4\,H_2SO_4 \rightarrow Fe_2(SO_4)_3 + 2\,CoSO_4 + Li_2SO_4 + 4\,H_2O. \quad (2)$$

The leaching of $LiNiO_2$ can be stated as follows:

$$4\,LiNiO_2(s) + 6\,H_2SO_4(aq) \rightarrow 2\,Li_2SO_4(aq) + 4\,NiSO_4(aq) + 6\,H_2O(aq) + O_2(g). \quad (3)$$

The retrieval of metals from spent LIBs typically involves an acid dissolution process that employs sulfuric acid, generated via bio-oxidation facilitated by microorganisms. Regardless of the energy source utilized, acid dissolution has been identified as the primary method for lithium (Li) extraction. However, when FeS_2 are used as energy carriers for cobalt (Co) bioleaching, a combination of acid dissolution and Fe^{2+} reduction takes place due to the physicochemical interaction between FeS_2 and Fe^{3+}. This interaction promotes the conversion of insoluble Co^{3+} into soluble Co^{2+} by Fe^{2+}, leading to the release of Co^{2+} from used batteries through acid dissolution. Additionally, the reduction attack of Fe^{2+} on the Co^{3+} moiety has been recognized as an efficient mechanism for the release of Co from spent LIBs. Therefore, by harnessing these unique mechanisms, the extraction of valuable metals and minerals from spent batteries can be carried out in an environmentally sustainable and cost-effective manner (Wu et al., 2019; Xin et al., 2009).

$$Fe^{2+} + Co^{3+} \rightarrow Fe^{3+} + Co^{2+}(CoO) \quad (4)$$

Through the interaction between Fe^{2+} and sulfuric acid, the structural integrity of the components within used batteries was compromised, leading to the release of metal ions. Xin et al. demonstrated that, for the dissolution of Co and Li from spent LIBs, the non-contact method was found to be indispensable, while the contact mechanism proved to be insufficient. It is critical to note that, unlike bioleaching of sulfide minerals, primarily used in this case, the presence of Fe^{3+} in the dissolution of batteries not only hinders but also reduces the recovery. Therefore, attention must be given to selecting bacterial species and energy sources (Xin et al., 2009).

Wu et al. examined the effect of extracellular polymeric in the leaching of LIBs, which were found to be a factor in improving Li and Co^{2+} recovery. In order to identify the specific constituent of extracellular polymeric substances (EPS) responsible for promoting the leaching of $LiCoO_2$, an analysis was conducted on protein, polysaccharide, and uronic acid components. It was shown that cysteine might increase the leaching of Li and Co^{2+}, but not glucose or uronic acid. This conclusion could be explained by the reducibility of cysteine's sulfhydryl groups, reducing the Co^{3+} to Co^{2+} (Wu et al., 2019).

3.2 Microorganisms

Mesophilic microorganisms have the ability to solubilize metallic components in LIBs. *Acidithiobacillus* has a better tolerance for metal toxicity, and it dominates crucial studies on LIB bioleaching (Jafari et al., 2018; Naseri et al., 2019; Quatrini & Johnson, 2019; Rawlings, 1997). They aid in metal dissolving by creating biogenic H_2SO_4 and ferric ion, both of which aid in metal leaching (Xin et al., 2009; Zeng et al., 2012). Among the microorganisms involved in the bioleaching of LIBs, *Acidithiobacillus ferrooxidans*, *Acidithiobacillus thiooxidans*, and *Leptospirillum ferrooxidans* have been extensively studied, particularly in mesophilic conditions. Additionally, certain moderate thermophilic bacteria, such as the sulfur-oxidizing *Acidithiobacillus caldus* and iron-oxidizing *Leptospirillum ferriphilum* have also been investigated in this context. (Roy et al., 2021a). The fundamental reactions of the iron and sulfur-oxidizing microorganisms are as follows:

$$S^0 + H_2O + 1.5O_2 \rightarrow SO_4^{2-} + 2H^+, \tag{5}$$

$$4Fe^{2+} + O_2 + 4H^+ \rightarrow 4Fe^{3+} + 2H_2O. \tag{6}$$

By increasing the temperature of the reaction and utilizing thermophilic microorganisms, it is possible to enhance the kinetics of the process. Thermophilic bacteria exhibit the ability to thrive and sustain metabolic activity at elevated temperatures, with different categories including moderately thermophilic bacteria (40–50 °C), thermophilic bacteria (50–75 °C), and extreme thermophilic bacteria (75–120 °C). One of the most significant disadvantages of bioleaching is its low kinetics which can be solved by utilizing thermophilic bacteria as they increase the bioleaching rate, namely *Acidithiobacillus caldus, Leptospirillum ferriphilum*, and *Sulfobacillus spp*. These particular strains of bacteria are of utmost significance in the process of LIBs bioleaching due to their ability to thrive in moderate temperatures and facilitate the leaching of metals from ores. *Acidithiobacillus caldus*, in particular, is known for its ability to oxidize sulfur and ferrous ions, while *Leptospirillum ferriphilum* is recognized for its capacity to oxidize iron and sulfur compounds. *Sulfobacillus spp.* are also valued for their role in sulfur oxidation. These microorganisms play an essential role in the LIBs bioleaching process by breaking down the ore and allowing for the extraction of valuable metals and *Ferroplasma spp.*, which were used as a mixed culture (Ghassa et al., 2020). Extreme thermophiles, like *Sulfolobus acidocaldarius*, *Sulfolobus solfataricus*, *Sulfolobus brierley*, and *Sulfolobus ambioalous* have been identified from volcanic springs. They can thrive at temperatures of 75–80 °C with pH values of 1–3. Due to their propensity to grow at elevated temperatures, these extreme thermophiles have a higher rate of metal bioleaching than moderate thermophiles and mesophiles. However, no study has been published so far describing the use of extreme thermophiles for LIB bioleaching (Roy et al., 2021a).

3.3 Effective Parameters

3.3.1 pH

The initial pH adjustment significantly affects recovery as the recovery of Li and Co^{2+} are virtually identical to the amount of acid usage. The greater the acid usage, the greater the amount of Li and Co recovered. It was determined that acid consumption is crucial for the leaching of $LiCoO_2$ due to its effect on cell development, and that adjusting the pH considerably enhanced bioleaching performance (Zeng et al., 2012). On the other hand, in the pH range from 2.5 to 4, the high initial pH causes the pH drop to start later, but in the end, in each case, the final pH is between 2 and 2.5 (Mishra et al., 2008). Moreover, some researchers showed that cobalt dissolution had only a weak connection with solution pH (Li et al., 2013), but both Co and Li would dissolve at lower pH (Boxall et al., 2018). The optimal pH range for acidophilic bacteria is 1.5–2.5, and this range has a high rate of oxidation of ferrous salt and elemental sulfur (Boxall et al., 2018; Khatri et al., 2019; Niu et al., 2014).

3.3.2 ORP

To date, there has been no published research on the ideal oxidation–reduction potential (ORP) range required to maximize metal leaching efficiency during bioleaching LIBs. However, it has been observed that cobalt dissolution can be improved by increasing the redox potential. In other words, a higher ORP range could potentially enhance the leaching of cobalt from spent LIBs. While further research is needed to determine the optimal ORP range for maximizing the efficiency of metal leaching during LIB bioleaching, these initial findings suggest that controlling the redox potential could improve the overall yield of valuable metals and minerals during the recycling process (Li et al., 2013). As previously noted, the bioleaching of cobalt from spent LIBs involves a combination of acid dissolution and Fe^{2+} reduction, unlike the leaching process for lithium. As a result, the oxidation and reduction conditions are more likely to impact the dissolution of cobalt. Cobalt exhibits high solubility in situations characterized by high pH levels and reducing conditions. However, its solubility decreases significantly in acidic conditions unless the ORP is simultaneously increased. This indicates that a higher oxidizing environment is necessary to retain cobalt in solution compared with other leaching tests. As such, fine regulation of the ORP range is crucial to ensure optimal cobalt dissolution during LIB bioleaching, and it is an area that requires further exploration and research. Enhancing the solubilization of cobalt through the utilization of combined bioreagents can potentially be achieved by augmenting the concentration of ferric iron oxidant or introducing an additional oxidizing agent to sustain the required oxidation–reduction potential (ORP) conditions conducive to effective cobalt leaching (Boxall et al., 2018).

3.3.3 Bacteria Energy Source

For Li extraction, sulfur and a sulfur-oxidizing bacteria (like *Acidithiobacillus thiooxidans*) are required, indicating that the metal would be released via acid solution caused by biogenic H_2SO_4; however, the mixed energy source which are Fe^{2+} and S, and the maximum dissolving yield for Co, Ni, and Mn could be found in a mixed culture system, demonstrating that these metals mobilize via a combination of Fe^{2+} reduction and acid dissolution (Zeng et al., 2012). Generally, the highest Li recovery is possible with a lower pH and a greater S concentration, and the highest Co recovery is possible with a lower pH and a greater $FeSo_4$ concentration (Boxall et al., 2018). The higher extraction efficiency of Ni, Co, and Mn could be obtained because of stronger reduction caused by highly concentrated Fe^{2+} (Wang et al., 2018). It should also be noted that, especially at pH higher than 2, the metal dissolution could become slower as the Fe^{2+} ion concentration increases. Increased Fe^{2+} concentrations can decrease solubility due to Fe^{3+} co-precipitation with the metals in the residues (Mishra et al., 2008). The impact of Fe^{2+} on moderate thermophilic acidophilic microorganisms would be different as these cultures have a limited ability to oxidize Fe^{2+} compared with mesophilic acidophilic cultures. The highest ORP values could be seen in low ferrous ion concentration tests. Due to the depleted ferrous ion concentration, practically all Fe^{2+} ions would be oxidized to Fe^{3+}. Cobalt recoveries with these cultures were the same as control tests, indicating that bacteria did not affect Co dissolving in the presence of ferrous ions. As mentioned, ferrous salt improves bioleaching dissolution. However, to avoid the use of this chemical and to reduce the expense of the process, iron sulfate heptahydrate can be substituted with iron scrap. According to Eq. 7, iron is dissolved in sulfuric acid, forming $FeSO_4$, which enhances Co dissolution according to Eq. 2.

$$Fe + H_2SO_4 + 0.5\,O_2 \rightarrow FeSO_4 + H_2O \tag{7}$$

Iron scrap boosts Li recovery by eliminating the Co layer from the surface of the particles and enhancing acid diffusion. Therefore, it may be concluded that $FeSO_4.7H_2O$ can be substituted for iron scrap. Nonetheless, this modification lengthens the procedure. Although iron scrap addition does not influence Li or Ni recovery, it can significantly boost cobalt recovery (Ghassa et al., 2020).

3.3.4 Temperature

Temperature is an important parameter because although it does not significantly affect the amount of ΔG, it affects the growth and activity of bacteria. For each culture, an optimum temperature should be found and optimized, e.g., for mixed cultures of *Alicyclobacillus spp.* and *Sulfobacillus sp.*, increasing the temperature from 30 to 35 °C boosts their metabolism, but further increasing it to 40 °C reduces their activity, resulting in a decrease in efficacy (Niu et al., 2014).

3.3.5 Pulp Density

One of the primary drawbacks of LIBs bioleaching technologies is the drastic decline in the recovery with pulp densities higher than 1%. High toxicity linked with significant metal release dosages or deteriorated factors such as ORP or pH, which injured the growth and activity of bacteria, were frequently the underlying causes of a decrease in bioleaching efficiency (Xin et al., 2012a). Numerous issues arise as a result of the high solid percentage, which is that because of the increased metal toxicity, limited air incursion due to the increased viscosity, and decreased dissolved oxygen, the recovery of bioleaching would remain constant in a specific pulp density (Roy et al., 2021c).

4 Summary of Previous Studies

Mishra et al. studied the bioleaching of LIBs using *Acidithiobacillus ferrooxidans*. Cobalt bio-dissolution was shown to be faster than lithium. The Fe^{2+} ion was used in the leaching tests to promote cell development in the lixiviant, but the metal dissolution became slower as the Fe^{2+} ion concentration increased. Increased Fe^{2+} concentrations decreased solubility due to Fe^{3+} co-precipitation with the metals in the remaining. Furthermore, the alteration in the solid-to-liquid ratio (w/v) exerted an influence on the dissolution of metals, causing a cessation in cell growth due to the heightened metal concentration present within the waste sample (Mishra et al., 2008).

In a study conducted by Zeng et al., the bioleaching extraction of valuable metals from three frequently discarded electric vehicle LIBs cathodes was examined at a pulp density of 1%. The *Acidithiobacillus thiooxidans* bacteria demonstrated the highest extraction efficiency for Li, suggesting that the metal was mobilized through acid solution caused by biogenic H_2SO_4. However, it was observed that the dissolution rate of Co, Ni, and Mn was most significant in the mixed energy source-mixed culture system, indicating that the mobilization of these metals resulted from a synergistic interplay between Fe^{2+} reduction and acid dissolution mechanisms. In addition, Li extraction was accomplished using a non-contact technique, but Co, Ni, and Mn mobilization needed contact between the cathodes and cells. The extraction efficiency of four significant metals from the resistive $LiNi_xCo_yMn_{1-x-y}O_2$ compound surpassed a remarkable 95% on average. The relatively elevated extraction rates of the valuable metals imply that cost-effective autotrophic bioleaching can be employed to retrieve used EV LIBs (Zeng et al., 2012).

In a study conducted by Li et al., the impact of solution pH and redox potential on the bioleaching of LIBs was investigated using *Acidithiobacillus ferrooxidans*. The researchers conducted a series of bioleaching tests with varying starting pH and ferrous ion concentrations to assess their effects on the bioleaching process. These findings provide valuable insights into the factors that significantly impact the efficiency of metals dissolution during LIB bioleaching and can help guide the development of efficient recycling processes for spent batteries. Nonetheless, cobalt dissolution was enhanced at greater redox potentials. Their cyclic voltammograms study revealed that dissolution rates increase above 0.4 V and rapidly decline below 1.3 V (Li et al., 2013). Boxall et al. also examined the parameters affecting the bioleaching of LIBs. They could recover 99.2% of Li, 50.4% of Co, and 89.4% of Ni at pH = 1.5, $FeSO_4 = 36.7$ g/L, and $S^0 = 5$ g/L as the optimal condition with a mixture of *Acidithiobacillus ferrooxidans* and *Acidithiobacillus thiooxidans*. Generally, the highest Li recovery was possible with a lower pH and a greater S concentration, and the highest Co recovery was possible with a lower pH and a greater $FeSO_4$ concentration (Boxall et al., 2018).

In a recent study, Wang et al. discovered that an extracellular polymeric substance (EPS) containing polysaccharides, proteins, and lipids can play a crucial role in the adhesion of LIBs and cells. The EPS facilitated the adhesion of cells to the batteries through minor electrostatic and dominant hydrophobic forces. Moreover, the presence of EPS resulted in the concurrent adsorption of Fe^{3+} and Fe^{2+} to amounts of 0.9 g/g and 1.6 g/g, respectively. This resulted in a high density of these ions bonded with other molecules. The enrichment of Fe^{3+} facilitated the attachment of cells to the cathodes, with the EPS serving as an intermediary between them. The higher extraction efficiency of Ni, Co, and Mn was attributed to the stronger reduction caused by highly concentrated Fe^{2+}. Furthermore, the EPS increased the electronic potential while reducing the electronic resistance, resulting in superior electron transfer from Fe^{2+}. These findings provide valuable insights into the role of EPS in the bioleaching process and can guide the development of more efficient and effective methods for the recycling of spent LIBs (Wang et al., 2018).

Hartono et al. tried to research the possibility of using bacteria strains which can reach 62.83% of lithium recovery with 2 mg/mL pulp density, initial pH of 7, the temperature of 30 °C, stirring rate of 120 rpm, and battery/soil mass ratio of 100 g/100 g. To improve their results, they suggested an adaptation environment with LiCl solution (Hartono et al., 2017). In another study that focused on the function of local bacteria, Cai et al. tried to present two bioleaching consortia that were grown in neutral activated sludge for 60 days and were characterized as phylogenetically distinct from documented bioleaching bacteria which could recover 69.46% of Li and 67.6% of Mn in 7 days. Three mixotrophs and two chemoautotrophs, three of which were unique *Sulfobacillus* and *Leptospirillum* species, were domesticated for a further 100 days utilizing cathodic materials. The findings unveiled the presence of novel bioleaching bacteria that exhibited a reduced leaching cycle and enhanced resistance to organic compounds when enriched from non-acidic environments. These results suggest a significant prospect for the recovery of metals from used LIBs or similar organic-rich settings (Cai et al., 2021).

Niu et al. investigated the bioleaching behavior of spent at pulp densities and its thermodynamics and kinetics ranging from 1 to 4%. It was found that bioleaching processes have a promising capability. It has an enormous negative value of ΔG, 12.7–11.4 times greater than leaching using $FeSO_4/H_2O_2$ as reactants, the product layer diffusion model provided the most accurate description of the bioleaching behavior of Co and Li. Temperature is a critical parameter in the bioleaching process, as it impacts the growth and activity of bacteria, although it does not significantly affect the amount of ΔG. Increasing the temperature from 30 to 35 °C enhances the effectiveness of the process. However, increasing the temperature further to 40 °C can hinder bacterial activity and decrease the efficacy of the bioleaching process. The results obtained from the study indicated that there was a decrease in the bioleaching efficiency for cobalt from 52 to 10% and for lithium from 80 to 37% as the pulp density increased from 1 to 4%. Nevertheless, by implementing process modifications such as the adjustment of media acidity to pH 2.0, the utilization of mixed energy substrates and temperature maintenance at 35 °C, a maximum extraction efficiency of 89% for Li and 72% for Co could be achieved at 2% pulp density. These findings highlight the significance of optimizing process parameters to achieve optimal bioleaching performance and improve the overall efficiency of the recycling process for spent LIBs (Niu et al., 2014).

In addition to this, Jegan Roy et al. concentrated their efforts on improving the efficiency of bioleaching used LIBs that included nickel, manganese, and cobalt (NMC). They were able to improve bioleaching in a similar manner by altering the amount of H_2SO_4 and ferric ions in the *Acidithiobacillus ferrooxidans* cultures, in addition to providing a cyclic technique. This allowed them to extract more metals from the waste. The strategy that was suggested comprised supplementing the powder with new bacterial culture each time the leached liquid was changed after a period of twenty-four hours. As a result of the increased metal toxicity, reduced dissolved oxygen, and limited air intrusion resulting from the heightened viscosity, the bioleaching process was conducted at a specific pulp density. This decision was based on the notion that the recovery of bioleaching would remain stable at a particular pulp density. After 72 h of processing at a high pulp density of 100 g/L, researchers were able to achieve maximum recoveries of 90% for Ni, 92% for Mn, 82% for Co, and 89% for Li. These findings emphasize the significance of achieving an optimal pulp density in the bioleaching procedure to enhance the retrieval of valuable metals and minerals from used LIBs. The findings hold promise for facilitating the creation of recycling techniques that are more productive and proficient in managing expended batteries, leading to a reduction in the ecological consequences of battery waste and promoting the principles of a circular economy (Roy et al., 2021c). In a work quite similar to this one, Roy et al. investigated the use of *Acidithiobacillus ferrooxidans* for the purpose of bioleaching a mixture of $LiCoO_2$-based LIBs at a high pulp density. They showed that the recovery of bioleaching would improve if they were to include $FeSO_4$ in the nutrient media and then increase its concentration there up to 150 g/L. The effectiveness of the bioleaching process was significantly improved due to several factors, including the replenishment of bacterial culture three times and the presence of a high concentration of biogenic H_2SO_4 (0.52 M) and Fe^{3+} (36.86 g/L) in the culture. Analysis revealed that after 72 h of bioleaching, utilizing a pulp density of 100 g/L, the maximum recovery of 94% for Co and 60% for Li could be achieved. These results highlight the importance of optimizing the bioleaching process by adjusting critical parameters such as the concentration of sulfuric acid and ferrous ions and the duration of the process (Roy et al., 2021b).

In order to circumvent the limitations caused by the inhibition of microorganisms at low pulp densities and the scarcity of substrates, Boxall et al. investigated the feasibility of utilizing a non-contact indirect bioleaching approach to extract valuable metals from LIBs. Their objective was to achieve high leaching yields while operating at room temperature. However, when the number of leaching stages was raised to four within an hour, there was a significant enhancement in the leaching yields of all metals. These findings offer a substantial basis for the establishment of practical leaching techniques for battery waste that require less usage of inorganic acid in the future. These outcomes may enable the development of more sustainable and efficient approaches to recycle valuable metals from spent batteries, ultimately reducing the environmental impact of battery waste. The leach yields achieved in this study are much lower compared with the leach yields produced in typical hydrometallurgical procedures. The highest leaching efficiency was reported in trials utilizing strong inorganic acids such as sulfuric and hydrochloric acids. However, there would be significant energy savings associated with leaching at an ambient temperature, as well as the avoidance of releasing dangerous compounds to the environment, which would result in wastes requiring additional processing farther downstream before they could be disposed of Boxall et al. (2018).

The bio-electro-hydrometallurgical platform shows efficient and sustainable method for recycling of spent LIBs. By integrating electrokinetics, bioleaching, and selective adsorption, the researchers were able to achieve an optimal separation and purification of Co, Li, and Mn. The use of PC-88A/TOA-modified granular activated carbon as an adsorbent was particularly effective in selectively capturing the target metals. The study's detailed analysis of the factors affecting the dissolution of active cathode materials provides valuable insights into the recycling process, which can be optimized by adjusting the mass ratio of cathode active materials to $S+FeS_2$, the ratio of total solids cathode active materials$+S+FeS_2$ to medium, processing duration, and Fe^{+3} concentration. The electrokinetic device illustrated in Fig. 3 provides a clearer understanding of the platform's design and operation.

Domestication studies indicated the possible sulfur-oxidizing bacteria strain's adaptability to the rigorous electrokinetics environment, even though a high pulp slowed microbe growth and metabolism. The maximum metals recovery was obtained using a 15% mass ratio, a 40% S-L ratio, a 12-day processing duration, and a 0.2 M ferric ion concentration. The highest recovery values for Co, Li, and Mn were 91.45%, 93.64%, and 87.92%, respectively. The

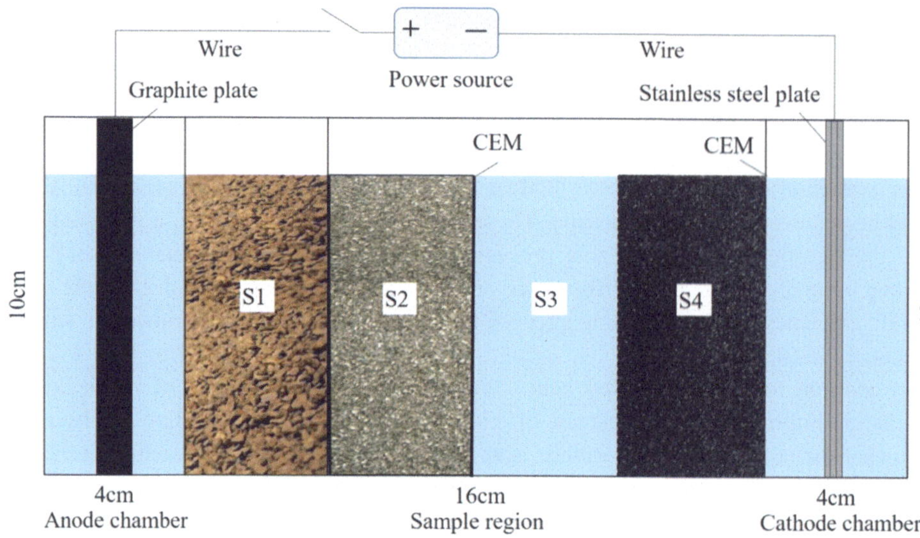

Fig. 3 Bio-electro-hydrometallurgical device, featuring four sections denoted as S1 (haematite-stacking area), S2 (bioleaching area), S3 (buffer area), and S4 (GAC-stacking zone) (Xin et al., 2016)

relative proportions of cathode active materials to $S + FeS_2$ were found to significantly impact the overall dissolution of cathode active materials. The study revealed that a bio-electrokinetic system, combining bioleaching, selective adsorption using granular activated carbon, and electrokinetics, proved effective in the Co, Mn, Li recovery from the cathodic active materials of spent LIBs (Huang et al., 2019a).

Khatri et al. compared the efficiency of hydro and bio-hydrometallurgical procedures to extract multiple metals from LIBs. The study also examined the impact of cell presence and absence at higher pulp densities. The study findings demonstrated that the modified acidophilic iron-oxidizing consortium displayed the highest rates of metal extraction when the process conditions were optimized. These optimal conditions involved a pH level of 2, a two-step bioleaching process using Leptospirillum ferriphilum, a ferrous iron concentration of 9 g/L (Khatri et al., 2019) (Table 3).

5 Bioleaching of Waste LIBs with Heterotrophic Bacteria and Fungi

Heterotrophic microorganisms have been extensively studied and utilized in the spent LIBs recycling due to their ability to breakdown and consume organic matter for energy. While both heterotrophic bacteria and fungi can generate organic acids and aid in the bioleaching process, fungi have been predominantly favored in the bioleaching of spent LIBs (Alavi et al., 2021; Bahaloo-Horeh & Mousavi, 2017; Bahaloo-Horeh et al., 2018; Biswal et al., 2018; Horeh et al., 2016). Fungi have been recognized for their effectiveness in bioleaching owing to their shorter lag phase and faster leaching rate. In addition, fungi have a greater ability to tolerate hazardous compounds, can thrive in both alkaline and acid-consuming environments, and can excrete valuable metabolites such as EPS, proteins, exo-polysaccharides, organic acids, and complexing agents like siderophores. These metabolites can be effectively utilized to solubilize metals from various sources. The utilization of fungi presents a sustainable and promising alternative to conventional bioleaching methods and can potentially pave the way for the development of more sustainable and efficient recycling processes for spent LIBs. By connecting the potential of fungi and their metabolites, we can create a more eco-friendly and resource-efficient approach to metal recovery and promote the circular economy (Pollmann et al., 2016). Fungi naturally secrete organic acids, which can play a critical role in the chelation of metal ions. Chelation is a process in which a metal ion forms a complex with a ligand, which is a molecule capable of binding to the metal ion. In the context of bioleaching, organic acids act as ligands, binding to metal ions and making them more soluble and accessible for extraction by the microorganisms (Wu & Ting, 2006). Fungi, such as *Penicillium simplicissimum*, *Penicillium chrysogenum*, and *Aspergillus niger* have been used to recover heavy metals from different sources (Deng et al., 2013; Faraji et al., 2018; Harwood et al., 2017; Horeh et al., 2016; Ku et al., 2016; Xia et al., 2018). *Aspergillus niger* has demonstrated high performance in the bioleaching of LIBs among fungal species because of its ability to grow even in highly alkaline media and produce numerous chelating agents and organic acids (Akcil et al., 2015).

Bahaloo-Horeh et al. utilized *Aspergillus niger* fungus to recover valuable metals from spent LIBs. The researchers explored several bioleaching techniques, included one-step, two-step, and spent medium bioleaching. The findings indicated that spent medium was the most effective

Table 3 Summary of LIB bioleaching with autotrophic microorganisms

Culture	Culture media	Temperature and initial pH	Pulp density	Method	Bioleaching efficiency	Description	Refs.
Sulfur-oxidizing bacteria obtained from acidic drainage in an abandoned mine pit in Yunnan Province	9K without iron sulfide	pH = 2,3,4, and 5	Pyrite was mixed with medium to prepare 5%, 10%, 20%, and 30% (wt%) pyrite pulp	1-step	Co = 91.45% Li = 93.64% Mn = 87.92%	The ideal condition for the process involves a mass ratio of 15%, an S/L ratio of 40%, a processing time of 12 days, and a ferric concentration of 0.2 M	Huang et al. (2019a)
Alicyclobacillus sp./ Sulfobacillus sp.	Basic medium	pH = 1–2 35–40 °C	1%, 2%, and 4%	1-step	Li = 89% Co = 72%	Mixed energy substrates = 4.0 g/L	Niu et al. (2014)
Divergent bacteria consortium from neutrally activated sludge	9K	pH = 2.0	–	1-step	Li = 69.46% Mn = 67.60%	The optimal conditions for the experiment involved the addition of cathodic materials (NCM523 and $LiMn_2O_4$) into individual fresh media at a final concentration of 780.0 mg/L for NCM523 and 164.5 mg/L for $LiMn_2O_4$	Cai et al. (2021)
Acidithiobacillus caldus, Leptospirillum ferriphilum, Sulfobacillus spp., and *Ferroplasma spp*	9k	45 °C pH = 1.8	10 g/L	1-step	Co = 99.9% Ni = 99.7% Li = 84%	The bioleaching effectiveness of the test conducted with scrap iron is almost identical to that of the test conducted with iron sulfate heptahydrate. However, the period needed for leaching increased from two to six days	Ghassa et al. (2020)
Acidithiobacillus ferrooxidans	Modified 9K	pH = 2.0 30 °C	Starting from 5 g/L to 100 g/L	2-step	Co = 94% Li = 60%	Best metals recovery achieved at a pulp density of 100 g/L	Roy et al. (2021b)
Local source microorganisms	3 g/l of meat extract, 5 g/l of peptone, and 5 g/l of NaCl	pH = 7 30 °C	2,3,4 mg/mL	1-step	Li = 62.83%	Soil/battery mass ratio = 100 g/100 g Pulp density = 2 mg/mL	Hartono et al. (2017)
Acidithiobacillus ferrooxidans	9K	pH = 1.5 35 °C	1%	2-step	Co = 47.6%	Highest cobalt dissolution percentage of 48.2% was achieved at the end of experiments when initial Fe2 + concentration = 45 g/L	Li et al. (2013)
Acidithiobacillus ferrooxidans	9K	pH = 2.9 35 °C	1%	2-step	Co-recovery in the presence of Ag = 98.4% absence = 43.1%	$AgNO_3$ = 0.02 g/L	Zeng et al. (2013a)
Acidithiobacillus thiooxidans, Acidithiobacillus ferrooxidans, and *Leptospirillum ferriphilum*	Mineral salt medium	30 °C	4%	Spent medium	Li = 100% Ni = about 60% Co = about 43% Mn = about 45%	Best result: with leachate + Intact cell + EPS;	Wang et al. (2018)

(continued)

Table 3 (continued)

Culture	Culture media	Temperature and initial pH	Pulp density	Method	Bioleaching efficiency	Description	Refs.
Leptospirillum ferriphilum-dominated consortium	Medium 9K	pH = 2.0 31 ± 3 °C	10 g/L	2-step	Li = about 22% Co = about 60% Ni = about 78% Zn = about 70% Cu = about 55%	Ferrous iron = 9 g/L1	Khatri et al. (2019)
Acidithiobacillus thiooxidans and *Leptospirillum ferriphilum*	Basic medium	30 °C pH = 1	1% pulp density	1-step	LiFeO4 (Li = about 100%) $LiMn_2O_4$ (Li and Mn = about 100%) $LiNi_xCo_yMn_{1-x-y}O_2$ (Li, Mn, Ni, and Co = more than 95%, with adjusted pH at 1)	16.0 g/L mixed energy source (elemental sulfur: pyrite at 1:1 by weight)	Xin et al. (2016)
Acidithiobacillus ferrooxidans	Iron-free 9K medium	pH = 2.5 30 °C	0.5 g/L	1-step	Li = about 10% Co = about 70%	Fe^{2+} = 3 g/L and elemental sulfur = 1%	Mishra et al. (2008)
Sulfur-oxidizing and iron-oxidizing bacteria	The basic medium	30 °C pH = 1	1%	1-step and 2-step	Cobalt dissolution = about 1400 mg/L Li dissolution = 300 mg/L	Fe^{2+}–H_2SO_4 combined system	Xin et al. (2009)
Acidithiobacillus ferrooxidans	Modified 9K media based on the basal salts	pH = 2 30 °C	100 g/L	2-step	Ni = 90% Mn = 92% Co = 82% Li = 89%	During the bioleaching process, the bacterial culture was supplemented for a total of three cycles	Roy et al. (2021c)
Acidithiobacillus ferrooxidans and Acidithiobacillus thiooxidans	Basal salts medium	25 °C pH = 1.8	10%	Spent medium	Co = 53.2% Li = 60% Ni = 48.7% Mn = 81.8% Cu = 74.4%	In the leaching process, the utilization of both biogenic iron and biogenic acid in sequential batch leaching was employed in multiple stages, specifically four sequential leaching stages, each lasting one hour	Boxall et al. (2018)
Acidithiobacillus *ferrooxidans and Acidithiobacillus thiooxidans*	9K	pH = 1.5 32 °C	40 g/L	2-step	Li = 99.2% Co = 50.4% Ni = 89.4%	Iron sulfate concentrate = 36.7 g/L sulfur concentration = 5g/L and initial pH of 1.5 for the best inoculum ratio of 3/2 (A.f/A.t)	Heydarian et al. (2018)
Acidithiobacillus ferrooxidans	9K	35 °C pH = 2.9	1%	2-step	Co = 99.9% 43.1% (in the absence of copper)	Maximum Co recovery was achieved in the presence of 0.75 g/L copper	Zeng et al. (2012)

(continued)

bioleaching method for maximizing metal recovery from spent LIBs. At a pulp density of 1%, the maximum recovery efficiency in spent medium bioleaching was Cu 95%, Li 70%, Mn 65%, Al 45%, Co 45%, and Ni 38%, respectively. Citric acid was found to play a crucial role in the bioleaching efficiency of *Aspergillus niger*, surpassing other organic acids such as gluconic, oxalic, and malic acid. The findings can help guide the development of more efficient and sustainable recycling processes for spent batteries, which can promote the cost-effective process and reduce the environmental impact of battery waste. Furthermore, bioleaching was more effective at removing heavy metals than chemical leaching. The authors proved that the bio-hydrometallurgical route is more effective for recovering heavy metals from spent LIBs (Horeh et al., 2016). Another study, Biswal and colleagues recycled spent medium bioleaching with *Aspergillus niger* SG1 and MM1 strains at a pulp density of 0.25% (w/v) to remove Co and Li from spent LIBs. Their findings showed that Aspergillus niger strain MM1 was highly effective in dissolving Co (82%) and Li (100%) during the bioleaching of LIBs (Biswal et al., 2018). The investigation also uncovered that *Aspergillus niger's* adaptation to heavy metals resulted in an enhancement in the production of organic acids and an increase in metal leaching efficiency when comparing adapted fungi to unadapted fungi. These results emphasize the promising potential of *Aspergillus niger* and spent medium bioleaching as an effective and sustainable approach for spent LIBs recycling. By considering the unique capabilities of *Aspergillus niger* and optimizing the bioleaching process, we can create a more efficient and sustainable approach to metal recovery, which can help promote the circular economy and reduce the environmental impact of battery waste (Bahaloo-Horeh et al., 2018). A study conducted by Alavi et al. analyzed the effectiveness of one-step bioleaching, two-step bioleaching, and spent medium bioleaching methods for the recycling of metals from LIBs, utilizing a mixed fungal culture of *Aspergillus niger* and *Aspergillus tubingensis*. According to the findings, the spent medium bioleaching method proved to be the most. The investigation also demonstrated that oxalic acid was released in higher quantities than citric acid, although citric acid played a critical role in the bioleaching process. These results emphasize the importance of understanding the role of different organic acids and other compounds in the bioleaching process. By optimizing the presence and concentration of these compounds, we can improve the efficiency and effectiveness of the bioleaching process (Alavi et al., 2021). The outcomes of this study demonstrated the potential of *Aspergillus niger* and other fungal cultures in spent LIBs recycling. By utilizing spent medium bioleaching and optimizing organic acid secretion, particularly citric acid, we can develop more efficient and sustainable methods for metal recovery. These findings can potentially foster the development of innovative and effective solutions for reducing the ecological impact of battery waste. The use of fungal-based bioleaching methods can help promote a circular economy and minimize the environmental harm caused by metal-containing waste.

5.1 Mechanisms

Heterotrophic microorganisms are known for their remarkable capacity to produce organic acids and chelating compounds, which play a significant role in metal bioleaching. Acidolysis, complexolysis, bioaccumulation, redoxolysis, and biosorption have been reported as the main mechanisms by which organic acids react with metals through bioleaching with heterotrophic cultures (Le et al., 2006; Simate et al., 2010). Organic substances are used as an energy source by heterotrophic bacteria and fungus through their metabolism. Fungi secretes numerous organic acids (gluconic, citric, oxalic, etc.) while growing on organic supplements (Sierra-Alvarez, 2007; Xu et al., 2014). The process of transforming glucose or sucrose into citric acid consists of a sequence of enzymatic reactions that occur in two separate cellular compartments, namely the cytosol and the mitochondrion. The first step involves the transportation of glucose through the glycolysis pathway into the cytosol, where it is transformed into pyruvate (as shown in Fig. 4). The pyruvate molecule undergoes two distinct fates in the citric acid production process. One pyruvate molecule undergoes decarboxylation to yield acetyl-CoA via the mitochondrial pyruvate dehydrogenase complex. Meanwhile, the other pyruvate molecule is carboxylated to form oxalo-acetic acid by pyruvate carboxylase in the cytosol. Afterward, the oxalo-acetic acid is transported to the mitochondrion, facilitated by malate and reacts with acetyl-CoA to generate citric acid. This intricate process demonstrates the complex biochemical pathways involved in the conversion of glucose or sucrose to citric acid. Understanding these pathways and their regulation can potentially aid in the development of more efficient and sustainable methods for citric acid production. The final product is then transported out of the mitochondrion and eventually exits the cell. Oxalo-acetase, an enzyme that catalyzes the hydrolysis of oxalo-acetic acid to oxalic acid and acetic acid, can produce oxalic acid from oxalo-acetic acid. Overall, the production of citric acid from glucose or sucrose involves a complex series of biochemical reactions that occur in specific cellular compartments requiring the coordinated action of multiple enzymes (Magnuson & Lasure, 2004). Throughout the growth phase of fungi, an array of organic acids including citric acid, lactic acid, gluconic acid, and oxalic acid, as well as enzymes, are secreted. These

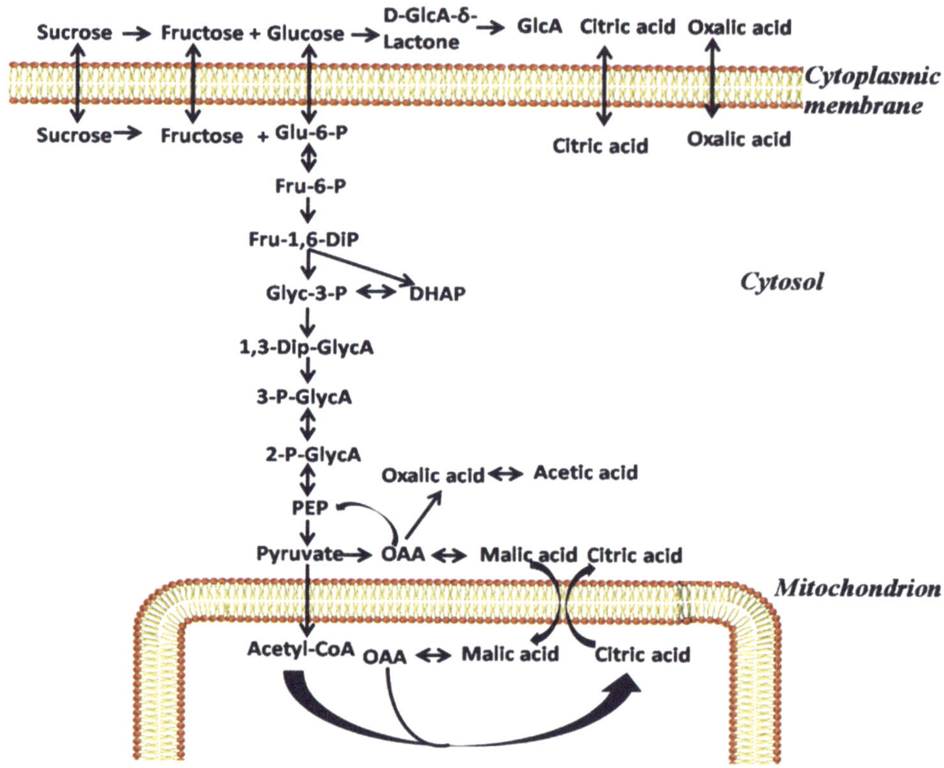

Fig. 4 Production of organic acids including gluconic acid, citric acid, and oxalic acid, from glucose or sucrose (Srichandan et al., 2019)

organic acids serve as bio-lixiviants, as supported by the following evidence (Xia et al., 2018), and are utilized for the oxidation of metals from spent LIBs via the O_2/H_2O redox pair. Following this, the metal ions are protonated and subsequently complexed with the organic acids. The protonation lead to the release and movement of free metal cations that are generated through acidolysis, resulting in the dissolution of metals at an acidic pH (as seen in Eqs. 8–12). This process highlights the crucial role of organic acids in bioleaching and their potential to be harnessed for metal recovery from various sources. By optimizing the bioleaching process and the presence of these organic acids, we can develop more efficient and sustainable methods for metal recovery from spent LIBs, which can contribute to the circular economy and reduce the environmental impact of battery waste (Bahaloo-Horeh & Mousavi, 2017; Bahaloo-Horeh et al., 2018; Biswal et al., 2018). Further research in this area can potentially lead to the development of innovative and effective solutions for sustainable resource management and metal recovery. The principal mechanism of fungal bioleaching is complexolysis, wherein metal cations interact with organic acid anions to generate complexes.

The *Aspergillus niger* and *Penicillium simplicissimum* fungi were found to be the most effective strains to bioleach metals from spent LIBs by creating a variety of organic acids as metabolites in sucrose ($C_{12}H_{22}O_{11}$) medium.

$$C_{12}H_{22}O_{11} + H_2O \rightarrow 2C_6H_{12}O_6, \tag{8}$$

$$\begin{aligned} &C_6H_{12}O_{11}(\text{Glucose}) + 4.5O_2 \\ &\rightarrow 3C_2H_2O_4(\text{Oxalic} - \text{acid}) + 3H_2O, \end{aligned} \tag{9}$$

$$C_2H_2O_4(\text{Oxalic} - \text{acid}) \rightarrow (C_2HO_4)^- + H^+, \tag{10}$$

$$(C_2HO_4)^- + Me^+(\text{metal}) \rightarrow Me(C_2HO_4)^-, \tag{11}$$

$$\begin{aligned} &7C_2H_2O_4 + 2LiCoO_2 \rightarrow 2LiC_2HO_4 + 2Co(HC_2O_4)_2 \\ &+ 4H_2O + 2CO_2. \end{aligned} \tag{12}$$

5.1.1 Acidolysis

Acidolysis is a widely utilized fast leaching method for fungi and other heterotrophic microorganisms. The leaching process involving fungi is an indirect method, whereby the oxygen atoms of the metal compound are protonated by the organic acids secreted by the fungi. This is exemplified by Eq. 13, where protons attach to the solid surface and react. The detachment of metal from the solid surface is facilitated by the interaction between oxygen, protons, and water. This process highlights the important role of organic acids and the protonation process in the bioleaching of metals from various sources. The potential of organic acid-based bioleaching methods can help create a more sustainable and environmentally friendly approach to metal recovery and resource management (Srichandan et al., 2019).

$$MeO + 2H^+ \rightarrow Me^{2+} + H_2O \tag{13}$$

MeO is a metal oxide such as NiO. Low pH has been shown to benefit acidolysis by weakening chemical bonds in the waste matrix, allowing metals to dissolve more quickly (Bahaloo-Horeh et al., 2018; Biswal et al., 2018).

5.1.2 Complexolysis

Complexolysis is a process that involves the mobilization of metal ions through the acidolysis mechanism, which is stabilized by complexation with organic acids and other chelating agents. Through complexolysis, metals are solubilized and mobilized in a form that is readily available for extraction. Fungi release organic acids and amino acids, playing an essential role in complexolysis (Srichandan et al., 2019). Heterotrophic microorganisms such as *Actinomycetes* and fungi can increase the solubility of metals by forming metal-complexing ligands in reaction with organic acids and amino acids, although fungi restricted amino acid excretion (Srichandan et al., 2019). The molecules of organic acids possess the unique capability to generate intricate metal chelators, which assist in the dissolution of metal ions. Chelators possess the capability of generating stronger bonds with metal ions when compared with the lattice bonds that are formed between solid particles and metal ions. This attribute contributes to an enhancement of the bioleaching process, as it facilitates the solubilization of the metal ions (Islam et al., 2020). When metal ions interact with organic ligands, they can form durable complexes. The toxicity and stability of chelating agents in a solution are greatly affected by the specific organic ligands and metal ions involved in the process. Recent studies have revealed that certain types of bacteria and fungi possess the capacity to excrete iron-chelating compounds with low molecular weight, which serve as the principal agent for dissolving Iron (III) (Islam et al., 2020). Additionally, studies have revealed that amino acids and organic acids produced by fungi secrete protons that augment the capacity of metal solubilization complexing. These mechanisms contribute to the efficient bioleaching of metals and highlight the potential of organic acid molecules in the development of sustainable and eco-friendly methods for metal recovery. Equations (14) and (15) illustrate a complexolysis and chelation reactions involving metal ions with citric acid (Srivastava et al., 2020):

$$Me^{+2} + C_6H_8O_7 \rightarrow Me(C_6H_5O_7)^- + 3H^+, \tag{14}$$

$$Ni^{+2} + C_6H_8O_7 \rightarrow Ni(C_6H_8O_7)^- + 3H^+. \tag{15}$$

Under weaker acidic conditions, the complexolysis process is the main mechanism. Complexolysis has two essential functions (Brandl & Faramarzi, 2006):

- Improve the solubility of metal ions that have already been solubilized by acidolysis (with the exception of metal-oxalate complexes such as Ni oxalate, which have low solubility),
- Facilitate the removal of metal species from the surface by polarizing the critical bonds through ligand exchange.

5.1.3 Redoxolysis

Fungi use the redoxolysis mechanism, to increase metals' mobility through microbial oxidations and reduction processes to obtain energy from minerals (Mishra & Rhee, 2014). In fungal leaching of manganese, solubilization happens due to enzymatic reduction of oxidized manganese, as indicated in Eq. (16) (Asghari et al., 2013).

$$MnO_2 + 2e^- + 4H^+ \rightarrow Mn^{+2} + 2H_2O \tag{16}$$

5.1.4 Bioaccumulation

The bioaccumulation mechanism happens when the soluble metal ions are transported through the cell membrane, causing solid particles to accumulate or precipitate in vacuoles (Brandl et al., 1999). Fungi possess a cell wall that harbors numerous functional groups, such as hydroxyl, carboxyl, amino, phosphate, and sulfate groups, which have the potential to bind to metal ions. This feature of fungi opens up the possibility of using them in the bioleaching process for the recovery of metals from spent batteries, as they can act as efficient biosorbents for metal ions. By exploiting the unique properties of fungi, it may be possible to develop more sustainable and economical methods for metal extraction and recycling (Kapoor & Viraraghavan, 1995). The bioaccumulation of metal ions in the mycelia of fungi is facilitated through active metabolic reactions and passive adsorption (Dusengemungu et al., 2021). Certain fungi, such as *Penicillium and Aspergillus* have been found to possess a high ability to accumulate metal ions (Asghari et al., 2013). Arya and Kumar (2020) observed that *Aspergillus niger* was able to bioaccumulate 77% of the total solubilized lithium into its biomass during lithium recovery (Arya & Kumar, 2020). This finding suggests that the bioaccumulation of Li is a significant mechanism for its recovery. This process promotes the solubilization of Li^+ cations that amass within the cells, improving the equilibrium and ultimately resulting in a higher efficiency of lithium dissolution. The bioaccumulation of metal ions in fungi provides a sustainable and eco-friendly solution for metal recovery, as it reduces the need for complex and costly chemical processes. These results highlight the potential of fungi-based approaches for LIBs recycling. Further research in this area can lead to the development of innovative and sustainable methods for metal recovery, contributing to the promotion of circular economy principles and reducing the environmental impact of metal waste. The recovery of metals from fungal biomass would be a future research challenge. Examples of precipitation reactions are given in Eqs. (17) and (18).

$$Me^{2+} + C_2H_2O_4(\text{Oxalic} - \text{acid}) \rightarrow Me(C_2O_4)(s) + 2H^+, \tag{17}$$

$$Me^{2+} + SO_4^{2-} \rightarrow MeSO_4(s). \quad (18)$$

Me^{2+} is a metal cation (Dusengemungu et al., 2021).

5.1.5 Biosorption

Studies have shown that biosorption is a phenomenon that occurs during fungal bioleaching (Dusengemungu et al., 2020; Iram et al., 2015; Ong et al., 2017). The process consists of the dissolution of metal from the leached material by the organic acids generated by fungi. Metal ions present in the leaching solution are adsorbed by the fungal biomass, which results in a reduction of the amount of metal present in the solution. The biosorption process involves various reactions, including ion exchange, complexation, adsorption, and precipitation. It is sometimes described as an independent metabolism accumulation of metals.

The effectiveness of metal biosorption can be influenced by a number of factors, such as the type and amount of biosorbent used, the surface area of the biomass, and various physicochemical factors like pH, temperature, and ion concentration (Işıldar et al., 2019). Biosorption presents a promising approach for the removal of metals from waste streams, contributing to the promotion of circular economy principles and reducing the environmental impact of metal waste.

5.2 Parameters Affecting Heterotrophic Bioleaching

Heterotrophic microorganisms like fungi excrete organic acids that play an essential role during bioleaching. However, numerous operational characteristics such as growth, medium pH, energy source, substrate concentration, etc., significantly impact the type and amount of organic acid produced by heterotrophic microorganisms. As a result, parameter optimization is critical because maximal leaching occurs when the settings are tuned for fungus growth. Table 4 has detailed different parameters' effects on bioleaching of spent LIBs with heterotrophic microorganisms.

5.2.1 PH

During fungal growth, pH plays an important role. First, germination from conidiophores occurs at a higher initial pH (>5). During this phase, any changes in pH could be harmful. Due to organic acids and ammonium formation, the medium's pH tends to reduce (2.5) after germination (Moh et al., 2005). The optimal pH value of *Aspergillus niger* is around 5.0 in bioleaching of spent LIBs. In the bioleaching process, the pH rises in proportion to the pulp density (S/L) due to the alkaline nature of LIBs powder (Bahaloo-Horeh et al., 2018).

5.2.2 Temperature

Temperature is a pivotal parameter during bioleaching because higher temperatures result in a faster reaction rate. Hence even minor temperature changes might affect microorganism growth and the bioleaching process. Reducing temperatures during the growth phase diminishes the likelihood of achieving a successful collision between enzymes and substrates. Conversely, elevated temperatures lead to the denaturation of essential enzymes involved in the cell cycle. (Walker & White, 2018). As a result, an appropriate temperature range is necessary for proper fungal growth. Temperatures of 30–32 °C are ideal for the growth and bioleaching of spent LIBs by *Aspergillus niger*.

5.2.3 Pulp Density

The concentration of pulp plays a critical role in the bioleaching process of LIBs. High pulp densities can impede the leaching rate by elevating the viscosity of the leaching solution. This, in turn, restricts the distribution of dissolved oxygen and air to the microorganisms, thereby reducing the effectiveness of the bioleaching process. This can prevent penetration of oxygen into medium, negatively impacting the metabolism of microorganisms and reducing the efficiency of metal leaching. In situations where the viscosity is high, the diffusion of oxygen and nutrients into the microbial cells is impeded, leading to a decline in their metabolic activity. Therefore, the optimization of pulp density is critical for the development of efficient and sustainable methods for metal recovery from spent LIBs. The understanding of the impact of pulp density on the bioleaching process can contribute to the development of innovative solutions for the reduction of the environmental impact of battery waste (Naseri et al., 2020).

5.2.4 Nutrient or Source of Energy

The filamentous fungus can bioleach metals faster than bacteria due to their shorter lag and exponential growth phases as long as nutrients are available (Dusengemungu et al., 2021; Horeh et al., 2016; Moh et al., 2005). The concentration and nature of the carbon source utilized are pivotal to the conversion of organic carbon into acids. Fungi are capable of utilizing an assortment of carbon sources, including glucose, sucrose, fructose, galactose, and molasses, among other options (Natarajan & Das, 2003; Walker & White, 2018). Due to its cost-effectiveness, sucrose has become the most commonly employed carbon source in various industries (Hamad et al., 2015). In addition, one sucrose molecule yields two simple sugar molecules (Walker & White, 2018). The production of organic acids is dependent on the types and concentrations of sugars present in the system. Furthermore, various nutrients, including nitrogen, phosphorus, magnesium, and others are required for fungi to grow and for the production of organic acids. Since

Table 4 Summary of LIB bioleaching with heterotrophic microorganisms

Device	Microorganism	Culture media	Organic acid secretion	Temp (°C) and Pulp density	Bioleaching efficiency	Description and methods	Ref
Mobile phone batteries	*Aspergillus niger* (PTCC 5210)	Potato dextrose agar (PDA) plates and incubated for 7 days at 30 °C	Citric acid Gluconic acid Malic acid Oxalic acid	30 °C with a 2% pulp density	Cu 100%, Li 100%, Mn 77%, Al 75%, Co 64%, Ni 54%	Among citric acid, oxalic acid, and malic acid, citric acid, with its three carboxyl groups, demonstrates superior efficacy as a metal leaching agent compared with oxalic acid and malic acid, which possess only two carboxyl groups. Conversely, gluconic acid, which features just one carboxyl group, exhibits the lowest capacity for metal leaching. The approach involves utilizing a spent medium	Bahaloo-Horeh & Mousavi (2017)
Mobile phone Li battery	*Aspergillus niger* (PTCC 5210)	3.9% (w/v) potato dextrose agar (PDA) slant	Citric acid Oxalic acid	30 **°C** with a pulp density of 1%	Cu 100%, Li 95%, Mn 70%, Al 65%, Co 45%, Ni 38%	Demonstrated that bioleaching outperforms chemical leaching in terms of achieving superior removal rates for heavy metals. The methodology involves utilizing a spent medium	Horeh et al. (2016)
Mobile phone battery	*Adapted Aspergillus niger* (PTCC 5010)	Potato dextrose agar (PDA) plates and incubated for 7 days at 30° C	Malic acid, Gluconic acid, Oxalic acid, Citric acid	30 **°C** with a pulp density of 1%	Li 100%, Cu 94%, Mn 72%, Al 62%, Ni 45%, Co 38%	The ability of *Aspergillus niger* to acclimate to heavy metals has shown promising outcomes in enhancing the production of organic acids and improving the efficiency of metal leaching when compared with non-adapted fungi. The methodology involves employing a 1-step bioleaching process using adapted *Aspergillus niger*	Bahaloo-Horeh et al. (2018)

(continued)

Table 4 (continued)

Device	Microorganism	Culture media	Organic acid secretion	Temp (°C) and Pulp density	Bioleaching efficiency	Description and methods	Ref
Different types of spent LIB	*Aspergillus niger* isolated	3.9% (w/v) potato dextrose agar (PDA)	Citric acid Oxalic acid Gluconic acid	30 °C with 0.25% pulp density	Li 100% Co 82%	Remarkable efficacy of fungal leaching in the extraction of Co and Li from intricate blends of spent LIB powder. The extraction rates of these metals through fungal leaching have been found to be comparable or even surpassing those achieved through conventional inorganic acid leaching methods, especially when it comes to Li removal. The methodology encompasses the utilization of a spent medium as a key component	Biswal et al. (2018)
Cellphones batteries	*Mixed fungal medium* of *Aspergillus niger*—and *Aspergillus tubingensis*	Dextrose agar (SDA)	Oxalic acid was secreted more than citric acid in the presence of battery power, but citric acid has a notable role in the bioleaching	30 °C with 1% pulp density	Al 80%, Co 60%, Mn 90%, Li 85%, Ni 70%	Through the implementation of bioleaching in conjunction with vinasse, a significantly enhanced yield was attained in comparison with impure sucrose and sucrose alone. The utilization of a spent medium is among the methods employed in this approach	Alavi et al., 2021)

fungi are unable to fix nitrogen, an external nitrogen source is necessary. Ammonium sulfate is a favorable nitrogen source, as it also provides a usable sulfur source (Walker & White, 2018). Optimizing the carbon and nutrient sources is essential in creating efficient and sustainable approaches to the production of organic acids by fungi. Using low-cost agricultural by-products as carbon sources for fungi is the optimal and effective choice for producing organic acids. The understanding of the impact of carbon and nutrient sources on the production of organic acids by fungi can contribute to the development of innovative and sustainable bioprocesses for the production of organic acids, promoting circular economy principles and reducing the dependence on non-renewable resources.

6 Methods for Process Intensification

6.1 Catalysis for Bioleaching of LIBs

Although bioleaching has several merits and environmental advantages in comparison with conventional hydrometallurgical processes, its relatively slow dissolution kinetics can be considered one of its major demerits. This slow leaching rate is the key underlying reason that hinders the large-scale application of bioleaching in industrial plants (Abdollahi et al., 2021). Researchers have evaluated the efficiency of various metal ions such as Ag^+, Hg^{2+}, Bi^{3+}, Cu^{2+}, Co^{2+}, etc., with non-metallic compounds such as activated carbon and polyethylene glycol as catalysts for the bioleaching process. Poor bioleaching yields and the necessity for large amounts of catalyst to achieve substantial recovery yields are two disadvantages of non-metallic catalysts (Bahaloo-Horeh et al., 2019). Metal ions have gotten the most attention out of all the other catalysts. Catalytic metallic ions can increase the efficiency of metal removal and substrate oxidation by speeding up electron transport (Niu et al., 2015). Due to their remarkable catalytic characteristics, metal ions are predicted to substantially impact the development of large-scale bioleaching technologies shortly. Ag^+ has gotten the most attention out of all of these metal ions. Various Ag compounds have been utilized as a source of silver ions in the bioleaching media, including silver sulfate (Ag_2SO_4), silver nitrate ($AgNO_3$), and silver chloride (AgCl). While Ag^+ ions have been substantiated to be beneficial in ameliorating the leaching kinetics and metals recovered from minerals and secondary sources, the majority of Ag^+ bioleaching research has focused on Cu minerals like chalcopyrite (Abdollahi et al., 2015; Pathak et al., 2017). Silver ions can also be employed in a $LiCoO_2$ bioleaching driven by *Acidithiobacillus ferrooxidans*, according to Zeng et al., The silver ion plays a key role in the creation of the $AgCoO_2$ intermediate product, which is then oxidized by Fe^{3+} ions to Co^{2+}, then Ag^+ is released into the medium, where it may be reused (Zeng et al., 2013b). The following equations can define the catalytic mechanism of Ag^+ ions on LIB bioleaching:

$$Ag^+ + LiCoO_2 \rightarrow AgCoO_2 + Li^+, \tag{19}$$

$$AgCoO_2 + 3Fe^{3+} \rightarrow 3Fe^{2+} + Ag^+ + O_2 + Co^{2+}. \tag{20}$$

Due to the bacterial metabolism, the Fe^{2+} ions generated are oxidized back to Fe^{3+} ions:

$$4Fe^{2+} + O_2 + 4H^+ \rightarrow 4Fe^{3+} + 2H_2O. \tag{21}$$

According to Zeng et al., adding 0.02 g/L silver ions to the culture medium can result in 98% Co extraction from spent LIBs in seven days, whereas Co dissolution was only 43% in the absence of Ag^+ (Zeng et al., 2013b). Application of Cu^{2+} ions instead of Ag^+ ions to improve Co solubility in LIBs utilizing *Acidithiobacillus ferrooxidans* was also reported by Zeng et al. In just six days, the extraction of Co was increased to 99.9% by utilizing 0.75 g/L of Cu^{2+} ions; conversely, in the absence of copper ions, the dissolving efficiency was 43.1% after ten days. The catalytic behavior of copper ions is attributed to the formation of $CuCo_2O_4$ intermediates on the surface of $LiCoO_2$ due to cationic exchange interactions, according to the authors (Zeng et al., 2012). As illustrated in the following equations, the Fe^{3+} ions may easily dissolve the intermediate product ($CuCo_2O_4$), leading to a greater Co recovery yield:

$$Cu^{2+} + 2LiCoO_2 \rightarrow CuCo_2O_4 + 2Li^+, \tag{22}$$

$$CuCo_2O_4 + 6Fe^{3+} \rightarrow 6Fe^{2+} + Cu^{2+} + 2O_2 + 2Co^{2+}. \tag{23}$$

The noteworthy point is that with the equal mass of catalyst, Ag^+ ions can provide a superior yield compared with Cu^{2+} ions. The application of metal ions as the catalyst for spent LIB bioleaching has various challenges like the considerable cost of various metal ions such as Ag, Hg, Bi, and Ru for the large-scale application, and the metal toxicity of high dosage catalysts for microorganisms, metal recovery, and environmental safety. For the first issue, more research needs to be conducted to find more economically suitable catalysts with the same effectiveness. Metal toxicity of microorganisms is also a critical issue when using catalysts. *Acidithiobacillus ferrooxidans*, for example, have been shown to be adversely impacted by a minimal concentration of Ag^+ of 0.1 mg/L. In the bacterial cell, Ag^+ ions can take the role of Fe^{2+} ions in the active site of oxidizing enzymes. Identifying and employing resistant microorganisms capable of tolerating larger concentrations of metallic ions are critical in dealing with these problems (Pathak et al., 2017).

6.2 Sonobioleaching

Sonobioleaching using ultrasound is another method to put a curb on slow bioleaching kinetics. In this approach, metal dissolution utilizing metabolites generated by bacteria is enhanced by employing ultrasonic wave (Anjum et al., 2014). Low quantities of acid cause a superior boost of the leaching rate in the sonobioleaching process; this is favorable for the bioleaching approach since bioprocesses produce relatively low metabolite concentrations. The sonobioleaching technique also facilitates bio-lixiviant penetration into solid particles (Anjum et al., 2014; Vyas & Ting, 2018). Ultrasonication improves bacterial metabolic activity and changes the permeability of cell membranes by increasing agitation at the macroscopic and microscopic levels. Low-frequency ultrasonication improves microbial growth, while high-frequency ultrasonication promotes cell wall breakdown. The ideal frequency of 40 kHz at 1.5 W enhanced *Aspergillus niger* growth; frequencies below and beyond this limit reduced ultrasonic efficiency. As reported by Vargas et al., metabolic activities of *Aspergillus niger* can be increased by 28% after exposure to sonication of more than 4 min (Gu et al., 2018; Vargas et al., 2004). Despite the promising potential of sonobioleaching, unfortunately, its application for LIB recycling has never been studied before; thus, it can be an interesting topic for future investigations.

6.3 High Pulp Density Bioleaching

The importance of pulp density in bioleaching cannot be overstated, as it significantly impacts the efficiency, kinetics, and economic feasibility of the process. Oxygen and carbon dioxide restrictions, excessive shear force, and the accumulation of leached metal ions are cited as primary challenges when aiming for optimal bioleaching operations, particularly in relation to high pulp density scenarios (Gu et al., 2018; Wang et al., 2014b). Increasing the pulp density from 10 g/L to 20 g/L presents a notable reduction of 50% in both leaching media volume and reactor size, leading to a substantial cost reduction in bioleaching processes. Typically, the pulp density in the bioleaching of low-grade ores is maintained at 10% or higher (Rohwerder et al., 2003). Evaluating the impacts of high pulp density on bacterial behavior is essential because pulp density should be maximized to increase the potential of LIB bioleaching's commercial application (Norris, 1997). There are only a few studies that evaluated the potential of LIB bioleaching at substantial pulp densities. As reported by Niu et al., the extraction efficiency of Co and Li from spent LIBs is significantly influenced by the solid/liquid ratio. By increasing solid content from 1 to 4%, Co extraction declined from 52 to 10% for and from 80 to 37% for Li (Niu et al., 2014). Roy et al. used the autotrophic bacteria *Acidithiobacillus ferrooxidans* to explore the bioleaching of a combination of $LiCoO_2$-based LIBs at elevated pulp density (100 g/L). At this pulp density, 60% lithium and 94% cobalt recovery were achieved in 72 h by increasing biogenic acid generation in the culture media and refilling the bacterial culture for three cycles (Roy et al., 2021b). Regardless of LIB bioleaching, there are some studies on spent Zn-Mn batteries with the purpose of maximizing the pulp density for industrial applications, in which bioleaching at 10% pulp density was employed with promising results (Niu et al., 2015; Xin et al., 2012a).

As previously mentioned, high pulp density and employment of metal ion catalysts can significantly reduce the microorganism metabolism due to metal toxicity and other inhibitory impacts. However, there are various approaches to put curb on these issues. Metals extraction from LIBs throughout bioleaching impedes the metabolic pathways of microorganisms that produce metabolites. Additionally, extracted metal ions can disintegrate proteins and nucleic acid in cells. The toxicity of microorganisms to LIBs is influenced by the concentration of heavy metals. During the bioleaching process, it is important for the microorganisms to maintain a stable population with efficient functioning and the ability to sustain an acidic medium with inhibitory factors. Adapting microorganisms to high metal content and employing adapted bacteria or fungi is crucial for maximizing the pulp density in LIB bioleaching. Through a progressive sub-culturing procedure, bacteria can be adapted to the LIB by gradually exposing them to larger solid content (Roy et al., 2021a; Sethurajan & Gaydardzhiev, 2021; Srichandan et al., 2019). Heydarian et al. utilized a mixed culture of adapted acidophilic bacteria of *Acidithiobacillus thiooxidans* and *Acidithiobacillus ferrooxidans* for waste laptops' LIB bioleaching. The adaptation period was 128, and the bacterial tolerance threshold of LIBs reached to around 40 g/L. Using an adapted bacterial culture, the LIBs bioleaching efficiency reached 99.2% for lithium, 89.4% for nickel, and 50.4% for cobalt (Heydarian et al., 2018).

Another method is using synthesized biology technologies like genetic engineering to improve bioleaching microorganisms' resistance and resilience to different stress factors prevalent in harsh conditions, increasing bioleaching efficiency (Baker-Austin & Dopson, 2007). For the bioleaching of LIBs, synthetic biology techniques may also change the metabolic pathways of new bacteria (Dopson &

Holmes, 2014). According to Gumulya et al., there are four possible pathways for industrial bioleaching microorganisms: acid tolerance, thermotolerance, osmotolerance, and, most importantly, metal tolerance (Gumulya et al., 2018). Synthesis of metal tolerance microorganisms, coupled with proper adaptation, can bring about the desirable efficiency at maximum pulp density, which subsequently makes the industrial LIB bioleaching more economically reasonable.

6.4 Kinetics Studies

To unravel the intricacies of the leaching process, an essential step involves delving into kinetic modeling. The aim of kinetic modeling is to identify the optimal equation that captures the rate of the process and enables the extraction of kinetic parameters. This knowledge is crucial for tasks such as plant design, real-time optimization for automated control, determining operating conditions in large-scale industrial settings, and maximizing leaching yields. In the realm of hydrometallurgy, the widely utilized shrinking core model provides valuable insights into comprehending the dissolving mechanism of solid materials (Baker & Bishop, 1997; Dickinson & Heal, 1999). Based on this model, the reaction can be divided into three essential stages: the transfer of bio-generated metabolites from the solution to the solid surface, the diffusion of metabolites within the solid particles, and the chemical reaction occurring on the solid surface. The initial phase is not the determining factor when the stirring is appropriately executed, thereby indicating that the process is predominantly influenced by the diffusion of reactants through the particle surface or the chemical reaction itself. In cases where the rate of dissolution is governed by the chemical reaction transpiring on the particle surface, the following equation can be employed as a suitable representation:

$$1 - (1 - X)^{\frac{1}{3}} = k_t. \tag{24}$$

The dissolution kinetics may be described using the following equation, assuming that diffusion through the product layer is the rate-limiting step:

$$1 - \frac{2}{3}X - (1 - X)^{\frac{2}{3}} = k_t. \tag{25}$$

In both of these equations, X represents the reacted metal fraction, t represents the bioleaching duration, and k_t represents the kinetic constant (Levenspiel, 1998).

Over the past few years, other models have been introduced and modified for other types of mechanisms:

Diffusion through the product layer controls the dissolution rate (İkiz et al., 2006):

$$k_t = \left[1 - (1 - X)^{\frac{1}{3}}\right]^2 \tag{26}$$

The surface chemical reaction with the shrinking core limits the reaction (Padilla et al., 2008):

$$k_t = 1 - (1 - 0.45X)^{\frac{1}{3}} \tag{27}$$

Mixed control model by shrinking core model (diffusion control; chemical reaction control) (Ghassa et al., 2017):

$$k_t = 1 - (1 - X)^{\frac{2}{3}} \tag{28}$$

Mixed control model (surface reaction control; and diffusion through sulfur layer) (Sokić et al., 2009):

$$k_t = -\ln(1 - X) \tag{29}$$

Few researchers studied the kinetics of bioleaching to recycle both valuable and environmentally hazardous metals from spent LIBs, which indicates a lack of awareness in this area. Low kinetic is one of the important disadvantages in bioleaching process of LIBs. Niu et al. investigated the kinetic behavior of Li and Co bioleaching from spent LIBs with *Alicyclobacillus sp.* and *Sulfobacillus sp.* at 2% pulp density. The product layer diffusion model had the highest correlation coefficients, noting that other models also had R^2 higher than 90%. The intrinsic mechanisms behind the phenomena remained unknown (Niu et al., 2014). The study's kinetic results provided insight into how mesophilic sulfur-oxidizing bacteria-modified granular-activated carbon adsorbs Co. The process involves external diffusion initially, followed by surface adsorption and culminating in chemical immobilization (Huang et al., 2019a).

6.5 Bioleaching with Mixed Cultures

Bioleaching with mixed cultures rather than pure strains has gained a lot of attention in recent years, mostly due to the positive synergistic effect of microorganisms that leads to higher efficiency and faster kinetics (Liao et al., 2021; Xia et al., 2018). It has been substantiated that single microorganism bioleaching is less effective than mixed culture bioleaching regarding the metal dissolution from various sources. Mixed mesophilic cultures constituted from *Acidithiobacillus ferrooxidans*, *Leptospirillum ferrooxidans*, and *Acidithiobacillus thiooxidans* have been widely employed for metal leaching from wastes, minerals, and other sources. Mixed moderate thermophilic culture comprised of microorganisms like *Sulfobacillus thermosulfidooxidans*, *Leptospirillum ferriphilum*, *Acidithiobacillus caldus*, and *Ferroplasma spp.* at the optimum temperature of 45 °C have shown promising results for extraction of Co from cobaltite bearing ore (Abdollahi et al., 2021). In comparison with a single culture of *Acidithiobacillus thiooxidans*, a mixed culture demonstrates a greater extraction of manganese (insoluble Mn^{4+} to soluble Mn^{2+}) by Fe^{2+}. The decline in ORP throughout the bioleaching process in a mixed culture medium is indicated by the large fluctuations in ORP. Furthermore, mixed culture produces more ferric ions than pure strain iron-oxidizing microbes due to the lower pH value. More ferric ions result in a more extensive formation of Fe^{2+}, which speeds up the reduction of Mn^{4+} and the recovery of Mn^{2+}. Consequently, using mixed culture yields the highest manganese extraction (Xin et al., 2009, 2012b).

In a study by Alavi et al., spent cellphone LIB bioleaching was employed with mixed-fungus culture (*Aspergillus tubingensis* and *Aspergillus niger*) by 1-step, 2-step, and spent medium approaches. Their investigations revealed that spent medium bioleaching in the presence of vinasse with 1% solid percentage and at 140 rpm and 30 °C provides the highest metal yields of approximately 82% Al, 62% Co, 98% Mn, 91% Li, and 81% Ni (Alavi et al., 2021). The investigation conducted by Xin et al. showed that the maximum dissolving yield for Co, Ni, and Mn were obtained in the mixed culture system (*Acidithiobacillus thiooxidans* and *Leptospirillum ferriphilum*), implying that these ions were mobilized by a combination of Fe^{2+} reduction and acid dissolution. The extraction efficiency of four important metals from refractory $LiNi_xCo_yMn_{1-x-y}O_2$ in a mixed system was above 95% on average (Xin et al., 2016). Employing mixed moderate thermophilic culture in the presence of iron scrap as a reductant also showed promising results in a study by Ghassa et al. (Ghassa et al., 2020). A mixture of bio-metabolites can be employed to undertake bioleaching. Bioleaching using a mixture of ferric ion and biogenic acid was shown to improve leaching efficiency in terms of pH and ORP compared with solo bioreagents. When the bioreagents were combined, their pH was lower than that of the solo bioreagents. This suggests that a more oxidizing environment was created, which led to more effective metal leaching (Boxall et al., 2018).

7 Methods of Metal Recovery

7.1 Solvent Extraction

Solvent extraction stands as a prominent method for purifying and extracting metals from the leachate of LIBs. The dissimilar solubility behaviors exhibited by various solutes when in the presence of organic solvents within a two-phase system determine the separation process during solvent extraction. An appropriate solvent should demonstrate selectivity toward the desired chemical and exhibit significant potential for recycling. Bioleaching-solvent extraction-electrowinning route has gained much popularity in past years, and many studies have been conducted in this regard (Akbari & Ahmadi, 2019; Irrgang et al., 2021). Various organic extractants such as MextralVR 5640H, MextralVR 272 P, P-204, P-507, D2EHPA, Cyphos IL 102, and Cyanex 272 have been utilized to recover metals from LIB's PLS (Kang et al., 2010; Keller et al., 2021; Lei et al., 2022; Punt et al., 2021; Shuya et al., 2020; Torkaman et al., 2017; Xu et al., 2020). Existing Ni, Co, Mn, and Li together in the cathode of LIBs (such as in NCM cells) can make separation more difficult and may lead to co-extraction. However, there are ways to separate these metals: using Cyanex 272 can separate Ni and Co, while using D2EHPA can separate Mn and Co when used together, and precipitation can separate Li and Ni (Chen & Ho, 2018). Solvent extraction often involves the use of diluents like paraffin, naphthenes, and alkyl aromatics to thin out the viscous extractants. However, the use of hazardous and flammable organic diluents remains a concern (Shamsuddin, 2021). This should also be taken into account that returning aqueous raffinate in the bioleaching-solvent extraction cycle might contain a small proportion of organic matters that can significantly hinder the bioleaching process. As reported by Saneie et al., the presence of an organic phase in the returning raffinate can substantially inhibit the bio-oxidation and metabolism of bioleaching microorganisms resulting in lower leaching kinetics. Thus, eliminating organic extractants from the aqueous phase is crucial (Saneie et al., 2021).

7.2 Chemical and Biological Precipitation

The differential in solubility of several chemical species in a mixture in the presence of selectively applied chemical compounds is the basis for precipitation. The fluctuation of solubility as pH changes is utilized in the separation's favor. At pH = 2 and temperatures within 40 and 50 °C, potassium permanganate can be employed as a reagent to precipitate manganese as manganese dioxide. It is reported that dimethylglyoxime (DMG) can precipitate nickel in the presence of ammonia at pH 8 to 11 (Vanitha & Balasubramanian, 2013). Typically, Fe, Cu, and Al are removed at the start of the precipitation process. Precipitation is a low-cost approach to separate them from other ions. At low pHs (3–6), they are easily eliminated by precipitation using NaOH or other hydroxides (Zou et al., 2013). At higher pHs (8–12), Co, Ni, and Mn can also be selectively precipitated as hydroxides. These metals can be eliminated in the form of carbonates by adding Na_2CO_3, sulfides by adding $(NH_4)_2S$ at a pH of 6 to 10, and other ways. Lithium may be recovered in three forms: carbonate

(Li_2CO_3), fluoride (LiF), and phosphate (LiP) (Li_3PO_4) (Duarte Castro et al., 2021). In a recent study, Biswal et al. employed bioleaching followed by precipitation for metal recovery from waste LIBs. More than 88% of Co from the fungal PLS recovered in the form of cobalt oxalate, cobalt sulfides, and cobalt hydroxides, and about 74% of Li in the lithium carbonate form (Biswal et al., 2018).

In the bioprecipitation technique, the reagents for precipitation are produced by different microorganisms. One of the most remarkable biochemical reactions for recovering metal ions from PLS is biogenic sulfide precipitation. Peculiar bacteria classified as "sulfate-reducing bacteria" (SRB) can produce biogenic sulfides. SRBs use sulfate ions in their metabolism and convert them to sulfides via sulfur. The following reactions demonstrate the process's basic transformations:

$$\text{Organic matter} + SO_4^{2-} \rightarrow H_2S + HCO_3^-, \tag{30}$$

$$Me^{2+} + HS^- \rightarrow MeS_{(s)} + H^+, \tag{31}$$

where Me^{2+} is the metal cation. This approach has shown promising results and a decent industrial application potential (Meshram et al., 2014). Bioprecipitation has been successfully employed in previous studies for the separation of metals from polymetallic PLS, either synthetic or real leachates (Esposito et al., 2006; Sethurajan et al., 2017). The microbial consortium consisted of a variety of bacteria, with *Desulfovibrio spp.* being the most prevalent was utilized by Calvert et al. for selective separation of metals from LIB leachate. At pH around 5, more than 92% of Al and Cu was removed in the sulfide form, and at pH = 10, about 99% of cobalt was precipitated with the co-precipitation of almost all Cd, Ni, Mn, and Zn (Calvert et al., 2019).

7.3 Other Methods

There are various other approaches for metal reclamation from LIB's PLS, including electrodeposition, adsorption, and bioelectrochemical processes. Metals can be extracted from a solution through electrochemical deposition due to variations in electric potential. Recovery of Co from LIB leachate with electrodeposition has been studied by a few researchers (Duarte Castro et al., 2021; Quintero-Almanza et al., 2019). Wang et al. used manganese-type lithium ion sieves to conduct lithium extraction via adsorption and isolated 99.9% of Li from the PLS, including Li, Ni, and Co (Wang et al., 2017). Bioelectrochemical processes, often known as BES, are a subclass of techniques for recovering metals from aqueous solutions that are still in their infancy, although they are already gaining attention from researchers and developers. They were initially designed and put to use to improve the quality of wastewater, and it wasn't until recently that they were put to use for the recovery of metals from PLS (Nancharaiah et al., 2015). Investigations are currently being carried out in order to gain a deeper comprehension of the fundamentals behind the bio-recovery of cobalt from synthetic solutions by means of bioelectrochemical systems. The BES's performance in treating actual leachates, on the other hand, has yet to be shown (Huang et al., 2019b).

8 Future Prospectives and Conclusions

In the mineral processing and extractive metallurgy industry, bioleaching is a sustainable and environmentally friendly method for the extraction of metals from primary and secondary resources. This method has the potential to help save non-renewable energy sources and reduce greenhouse gas emissions. The lithium cobalt oxide ($LiCoO_2$) serves as the primary and most significant component of LIBs. These batteries are complex, multi-metallic compounds with high concentrations of lithium, cobalt, manganese, nickel, copper, and aluminum (cathodic material). The elemental makeup of LIBs might change significantly depending on its application. Autotrophs and heterotrophs are the two types of microorganisms employed for the LIB bioleaching process. Through acidolysis and redoxolysis, metabolites generated by microorganisms, such as biogenic Fe^{3+} and H_2SO_4 in the case of autotrophs, and weak organic acid in the case of heterotrophs can leach metals from the active cathode materials. The time it takes to produce microorganisms can be significantly reduced by improving the growth conditions. In this chapter, we categorized the most influential parameters of the LIB bioleaching process with both autotrophic and heterotrophs microorganisms and their most dominant mechanisms. The most important drawbacks of the LIB bioleaching that prevents its industrial application were mentioned, and several approaches for process enhancement, including admission of catalysts, employing adapted or mixed cultures, sonobioleaching, and genetic engineering were summarized. With advancements in process kinetics, higher pulp density and improved microbial tolerance, there is potential for the application of bioleaching techniques in the industrial sector. The utilization of bioleaching has already yielded a notable leaching efficiency of 80–90%, indicating promising prospects for large-scale recycling of waste LIBs in the near future. The high pulp density of 100 g/L was used in this study. In spite of the numerous studies and new research that have been conducted recently, LIB bioleaching is still in its infancy. More study is necessary to improve the process's efficiency, kinetics, and selectivity. In subsequent research, the primary focus should be on determining how to best optimize operating parameters in order to make it easier to process large pulp densities and open the door to potential for industrial expansion. Furthermore, the biological recovery of key metals from the PLS is of significant importance for comprehensive recycling and regeneration and should be explored so that the necessary metals can be recovered selectively. Finally, the proper disposal and treatment of created tailings, process effluents, and sludge must be established.

References

Abdollahi, H., Noaparast, M., Shafaei, S. Z., Manafi, Z., Muñoz, J. A., & Tuovinen, O. H. (2015). Silver-catalyzed bioleaching of copper, molybdenum and rhenium from a chalcopyrite–molybdenite concentrate. *International Biodeterioration and Biodegradation, 104*, 194–200.

Abdollahi, H., Saneie, R., Shafaei, S. Z., Mirmohammadi, M., Mohammadzadeh, A., & Tuovinen, O. H. (2021). Bioleaching of cobalt from magnetite-rich cobaltite-bearing ore. *Hydrometallurgy, 204*, 105727. https://doi.org/10.1016/j.hydromet.2021.105727

Abhilash, Pandey, B. D., & Natarajan, K. A. (2015). Microbiology for minerals, metals, materials and the environment. CRC Press. https://doi.org/10.1201/b18124.

Akbari, S., & Ahmadi, A. (2019). Recovery of copper from a mixture of printed circuit boards (PCBs) and sulphidic tailings using bioleaching and solvent extraction processes. *Chemical Engineering and Processing-Process Intensification, 142*, 107584. https://doi.org/10.1016/j.cep.2019.107584

Akcil, A., Vegliò, F., Ferella, F., Okudan, M. D., & Tuncuk, A. (2015). A review of metal recovery from spent petroleum catalysts and ash. *Waste Management, 45*, 420–433. https://doi.org/10.1016/j.wasman.2015.07.007

Alavi, N., Partovi, K., Majlessi, M., Rashidi, M., & Alimohammadi, M. (2021). Bioleaching of metals from cellphones batteries by a co-fungus medium in presence of carbon materials. *Bioresource Technology Reports, 15*, 100768. https://doi.org/10.1016/j.biteb.2021.100768

Ambrose, H., & Kendall, A. (2020). Understanding the future of lithium: Part 1, resource model. *Journal of Industrial Ecology, 24*, 80–89. https://doi.org/10.1111/jiec.12949

Anjum, F., Shahid, M., Bukhari, S., & Potgieter, J. H. (2014). Combined ultrasonic and bioleaching treatment of hospital waste incinerator bottom ash with simultaneous extraction of selected metals. *Environmental Technology, 35*, 262–270. https://doi.org/10.1080/09593330.2013.824992

Arya, S., & Kumar, S. (2020). Bioleaching: urban mining option to curb the menace of E-waste challenge. *Bioengineered, 11*, 640–660. https://doi.org/10.1080/21655979.2020.1775988

Asghari, I., Mousavi, S. M., Amiri, F., & Tavassoli, S. (2013). Bioleaching of spent refinery catalysts: A review. *Journal of Industrial and Engineering Chemistry, 19*, 1069–1081. https://doi.org/10.1016/j.jiec.2012.12.005

Bahaloo-Horeh, N., Vakilchap, F., & Mousavi, S. M. (2019). Biohydrometallurgical methods for recycling spent lithium-ion batteries. https://doi.org/10.1007/978-3-030-31834-5_7.

Bahaloo-Horeh, N., & Mousavi, S. M. (2017). Enhanced recovery of valuable metals from spent lithium-ion batteries through optimization of organic acids produced by *Aspergillus Niger*. *Waste Management, 60*, 666–679. https://doi.org/10.1016/j.wasman.2016.10.034

Bahaloo-Horeh, N., Mousavi, S. M., & Baniasadi, M. (2018). Use of adapted metal tolerant *Aspergillus niger* to enhance bioleaching efficiency of valuable metals from spent lithium-ion mobile phone batteries. *Journal of Cleaner Production, 197*, 1546–1557. https://doi.org/10.1016/j.jclepro.2018.06.299

Baker, P. G., & Bishop, P. L. (1997). Prediction of metal leaching rates from solidified/stabilized wastes using the shrinking unreacted core leaching procedure. *Journal of Hazardous Materials, 52*, 311–333. https://doi.org/10.1016/S0304-3894(96)01814-6

Baker-Austin, C., & Dopson, M. (2007). Life in acid: pH homeostasis in acidophiles. *Trends in Microbiology, 15*, 165–171. https://doi.org/10.1016/j.tim.2007.02.005

Bardi, U., Jakobi, R., & Hettiarachchi, H. (2016). Mineral resource depletion: A coming age of stockpiling? *BioPhysical Economics and Resource Quality, 1*, 4. https://doi.org/10.1007/s41247-016-0004-x

Bharadwaj, A., & Ting, Y. P. (2013). Bioleaching of spent hydrotreating catalyst by acidophilic thermophile Acidianus brierleyi: Leaching mechanism and effect of decoking. *Bioresource Technology, 130*, 673–680. https://doi.org/10.1016/j.biortech.2012.12.047

Biswal, B. K., Jadhav, U. U., Madhaiyan, M., Ji, L., Yang, E. H., & Cao, B. (2018). Biological leaching and chemical precipitation methods for recovery of co and li from spent lithium-ion batteries. *ACS Sustainable Chemistry and Engineering., 6*, 12343–12352. https://doi.org/10.1021/acssuschemeng.8b02810

Bosecker, K. (1997). Bioleaching: Metal solubilization by microorganisms. *FEMS Microbiology Reviews, 20*, 591–604. https://doi.org/10.1016/S0168-6445(97)00036-3

Boxall, N. J., Cheng, K. Y., Bruckard, W., & Kaksonen, A. H. (2018). Application of indirect non-contact bioleaching for extracting metals from waste lithium-ion batteries. *Journal of Hazardous Materials, 360*, 504–511. https://doi.org/10.1016/j.jhazmat.2018.08.024

Brandl, H., Bosshard, R., & Wegmann, M. (1999). Computer-munching microbes: Metal leaching from electronic scrap by bacteria and fungi, Process. *Metallurgy, 9*, 569–576. https://doi.org/10.1016/S1572-4409(99)80146-1

Brandl, H., & Faramarzi, M. A. (2006). Microbe-metal-interactions for the biotechnological treatment of metal-containing solid waste. *China Particuology, 4*, 93–97. https://doi.org/10.1016/S1672-2515(07)60244-9

Cai, X., Tian, L., Chen, C., Huang, W., Yu, Y., Liu, C., Yang, B., Lu, X., & Mao, Y. (2021). Phylogenetically divergent bacteria consortium from neutral activated sludge showed heightened potential on bioleaching spent lithium-ion batteries. *Ecotoxicology and Environmental Safety, 223*, 112592. https://doi.org/10.1016/j.ecoenv.2021.112592

Calvert, G., Kaksonen, A. H., Cheng, K. Y., Van Yken, J., Chang, B., Boxall, N. J. (2019). Recovery of metals from waste lithium ion battery leachates using biogenic hydrogen sulfide. *Minerals 9*. https://doi.org/10.3390/min9090563.

Chen, W. -S., & Ho, H. -J. (2018). Recovery of valuable metals from lithium-ion batteries NMC cathode waste materials by hydrometallurgical methods. *Metals 8* (2018). https://doi.org/10.3390/met8050321.

Chen, K., & Xue, D. (2014). Anode performances of mixed $LiMn_2O_4$ and carbon black toward lithium-ion battery. *Functional Materials Letters, 07*, 1450017. https://doi.org/10.1142/S1793604714500179

Chen, M., Ma, X., Chen, B., Arsenault, R., Karlson, P., Simon, N., & Wang, Y. (2019). Recycling end-of-life electric vehicle lithium-ion batteries. *Joule., 3*, 2622–2646. https://doi.org/10.1016/j.joule.2019.09.014

Deng, X., Chai, L., Yang, Z., Tang, C., Wang, Y., & Shi, Y. (2013). Bioleaching mechanism of heavy metals in the mixture of contaminated soil and slag by using indigenous Penicillium chrysogenum strain F1. *Journal of Hazardous Materials, 248–249*, 107–114. https://doi.org/10.1016/j.jhazmat.2012.12.051

Deng, Z., Mo, Y., & Ong, S. P. (2016). Computational studies of solid-state alkali conduction in rechargeable alkali-ion batteries. *NPG Asia Materials, 8*, 1–15. https://doi.org/10.1038/am.2016.7

Dickinson, C. F., & Heal, G. R. (1999). Solid–liquid diffusion controlled rate equations. *Thermochimica Acta, 340–341*, 89–103. https://doi.org/10.1016/S0040-6031(99)00256-7

Dopson, M., & Holmes, D. S. (2014). Metal resistance in acidophilic microorganisms and its significance for biotechnologies. *Applied Microbiology and Biotechnology, 98*, 8133–8144. https://doi.org/10.1007/s00253-014-5982-2

Duarte Castro, F., Vaccari, M., & Cutaia, L. (2021). Valorization of resources from end-of-life lithium-ion batteries: A review. *Critical Reviews in Environmental Science and Technology*, pp. 1–44. https://doi.org/10.1080/10643389.2021.1874854.

Dusengemungu, L., Kasali, G., Gwanama, C., Ouma, K. O. (2020). Recent advances in biosorption of copper and cobalt by filamentous fungi. *Frontiers in Microbiology 11*. https://doi.org/10.3389/fmicb.2020.582016.

Dusengemungu, L., Kasali, G., Gwanama, C., & Mubemba, B. (2021). Overview of fungal bioleaching of metals. *Environmental Advances, 5*, 100083. https://doi.org/10.1016/j.envadv.2021.100083

Esposito, G., Veeken, A., Weijma, J., & Lens, P. N. L. (2006). Use of biogenic sulfide for ZnS precipitation. *Separation and Purification Technology, 51*, 31–39. https://doi.org/10.1016/j.seppur.2005.12.021

Faraji, F., Golmohammadzadeh, R., Rashchi, F., & Alimardani, N. (2018). Fungal bioleaching of WPCBs using *Aspergillus niger*: Observation, optimization and kinetics. *Journal of Environmental Management, 217*, 775–787. https://doi.org/10.1016/j.jenvman.2018.04.043

Fergus, J. W. (2010). Recent developments in cathode materials for lithium ion batteries. *Journal of Power Sources, 195*, 939–954. https://doi.org/10.1016/j.jpowsour.2009.08.089

Ghassa, S., Farzanegan, A., Gharabaghi, M., & Abdollahi, H. (2020). Novel bioleaching of waste lithium ion batteries by mixed moderate thermophilic microorganisms, using iron scrap as energy source and reducing agent. *Hydrometallurgy, 197*, 105465. https://doi.org/10.1016/j.hydromet.2020.105465

Ghassa, S., Noaparast, M., Shafaei, S. Z., Abdollahi, H., Gharabaghi, M., & Boruomand, Z. (2017). A study on the zinc sulfide dissolution kinetics with biological and chemical ferric reagents. *Hydrometallurgy, 171*, 362–373. https://doi.org/10.1016/j.hydromet.2017.06.012

Golmohammadzadeh, R., Faraji, F., & Rashchi, F. (2018). Recovery of lithium and cobalt from spent lithium ion batteries (LIBs) using organic acids as leaching reagents: A review. *Resources, Conservation and Recycling, 136*, 418–435. https://doi.org/10.1016/j.resconrec.2018.04.024

Gu, T., Rastegar, S. O., Mousavi, S. M., Li, M., & Zhou, M. (2018). Advances in bioleaching for recovery of metals and bioremediation of fuel ash and sewage sludge. *Bioresource Technology, 261*, 428–440. https://doi.org/10.1016/j.biortech.2018.04.033

Gumulya, Y., Boxall, N. J., Khaleque, H. N., Santala, V., Carlson, R. P., & Kaksonen, A. H. (2018). In a quest for engineering acidophiles for biomining applications: Challenges and opportunities. *Genes 9*. https://doi.org/10.3390/genes9020116.

Habib, K., Hamelin, L., & Wenzel, H. (2016). A dynamic perspective of the geopolitical supply risk of metals. *Journal of Cleaner Production, 133*, 850–858. https://doi.org/10.1016/j.jclepro.2016.05.118

Hamad, H. O., Alma, M. H., Ismael, H. M., & Göçeri, A. (2015). The effect of some sugars on the growth of *Aspergillus niger*. *Kahramanmaraş Sütçü İmam Üniversitesi Doğa Bilimleri Dergisi 17*, 7. https://doi.org/10.18016/ksujns.28479.

Hamidah, N. L., Nugroho, G., & Wang, F. M. (2015). *Electrochemical Analysis of Electrolyte Additive Effect on Ionic Diffusion for High-Performance Lithium Ion Battery*. https://doi.org/10.1007/s11581-015-1505-0

Hartono, M., Astrayudha, M. A., Petrus, H. T. B. M., Budhijanto, W., & Sulistyo, H. (2017). Lithium recovery of spent lithium-ion battery using bioleaching from local sources microorganism. *Rasayan Journal of Chemistry, 10*, 897–903. https://doi.org/10.7324/RJC.2017.1031767

Harwood, C. V., Ph, D., Cunningham, C. J., Ph, D., Scott, K., Ph, D. (2017). Bioleaching potential of filamentous fungi to mobilize lithium and cobalt from spent rechargeable li-ion batteries by aldo lobos a thesis submitted in partial fulfillment of the requirements for the degree of Master of Science in integrative biology with

He, L.-P., Sun, S.-Y., Song, X.-F., & Yu, J.-G. (2015). Recovery of cathode materials and Al from spent lithium-ion batteries by ultrasonic cleaning. *Waste Management, 46*, 523–528. https://doi.org/10.1016/j.wasman.2015.08.035

Heydarian, A., Mousavi, S. M., Vakilchap, F., & Baniasadi, M. (2018). Application of a mixed culture of adapted acidophilic bacteria in two-step bioleaching of spent lithium-ion laptop batteries. *Journal of Power Sources, 378*, 19–30. https://doi.org/10.1016/J.JPOWSOUR.2017.12.009

Horeh, N. B., Mousavi, S. M., & Shojaosadati, S. A. (2016). Bioleaching of valuable metals from spent lithium-ion mobile phone batteries using *Aspergillus niger*. *Journal of Power Sources, 320*, 257–266. https://doi.org/10.1016/J.JPOWSOUR.2016.04.104

Huang, L., Zhou, Q., & Quan, X. (2019). Recovery of metals from wastes using bioelectrochemical systems BT—bioelectrochemistry stimulated environmental remediation: From bioelectrorespiration to bioelectrodegradation. In A.-J. Wang, B. Liang, Z.-L. Li, & H. -Y. Cheng (Eds.) (pp. 121–156). Springer. https://doi.org/10.1007/978-981-10-8542-0_6.

Huang, T., Liu, L., & Zhang, S. (2019a). Recovery of cobalt, lithium, and manganese from the cathode active materials of spent lithium-ion batteries in a bio-electro-hydrometallurgical process. *Hydrometallurgy, 188*, 101–111. https://doi.org/10.1016/j.hydromet.2019.06.011

İkiz, D., Gülfen, M., Aydın, A. O. (2006). Dissolution kinetics of primary chalcopyrite ore in hypochlorite solution. *Minerals Engineering 19*, 972–974. https://doi.org/10.1016/j.mineng.2005.09.047.

Iram, S., Shabbir, R., Zafar, H., & Javaid, M. (2015). Biosorption and Bioaccumulation of Copper and Lead by Heavy Metal-Resistant Fungal Isolates. *Arabian Journal for Science and Engineering, 40*, 1867–1873. https://doi.org/10.1007/s13369-015-1702-1

Irrgang, N., Monneron-Enaud, B., Möckel, R., Schlömann, M., & Höck, M. (2021). Economic feasibility of the co-production of indium from zinc sulphide using bioleaching extraction in Germany. *Hydrometallurgy, 200*, 105566. https://doi.org/10.1016/j.hydromet.2021.105566

Işıldar, A., van Hullebusch, E. D., Lenz, M., Du Laing, G., Marra, A., Cesaro, A., Panda, S., Akcil, A., Kucuker, M. A., & Kuchta, K. (2019). Biotechnological strategies for the recovery of valuable and critical raw materials from waste electrical and electronic equipment (WEEE)—A review. *Journal of Hazardous Materials, 362*, 467–481. https://doi.org/10.1016/j.jhazmat.2018.08.050

Islam, A., Ahmed, T., Awual, M. R., Rahman, A., Sultana, M., Aziz, A. A., Monir, M. U., Teo, S. H., & Hasan, M. (2020). Advances in sustainable approaches to recover metals from e-waste-a review. *Journal of Cleaner Production, 244*, 118815. https://doi.org/10.1016/j.jclepro.2019.118815

Jafari, M., Abdollahi, H., Shafaei, S. Z., Gharabaghi, M., Jafari, H., Akcil, A., & Panda, S. (2019). Acidophilic bioleaching: A review on the process and effect of organic–inorganic reagents and materials on its efficiency. *Mineral Processing and Extractive Metallurgy Review, 40*, 87–107. https://doi.org/10.1080/08827508.2018.1481063

Jafari, S., Aghaei, S. S., Afifi-Sabet, H., Shams-Ghahfarokhi, M., Jahanshiri, Z., Gholami-Shabani, M., Shafiei-Darabi, S., & Razzaghi-Abyaneh, M. (2018). Exploration, antifungal and antiaflatoxigenic activity of halophilic bacteria communities from saline soils of Howze-Soltan playa in Iran. *Extremophiles, 22*, 87–98. https://doi.org/10.1007/s00792-017-0979-2

Johnson, D. B. (2014). Biomining—biotechnologies for extracting and recovering metals from ores and waste materials. *Current Opinion in Biotechnology, 30*, 24–31. https://doi.org/10.1016/j.copbio.2014.04.008

Johnson, D. (2018). The Evolution current status, and future prospects of using biotechnologies in the mineral extraction and metal recovery sectors. *Minerals, 8*, 343. https://doi.org/10.3390/min8080343

Kang, J., Senanayake, G., Sohn, J., & Shin, S. M. (2010). Recovery of cobalt sulfate from spent lithium ion batteries by reductive leaching and solvent extraction with Cyanex 272. *Hydrometallurgy, 100*, 168–171. https://doi.org/10.1016/j.hydromet.2009.10.010

Kapoor, A., & Viraraghavan, T. (1995). Fungal biosorption—an alternative treatment option for heavy metal bearing wastewaters: A review. *Bioresource Technology, 53*, 195–206. https://doi.org/10.1016/0960-8524(95)00072-M

Keller, A., Hlawitschka, M. W., & Bart, H. -J. (2021). Application of saponified D2EHPA for the selective extraction of manganese from spend lithium-ion batteries. *Chemical Engineering and Processing-Process Intensification* 108552. https://doi.org/10.1016/j.cep.2021.108552.

Khatri, B. R., Tipre, D. R., & Dave, S. R. (2019). Comparison of hydro- and biohydrometallurgical extraction of metals from waste li-ion batteries of cell phone. *Journal of Sustainable Metallurgy, 5*, 250–261. https://doi.org/10.1007/s40831-019-00223-z

Kim, J.H., Gibb, H.J., & Howe, P.D. (2006). Cobalt and inorganic cobalt compounds, IPCS Concise Int. Chem. Assess. Doc, pp. 1–82.

Kim, S., Bang, J., Yoo, J., Shin, Y., Bae, J., Jeong, J., Kim, K., Dong, P., & Kwon, K. (2021). A comprehensive review on the pretreatment process in lithium-ion battery recycling. *Journal of Cleaner Production, 294*, 126329. https://doi.org/10.1016/j.jclepro.2021.126329

Ku, H., Jung, Y., Jo, M., Park, S., Kim, S., Yang, D., Rhee, K., An, E.-M., Sohn, J., & Kwon, K. (2016). Recycling of spent lithium-ion battery cathode materials by ammoniacal leaching. *Journal of Hazardous Materials, 313*, 138–146. https://doi.org/10.1016/J.JHAZMAT.2016.03.062

Kwon, S. J., Lee, S. E., Lim, J. H., Choi, J., & Kim, J. (2018). Performance and life degradation characteristics analysis of NCM LIB for bess. *Electronics, 7*, 1–19. https://doi.org/10.3390/electronics7120406

Lai, X., Huang, Y., Gu, H., Deng, C., Han, X., Feng, X., & Zheng, Y. (2021). Turning waste into wealth: A systematic review on echelon utilization and material recycling of retired lithium-ion batteries. *Energy Storage Materials, 40*, 96–123. https://doi.org/10.1016/j.ensm.2021.05.010

Lavoie, Y., Danet, F., & Lombard, B., Lithium-ion batteries for industrial applications. In *2017 petroleum and chemical industry technical conference* (pp. 283–290). IEEE. https://doi.org/10.1109/PCICON.2017.8188747.

Le, L., Tang, J., Ryan, D., & Valix, M. (2006). Bioleaching nickel laterite ores using multi-metal tolerant *Aspergillus foetidus* organism. *Minerals Engineering, 19*, 1259–1265. https://doi.org/10.1016/j.mineng.2006.02.006

Lei, S., Sun, W., & Yang, Y. (2022). Solvent extraction for recycling of spent lithium-ion batteries. *Journal of Hazardous Materials, 424*, 127654. https://doi.org/10.1016/j.jhazmat.2021.127654

O. Levenspiel, Chemical reaction engineering, John wiley & sons, 1998.

Li, L., Zeng, G. S., Luo, S. L., Deng, X. R., & Xie, Q. J. (2013). Influences of solution pH and redox potential on the bioleaching of LiCoO2 from spent lithium-ion batteries. *Journal of the Korean Society for Applied Biological Chemistry, 56*, 187–192. https://doi.org/10.1007/s13765-013-3016-x

Liang, Y., Su, J., Xi, B., Yu, Y., Ji, D., Sun, Y., Cui, C., & Zhu, J. (2017). Life cycle assessment of lithium-ion batteries for greenhouse gas emissions. *Resources, Conservation and Recycling, 117*, 285–293. https://doi.org/10.1016/j.resconrec.2016.08.028

Liao, X., Ye, M., Li, S., Liang, J., Zhou, S., Fang, X., Gan, Q., & Sun, S. (2021). Simultaneous recovery of valuable metal ions and tailings toxicity reduction using a mixed culture bioleaching process. *Journal of Cleaner Production, 316*, 128319. https://doi.org/10.1016/j.jclepro.2021.128319

P. Liu, L. Xiao, Y. Chen, H. Chen, Highly enhanced electrochemical performances of LiNi0.815Co0.15Al0.035O2 by coating via conductively LiTiO2 for lithium-ion batteries, Ceram. Int. 45 (2019) 18398–18405. https://doi.org/10.1016/j.ceramint.2019.06.055.

Magnuson, J. K., Lasure, L. L. (2004). Organic acid production by filamentous fungi. In *Advances in fungal biotechnology for industry, agriculture, and medicine* (pp. 307–340). Springer. https://doi.org/10.1007/978-1-4419-8859-1_12.

Makuza, B., Tian, Q., Guo, X., Chattopadhyay, K., & Yu, D. (2021). Pyrometallurgical options for recycling spent lithium-ion batteries: A comprehensive review. *Journal of Power Sources, 491*, 229622. https://doi.org/10.1016/j.jpowsour.2021.229622

Mekonnen, Y., Sundararajan, A., & Sarwat, A. I. (2016). A review of cathode and anode materials for lithium-ion batteries. In SoutheastCon 2016, IEEE, pp. 1–6. https://doi.org/10.1109/SECON.2016.7506639.

Meshram, P., Pandey, B. D., & Mankhand, T. R. (2014). Extraction of lithium from primary and secondary sources by pre-treatment, leaching and separation: A comprehensive review. *Hydrometallurgy, 150*, 192–208. https://doi.org/10.1016/j.hydromet.2014.10.012

R. Methekar, S. Anwani, Manufacturing of Lithium Cobalt Oxide from Spent Lithium-Ion Batteries: A Cathode Material, in: 2019: pp. 233–241. https://doi.org/10.1007/978-981-13-1966-2_20.

Mishra, D., Kim, D. J., Ralph, D. E., Ahn, J. G., & Rhee, Y. H. (2008). Bioleaching of metals from spent lithium ion secondary batteries using *Acidithiobacillus ferrooxidans*. *Waste Management, 28*, 333–338. https://doi.org/10.1016/j.wasman.2007.01.010

Mishra, D., & Rhee, Y. H. (2014). Microbial leaching of metals from solid industrial wastes. *Journal of Microbiology, 52*, 1–7. https://doi.org/10.1007/s12275-014-3532-3

Moh, K., Aung, M., & Ting, Y. (2005). Bioleaching of spent fluid catalytic cracking catalyst using *Aspergillus niger 116*, 159–170. https://doi.org/10.1016/j.jbiotec.2004.10.008.

Nancharaiah, Y. V., Venkata Mohan, S., & Lens, P. N. L. (2015). Metals removal and recovery in bioelectrochemical systems: A review. *Bioresource Technology, 195*, 102–114. https://doi.org/10.1016/j.biortech.2015.06.058

Naseri, T., Bahaloo-Horeh, N., & Mousavi, S. M. (2019). Bacterial leaching as a green approach for typical metals recovery from end-of-life coin cells batteries. *Journal of Cleaner Production, 220*, 483–492. https://doi.org/10.1016/j.jclepro.2019.02.177

Naseri, T., Bahaloo-horeh, N., & Mousavi, S. M. (2020). Bacterial leaching as a green approach for typical metals recovery from end-of-life coin cells batteries. *Journal of Cleaner Production, 220*, 483–492. https://doi.org/10.1016/j.jclepro.2019.02.177

Natarajan, K. A., & Das, A. (2003). Surface chemical studies on "Acidithiobacillus" group of bacteria with reference to mineral flocculation. *International Journal of Mineral Processing, 72*, 189–198. https://doi.org/10.1016/S0301-7516(03)00098-X

Niu, Z., Huang, Q., Wang, J., Yang, Y., Xin, B., & Chen, S. (2015). Metallic ions catalysis for improving bioleaching yield of Zn

and Mn from spent Zn-Mn batteries at high pulp density of 10%. *Journal of Hazardous Materials, 298*, 170–177.
Niu, Z., Zou, Y., Xin, B., Chen, S., Liu, C., & Li, Y. (2014). Process controls for improving bioleaching performance of both Li and Co from spent lithium ion batteries at high pulp density and its thermodynamics and kinetics exploration. *Chemosphere, 109*, 92–98. https://doi.org/10.1016/j.chemosphere.2014.02.059
Norris, P. R. (1997). Thermophiles and bioleaching BT—biomining: Theory, microbes and industrial processes. In D. E. Rawlings (Ed.) (pp. 247–258). Springer. https://doi.org/10.1007/978-3-662-06111-4_12.
Olivetti, E. A., Ceder, G., Gaustad, G. G., & Fu, X. (2017). Lithium-Ion battery supply chain considerations: Analysis of potential bottlenecks in critical metals. *Joule., 1*, 229–243. https://doi.org/10.1016/j.joule.2017.08.019
Ong, G. H., Ho, X. H., Shamkeeva, S., Fernando, A. S. M. S., & Wong, L. S. (2017). Biosorption study of potential fungi for copper remediation from Peninsular Malaysia. *Remediation Journal, 27*, 59–63. https://doi.org/10.1002/rem.21531
Ordoñez, J., Gago, E. J., & Girard, A. (2016). Processes and technologies for the recycling and recovery of spent lithium-ion batteries. *Renewable and Sustainable Energy Reviews, 60*, 195–205. https://doi.org/10.1016/j.rser.2015.12.363
Padilla, R., Pavez, P., & Ruiz, M. C. (2008). Kinetics of copper dissolution from sulfidized chalcopyrite at high pressures in H2SO4–O2. *Hydrometallurgy, 91*, 113–120.
Pathak, A., Morrison, L., & Healy, M. G. (2017). Catalytic potential of selected metal ions for bioleaching, and potential techno-economic and environmental issues: A critical review. *Bioresource Technology, 229*, 211–221. https://doi.org/10.1016/j.biortech.2017.01.001
Pollmann, K., Kutschke, S., Matys, S., Kostudis, S., Hopfe, S., & Raff, J. (2016). Novel biotechnological approaches for the recovery of metals from primary and secondary resources. *Minerals, 6*, 54. https://doi.org/10.3390/min6020054
Punt, T., Akdogan, G., Bradshaw, S., & van Wyk, P. (2021). Development of a novel solvent extraction process using citric acid for lithium-ion battery recycling. *Minerals Engineering, 173*, 107204. https://doi.org/10.1016/j.mineng.2021.107204
Quatrini, R., & Johnson, D. B. (2019). Acidithiobacillus ferrooxidans. *Trends in Microbiology, 27*, 282–283. https://doi.org/10.1016/j.tim.2018.11.009
Quintero-Almanza, D., Gamiño-Arroyo, Z., Sánchez-Cadena, L. E., Gómez-Castro, F. I., Uribe-Ramírez, A. R., Aguilera-Alvarado, A. F., & Ocampo Carmona, L. M. (2019). Recovery of cobalt from spent lithium-ion mobile phone batteries using liquid–liquid extraction. *Batteries 5*. https://doi.org/10.3390/batteries5020044.
Rawlings, D. E. (1997). Mesophilic, autotrophic bioleaching bacteria: Description, physiology and role. *Biomining*, pp. 229–245. https://doi.org/10.1007/978-3-662-06111-4_11.
Rohwerder, T., Gehrke, T., Kinzler, K., & Sand, W. (2003). Bioleaching review part A. *Applied Microbiology and Biotechnology, 63*, 239–248. https://doi.org/10.1007/s00253-003-1448-7
Roy, J. J., Cao, B., & Madhavi, S. (2021a). A review on the recycling of spent lithium-ion batteries (LIBs) by the bioleaching approach. *Chemosphere, 282*, 130944. https://doi.org/10.1016/j.chemosphere.2021.130944
Roy, J. J., Madhavi, S., & Cao, B. (2021b). Metal extraction from spent lithium-ion batteries (LIBs) at high pulp density by environmentally friendly bioleaching process. *Journal of Cleaner Production, 280*(2021), 124242. https://doi.org/10.1016/j.jclepro.2020.124242
Roy, J. J., Srinivasan, M., & Cao, B. (2021c). Bioleaching as an eco-friendly approach for metal recovery from spent nmc-based lithium-ion batteries at a high pulp density. *ACS Sustainable Chemistry and Engineering, 9*, 3060–3069. https://doi.org/10.1021/acssuschemeng.0c06573
Saneie, R., Abdollahi, H., Shafaei, S. Z., & Mohammadzadeh, A. (2021). Inhibitory effect of solvent extractants on growth and metabolism of acidophiles. *Mineral Processing and Extractive Metallurgy Review*, pp. 1–12. https://doi.org/10.1080/08827508.2021.1971664.
Sethurajan, M., & Gaydardzhiev, S. (2021). Bioprocessing of spent lithium ion batteries for critical metals recovery—A review. *Resources, Conservation and Recycling, 165*, 105225. https://doi.org/10.1016/j.resconrec.2020.105225
Sethurajan, M., Lens, P. N. L., Rene, E. R., van de Vossenberg, J., Huguenot, D., Horn, H. A., Figueiredo, L. H. A., & van Hullebusch, E. D. (2017). Bioleaching and selective biorecovery of zinc from zinc metallurgical leach residues from the Três Marias zinc plant (Minas Gerais, Brazil). *Journal of Chemical Technology and Biotechnology, 92*, 512–521. https://doi.org/10.1002/jctb.5026
Shamsuddin, M. (2021). Hydrometallurgy BT—physical chemistry of metallurgical processes, Second Edition. In M. Shamsuddin (Ed.) (pp. 429–529). Springer International Publishing. https://doi.org/10.1007/978-3-030-58069-8_11.
Shin, S.M., Kim, N.H., Sohn, J.S., Yang, D.H., & Kim, Y.H. (2005) Development of a metal recovery process from Li-ion battery wastes. *Hydrometallurgy 79*, 172–181. https://doi.org/10.1016/j.hydromet.2005.06.004
Shuya, L., Yang, C., Xuefeng, C., Wei, S., Yaqing, W., & Yue, Y. (2020). Separation of lithium and transition metals from leachate of spent lithium-ion batteries by solvent extraction method with Versatic 10. *Separation and Purification Technology, 250*, 117258. https://doi.org/10.1016/j.seppur.2020.117258
Sierra-Alvarez, R. (2007). Fungal bioleaching of metals in preservative-treated wood. *Process Biochemistry, 42*, 798–804. https://doi.org/10.1016/j.procbio.2007.01.019
Simate, G. S., Ndlovu, S., & Walubita, L. F. (2010). The fungal and chemolithotrophic leaching of nickel laterites—Challenges and opportunities. *Hydrometallurgy, 103*, 150–157. https://doi.org/10.1016/j.hydromet.2010.03.012
Sokić, M. D., Marković, B., & Živković, D. (2009). Kinetics of chalcopyrite leaching by sodium nitrate in sulphuric acid. *Hydrometallurgy, 95*, 273–279. https://doi.org/10.1016/j.hydromet.2008.06.012
Sommer, P., Rotter, V. S., & Ueberschaar, M. (2015). Battery related cobalt and REE flows in WEEE treatment. *Waste Management, 45*, 298–305. https://doi.org/10.1016/j.wasman.2015.05.009
Srichandan, H., Mohapatra, R. K., Parhi, P. K., & Mishra, S. (2019). Bioleaching approach for extraction of metal values from secondary solid wastes: A critical review. *Hydrometallurgy, 189*, 105122. https://doi.org/10.1016/j.hydromet.2019.105122
Srichandan, H., Pathak, A., Singh, S., Blight, K., Kim, D.-J., & Lee, S. W. (2014). Sequential leaching of metals from spent refinery catalyst in bioleaching–bioleaching and bioleaching–chemical leaching reactor: Comparative study. *Hydrometallurgy, 150*, 130–143. https://doi.org/10.1016/j.hydromet.2014.09.019
Srivastava, R. R., Ilyas, S., Kim, H., Choi, S., Trinh, H. B., Ghauri, M. A., & Ilyas, N. (2020). Biotechnological recycling of critical metals from waste printed circuit boards. *Journal of Chemical Technology and Biotechnology, 95*, 2796–2810. https://doi.org/10.1002/jctb.6469
Subhan, A., Oemry, F., Khusna, S. N., & Hastuti, E. (2019). Effects of activated carbon treatment on Li4Ti5O12 anode material synthesis for lithium-ion batteries. *Ionics (kiel), 25*, 1025–1034. https://doi.org/10.1007/s11581-018-2633-0
Tanskanen, P. (2013). Management and recycling of electronic waste. *Acta Materials, 61*, 1001–1011. https://doi.org/10.1016/j.actamat.2012.11.005

Tarascon, J.-M., & Armand, M. (2001). Issues and challenges facing rechargeable lithium batteries. *Nature, 414*, 359–367. https://doi.org/10.1038/35104644

Torkaman, R., Asadollahzadeh, M., Torab-Mostaedi, M., & Maragheh, M. G. (2017). Recovery of cobalt from spent lithium ion batteries by using acidic and basic extractants in solvent extraction process. *Separation and Purification Technology., 186*, 318–325. https://doi.org/10.1016/j.seppur.2017.06.023

Vakilchap, F., Mousavi, S. M., & Shojaosadati, S. A. (2016). Role of *Aspergillus niger* in recovery enhancement of valuable metals from produced red mud in Bayer process. *Bioresource Technology, 218*, 991–998. https://doi.org/10.1016/j.biortech.2016.07.059

Vanitha, M., & Balasubramanian, N. (2013). Waste minimization and recovery of valuable metals from spent lithium-ion batteries—a review. *Environmental Technology Review, 2*, 101–115. https://doi.org/10.1080/21622515.2013.853105

Vargas, L. H. M., Pião, A. C. S., Domingos, R. N., & Carmona, E. C. (2004). Ultrasound effects on invertase from *Aspergillus niger*. *World Journal of Microbiology and Biotechnology, 20*, 137–142. https://doi.org/10.1023/B:WIBI.0000021726.69233.12

Versaci, D., Nasi, R., Zubair, U., Amici, J., Sgroi, M., Dumitrescu, M. A., Francia, C., Bodoardo, S., & Penazzi, N. (2017). New eco-friendly low-cost binders for Li-ion anodes. *Journal of Solid State Electrochemistry, 21*, 3429–3435. https://doi.org/10.1007/s10008-017-3665-5

Vikström, H., Davidsson, S., & Höök, M. (2013). Lithium availability and future production outlooks. *Applied Energy, 110*, 252–266. https://doi.org/10.1016/j.apenergy.2013.04.005

Vyas, S., & Ting, Y. -P. (2018). A review of the application of ultrasound in bioleaching and insights from sonication in (Bio) chemical processes. *Resources 7*. https://doi.org/10.3390/resources7010003.

Walker, G. M., White, N. A. (2018). Introduction to fungal physiology, pp. 1–35.

Wang, H., Huang, K., Zhang, Y., Chen, X., Jin, W., Zheng, S., Zhang, Y., & Li, P. (2017). Recovery of lithium, nickel, and cobalt from spent lithium-ion battery powders by selective ammonia leaching and an adsorption separation system. *ACS Sustainable Chemistry and Engineering, 5*, 11489–11495. https://doi.org/10.1021/acssuschemeng.7b02700

Wang, J., Tian, B., Bao, Y., Qian, C., Yang, Y., Niu, T., & Xin, B. (2018). Functional exploration of extracellular polymeric substances (EPS) in the bioleaching of obsolete electric vehicle $LiNi_xCo_yMn_{1-x-y}O_2$ Li-ion batteries. *Journal of Hazardous Materials, 354*, 250–257. https://doi.org/10.1016/j.jhazmat.2018.05.009

Wang, X., Chen, J., Yan, X., Wang, X., Zhang, J., Huang, J., & Zhao, J. (2015). Heavy metal chemical extraction from industrial and municipal mixed sludge by ultrasound-assisted citric acid. *Journal of Industrial and Engineering Chemistry, 27*, 368–372. https://doi.org/10.1016/j.jiec.2015.01.016

Wang, X., Gaustad, G., Babbitt, C. W., Bailey, C., Ganter, M. J., & Landi, B. J. (2014a). Economic and environmental characterization of an evolving Li-ion battery waste stream. *Journal of Environmental Management, 135*, 126–134. https://doi.org/10.1016/j.jenvman.2014.01.021

Wang, Y., Su, L., Zeng, W., Wan, L., Chen, Z., Zhang, L., Qiu, G., Chen, X., & Zhou, H. (2014b). Effect of pulp density on planktonic and attached community dynamics during bioleaching of chalcopyrite by a moderately thermophilic microbial culture under uncontrolled conditions. *Minerals Engineering, 61*, 66–72. https://doi.org/10.1016/j.mineng.2014.03.012

Windisch-Kern, S., Gerold, E., Nigl, T., Jandric, A., Altendorfer, M., Rutrecht, B., Scherhaufer, S., Raupenstrauch, H., Pomberger, R., Antrekowitsch, H., & Part, F. (2022). Recycling chains for lithium-ion batteries: A critical examination of current challenges, opportunities and process dependencies. *Waste Management, 138*, 125–139. https://doi.org/10.1016/j.wasman.2021.11.038

Wu, H.-Y., & Ting, Y.-P. (2006). Metal extraction from municipal solid waste (MSW) incinerator fly ash—Chemical leaching and fungal bioleaching. *Enzyme and Microbial Technology, 38*, 839–847. https://doi.org/10.1016/j.enzmictec.2005.08.012

Wu, W., Liu, X., Zhang, X., Li, X., Qiu, Y., Zhu, M., & Tan, W. (2019). Mechanism underlying the bioleaching process of $LiCoO_2$ by sulfur-oxidizing and iron-oxidizing bacteria. *Journal of Bioscience and Bioengineering, 128*, 344–354. https://doi.org/10.1016/j.jbiosc.2019.03.007

Xia, M., Bao, P., Liu, A., Wang, M., Shen, L., Yu, R., Liu, Y., Chen, M., Li, J., Wu, X., Qiu, G., & Zeng, W. (2018). Bioleaching of low-grade waste printed circuit boards by mixed fungal culture and its community structure analysis. *Resources, Conservation and Recycling, 136*, 267–275. https://doi.org/10.1016/j.resconrec.2018.05.001

Xiang, Y., Wu, P., Zhu, N., Zhang, T., Liu, W., Wu, J., & Li, P. (2010). Bioleaching of copper from waste printed circuit boards by bacterial consortium enriched from acid mine drainage. *Journal of Hazardous Materials, 184*, 812–818. https://doi.org/10.1016/j.jhazmat.2010.08.113

Xin, B., Jiang, W., Aslam, H., Zhang, K., Liu, C., Wang, R., & Wang, Y. (2012b). Bioleaching of zinc and manganese from spent Zn-Mn batteries and mechanism exploration. *Bioresource Technology, 106*, 147–153. https://doi.org/10.1016/j.biortech.2011.12.013

Xin, B., Jiang, W., Li, X., Zhang, K., Liu, C., Wang, R., & Wang, Y. (2012a). Analysis of reasons for decline of bioleaching efficiency of spent Zn-Mn batteries at high pulp densities and exploration measure for improving performance. *Bioresource Technology, 112*, 186–192. https://doi.org/10.1016/j.biortech.2012.02.133

Xin, B., Zhang, D., Zhang, X., Xia, Y., Wu, F., Chen, S., & Li, L. (2009). Bioleaching mechanism of Co and Li from spent lithium-ion battery by the mixed culture of acidophilic sulfur-oxidizing and iron-oxidizing bacteria. *Bioresource Technology, 100*, 6163–6169. https://doi.org/10.1016/j.biortech.2009.06.086

Xin, Y., Guo, X., Chen, S., Wang, J., Wu, F., & Xin, B. (2016). Bioleaching of valuable metals Li Co, Ni and Mn from spent electric vehicle Li-ion batteries for the purpose of recovery. *Journal of Cleaner Production, 116*, 249–258. https://doi.org/10.1016/j.jclepro.2016.01.001

Xu, J., Thomas, H. R., Francis, R. W., Lum, K. R., Wang, J., & Liang, B. (2008). A review of processes and technologies for the recycling of lithium-ion secondary batteries. *Journal of Power Sources, 177*, 512–527. https://doi.org/10.1016/J.JPOWSOUR.2007.11.074

Xu, L., Chen, C., & Fu, M.-L. (2020). Separation of cobalt and lithium from spent lithium-ion battery leach liquors by ionic liquid extraction using Cyphos IL-101. *Hydrometallurgy, 197*, 105439. https://doi.org/10.1016/j.hydromet.2020.105439

Xu, T.-J., Ramanathan, T., & Ting, Y.-P. (2014). Bioleaching of incineration fly ash by *Aspergillus niger*—precipitation of metallic salt crystals and morphological alteration of the fungus. *Biotechnology Reports, 3*, 8–14. https://doi.org/10.1016/j.btre.2014.05.009

Xu, W., Wang, Z., Shi, L., Ma, Y., Yuan, S., Sun, L., Zhao, Y., Zhang, M., & Zhu, J. (2015). Layer-by-layer deposition of organic-inorganic hybrid multilayer on microporous polyethylene separator to enhance the electrochemical performance of lithium-ion battery. *ACS Applied Materials and Interfaces, 7*, 20678–20686. https://doi.org/10.1021/acsami.5b05457

Yu, Z., Han, H., Feng, P., Zhao, S., Zhou, T., Kakade, A., Kulshrestha, S., Majeed, S., & Li, X. (2020). Recent advances in the recovery of metals from waste through biological processes. *Bioresource Technology, 297*, 122416. https://doi.org/10.1016/j.biortech.2019.122416

Zeng, G., Deng, X., Luo, S., Luo, X., & Zou, J. (2012). A copper-catalyzed bioleaching process for enhancement of cobalt dissolution from spent lithium-ion batteries. *Journal of Hazardous Materials, 199–200*, 164–169. https://doi.org/10.1016/j.jhazmat.2011.10.063

Zeng, G., Luo, S., Deng, X., Li, L., & Au, C. (2013a). Influence of silver ions on bioleaching of cobalt from spent lithium batteries. *Minerals Engineering, 49*, 40–44. https://doi.org/10.1016/j.mineng.2013.04.021

Zeng, X., Li, J., & Liu, L. (2015). Solving spent lithium-ion battery problems in China: Opportunities and challenges. *Renewable and Sustainable Energy Reviews, 52*, 1759–1767. https://doi.org/10.1016/j.rser.2015.08.014

Zeng, X., Li, J., & Singh, N. (2014). Recycling of spent lithium-ion battery: A critical review. *Critical Reviews in Environmental Science and Technology, 44*, 1129–1165. https://doi.org/10.1080/10643389.2013.763578

X. Zheng, W. Gao, X. Zhang, M. He, X. Lin, H. Cao, Y. Zhang, Z. Sun, Spent lithium-ion battery recycling – Reductive ammonia leaching of metals from cathode scrap by sodium sulphite, Waste Manag. 60 (2017) 680–688. https://doi.org/10.1016/j.wasman.2016.12.007.

Zheng, X., Zhu, Z., Lin, X., Zhang, Y., He, Y., Cao, H., & Sun, Z. (2018). A mini-review on metal recycling from spent lithium ion batteries. *Engineering, 4*, 361–370. https://doi.org/10.1016/j.eng.2018.05.018

Zou, H., Gratz, E., Apelian, D., & Wang, Y. (2013). A novel method to recycle mixed cathode materials for lithium ion batteries. *Green Chemistry, 15*, 1183–1191. https://doi.org/10.1039/C3GC40182K

Bio-Beneficiation: Relevance to Mineral Processing

Hadi Abdollahi, Sina Ghassa, Ehsan Ebrahimi, Amirhossein Mohammadzadeh, Morteza Shahbaznejad and Roozbeh Saneie

Abstract

The increasing demand for various minerals and the limited high-grade mineral resources, concerning economic and environmental issues, has led to increasing research attention to the bio-beneficiation of low-grade mineral resources. Bio-beneficiation can be defined as employing microorganisms (including bacteria, fungi, algae, and yeast) in mineral processing and related industries. The high potential of microorganisms and their metabolites, especially extracellular polymeric substances (EPS), has been substantiated in bio-beneficiation processes. The bio-beneficiation is generally divided into two main categories: including bioflotation and bioflocculation. The bioflotation uses the microorganisms and their products (such as biomass and EPS) as flotation reagents. Microorganisms and their biomass can be applied as collectors, depressants, dispersants, and frothers in minerals floatation. Bioflocculation is another application of biotechnology in minerals processing. Almost all mineral processing techniques should be carried out in a wet environment. Therefore, dewatering and water recycling are essential steps in mineral processing plants. Microorganisms can be used as flocculants to decrease the dewatering time or minerals separation with high efficiency. Comprehensive information about flocculation, classification of flocculants, and the application of EPS as bioflocculant for selective bioflocculation of different minerals are provided in this chapter. The mechanism of the bioflocculation process and the effect of different parameters on the bioflocculation of various minerals have been investigated. Moreover, recent research on the application of bioflocculation in the selective separation or removal of minerals using different types of microorganisms has been briefly reviewed. Employing microorganisms for both bioflotation and bioflocculation can be a great strategy to save the environment and decrease process costs.

Keywords

Bioflocculation · Bioflotation · Mineral processing · Bio-beneficiation

1 Introduction

Nowadays, microorganisms are widely used in bioleaching of sulfide minerals, recycling mineral processing plant tailings, and treating hazardous wastes (Johnson & Hallberg, 2005; Olson et al., 2003). With rising demand for minerals and declining high-grade resources, researchers have increasingly focused on the beneficiation of low-grade mineral resources to meet the demand of global markets. Also, considering the resources and economic issues, today's industrial approach is further decreasing the effect on the environment of mineral processing activities, resulting in considerable developments in the application of biotechnology in mineral processing procedures. In bio-beneficiation, which includes bioflocculation and bioflotation processes, microorganisms and their metabolites may be used as collectors, regulators, flocculants, depressants, and surface modifiers for the flotation or flocculation of selected minerals (Chapelle, 2000). The role of microorganisms and their metabolites in the bio-beneficiation industry for the selective removal or separation of gangue or valuable minerals has been fully established. The effectiveness of selective separation of minerals in the refining processes has been substantiated. Fundamentally, in bio-beneficiation, the role of microorganisms is to alter the physicochemical properties of minerals' surfaces to

H. Abdollahi (✉) · S. Ghassa · E. Ebrahimi · A. Mohammadzadeh · M. Shahbaznejad · R. Saneie
School of Mining Engineering, College of Engineering, University of Tehran, P.O. Box 1439957131, Tehran, Iran
e-mail: h_abdollahi@ut.ac.ir

S. Panda et al. (eds.), *Biotechnological Innovations in the Mineral-Metal Industry*, Advances in Science, Technology & Innovation, https://doi.org/10.1007/978-3-031-43625-3_6

achieve the highest separation efficiency (Groudev, 1987). Although biohydrometallurgical methods have been extensively explored and commercialized for ores like low-grade sulfides—particularly copper, uranium, lead, and zinc—as well as precious metals like gold and silver, bio-beneficiation methods are still evolving and require deeper investigations. (Hanumantha Rao & Forssberg, 2001; Holmes & Smith, 1995; Smith et al., 1991). The primary purpose of this chapter is to present recent reports on the applications of bio-beneficiation in mineral processing, address the mechanism of bioflocculation and bioflotation, and the factors affecting the process.

2 Absorption and Modification of Mineral Surfaces by Microorganisms

The absorption of microorganisms on solid surfaces is essential in nature for their growth. In most natural and artificial systems, the metabolic activity of microorganisms, especially bacteria, is associated with solid surfaces such as soil, minerals, and tissues. Surface biofilms can be employed in minerals' bioprocessing (Somasundaran et al., 2005). Production of metabolites by microorganisms and accumulation of their EPS is of significant importance in microorganism-mineral interactions (Deo & Natarajan, 1998). Interactions of microorganisms and minerals have the following results:

- Alteration in the chemical properties of the minerals' surface
- Reaction of bacterial cells and metabolic products with minerals
- Production of surface-active chemicals.

These interactions lead to surface modification, alteration of minerals' surface properties, selective dissolution of different components from the mineral's surface, and bioaccumulation of soluble metal ions. In addition to metabolic products, cell components of microorganisms such as bacterial cell walls and membranes can participate in these surface reactions (Somasundaran et al., 2005).

3 Bio-Beneficiation Mechanisms

Attachment between microorganisms and mineral particles is due to biological interactions like growth, metabolism, EPS secretion (Grossart et al., 2006; Simon et al., 2002), which may lead to the minerals particles' surface modifications (Kiørboe, 2001; Tang et al., 2014). In particular, through bioflocculation processes, the application of EPS has been proved (More et al., 2014; Sheng et al., 2010; Tang & Maggi, 2016). In fact, in the process of bioflocculation, the extracellular polymeric substances excreted by microorganisms are used as bioflocculants. Investigations have shown that increasing the content of EPS increases the capacity of bioflocculation (Badireddy et al., 2010). The collision between particles is the most important subject in particle aggregation; only a successful collision can cause flocs to agglomerate together. This effective collision can be created by different forces, such as electrostatic forces, Van der Waals forces, hydrophilic and hydrophobic interactions, and the polymer bridging through chemical, physical, and biological processes. Since the role of EPS and microorganisms in bioflocculation is not understood well yet, bioaccumulation mechanisms in bioflocculation remain unclear (Lai et al., 2018). However, several mechanical theories, including Derjaguin, Landau, Verwey, and Overbeek (DLVO) and extended DLVO (XDLVO) theory, cationic bridging theory (Sobeck & Higgins, 2002), and polymer bridging or adsorption bridging theory (Bolto & Gregory, 2007), have been suggested as the most important possible mechanisms for bio-beneficiation.

3.1 DLVO and XDLVO Theory

The (DLVO) theory was proposed to define the stabilization of colloidal suspensions for the first time (Derjaguin et al., 1987; Lai et al., 2018). According to this theory, the flocculation of dispersed particles in water can be explained quantitatively. The total surface energy of a particle is calculated by considering Lifshitz–Van der Waals (W^{LW}) and electric double-layer energy (W^{WE}) (Rijnaarts et al., 1999), as shown in (Fig. 1a) and (Fig. 1b). Particle aggregation occurs when particles defeat a potential barrier. DLVO theory can also define the mechanism of bioflocculation. The total interaction energy (W^{TOT}) from extended DLVO (XDLVO) theory can be represented by Eq. (1) (Chia et al., 2011; Lai et al., 2018):

$$W^{TOT} = W^{LW} + W^{EL} + W^{AB}, \quad (1)$$

where W^{EL} is a practical function of the distance between two particles with an amplitude of the thickness of electric double layers, W^{LW} decreases with increasing distance between particles, and W^{AB} is the energy due to Lewis base acid interactions, and based on thermodynamic parameters, it can be attractive or repulsive (Lai et al., 2018). Liu et al. (2008) examined the formulas of this theory, the sizes, and contributions of W^{EL}, W^{LW}, and W^{AB} in detail (Liu et al., 2008). Their research shows that Van der Waals attraction forces dominate at small distances between particles, while at middle distances, electric double-layer repulsion and Lewis acid–base interactions may predominate. Either

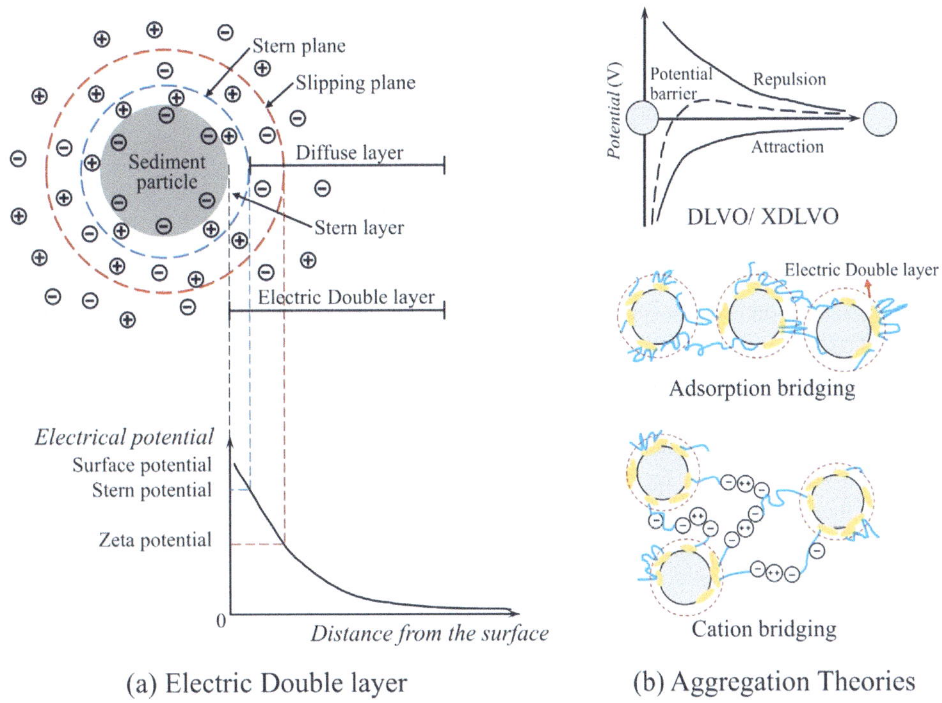

Fig. 1 **a** Electrical potential of double layer, **b** Three possible mechanisms for bio-beneficiation (Lai et al., 2018)

DLVO or XDLVO theories were used extensively in bioflocculation and bioflotation (Li et al., 2012, 2014; Liu et al., 2007, 2008, 2010).

3.2 Polymer Bridging (Adsorption Bridging) Theory

Polymers are heavy molecules with long chains in their structure that have many binding sites for attachment to particles. EPSs produced by microorganisms are known as natural biopolymers. In general, polymers can adsorb on the surface of particles' by formation loops and tails, improving the attachment between particles and bridging them together (Bolto & Gregory, 2007; Joon & Schlautman, 2015). As shown in (Fig. 1b), this mechanism is known as polymer bridging or the adsorption bridging in bio-beneficiation. Hocking et al. (1999) suggested that the longer polymers enable to interact with more than one particle. Generally, the flocs formed by the polymer bridging mechanism are stronger than the flocs produced by other mechanisms because, in the case of polymer bridging, the flocs are highly flexible that do not easily disappear in the face of shear stress (Gregory & Barany, 2011). The degree of the covered surface area by the adsorbed polymer is an important factor in the bridging chance because, in the polymer bridging mechanism, the particles connected by the bridge must have an unoccupied surface to connect the polymer parts (Biggs et al., 2000). At high adsorption levels, the available adsorption sites will be insufficient, while at low adsorption levels, the connection of the formed bridges to each other will be insufficient, both of which are unsuitable for the polymer bridging adsorption flocculation mechanism (Lai et al., 2018). Bolto and Gregory (2007) indicated that the optimal bridging flocculation occurs at a level lower than saturation surface coverage of the adsorbed polymers (Bolto & Gregory, 2007).

3.3 Cation Bridging Theory

This theory was applied in by Tezuka (1969) primarily (Tezuka, 1969). Interactions between EPS and solid particles play a critical role in the bioflocculation of colloids. The bioflocculation process is also influenced by the surface characteristics of EPS. In the compound of EPS, there are many functional groups with negative charges (Sheng et al., 2010). Consequently, the presence of cations (solid particles with positive surface charge) can act as a bridge that connects functional groups in EPS, leading to particle aggregation, resulting in bioflocculation of solids, as presented in (Fig. 1b) (Sobeck & Higgins, 2002; Wilén et al., 2008). Based on the cation bridging theory, the effectiveness of bioflocculation depends on EPS molecular structure, the weight of the molecule, the surface charge of EPS, and the positive charge density on a solid surface, so the presence of polyvalent cations can increase the efficiency of bioflocculation (Lai et al., 2018). DLVO and polyvalent cation bridging theories can examine the impact of cations on bioflocculation. The electric double layers are compacted when there is a high concentration of cations in a solution (Mietta, 2010) and thus, based on DLVO theory, increase the flocculation capacity. The polyvalent cation performs like a link between the negative charge of biopolymer and solid particles, thereby improving flocculation (Higgins & Novak, 1997a; Sobeck & Higgins, 2002). However, the actual effects of cations, especially the polyvalent cation in this mechanism, are not recognized well yet, and more profound researches are required. In general, the mechanism of the bioflocculation process is complex but can be described by XDLVO theories, adsorption bridging, and cationic bridging theory. Nevertheless, each has its limitations.

XDLVO theory evaluates the energy between particles, but the zeta potential of particles surface and charge redistribution by EPS are not factored into the equation in the XDLVO theory. In contrast, these factors could significantly impact the bioflocculation process (Sobeck & Higgins, 2002). The adsorption action between particles and EPS can be described by adsorption bridging theory, and the cation bridging theory clarifies why polyvalent cations can help bioflocculation. However, these two theories are conceptual models, and there is a limit to the quantitative evaluation of these theories (Lai et al., 2018).

3.4 Extracellular Polymeric Substances (EPS)

The metabolism of microorganisms leads to the secretion of extracellular polymeric substances (EPS). Wingender et al. (1999) proposed the "EPS" as a comprehensive and common term to denote various biomolecules produced by microbes, which include proteins, nucleic acids, polysaccharides, lipoproteins, and other biopolymer substances (Wingender et al., 1999). Wingender et al. (1999) suggested that all extracellular polymers that do not bind directly to peptidoglycan must be considered as EPS (Wingender et al., 1999). EPS are mainly macromolecular with high molecular weights (Liu & Fang, 2003; Wingender et al., 1999). Outside of the cell, EPS is categorized into two groups (Laspidou, 2002; Wingender et al., 1999):

1. Bound EPS, which includes coatings, capsular polymers, compact gels, polymers, and attached organic compounds
2. Soluble EPS, which contains soluble macromolecules, colloids, and sludge.

Bound EPS is completely limited to cells, while soluble EPS binding to cells is weak and easily dissolved in solution. In general, these two types of EPS can be detached by centrifugation. The bound EPS structure is generally represented by double layers (Fig. 2b) (Lin et al., 2014). The inner layer comprises tightly bound EPS (TB-EPSs), which have a special shape and are firmly placed adjacent to the cell surface, and the outer layer contains loosely bound EPS (LB-EPSs).

Macromolecules and polymer compounds produced by microorganisms are used as bioflocculants in bioflocculation processes. The volume of EPS production by microorganisms is mainly affected by the growth stage of microbes. Different food sources such as carbon, nitrogen, and other nutrients and environmental factors affect EPS production. Operating factors such as pH, temperature, metal ions, and oxygen content of the culture medium (Nichols et al., 2005; Sheng et al., 2010) also directly affect the amount of EPS and bioflocculants produced by microorganisms (Ye et al., 2011). Badireddy et al. (2010) suggested that bioflocculation capacity improves with the secretion of EPS. In the exponential growth stage of the microorganism, due to the low level of EPS content, the bioflocculation capacity is low, while the bioflocculation capacity significantly increases in the growth stabilization stage, with increasing EPS secretion (Badireddy et al., 2010). More et al. (2014) suggested that the optimal amount of EPS excreted by several microorganisms in the bioflocculation of kaolinite was in the range of 1–5000 mg/l depending on the composition

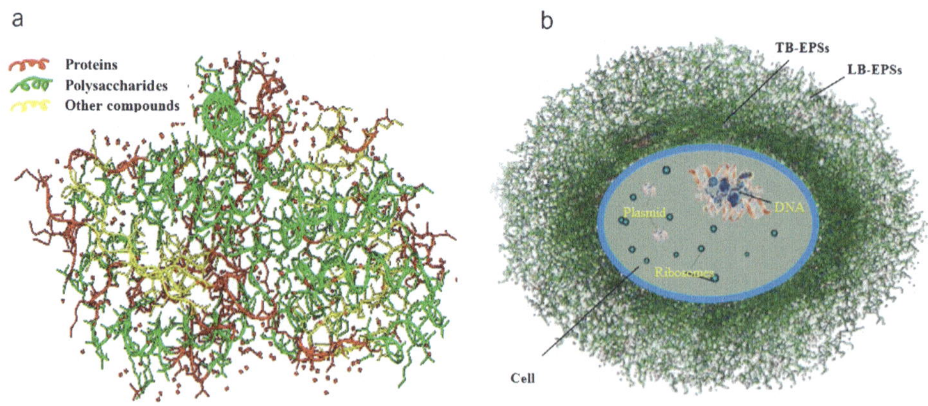

Fig. 2 **a** EPS compounds, **b** Different types of EPS sections: tightly bound EPS (TB-EPS), loosely bound EPS (LB-EPS), spatial patterns, and their position (Lin et al., 2014)

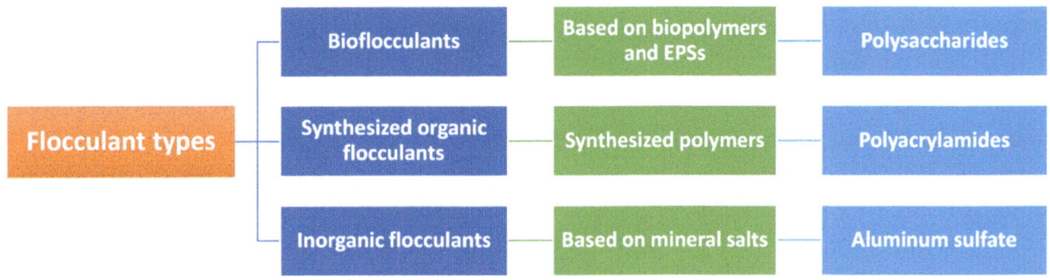

Fig. 3 Flocculant types with examples

of the EPS. It was indicated that the bioflocculation capacity should be determined by the EPS dosage and the characteristics of the EPS compounds (More et al., 2014; Tsuneda et al., 2003). Carbohydrates and proteins are regularly the main constituents of EPS (Fig. 2). (Fr/olund et al., 1995; Frølund et al., 1996). In addition, nucleic acids, humic substances, uronic acids, and some other components may be found in EPS (D'Abzac et al., 2010a, 2010b; Dignac et al., 1998; Frølund et al., 1996).

Numerous functional groups exist in the EPS compounds that significantly alter the surface properties of the flocs. The macromolecules in the EPS compound can interact with ultrafine particles by surface adsorption, hydrogen bonding, complex formation in surface, hydrophilic and hydrophobic interactions, protein-polysaccharide interactions, and electrostatic interaction, all of which indicate specific roles of EPS functional groups in bioflocculation (Higgins & Novak, 1997b; Parikh & Chorover, 2006; Wilén et al., 2003). The phosphate, carboxylate, and amine functional groups are involved in bacterial adsorption to the surface of the mineral (Parikh & Chorover, 2006). According to Lurie et al. (1997), humic compounds and other functional groups with extremely negative charge densities might cause upper repulsive forces between molecules (Lurie & Rebhun, 1997). In a study, Badireddy et al. (2010) showed that functional groups include carbohydrates and alcohols, which improve the bioflocculation process, while functional groups comprise carboxylate, carbonyl, acetal and in some cases reduce bioflocculation efficiency. Therefore, functional groups can play an essential role in bioflocculation by altering particle surface characteristics and floc interactions (Badireddy et al., 2010).

4 Bioflocculation

4.1 Flocculation

With the increasing complexity of ore compositions, excessive crushing and grinding of these resources for liberation in mineral processing plants have led to the formation of large volumes of fine particles. One of the most common methods for the separation of fine, very fine, and colloidal minerals is their selective flocculation using flocculants. The fine particles that must be separated from the liquid might vary in size, morphology, and chemical nature. Colloidal particles are larger than molecules, but they are so small that the gravitational forces are far less than the repulsive electric force between them. Under such conditions, the stability of Brownian motion keeps the particles suspended and dispersed in the liquid. Flocculation occurs when predominantly high molecular weight polymeric compounds cause particle aggregation and floc formation. The particular compounds used for this purpose are called "flocculants" (Hughes, 1990). Flocculants are divided into three main categories (Fig. 3) (Salehizadeh & Shojaosadati, 2001): (a) inorganic flocculants, such as aluminum sulfate salts, polyaluminum chloride, iron chloride, and iron sulfate, (b) synthesized organic flocculants, such as polyacrylamide and its derivatives, and polyethylenimine, (c) biological or bioflocculants, such as chitosan, sodium alginate, gelatin, and flocculants based on extracellular polymeric substances (EPS); bioflocculants are produced by microorganisms and generally containing polysaccharides, proteins, glycoproteins, and amino acids.

4.2 Biotechnology for Flocculation

Most synthetic flocculants are significantly toxic and harmful to animals, sea organisms, and humans (Campbell, 2002). For example, acrylamide monomer, a small amount of which can cause contamination, is carcinogenic and dangerous to humans (Gao et al., 2009). It was proved that aluminum salts might cause Alzheimer's in humans (Salehizadeh & Shojaosadati, 2001). Due to the mentioned concerns, today, the bioflocculants are preferable for flocculation process in various industrial plants such as water and wastewater treatment, decolorization of solutions, mining and mineral processing, pharmacy and serology, food productions, and many industries because most bioflocculants are non-toxic and fully environmentally friendly (Chen et al., 2014; Salehizadeh & Shojaosadati, 2001; Virk-Baker et al., 2014; Zhuang et al., 2012). Different types of microorganisms, including bacteria, fungi, microalgae, and their metabolites, have been applied as bioflocculants with different structures and properties. These biopolymers are active materials that are biodegradable and environmentally friendly with significant flocculation capabilities (Ben et al., 2018). Microbial bioflocculants produced during the growth of bacteria are different in the composition of polysaccharides, proteins, cellulose, sugar, and polyaminoacids. The type and amount of nutrients and culture medium conditions for the growth of microorganisms have an essential effect on the amount of bioflocculant produced. Thus, the amount of bioflocculant produced is directly affected by carbon and nitrogen sources in the culture medium, operating temperature, pH, inoculation ratio, and aeration rate (Salehizadeh & Yan, 2014). The absorption of bacteria on the minerals is very important for the surface modification of minerals in bio-beneficiation processes. Most bioflocculation and bioflotation studies have shown that the initial absorption of bacteria is mainly controlled by the physicochemical of the bacterial cell surface properties, which are associated with the arrangement of protein membranes and polysaccharides (Raichur et al., 1996).

4.3 Application of Microorganisms in Bioflocculation

4.3.1 Paenibacillus polymyxa

Paenibacillus polymyxa is a neutrophilic, heterotrophic, gram-positive, periflagellated bacterium found with some oxide minerals. It releases EPS which contains polysaccharides, proteins, organic acids such as formic acid, acetic acid, and oxalic acids (Murphy, 1952). The application of *P. polymyxa* to the beneficiation of different minerals has been reported in several works.

Oxide Minerals

Interaction between *Paenibacillus polymyxa* with oxide minerals causes chemical changes on minerals and bacteria surface (Deo & Natarajan, 1998; Vijayalakshmi & Raichur, 2002), for instance, based on Deo et al. (1998) investigations, after biotreatment by *P. polymyxa*, quartz and kaolinite became hydrophobic, while hematite and corundum converted hydrophilic. The EPS content of bacteria causes surface-chemical changes. As shown in (Fig. 4), adhesion of biopolymers extracted from *P. polymyxa* on minerals surfaces follows this order: extracellular proteins show high adhesion on kaolinite and quartz, while extracellular polysaccharides show high adhesion on corundum and hematite surface (Deo & Natarajan, 1998).

Extracted bioflocculants from bacterial EPS such as polysaccharides can selectively flocculate fine particles of hematite and corundum through a polymer bridging mechanism. Based on (Table 1), the settling rates of quartz and kaolinite saw to be reduced. In contrast, the settling rate of hematite and corundum enhanced due to interaction with *P. polymyxa* cells or the metabolite (Deo & Natarajan, 1998). Thus, selective bio-beneficiation of oxide minerals like hematite and corundum was accomplished by selective dispersion of kaolinite and quartz particles.

Coal

Paenibacillus polymyxa also is used to remove ash and quartz from coal samples by bioflocculation (Vijayalakshmi & Raichur, 2002). Vijayalakshmi et al. (2002) investigated the application of the *P. polymyxa* to remove ash from coal samples. As presented in (Fig. 5), the high ash coal and the bacterium have a very similar ZPC in the pH range of 2–3, and both show a negative charge over a wide range of pH values. The maximum adhesion of *P. polymyxa* on the coal samples occurs at pH 2, which is very close to the ZPC, while the minimum number of bacteria is attached to the quartz surface at this pH (Vijayalakshmi & Raichur, 2002). Settling studies in the presence of the *P. polymyxa* proved that coal samples flocculated at a higher rate than quartz, thus representative selective separation of coal from quartz by bioflocculation (Fig. 6). Similar results were demonstrated for removing ash from coal samples. According to results, almost 55–60% of the ash was removed from coal samples in a bioflocculation using *P. polymyxa* (Vijayalakshmi & Raichur, 2002).

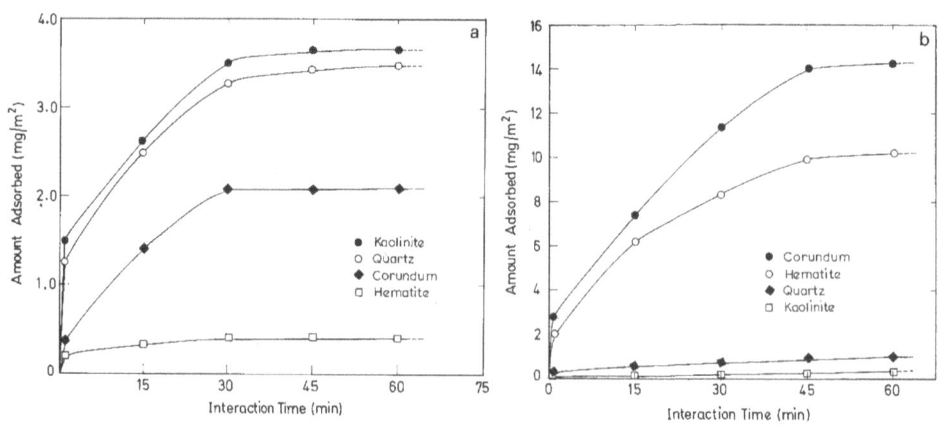

Fig. 4 *P. polymyxa* adsorption density **a** protein and **b** polysaccharides on the surface of different minerals (Deo & Natarajan, 1998)

Table 1 Bioflocculation of minerals interacted with cells and metabolite of *P. polymyxa* (5 min at 5% pulp density, size <38 μm) (Deo & Natarajan, 1998)

Mineral	pH	Weight settled(%) in 1 min		
		Control	Bacterial cells	Metabolite
Quartz	4–5	58	10	17.1
	7	45	18	–
	12	38	11	–
Kaolinite	4–5	85	40	1.65
	7	70	20	–
	12	62	18	–
Corundum	4–5	85	97	90
	7	82	96	–
	12	70	90	–
Hematite	4–5	84	99	92
	7	80	95	–
	12	70	98.8	–

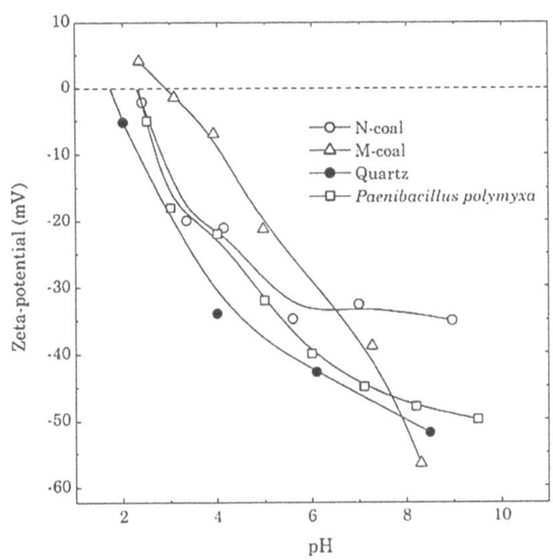

Fig. 5 Zeta potential of coal samples and the *P. polymyxa* (Vijayalakshmi & Raichur, 2002)

Sulfide Minerals

Santhiya et al. (2002) studied on adsorption of *P. polymyxa* and its EPSs on galena and sphalerite. Considering their results, selective bioflocculation and separation of galena from sphalerite in the presence of *P. polymyxa* at pH = 9–9.5 proved (Santhiya et al., 2001a, 2002). Adsorption studies showed that the amount of adsorbed bacteria cells onto galena against sphalerite was significant, and the adsorption density of the *P. polymyxa* cells onto the galena surface was not affected by pH. In contrast, the adsorption of the bacterial cells onto the sphalerite surface reduced with a rise in the amount of pH (Santhiya et al., 2001a). For galena-interacted cells, polysaccharides were the dominant bioflocculant. In contrast, for sphalerite-interacted cells, the protein was the main bioflocculation agent (Santhiya et al., 2002). The surface hydrophobicity investigations approved that the sphalerite was more hydrophobic while the galena was more hydrophilic after interaction with *P. polymyxa* cells. Thus, in addition to the hydrophilic character of galena-interacted cells, high polysaccharides and fewer protein concentrations could be the main reason for selective flocculation and separation of galena from sphalerite. Almost 95% of galena is separated from sphalerite through selective bioflocculation presented in (Table 2) (Santhiya et al., 2001a).

Patra et al. (2004) studied on adsorption of *P. polymyxa* and its EPSs on pyrite and sphalerite. According to their studies, selective bioflocculation and separation of pyrite from sphalerite were verified after interaction with either bacterial cells or extracellular proteins in the pH range of 8–9 (Patra & Natarajan, 2004a). Adsorption investigations showed that the amount of adsorbed bacteria cells onto the pyrite surface was significant compared to sphalerite. Patra et al. (2006) also examined selective bioflocculation and pyrite removal from galena in the presence of *P. polymyxa*, and it EPSs in the pH range of 6–7 after interaction with either bacterial cells or extracellular proteins. Adsorption investigations showed that the adsorption of bacterial cells onto pyrite was significant against galena. Selective flocculation results at pH = 8.5–9 verified that 91% of pyrite could be selectively separated from galena (Patra & Natarajan, 2006).

Patra & Natarajan (2004a, b) achieved selective separation of chalcopyrite and pyrite from oxide gangue minerals like quartz and calcite through bioflocculation after interaction with cells of *P. polymyxa* or proteins separated from its EPS (Patra & Natarajan, 2003, 2004b). Based on the results, *P. polymyxa* cells had a great affinity for chalcopyrite compared to quartz. Adhesion of *P. polymyxa* follows this order (Patra & Natarajan, 2004b): Chalcopyrite >> Calcite > Quartz.

In the case of quartz biotreatment, the author indicated that extracellular bacterial protein could be responsible for hydrophobicity increasing in quartz surface, which helped dispersion of quartz, while polysaccharides promoted quartz flocculation. They found that the dispersion performance of calcite interacted with *P. polymyxa* cells was similar to quartz. The author proved that extracellular protein was a predominant factor in the selective separation of quartz from chalcopyrite.

4.3.2 *Acidithiobacillus* Group

Acidithiobacillus group are gram-negative, acidophilic, mesophilic, bacteria such as *Acidithiobacillus ferrooxidans* and *Acidithiobacillus thiooxidans*, which are widely used in bioleaching and bio-beneficiation of

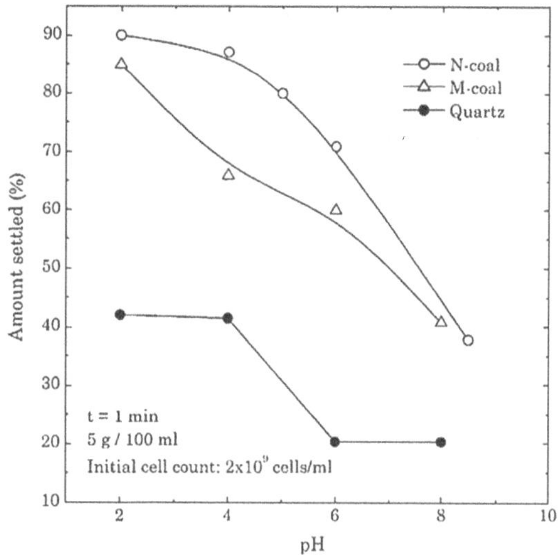

Fig. 6 Bioflocculation of coal and quartz by *P. polymyxa* (Vijayalakshmi & Raichur, 2002)

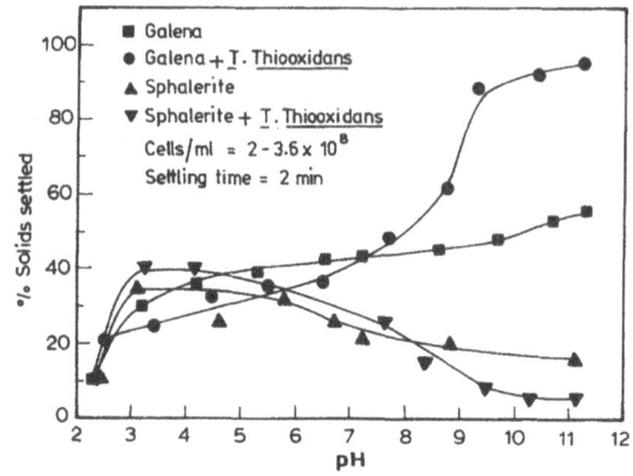

Fig. 7 Bioflocculation of galena and sphalerite as a function of pH in the absence and presence of *Acidithiobacillus thiooxidans* (Santhiya et al., 2000)

Table 2 Selective bioflocculation sphalerite from Galena using *P. polymyxa* cells (Santhiya et al., 2001a)

Experimental conditions	ZnS (%)		PbS (%)	
	Dispersed	Flocculated	Dispersed	Flocculated
Blank at pH 9–9.5	70.5	27.8	27.0	71.7
	70.3	29.5	27.0	72.3
With *Bacillus polymyxa* at pH 9–9.5	93.7	4.9	5.6	94.2
	94.3	4.6	5.8	94.9

sulfide minerals. Santhiya et al. (2000) examined selective bioflocculation and separation of galena from sphalerite by *Acidithiobacillus thiooxidans* (Santhiya et al., 2000). According to their studies, in the presence of *Acidithiobacillus thiooxidans*, 95% of galena flocculated while sphalerite dispersed (Santhiya et al., 2000). The effect of pH on the settling of galena and sphalerite is shown in (Fig. 7). In the presence of *Acidithiobacillus thiooxidans* cells, sphalerite sedimentation reduced from about 40% at pH = 3 to about 5% in the pH range of 10–11. In contrast, the percentage of galena flocculation improved from about 20% at pH = 2.5 to about 95% at pH = 11. Thus, galena can be separated from sphalerite in the presence of *Acidithiobacillus thiooxidans* in the pH range of 9–11 (Santhiya et al., 2000).

Natarajan et al. (2003) studied the selective separation of pyrite, chalcopyrite, and sulfur from quartz through selective flocculation/dispersion after biotreatment with *Acidithiobacillus ferrooxidans* and *Acidithiobacillus thiooxidans* (Natarajan & Das, 2003). According to Natarajan et al. (2003), after biotreatment with *Acidithiobacillus ferrooxidans* and *Acidithiobacillus thiooxidans* cells, the settling rates of pyrite and elemental sulfur improved. As a result, selective flocculation and separation of pyrite and sulfur from quartz were demonstrated (Natarajan & Das, 2003).

4.3.3 Bacillus subtilis

Bacillus subtilis is a gram-positive, neutrophilic, aerobic, periflagellated, and capsulated bacterium usually found in soil (Brock et al., 2006). The application of *Bacillus subtilis* for beneficiation of different minerals has been described in several approaches.

Iron Removal (Iron Oxides) from Kaolin Clays

Poorni and Natarajan (2013) used *Bacillus subtilis* and its EPSs to remove iron oxides such as hematite from kaolinite via selective bioflocculation. As the results indicated, in the presence of hematite, *Bacillus subtilis* secretion of extracellular polysaccharides (ECP) increased while kaolinite promoted the secretion of extracellular proteins (EP). Moreover, extracellular polysaccharides showed great affinity to the hematite surface, which caused the zeta potential of hematite to shift in the negative direction. In contrast, extracellular proteins showed great affinity to the kaolinite surface, which affected the zeta potential of kaolinite in a positive direction (Fig. 8). After biotreatment with *Bacillus subtilis*, hematite converted to more hydrophilic, and kaolinite exhibited higher surface hydrophobicity. As shown in (Table 3), almost 90% of iron could be removed from the kaolin clays after biotreatment with the ECP extracted from *Bacillus subtilis* (Poorni & Natarajan, 2013a).

Pyrite Removal from Galena

Sarvamangala et al. (2013) studied *Bacillus subtilis* and its EPSs to separate pyrite from galena. Adsorption investigations showed that the amount of adsorbed bacteria cells onto pyrite was significant compared to galena (Fig. 9).

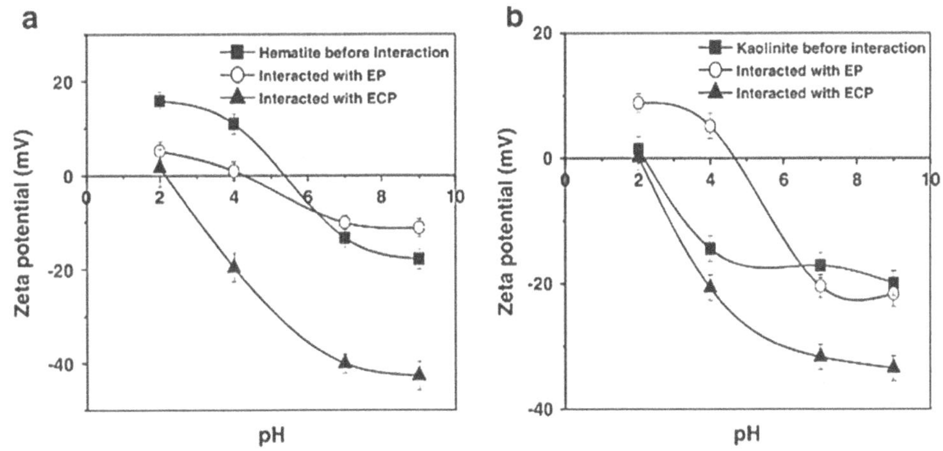

Fig. 8 Effect of extracellular protein (EP) and extracellular polysaccharide (ECP) of *Bacillus subtilis* on zeta potential of **a** hematite and **b** kaolinite (Poorni & Natarajan, 2013a)

Table 3 Settling behavior of hematite and kaolinite under different conditions in the presence of *Bacillus subtilis* (Poorni & Natarajan, 2013a)

Interaction conditions	Percent settled %			
	Hematite alone	Kaolinite alone	1:1 mineral mixture	
			Hematite	Kaolinite
Control (no bacterial interaction)	60	45	65	50
Solution-grown cells	70	35	75	40
Cell-free extract from solution-grown cells	75	20	70	35
Hematite-grown cells	95	12	95	10
Cell-free extract from hematite-grown cells	80	20	65	30
Kaolinite-grown cells	70	04	60	10
Cell-free extract from kaolinite-grown cells	60	20	70	30
EP from hematite-grown cells	80	50	82	25
EP from kaolinite-grown cells	75	08	80	10
ECP from hematite-grown cells	98	05	95	20
ECP from kaolinite-grown cells	90	10	75	20

After biotreatment with *Bacillus subtilis,* pyrite converted to more hydrophilic, and galena showed higher surface hydrophobicity. Furthermore, in the presence of galena, *Bacillus subtilis* secreted hydrophobic proteins, which enhanced dispersion of galena, while in the presence of pyrite; the bacteria secreted polysaccharides that enhanced the settling of pyrite. Therefore, selective bioflocculation and separation of pyrite from galena are evidenced in (Table 4) (Sarvamangala et al., 2013).

4.3.4 Bacillus licheniformis

Bacillus licheniformis is a rod-shaped, gram-positive, aerobic bacterium that usually can be isolated from natural sources, such as soil (Waldeck et al., 2006).

Quartz Removal from Kaolinite

Ghashoghchi et al. (2017) used *Bacillus licheniformis* cells and extracellular polymeric substances (EPS) to bio-flocculation of kaolin and quartz. It was indicated that extracellular protein secreted from *B. licheniformis* was more effective in quartz agglomeration, while extracellular polysaccharides secreted from *B. licheniformis* were more effective in kaolin flocculation. In the best state, the sedimentation of kaolin increased by 40% using bacterial cells and metabolites at pH = 7 and 3. Also, the sedimentation of quartz using the same bioflocculants was improved by about 50% at pH = 1–3 (Ghashoghchi et al., 2017). In (2019), Hosseini et al. (2020) also investigated the bioflocculation of quartz and kaolinite in the presence of *Bacillus*

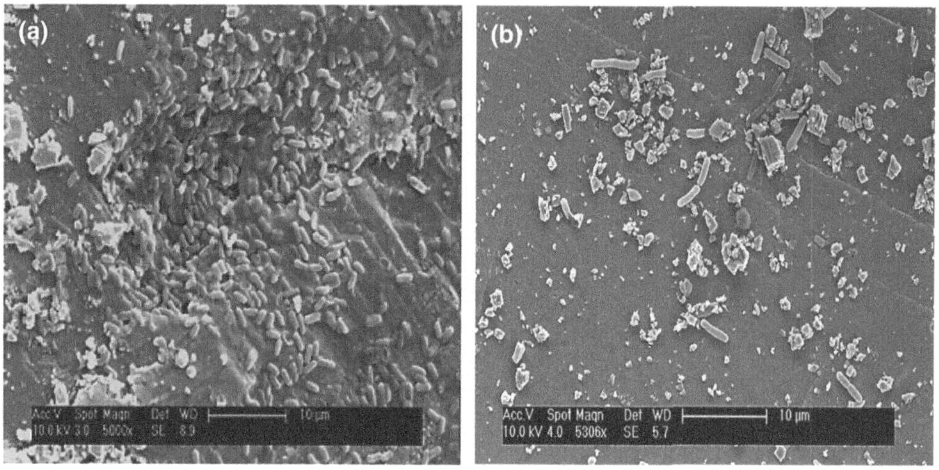

Fig. 9 SEM images of *B. subtilis* attached to **a** pyrite and **b** galena (Sarvamangala et al., 2013)

licheniformis cells and metabolites in basic conditions. The adsorption of bacterial cells and EPS on kaolinite was three times higher than adsorption on quartz. In the presence of bacterial cells and EPS, sedimentation of kaolinite was less than quartz at most of the pH values, and the selective bio-flocculation occurred at pH = 11 and 12. They indicated that in the presence of *Bacillus licheniformis* cells, 98.3% of the kaolinite was selectively separated from quartz (Hosseini et al., 2020).

Removal of Iron Oxides from Kaolinite and Quartz

Differential bioflocculation in the presence of *Bacillus licheniformis* and extracellular polymeric substances (EPS) was investigated by Hosseini et al. (2019) to separate kaolinite and quartz from hematite and goethite (Fig. 10). The best separation of kaolinite and quartz from iron oxides was observed at pH = 7. Based on this study, the application of bacterial protein was the best bioflocculant to remove hematite from kaolinite. However, the recovery of kaolinite was low. Thus, the authors suggested bacterial polysaccharides as bioflocculant, which resulted in 77.6% separation of hematite from the mixture with a kaolinite recovery of about 59.4%, respectively (Hosseini et al., 2019).

Table 4 Bioflocculation of pyrite and galena using *Bacillus subtilis* (Sarvamangala et al., 2013)

Interaction conditions	Percent settled %			
	Pyrite alone	Galena alone	1:1 mineral mixture	
			Pyrite	Galena
No bacterial interaction (control)	60	45	65	50
After interaction with cells	70	35	75	40
After interaction with cell-free metabolite	75	20	70	35
After interaction with EP (35 mg/g)	95	12	95	10

4.3.5 Other Microorganisms

In 1994, Schneiderl et al. (1994) studied the flocculation of hematite by the *Candida parapsilosis yeast*. The results revealed that *C. parapsilosis* and substances released from it improve flocculation of fine hematite suspensions (Schneider et al., 1994). Then in 1996, Raichur et al. (1996) used *Mycobacterium phlei* as a bioflocculant to remove sulfur and ash contents of coal by flotation and flocculation (Raichur et al., 1996). In 1999, Haas et al. (1999) investigated the application of *Corynebacterium xerosis* bacterium for fine fluorite flocculation. They indicated that *C. xerosis* cells improved the aggregation of fine fluorite particles (Haas et al., 1999). Padukone and Natarajan (2011) utilized *Saccharomyces cerevisiae yeast* and its metabolites for selective bioflocculation of quartz from calcite. Yeast cells showed a better affinity to the calcite surface. After biotreatment, the quartz surface was more hydrophobic, and calcite was hydrophilic. So this situation facilitated the selective separation of quartz from calcite (Padukone & Natarajan, 2011). The application of *Bacillus firmus* and its metabolites to bioflocculation of clay minerals was investigated by Karthiga Devi and Natarajan (2015). They also showed bioflocculant produced from *Bacillus firmus* could be used as absorbents to remove toxic Cr (VI) ions from aqueous solutions (Karthiga Devi & Natarajan, 2015). Selim & Rostom (2018) used *Bacillus cereus* to bioflocculation and separation of hematite from its mixture with silica. *Bacillus cereus* showed a higher affinity to hematite mineral surface compared with silica surface (Selim & Rostom, 2018). Camarate et al. (2021) used *Candida stellata yeast* for the selective separation of ultrafine hematite from quartz via the bioflocculation process (Camarate et al., 2021). The application of various microorganisms in minerals bioflocculation is reviewed in (Table 5).

Fig. 10 **a** Polysaccharide crystal formation on hematite. **b** Formation of kaolinite flocs after biotreatment with extracellular proteins. Attachment of *Bacillus licheniformis* on **c** goethite and **d** quartz particles (Hosseini et al., 2019)

5 Bioflotation

Froth flotation is used widely in mineral processing for selective separation of valuable minerals from gangue minerals. In this method, minerals with hydrophobic surfaces attach to the air bubble and selectively separate. The process is carried out in a wet environment, and different chemicals (including depressants, pH-adjusting reagents, dispersants, activators, collectors, and frothers) are used to modify the minerals surfaces (Gaudin, 1975).

To separate minerals with high recovery, conditioning should be carried out in several steps, and the following steps should be passed:

1. A mineral–water slurry with a pulp density of 25 to 35 should be made.
2. The pulp pH should be adjusted using acids and bases.
3. The dispersants are added to the pulp for spreading the particles.
4. Depressants are added to the pulp to make selected particles hydrophilic. The activators also can be added to the pulp in this stage. Activators are used for modifying the mineral's surface for better collectors' adsorption.
5. The collectors were added to the pulp to make targets minerals surface hydrophobic.
6. The frother is added to the pulp to produce stable bubbles.

After conditioning, hydrophobic minerals will be attached to the air bubbles due to Archimedes' force.

In the bioflotation method, microorganisms act as one of the mentioned chemical reagents. In other words, microorganisms can be used as depressants, dispersants, collectors, frothers, and even flocculants. Various microorganisms, such as autotrophic or heterotrophic bacteria, fungi, yeasts, and algae, can be used for this purpose. The applications of these microorganisms will be reviewed in the following sections.

5.1 Application of Microorganisms in Bioflotation

5.1.1 Acidithiobacillus ferrooxidans

Acidithiobacillus ferrooxidans is a gram-negative chemolithoautotrophic bacteria that has been used successfully in bioleaching and bio-beneficiation processes in recent years. This bacterium derives its energy from the oxidation of (Fe^{+2}) to (Fe^{3+}) and sulfur (S^0) to sulfuric acid (H_2SO_4) (Chandraprabha et al., 2004a; Dwyer et al., 2012; Pecina-Treviño et al., 2012).

For the first time, Misra et al. (1996) examined *Acidithiobacillus ferrooxidans* as pyrite depressants for coal desulfurization. This microorganism was used as an

Table 5 Application of microorganisms in minerals bioflocculation

Microorganism	Mineral bioflocculation	Bioflocculant	Optimal pH bioflocculation	Contact time (min)	Bioflocculation efficiency	References
Paenibacillus polymyxa	High-ash coals	Protein–Polysaccharide and 50 ml suspension bacterium	2–4	25–30	90%	Vijayalakshmi and Raichur (2002)
	Hematite, corundum, calcite from quartz	Protein–Polysaccharide	Corundum: 7 Hematite: 7 Calcite: 12	5	90–99.6%	Deo and Natarajan (1997)
	Hematite, corundum, from quartz and kaolinite	Protein–Polysaccharide	Corundum: 4–5 Hematite: 7 kaolinite:12	5	Corundum–quartz: 90–97% Hematite–quartz: 50–90% Hematite–kaolinite: 30–55%	Deo and Natarajan (1998)
	Hematite, corundum from quartz	Protein–Polysaccharide	4–7	15	–	Deo and Natarajan (1999)
	Sphalerite and galena	Protein–Polysaccharide	9–9.5	15	Galena: 94.9% Sphalerite: 4.6%	Santhiya et al., (2001a)
	Chalcopyrite from quartz	25 mg/l of extracellular bacterial proteins	8	20	95% of quartz could be removed 81% of calcite could be removed	Patra and Natarajan (2004b)
Acidithiobacillus thiooxidans	Galena and sphalerite	Bacterial cells (2–3 × 10^8 cell/ml)	11	120	95% of galena flocculated	Santhiya et al., (2000)
Acidithiobacillus ferrooxidans	Pyrite, chalcopyrite from quartz	Bacterial cells (1.8 × 10^8 cell/ml)	2	2	90% of quartz separation from pyrite, chalcopyrite is possible	Somasundaran et al., (2005)
Bacillus subtilis	Fine coal	Bacterial suspensions	2	5	99%	Vijayalakshmi and Raichur (2003)
	Pyrite	100 ml of bacterial cell suspension	6.0–7.5	3	98%	Sarvamangala et al., (2013)
					85%	
					80%	
	Galena				18%	
					26%	
					27%	
	Pyrite from galena				85% for pyrite and 21% for galena	
					74% for pyrite and 25% for galena	
					72% for pyrite and 26% for galena	
	Hematite and kaolinite	100 ml of bacterial cell suspension	7	5	95% for hematite and 10% for kaolinite	Poorni and Natarajan (2013a)
	Hematite and kaolinite	100 ml of bacterial cell suspension	7–8	5	95% for hematite and 20% for kaolinite	Poorni and Natarajan (2013b)
	Hematite and kaolinite	100 ml of bacterial cell suspension	7	5	95% for hematite and 15% for kaolinite	Poorni and Natarajan (2014)

(continued)

Table 5 (continued)

Microorganism	Mineral bioflocculation	Bioflocculant	Optimal pH bioflocculation	Contact time (min)	Bioflocculation efficiency	References
Bacillus licheniformis	Quartz	20% v/v of bacterial metabolite	1	8	92%	Ghashoghchi et al., (2017)
	Kaolinite	0.24 mg/ml polysaccharide. 20 g/l solid concentration	9	8	45%	
	Hematite and kaolin	0.24 mg/mL extracellular polysaccharides	7	8	59.4%	Hosseini et al., (2019)
	Hematite and quartz				69.2%	
	Goethite and kaolin				56.2%	
	Goethite and quartz				64.8%	
	Hematite	Bacterial cells (3.6 × 10^8 cell/ml)	5	8	100%	Sadeghizadeh et al., (2017)
	Goethite	Bacterial cells (3.6 × 10^8 cell/ml)	7	8	95%	
	Kaolinite from quartz	Bacterial cells (19 × 10^8 cell/ml)	12	6	98.3%	Hosseini et al., (2020)
	Kaolinite clays suspension	6ml from a 5 g/L bioflocculant solution	7 – 7.2	5	96%	Karthiga Devi and Natarajan (2015)
	Cr^{4+} removal	6ml from a 2 g/L bioflocculant solution	7	60–120	85%	
Bacillus firmus	Cr^{4+} removal	6ml from a 2 g/L bioflocculant solution	7	60–120	77%	
	Kaolinite clays suspension	6ml from a 5 g/L bioflocculant solution	9	5	92%	
Bacillus cereus	Hematite from quartz	10 ml of bacterial solution (10 × 10^{11} cell/ml). 10 g/L solid concentration	4.5	10	82%	Selim and Rostom (2018)
Corynebacterium xerosis	Fine fluorite particles	40 mg/L bacterial cell concentration	7	3	96%	Haas et al., (1999)
Yeast Cell Rupture	Hematite	cell as biofloculant (60 kg/t) or soluble cell fraction (1.2 kg/t)	7	3	98%	Schneider et al., (1994)
Mycobacterium phlei	Fine coal	250 ppm of bacteria solution	3.5	3	93%	Raichur et al., (1996)
Saccharomyces cerevisiae	1:1 mixture of quartz and calcite	100 ml of the metabolite from quartz-grown cells	7	3	92% for calcite and 21% for quartz	Padukone and Natarajan (2011)

alternative for cyanide that is extremely toxic. Bacteria had changed the surface properties of pyrite by attaching it to this mineral and making it hydrophilic. This research showed that the pyrite floatability using sodium isopropyl xanthate (PIX) collector was reduced from above 90% to less than 45% after bacterial treatment by increasing pH from 1 to 7 which means that pyrite can be depressed with *Acidithiobacillus ferrooxidans* at low pHs. Furthermore, the results indicated that pyrite depression is dependent on the bacterial counts and type and concentration of salts in culture media (Misra et al., 1996).

Amini et al. (2009a, b) also studied the effect of this type of bacteria on pyrite depression in coal flotation and compared results with sodium cyanide (Amini et al., 2009a). Again, the results showed that bacteria have a higher ability to depress the pyrite than sodium cyanide. The pyrite recovery decreased to less than 14% after ore treatment with *Acidithiobacillus ferrooxidans* (Amini et al., 2009b).

Nagaoka et al. (1999) conducted experiments on the *Acidithiobacillus ferrooxidans* depression ability on sulfide minerals, including pyrite (FeS_2), molybdenite (MoS_2), chalcocite (Cu_2S), millerite (NiS), and galena (PbS). They stated that in the absence of bacteria, all sulfide minerals showed high floatability. After bacterial contact with minerals surfaces, the floatability rate of pyrite decreased dramatically from 96 to 19%. The bacterial treatment also affected the flotation of millet and galena by reducing these minerals recoveries from 96 to 84% and 91% to 82%, respectively. On the other hand, bacterial treatment had almost no effect on the flotation of chalcocite and molybdenite. Since the ability of *Acidithiobacillus ferrooxidans* to inhibit pyrite flotation was confirmed, bioflotation was employed to separate pyrite from a mixture of minerals. The results once again demonstrated that pyrite could be successfully depressed after bacterial treatment (Nagaoka et al., 1999).

In a series of research, Chandraprabha et al. (2004) investigated the effect of *Acidithiobacillus ferrooxidans* and *Acidithiobacillus thiooxidans* on floatability pyrite, chalcopyrite, and arsenopyrite (Chandraprabha & Natarajan, 2006; Chandraprabha et al., 2004a, 2004b, 2005). In 2004, they stated that bacterial cells showed a greater tendency to bind with pyrite compared to arsenopyrite. The binding kinetics of *Acidithiobacillus ferrooxidans* on pyrite was high, and it reached equilibrium within 15 min, while the existence of insoluble arsenic on the surface of arsenopyrite prevented the initial binding of cells. They also examined the selective separation of pyrite and chalcopyrite in this research. In the kinetics and binding studies of *Acidithiobacillus ferrooxidans*, it was observed that the binding kinetics were fast for both pyrite and chalcopyrite and equilibrated in about 15 min. However, the binding rate for pyrite was higher. Moreover, the interaction of all three minerals with the cells caused the isoelectric points (IEP) of the minerals to reach higher pH values. However, this change was much more significant for pyrite than arsenopyrite and chalcopyrite. After minerals interaction with bacteria, the bacterial cell surface charges and minerals IEP showed similar changes, which were more significant for the pyrite reciprocal cells. In addition, selective separation of pyrite from arsenopyrite was carried out using potassium isopropyl xanthate as a collector at natural and acidic pH. The results stated that pyrite was depressed by sodium isopropyl xanthate as the collector after 5 min contact with bacterial cells. In similar conditions, there was no significant change in arsenopyrite recovery after interaction with bacterial cells for 5 min, in the presence of sodium isopropyl xanthate and copper sulfate ($CuSO_4$) as the activator. They stated that copper ions increase the recovery of both pyrite and arsenopyrite minerals; however, flotation improvement is more significant for arsenopyrite. This occurs because As^{3+} forms a stable arsenide complex with Cu^{2+} and Cu^{3+}, while Fe^{2+} and Fe^{3+} cannot form a stable formation as arsenide. In addition, the galvanic effect between pyrite and arsenopyrite reduces the adsorption of xanthates on the pyrite surface and increases on arsenopyrite, which improves the separation process between these two minerals. The results also showed that the *Acidithiobacillus ferrooxidans* could selectively remove pyrite from a mixture of pyrite and chalcopyrite. When pyrite and chalcopyrite are contacted separately with bacterial cells, pyrite was depressed, but chalcopyrite showed different behaviors in the flotation process using the xanthate (PIPX) collector. While if both minerals have interacted with bacteria simultaneously, the dissolution of copper from chalcopyrite activates the pyrite surface and disrupts the selective separation process. The dissolution of chalcopyrite in the presence and absence of cells and collectors was also investigated. The results indicated that the concentration of copper in the flotation environment was higher when pyrite and chalcopyrite were treated simultaneously. This clearly shows that the copper ions which were released from chalcopyrite migrate to the pyrite surface and activate it (Chandraprabha & Natarajan, 2006; Chandraprabha et al., 2004a, 2004b, 2005).

Hosseini et al. (2005) used *Acidithiobacillus ferrooxidans* for the bioflotation of two sulfide copper ores (sample A with higher copper grade and less iron and sample B with lower copper grade and more iron) as well as pyrite and chalcopyrite. According to the results, in all samples except chalcopyrite, the binding of *Acidithiobacillus ferrooxidans* increased with increasing cell numbers. The highest binding rate was for pyrite, then the B-sulfide ore sample, A-sulfide ore sample, and then chalcopyrite. Therefore, it was evident that the *Acidithiobacillus thiooxidans* is not interested in binding to copper sulfide while it binds selectively to pyrite. They said that because bacteria get their energy from the oxidation of iron and sulfur, they can be adsorbed on the surface of pyrite and prevent the adsorption of xanthate collectors. Bacteria do not absorb chalcopyrite due to the presence of copper, which is toxic to bacteria, and therefore, the collector can be absorbed on it and make it hydrophobic (Hosseini et al., 2005).

Rao et al. (1992) also showed that *Acidithiobacillus ferrooxidans* change the surface's chemical and flotation behavior of non-ferrous sulfides (sphalerite and galena) by direct microbial adhesion on the mineral surface. The initial cell concentration and duration of mineral interaction were the most effective mineral flotation parameters. The formation of insoluble lead sulfate causes galena to be depressed therefore not be floated, while in sphalerite, the formed zinc sulfate is soluble and does not depress. Only in high concentration may be the sphalerite depression occur. In addition, selective bioflotation of lead and zinc sulfides can be achieved with *Acidithiobacillus ferrooxidans* under acidic conditions (pH = 2) (Rao et al., 1992).

The pyrite depression in the bioflotation of high-grade pyrite and low-grade lead–zinc ore by *Acidithiobacillus ferrooxidans* was done by Mehrabani et al. (2010). In their research, the concentration effect of four parameters of PAX (collector), copper sulfate (activator), and bacteria and sodium cyanide (depressants) was investigated in three different levels. The optimal points of the process were to minimize the flotation rate of pyrite, which was achieved in four experiments. In these experiments, the concentration of collector and frother was lowest. In the results of these experiments, in the presence of bacteria, the recovery of pyrite decreased from 38.11 to 23.52%.

In comparison, the recovery of sphalerite increased from 65.91 to 74.03% and zinc grade from 15.22 to 20.84%. Also, under optimal conditions, the results obtained for pyrite recovery when using bacteria and sodium cyanide as depressants were similar. However, when using bacteria, the zinc grade increased by 3%, and the iron grade decreased by 4% (Mehrabani et al., 2010, 2011).

In 2018 and 2020, San Martin et al. studied the effect of *Acidithiobacillus ferrooxidans* on pyrite recovery in seawater. In the bacterial preparation of pyrite with *Acidithiobacillus ferrooxidans* for 15 min, at pH > 4, pyrite recovered in fresh water, pyrite decreased at pH > 8 in fresh. Thus, *Acidithiobacillus ferrooxidans* can act as a pyrite depressant even when inhibited by bacterial activity. At the same time, in seawater (pH = 8), pyrite is depressed but does not affect molybdenite or chalcopyrite (San Martín et al., 2018, 2020).

5.1.2 Acidithiobacillus thiooxidans

Acidithiobacillus thiooxidans is a gram-negative acidophilic mesophile and, along with *Acidithiobacillus ferrooxidans,* is the most-used bacteria for bioleaching. This bacterium is characterized as a sulfur-oxidizing bacterium (Jerez, 2019; Johnson, 2009). *Acidithiobacillus thiooxidans* produces sulfuric acid with a sulfur-oxidizing with sulfur-oxidizing enzymes and sulfite-oxidizing enzymes with sulfite as the critical intermediate (Chandraprabha & Natarajan, 2006), according to the Eqs. (2) and (3):

$$S^0 + O_2 + H_2O \rightarrow H_2SO_3 \tag{2}$$

$$H_2SO_3 + \frac{1}{2}O_2 \rightarrow H_2SO_4. \tag{3}$$

Since *Acidithiobacillus thiooxidans* utilize elemental sulfur or reduced sulfur, it can alter the surface of sulfide minerals. Three researchers reported using these bacteria for flotation of galena, sphalerite, pyrite, and chalcopyrite (Chandraprabha & Natarajan, 2006; Santhiya et al., 2000, 2001b).

Chandraprabha and Natarajan (2006) showed that the isoelectric point (IEP) for pure pyrite was standing at pH 3.25, and at the lower pHs, the electronegative character decreased. However, after bacterial treatment for 1 h and 12 h, IEP increased to 3.5 and 4.2, respectively. On the other hand, IEP for pure chalcopyrite IEP was on pH 2.4. Moreover, it could be shifted to 3 and 3.4 after 1 and 12 h interaction with bacteria, respectively. Adhesion kinetics of bacteria on chalcopyrite and pyrite surface was similar, and its equilibrium was reached after 80 min. However, the cell concentration on the pyrite surface was more than chalcopyrite (6.125×10^8 cells/ml compared to 6.25×10^8 cells/ml). In micro-flotation tests, recovery of both minerals was dropped after bacterial treatment. However, pyrite depression was more significant. While the recovery of pyrite flotation using 1 mM PIPX was 41% after bacterial treatment, the chalcopyrite recovery was 76% (Chandraprabha & Natarajan, 2006). In differential flotation tests with pyrite: chalcopyrite ratio of 1:1, the pyrite and chalcopyrite recoveries were 41% and 69%, respectively, after bacterial treatment and flotation at pH = 4.5, using 0.5 mM PIPX as the collector. The recoveries improved to 32% and 72% by increasing the pH to 6.5. Nevertheless, better separation was achieved by changing the conditioning sequence (addition of collector followed by interaction with cells). Thus, recoveries for pyrite and chalcopyrite were improved to 21% and 86%, respectively, at pH 4.5 and 19.3%, and 84.6% at pH 6.5 (Chandraprabha & Natarajan, 2006).

Santhiya et al. (2000) studied bacterial cell attachment on galena and sphalerite surfaces in three different pH (acidic, neutralized, and basic). The observations indicated that more cells were attached to the galena surface in comparison with sphalerite (about 10^9 cells/ml for galena compared to 5×10^7 for sphalerite). This difference in attachment concentration resulted from different solubility of these minerals in an oxidation acidic environment which is made in the presence of acidophilic microorganisms. The highest absorbed cell in minerals surface obtained with a bacterial count in culture media was 10^9 cells/ml. Electrostudies were performed at pHs between 2 and 2.5 for 1 h and 24 h bacterial treatment. Both minerals before interaction had an isoelectric point at pH = 2.2. After 1 h and 24 h interactions between minerals and bacteria, sphalerite IEPs shifted to pH = 2.6 and pH = 3.2, and galena IEPs shifted to pH = 3.1 and pH = 3.6, respectively, which means that bacterial treatment causes a more significant shift in galena IEP. As mentioned, these changes occurred due to the attachment of more bacterial cells to the galena surface. Microflotation tests were performed using PIPX as the collector and $CuSO_4$ as an activator for sphalerite. The result showed galena recovery is higher than 95% for all pHs and sphalerite recovery is higher than 90% in the pH range of 4.5–8. After 2 h interactions of minerals with cells at the pH of 2–2.5, the recovery of sphalerite stays at 90%, while galena gets fully depressed. In (1:1) differential flotation tests, both mineral recoveries stayed above 95% without cell interactions. However, after 2 h cell interactions at pH = 2–2.5, sphalerite recovery was higher than 90%, while most of the galena depressed (~90%) (Santhiya et al., 2000).

Santhiya et al. (2001) investigated galena IEP at pH = ~2.2. After bacterial treatments for 1 h and 24 h, the IEP shifted to pH = 3.0 and pH = 3.5, respectively. While IEP for pure sphalerite was ~2.3, it was shifted to 2.6 and 3.3, after 1 h and 24 h interaction with cells. Adsorption of bacteria on galena and sphalerite surfaces was measured for 24 h. The results showed that after 1 h, equilibrium is reached. The optimum adsorbed cell for galena was significantly higher than sphalerite, and it had more change in IEP after interaction with cells. It was interpreted that the sphalerite is more soluble in an oxidation environment than galena, making attachment more difficult. In differential tests, it was reported that galena and sphalerite recovery reached 95% and 98% in the presence of collector and activator, respectively. However, galena and sphalerite recovery stayed at 31% and 94% after bacterial treatment. In the absence of collector and activator recovery, 94% of galena were depressed, while 94% of sphalerite recovered, which shows the high potential of *Acidithiobacillus thiooxidans* for the separation of sphalerite and galena with high efficiency (Santhiya et al., 2001b).

5.1.3 Aspergillus niger

Aspergillus niger is heterotrophic fungi that produce large amounts of organic acids such as citric, gluconic, and oxalic acids. These products are able to complex and mobilize metals from non-sulfide minerals. As the tolerance of this microorganism is high and can produce critical lixiviants, they became one of the essential fungi in the bioleaching of metals and bioprocessing of minerals (Donati & Sand, 2007; Muddanna & Baral, 2019).

Gawel et al. (1997) investigated the potential of *Aspergillus niger* for magnesite depression in the presence of sodium oleate as the collector. The effects of bacterial pretreatment in 1, 7, and 14 days were studied. The results indicated that by increasing the bio-pretreatment time, the adsorption of sodium oleate on the magnesite surface decreased, and flotation recovery dropped. With 14 days of bio-pretreatment, recovery dropped from about 55% to 35%. This decrease probably happened because active surface sites were blocked by *Aspergillus niger* products (Gawel et al., 1997).

5.1.4 Bacillus subtilis

Bacillus subtilis is a mesophilic bacterium and is able to tolerate extreme conditions as it can form a protective endospore (Vasanthakumar et al., 2017). The produced EPS affects the aqueous phase conformation and adsorb on the mineral's surface. The adsorption would alter substrata's physicochemical properties (Sarvamangala et al., 2013).

Bacillus subtilis can be employed to decrease the sulfur and ash content in coal (Abdel-Khalek & El-Midany, 2013; El-Midany & Abdel-Khalek, 2014). Abdel-Khalek and El-Midany reduced coal ash and sulfur content from 6.65% and 3.3% to 1.95% and 0.92%, respectively. The zero points of charge (ZPC) of pure coal and bacterial-treated coal were found at pH = 3, but interacted coal had a more positive surface charge in higher pH values (Abdel-Khalek & El-Midany, 2013). El-Midany and Abdel-Khalek compared *Bacillus subtilis* with *Paenibacillus polymyxa* influence in coal flotation. Pure coal ZPC was found at pH = 2.5. After interaction with each bacteria, the ZPC point did not change and stayed the same, but its positivity in higher pH values increased (El-Midany & Abdel-Khalek, 2014).

In adsorption tests, both bacteria had maximum adsorption at pH 3, but *Bacillus subtilis* had faster kinetics and higher adsorption amounts. FTIR spectra analysis indicated that the main forces are hydrogen bonding and long-term hydrophobic, which means the adsorption nature is physical (El-Midany & Abdel-Khalek, 2014).

Bacillus subtilis shows high performance compared to *Paenibacillus polymyxa* for ash and sulfur removal. While the coal sample had 3.3% sulfur and 6.65% ash contents, *Bacillus subtilis* and *Paenibacillus polymyxa* decreased this amount to 0.92% sulfur and 1.95% ash content and 1.12% sulfur and 2.64% ash content, respectively (El-Midany & Abdel-Khalek, 2014).

5.1.5 Bacillus megaterium

Bacillus megaterium is a rod-shaped, neutrophilic, gram-positive, mesophilic bacterium, producing biotechnologically relevant vitamins and proteins that have become important in this industry. It can widely be found in numerous environments, such as soil to seawater, sediment. Its cell surface consists of components like peptidoglycan, teichoic and teichuronic acids, lipoproteins, lipopolysaccharides, surface proteins, polysaccharides, and polypeptides, which is essential in the microbe–mineral interactions (Vasanthakumar et al., 2013, 2014).

Vasanthakumar et al. (2014) studied the effect of *Bacillus megaterium* products on sphalerite and galena flotation as bio-collector. It was found that extracellular DNA is an essential parameter in sphalerite selective flotation since single-stranded DNA had a more significant bio-collector capacity compared to double-stranded DNA. With the addition of combined single-stranded DNA and non-DNA components, about 95% of sphalerite was recovered. It was found that the presence of calcium and phosphate components in the nutrient media is valuable for sphalerite selectively separating (Vasanthakumar et al., 2014).

Vasanthakumar et al. (2013) investigated the effect of the adaptation of *Bacillus megaterium* on the flotation and ZPC of sphalerite and galena. Adapted bacteria had less negative surface charge than un-adapted cells. Sphalerite-adapted cells obtained better flotation recovery in case of selectivity as un-adapted and galena-adapted ones (Vasanthakumar et al., 2013).

5.1.6 Paenibacillus polymyxa (P. polymyxa)

The main components of *P. polymyxa* EPS are polysaccharides, proteins, and organic acids (Patra & Natarajan, 2004a). These metabolic products are biodegradable, effective at extreme temperatures, and low toxic (Subramanian et al., 2003).

Patra et al. (2008) (Patra & Natarajan, 2006, 2008) investigated the role of extracellular *P. polymyxas* protein and polysaccharide in flotation. The results indicated that fractionated protein groups could selectively alter pyrite and chalcopyrite surfaces to hydrophilic while galena, quartz, and sphalerite hydrophobicity increased. Extracellular bacterial protein and extracellular polysaccharide absorbed with higher density in pyrite surface compared to galena, and galena could selectively be floated in the presence of extracellular polysaccharide. Also, Patra et al. (2004) (Patra & Natarajan, 2004a) selectively floated pyrite from sphalerite using *P. polymyxa*. The adsorption studies revealed higher adsorption density on pyrite surface at different pH, especially at pH = 9. ZPC of pyrite and sphalerite before and after interaction with cells (by 1×10^7 cells/ml) for 1 h, 2 h, and 24 h had no significant difference. Selective flotation of sphalerite at pH = 9 in the presence of extracellular bacterial protein was achieved, and recovery was 96%. Extracellular bacterial protein proved to work as a bioflocculant for pyrite, enhancing its depression.

Subramanian et al. (2003) investigated the role of *P. polymyxas* metabolite in the sphalerite-galena system. They observed a higher adsorption density of metabolites on the galena surface compared to sphalerite. ZPC of both minerals before and after interaction with *P. polymyxas* metabolite was measured. For sphalerite, the ZPC of pure sphalerite was around pH = 2.2, and after interaction with metabolite, it shifted to more basic pH values. In contrast, in galena's case, the ZPC remained the same after interaction with metabolite at about pH = 2.5. Results of micro-flotation showed that galena depressed after metabolite interaction, while 90% of sphalerite was recovered with the addition of $CuSO_4$ and PIPX. In differential flotation tests with only *P. polymyxas* metabolite, sphalerite was selectively floated at pH of 3.2–3.4 with 90% recovery, while 96% of galena was depressed. At the pH range of 9–9.5 with the addition of 10^{-6} M $CuSO_4$ and 10^{-4} M PIPX before metabolite interaction, about 95% of sphalerite recovered, while 95% of galena was depressed.

Patra et al. (2008) (Patra & Natarajan, 2008) investigated the effect of different types of EPS groups driven from *P. polymyxa,* using micro-flotation tests with a Halimond tube. Four different types of extracellular proteins (EP), comprised of various kinds of amino acids (protein groups), were studied in adsorption and flotation tests. The adsorption studies on different minerals are shown in (Figs. 11 and 12). As results indicate, quartz had 65%, 80%, 30%, and 40% recovery with groups A, B, C, and D proteins. All protein groups depressed pyrite and chalcopyrite, and galena had the highest recovery with group D. The results indicated that selective depression of pyrite and chalcopyrite in the presence of galena, quartz, and sphalerite could be achieved.

5.1.7 Bacterial Consortium

Govender et al. (2011) investigated the effects of different mixtures of bacteria and their EPS on chalcopyrite flotation. After parameter optimization, the efficiency of both produced EPS and bioleaching bacteria was investigated and compared. Five types of EPS are extracted from different bacteria, including *Acidithiobacillus caldus*, *Leptosprillium sp*, *Sulfobacillus sp*, *Ferroplasma sp*, *Acidianus sp*, *Metallosphaera sp*, *Sulfolobus sp*. The EPS constituents are mostly carbohydrates and proteins with a small amount of humic acid and uronic acid. The best recovery for chalcopyrite EPS flotation was achieved with 3.5×10^{-2} mg/g EPS concentration. The optimum cell concentration to be used for bioflotation was 1×10^{-6}. The flotation process was conditioned for 20 min with EPS or bacterial cell at pH = 9; after that, 1×10^{-5} M of SIBX was added, and conditioning continued for 5 more min. Obtained results are shown in (Fig. 13), as it is obvious that produced free EPS is more successful (Govender & Gericke, 2011).

Patra et al. (2008) (Patra & Natarajan, 2008) investigated the role of different EPS groups drove from *P. polymyxa* in micro-flotation tests with Halimond tube. The results

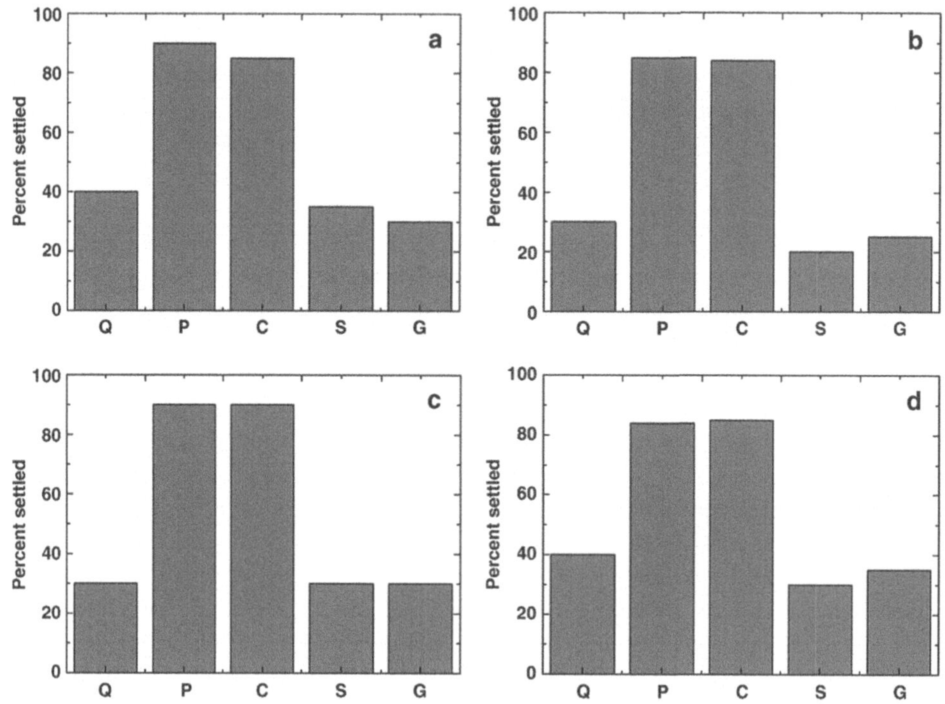

Fig. 11 Adsorption densities of protein fractions on various minerals **a** A group 0–20%, **b** B group 20–40%, **c** C group 40–60%, **d** D group 60–90% (Q: Quartz, P: Pyrite, C: Chalcopyrite, S: Sphalerite, G: Galena) (Patra & Natarajan, 2008)

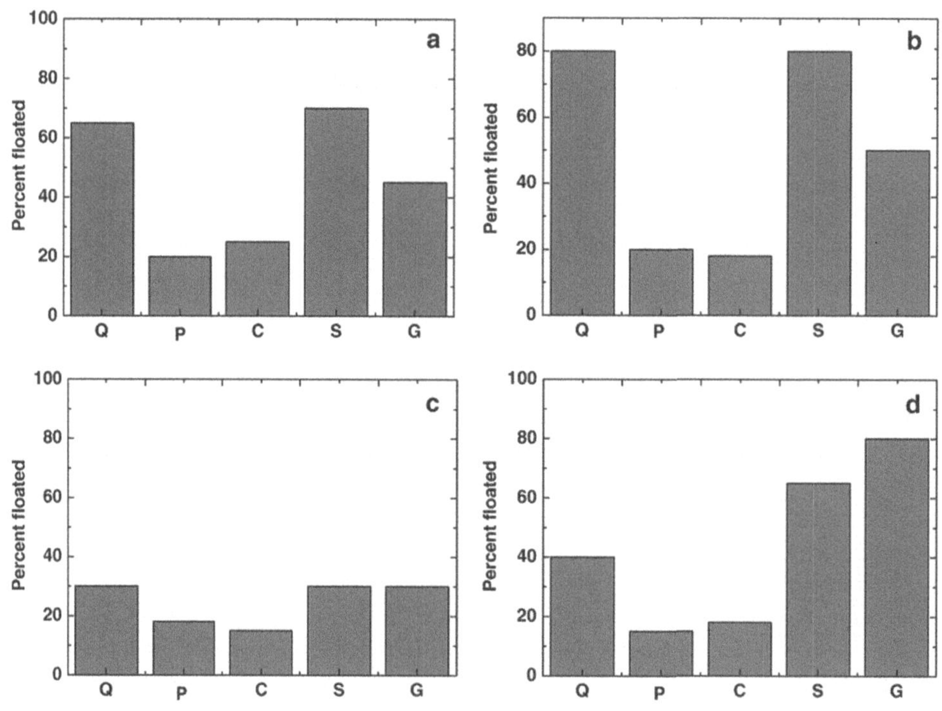

Fig. 12 Flocculation behavior of minerals after interaction with different protein fractions **a** A group 0–20%, **b** B group 20–40%, **c** C group 40–60%, **d** D group 60–90% (Q: Quartz, P: Pyrite, C: Chalcopyrite, S: Sphalerite, G: Galena) (Patra & Natarajan, 2008)

indicated that selective depression of pyrite and chalcopyrite in the presence of galena, quartz, and sphalerite could be achieved (Patra & Natarajan, 2008).

5.1.8 Leptosririllum ferrooxidans

Leptospirillum ferrooxidans is a chemolithotrophic acidophilic bacterium that can be found in abundance in acid mine drainage (AMD). This microorganism can bind selectively to sulfide minerals, having the ability to alter the properties of surfaces and alter their floatability. *Leptospirillum ferrooxidans* derive their energy from the oxidation of Fe^{2+} to Fe^{3+}.

In 2008 and 2011, Velinska et al. investigated the effect of *Leptospirillum ferrooxidans* on pyrite and chalcopyrite

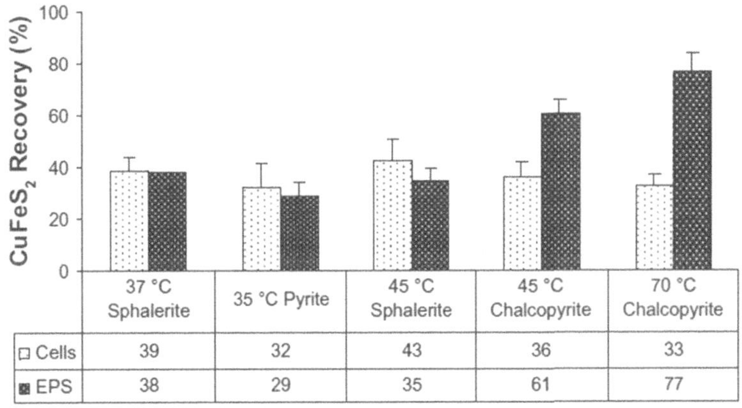

Fig. 13 Comparison of the efficiency of using bacterial cells with bound EPS and free

flotation. These researchers also investigated the thermodynamics and DLVO theory for bacterial attachment. The results indicated that bacterial treatment has a more significant effect on chalcopyrite than pyrite, both for bioflotation and bioflocculation. The DLVO theory also confirmed the results. At pH = 2, the ZPC of the chalcopyrite surface was lower due to the electrostatic repulsion forces associated with pyrite. In flotation experiments performed at pH = 4 and 0.5×10^{-4} mol/L xanthate as a collector, it was observed that the recovery of both minerals decreased in the presence of bacterial cells. However, this reduction was much higher for chalcopyrite in comparison to pyrite. Chalcopyrite flotation recovery was decreased from 95 to 25%, while pyrite recovery was reduced to 67% under similar conditions. These researches attributed the higher tendency of *Leptospirillum ferrooxidans* to bind the chalcopyrite surface rather than pyrite due to more defects and higher iron availability as a source of energy in chalcopyrite surface (Vilinska & Rao, 2008, 2011).

In another research, Bleeze et al. (2018) examined the effect of this microorganism in the separation of pyrite and chalcopyrite. The best separation of pyrite and chalcopyrite was obtained when *Leptospirillum ferrooxidans* were cultured in the presence of chalcopyrite before the addition of the PIPX collector. The results indicated that *Leptospirillum ferrooxidans* have an inhibitory effect on both minerals, while the presence of EPS in acidic conditions has a more inhibitory effect on pyrite than chalcopyrite (Bleeze et al., 2018).

Pacina et al. (2009) investigated the effect of *Leptospirillum ferrooxidans* on the flotation kinetics of chalcopyrite, pyrrhotite, and sphalerite. The results showed that the effect of this bacterium on the flotation rate of the studied sulfide minerals is directly related to the susceptibility of the ore to oxidation by the microorganism. The floatability of bacterially conditioned chalcopyrite increased by the addition of elemental sulfur to the treatment environment. The recovery of chalcopyrite was directly related to bacterial activity, while pH was not an adequate parameter. *Leptospirillum ferrooxidans* had a slightly depressing effect on pyrrhotite after more than 60 min conditioning. Also, as sphalerite had low sensitivity to oxidation, no change in flotation recovery was observed (Pecina et al., 2009).

Díaz-López et al. (2012) also studied the effect of *Leptospirillum ferrooxidans* on the flotation of chalcopyrite and pyrrhotite. The results indicated that the recovery of chalcopyrite in the presence of *Leptospirillum ferrooxidans* increased from 80 to 95%. However, in the case of pyrrhotite, the effect was different for each particle size, which made it a weak depressant because bacteria could not cover all the hydrophobic surfaces. In addition, chalcopyrite recovery was increased with a shorter contact time. In contrast, pyrrhotite conditioning needs high contact time. It was also stated that in both minerals, adsorption is a quick process that happens in the first 10–20 min, and that bacteria are more prone to pyrrhotite than chalcopyrite (Díaz-López et al., 2012).

5.1.9 Rhodococcus opacus

Rhodococcus opacus is a gram-positive bacterium that contains various organic compounds, including polysaccharides, carboxylic acids, lipid groups, and mycolic acids in its cell wall that exhibit amphoteric behavior at the cell surface.

Botero et al. (2007) investigated the effect of *Rhodococcus opacus* as a bio-collector of calcite and magnesite. The results showed that the bacterial treatment decreased the IEP for both minerals. In addition, the number of *Rhodococcus opacus* cells adsorbed on the surface of magnesite was 10 times higher compared to calcite. The interaction of *Rhodococcus apacus* with the surface of these two minerals was dependent on pH, and the optimal adsorption results of the two minerals were observed at a pH = ~7. The adsorption kinetics for both minerals was rapid and reached equilibrium after 5 min. The magnesite recovery at pH = ~5 and bacterial concentration of 100 ppm was 93%, while it was 55% at pH = ~7 and 220 ppm bacterial concentration (Botero et al., 2007).

In similar experiments, Merma et al. (2013) examined the effect of *Rhodococcus opacus* on quartz and apatite flotation. Bacteria altered the zeta potential of both minerals immediately after contact, which was more noticeable for apatite. It was also found that the bacterial cells' surface tension was influenced by pH and bacterial concentration. The lowest surface tension was found in the 0.15 g/l biomass and at a pH range of 3–7. However, bacterial cells were more inclined to the apatite surface. They also stated that the bioflotation of apatite and quartz particles depends on the pH and bacterial concentration. The best separation efficiency was obtained at pH = 5, and recoveries were 90% and 14% for apatite and quartz, respectively (Merma et al., 2013).

Kim et al. (2017) investigated the possibility of bioflotation of copper oxide (malachite) ores by *Rhodococcus opacus*. They observed that the agitator speed in the range of 800–1200 rpm did not influence malachite flotation. However, ionic strength and especially pH changes affect the recovery and grade of malachite. At pH = 7, the highest grade (98%) and recovery (93%) and pH = 11, the lowest grade (68%) and recovery (90%). The experimental results showed that bioflotation was appropriate for fine particle size and led to a higher degree of malachite freedom (5.17%) than the conventional process (1.43%) (Kim et al., 2017).

6 Future Prospective and Conclusion

Bio-beneficiation is the application of microorganisms (including bacteria, fungi, algae, and yeast) and their products (such as biomass and extracellular polymeric substances (EPS)) to facilitate the selective separation of gangue minerals from valuable minerals in mineral processing. Bio-beneficiation is divided into two main categories: bioflotation and bioflocculation. Microorganisms and their biomass can be used as collectors, depressants, dispersants, bioflocculants, and frothers. Fundamentally, the role of microorganisms in bio-beneficiation is to change the physicochemical properties of mineral surfaces to obtain the best separation efficiency. Several mechanisms have been proposed to explain mineral bio-beneficiation, including (DLVO) and extended DLVO (XDLVO) theory, cationic bridging theory, and polymer bridging or adsorption bridging theory. However, the actual role of microorganisms and their extracellular polymeric substances (EPS) in the bio-beneficiation process is not understood well yet. There is still a significant gap in the literature describing exact dominating bio-beneficiation mechanisms, so future studies in this area must be accomplished. Numerous studies on bioflotation and flocculation associated with the selective separation/removal of coal, oxide minerals, and sulfide minerals have been reported in this chapter, indicating the high potential of bio-beneficiation in mineral processing. Interactions of microorganisms and minerals cause alteration in the chemical properties of the minerals' surface, the reaction of bacterial cells and metabolic products with minerals, production of surface-active chemicals. So the intrinsic differences between minerals attaching to microorganisms and their products lead to their selective biological separation. Although biohydrometallurgical methods for various ores have been thoroughly studied and commercialized, bio-beneficiation has not yet been as widely used as biohydrometallurgical processes on industrial scales. So due to the general studies reviewed in this chapter, there is a noteworthy need for commercial-scale advancements in bio-beneficiation processes, and more attention should be directed to it.

References

Abdel-Khalek, M. A., & El-Midany, A. A. (2013). Application of *Bacillus subtilis* for reducing ash and sulfur in coal. *Environment and Earth Science, 70*, 753–760. https://doi.org/10.1007/s12665-012-2163-4

Amini, E., Hosseini, T. R., Oliazadeh, M., & Kolahdoozan, M. (2009a). Application of *Acidithiobacillus ferrooxidans* in coal flotation. *International Journal of Coal Preparation and Utilization, 29*, 279–288. https://doi.org/10.1080/19392690903558314

Amini, E., Oliazadeh, M., & Kolahdoozan, M. (2009b). Kinetic comparison of biological and conventional flotation of coal. *Minerals Engineering, 22*, 344–347. https://doi.org/10.1016/j.mineng.2008.09.008

Badireddy, A. R., Chellam, S., Gassman, P. L., et al. (2010). Role of extracellular polymeric substances in bioflocculation of activated sludge microorganisms under glucose-controlled conditions. *Water Research, 44*, 4505–4516. https://doi.org/10.1016/j.watres.2010.06.024

Ben, R. F., Mnif, W., & Siddeeg, S. M. (2018). Microbial flocculants as an alternative to synthetic polymers for wastewater treatment: A review. *Symmetry (basel), 10*, 1–19. https://doi.org/10.3390/sym10110556

Biggs, S., Habgood, M., Jameson, G. J., & Yan, Y. (2000). Aggregate structures formed via a bridging flocculation mechanism. *Chemical Engineering Journal, 80*, 13–22. https://doi.org/10.1016/S1383-5866(00)00072-1

Bleeze, B., Zhao, J., & Harmer, S. L. (2018). Selective attachment of *Leptospirillum ferrooxidans* for separation of chalcopyrite and pyrite through bio-flotation. *Minerals, 8*, 7–14. https://doi.org/10.3390/min8030086

Bolto, B., & Gregory, J. (2007). Organic polyelectrolytes in water treatment. *Water Research, 41*, 2301–2324. https://doi.org/10.1016/j.watres.2007.03.012

Botero, A. E. C., Torem, M. L., & Souza de Mesquita, L. M. (2007). Fundamental studies of *Rhodococcus opacus* as a biocollector of calcite and magnesite. *Minerals Engineering, 20*, 1026–1032. https://doi.org/10.1016/j.mineng.2007.03.017

Brock TD, Madigan MT, Martinko JM (2006) Biology of microorganisms, Eleventh edition. Biol Microorg Elev Ed 909

Camarate MC, Merma AG, Hacha RR, Torem ML (2021) Selective bioflocculation of ultrafine hematite particles from quartz using a biosurfactant extracted from Candida stellata yeast. Sep Sci Technol 1–12. https://doi.org/10.1080/01496395.2021.1881972

Campbell, A. (2002). The potential role of aluminium in Alzheimer's disease. *Nephrology, Dialysis, Transplantation, 17*, 17–20. https://doi.org/10.1093/ndt/17.suppl_2.17

Chandraprabha, M. N., & Natarajan, K. A. (2006). Surface chemical and flotation behaviour of chalcopyrite and pyrite in the presence of *Acidithiobacillus thiooxidans*. *Hydrometallurgy, 83*, 146–152. https://doi.org/10.1016/j.hydromet.2006.03.021

Chandraprabha, M. N., Natarajan, K. A., & Modak, J. M. (2004a). Selective separation of pyrite and chalcopyrite by biomodulation. *Colloids Surfaces B Biointerfaces, 37*, 93–100.

Chandraprabha, M. N., Natarajan, K. A., & Somasundaran, P. (2004b). Selective separation of arsenopyrite from pyrite by biomodulation in the presence of *Acidithiobacillus ferrooxidans*. *Journal of Colloid and Interface Science, 276*, 323–332.

Chandraprabha, M. N., Natarajan, K. A., & Somasundaran, P. (2005). Selective separation of pyrite from chalcopyrite and arsenopyrite by biomodulation using *Acidithiobacillus ferrooxidans*. *International Journal of Mineral Processing, 75*, 113–122.

Chapelle, F. H. (2000). *Ground-water microbiology and geochemistry* (p. 496). Wiley Sons .

Chen, Y., Geurts, M., Sjollema, S. B., et al. (2014). Acute toxicity of the cationic surfactant C12-benzalkonium in different bioassays: How test design affects bioavailability and effect concentrations. *Environmental Toxicology and Chemistry, 33*, 606–615. https://doi.org/10.1002/etc.2465

Chia, T. W. R., Nguyen, V. T., McMeekin, T., et al. (2011). Stochasticity of bacterial attachment and its predictability by the extended Derjaguin-Landau-Verwey-Overbeek Theory. *Applied and Environment Microbiology, 77*, 3757–3764. https://doi.org/10.1128/AEM.01415-10

D'Abzac, P., Bordas, F., Joussein, E., et al. (2010a). Characterization of the Mineral Fraction Associated to Extracellular Polymeric Substances (EPS) in Anaerobic Granular Sludges. *Environmental Science and Technology, 44*, 412–418. https://doi.org/10.1021/es901912g

D'Abzac, P., Bordas, F., Van Hullebusch, E., et al. (2010b). Extraction of extracellular polymeric substances (EPS) from anaerobic granular sludges: Comparison of chemical and physical extraction protocols. *Applied Microbiology and Biotechnology, 85*, 1589–1599. https://doi.org/10.1007/s00253-009-2288-x

Deo, N., & Natarajan, K. A. (1997). Interaction of *Bacillus polymyxa* with some oxide minerals with reference to mineral beneficiation and environmental control. *Minerals Engineering, 10*, 1339–1354. https://doi.org/10.1016/s0892-6875(97)00125-8

Deo, N., & Natarajan, K. A. (1998). Studies on interaction of *Paenibacillus polymyxa* with iron ore minerals in relation to beneficiation. *International Journal of Mineral Processing, 55*, 41–60. https://doi.org/10.1016/s0301-7516(98)00020-9

Deo, N., & Natarajan, K. A. (1999). Role of corundum-adapted strains of *Bacillus polymyxa* in the separation of hematite and alumina. *Minerals & Metallurgical Processing, 16*, 29–34. https://doi.org/10.1007/bf03403231

Derjaguin, B. V., Churaev, N. V., & Muller, V. M. (1987). The Derjaguin—Landau—Verwey—Overbeek (DLVO) Theory of stability of lyophobic colloids. In *Surface forces* (pp. 293–310). https://doi.org/10.1007/978-1-4757-6639-4_8

Díaz-López, C. V., Pecina-Treviño, E. T., & Orrantia-Borunda, E. (2012). A study of bioflotation of chalcopyrite and pyrrhotite mixtures in presence of *L. Ferrooxidans*. *Canadian Metallurgical Quarterly, 51*, 118–125. https://doi.org/10.1179/0008443312Z.00000000025

Dignac, M.-F., Urbain, V., Rybacki, D., et al. (1998). Chemical description of extracellular polymers: Implication on activated sludge floc structure. *Water Science and Technology, 38*, 45–53. https://doi.org/10.2166/wst.1998.0789

Donati ER, Sand W (2007) Microbial processing of metal sulfides. Microb. Process. Met. Sulfides 1–314

Dwyer, R., Bruckard, W. J., Rea, S., & Holmes, R. J. (2012). Bioflotation and bioflocculation review: Microorganisms relevant for mineral beneficiation. *Miner Process Extr Metall, 121*, 65–71.

El-Midany, A. A., & Abdel-Khalek, M. A. (2014). Reducing sulfur and ash from coal using *Bacillus subtilis* and *Paenibacillus polymyxa*. *Fuel, 115*, 589–595. https://doi.org/10.1016/j.fuel.2013.07.076

Fr/olund B, Griebe T, Nielsen PH (1995) Enzymatic activity in the activated-sludge floc matrix. Appl Microbiol Biotechnol 43:755–761. https://doi.org/10.1007/BF00164784

Frølund, B., Palmgren, R., Keiding, K., & Nielsen, P. H. (1996). Extraction of extracellular polymers from activated sludge using a cation exchange resin. *Water Research, 30*, 1749–1758. https://doi.org/10.1016/0043-1354(95)00323-1

Gao, Q., Zhu, X.-H., Mu, J., et al. (2009). Using Ruditapes philippinarum conglutination mud to produce bioflocculant and its applications in wastewater treatment. *Bioresource Technology, 100*, 4996–5001. https://doi.org/10.1016/j.biortech.2009.05.035

Gaudin AM (1975) Flotation. Vol. 2. Mc Graw Hill Co

Gawel, J., Maliszewska, I., & Sadowski, Z. (1997). The effect of biopretreatment on the flotation recovery of magnesite tailings. *Minerals Engineering, 10*, 813–824. https://doi.org/10.1016/s0892-6875(97)00059-9

Ghashoghchi, R. A., Hosseini, M. R., & Ahmadi, A. (2017). Effects of microbial cells and their associated extracellular polymeric substances on the bio-flocculation of kaolin and quartz. *Applied Clay Science, 138*, 81–88. https://doi.org/10.1016/j.clay.2017.01.002

Govender, Y., & Gericke, M. (2011). Extracellular polymeric substances (EPS) from bioleaching systems and its application in bioflotation. *Minerals Engineering, 24*, 1122–1127. https://doi.org/10.1016/j.mineng.2011.02.016

Gregory, J., & Barany, S. (2011). Adsorption and flocculation by polymers and polymer mixtures. *Advances in Colloid and Interface Science, 169*, 1–12. https://doi.org/10.1016/j.cis.2011.06.004

Grossart, H. P., Kiørboe, T., Tang, K. W., et al. (2006). Interactions between marine snow and heterotrophic bacteria: Aggregate formation and microbial dynamics. *Aquatic Microbial Ecology, 42*, 19–26. https://doi.org/10.3354/ame042019

Groudev, S. N. (1987). Use of heterotrophic microorganisms in mineral biotechnology. *Acta Biotechnologica, 7*, 299–306. https://doi.org/10.1002/abio.370070404

Haas, S. R., Nascimento, F. R., Schneider, I. A. H., & Gaylarde, C. (1999). Flocculation of fine fluorite particles with *Corynebacterium xerosis*. *Revista De Microbiologia, 30*, 225–230. https://doi.org/10.1590/S0001-37141999000300007

Hanumantha Rao, K., & Forssberg, K. S. (2001). Minerals bioprocessing IV. *Minerials Engineering, 4*, 1127–1141.

Higgins, M. J., & Novak, J. T. (1997a). The effect of cations on the settling and dewatering of activated sludges: Laboratory results. *Water Environment Research, 69*, 215–224. https://doi.org/10.2175/106143097X125371

Higgins, M. J., & Novak, J. T. (1997b). Characterization of Exocellular Protein and Its Role in Bioflocculation. *Journal of Environmental Engineering, 123*, 479–485. https://doi.org/10.1061/(ASCE)0733-9372(1997)123:5(479)

Hocking, M. B., Klimchuk, K. A., & Lowen, S. (1999). Polymeric flocculants and flocculation. *Journal of Macromolecular Science Part C Polymer Review, 39*, 177–203. https://doi.org/10.1081/MC-100101419

Holmes, D. S., & Smith, R. W. (1995). Minerals bioprocessing-II. *Minerals Metals & Materials Society, 4*, 1127–1141.

Hosseini, M. R., Bahrami, A., Ahmadi, A., et al. (2019). Application of differential bio-flocculation in the removal of hematite and goethite from kaolin and quartz. *Chemical Engineering*

Communications, 206, 815–827. https://doi.org/10.1080/00986445.2018.1530990

Hosseini, M. R., Sadeghieh, S. M., Azizinia, M. R., & Tabatabaei, S. H. (2020). Biological separation of quartz from kaolinite using *Bacillus licheniformis*. *Separation Science and Technology, 55*, 2061–2071. https://doi.org/10.1080/01496395.2019.1617738

Hosseini, T. R., Kolahdoozan, M., Tabatabaei, Y. S. M., et al. (2005). Bioflotation of Sarcheshmeh copper ore using *Thiobacillus ferrooxidans* bacteria. *Minerals Engineering, 18*, 371–374.

Hughes MA (1990) Coagulation and flocculation Solid. In: Liquid separation 3th. L. Savrosky

Jerez CA (2019) Bioleaching and biomining for the industrial recovery of metals. Compr Biotechnol 758–771. https://doi.org/10.1016/B978-0-12-809633-8.09185-8

Johnson DB (2009) Extremophiles: Acidic Environments. Encycl Microbiol 107–126. https://doi.org/10.1016/B978-012373944-5.00278-9

Johnson, D. B., & Hallberg, K. B. (2005). Acid mine drainage remediation options: A review. *Science of the Total Environment, 338*, 3–14. https://doi.org/10.1016/j.scitotenv.2004.09.002

Joon, L.B., & Schlautman, M. (2015). Book of abstracts. In *Adsorption and flocculation mediated by polymeric substances in cohesive sediment suspensions: Experimental study* (pp. 1–28).

Karthiga Devi, K., & Natarajan, K. A. (2015). Production and characterization of bioflocculants for mineral processing applications. *International Journal of Mineral Processing, 137*, 15–25. https://doi.org/10.1016/j.minpro.2015.02.007

Kim, G., Choi, J., Silva, R. A., et al. (2017). Feasibility of bench-scale selective bioflotation of copper oxide minerals using *Rhodococcus opacus*. *Hydrometallurgy, 168*, 94–102. https://doi.org/10.1016/j.hydromet.2016.06.029

Kiørboe, T. (2001). Formation and fate of marine snow: small-scale processes with large- scale implications. *Scientia Marina, 65*, 57–71. https://doi.org/10.3989/scimar.2001.65s257

Lai, H., Fang, H., Huang, L., et al. (2018). A review on sediment bioflocculation: Dynamics, influencing factors and modeling. *Science of the Total Environment, 642*, 1184–1200. https://doi.org/10.1016/j.scitotenv.2018.06.101

Laspidou, C. (2002). A unified theory for extracellular polymeric substances, soluble microbial products, and active and inert biomass. *Water Research, 36*, 2711–2720. https://doi.org/10.1016/S0043-1354(01)00413-4

Li, H., Wen, Y., Cao, A., et al. (2012). The influence of additives (Ca^{2+}, Al^{3+}, and Fe^{3+}) on the interaction energy and loosely bound extracellular polymeric substances (EPS) of activated sludge and their flocculation mechanisms. *Bioresource Technology, 114*, 188–194. https://doi.org/10.1016/j.biortech.2012.03.043

Li, H., Wen, Y., Cao, A., et al. (2014). The influence of multivalent cations on the flocculation of activated sludge with different sludge retention times. *Water Research, 55*, 225–232. https://doi.org/10.1016/j.watres.2014.02.014

Lin, H., Zhang, M., Wang, F., et al. (2014). A critical review of extracellular polymeric substances (EPSs) in membrane bioreactors: Characteristics, roles in membrane fouling and control strategies. *Journal of Membrance Science, 460*, 110–125. https://doi.org/10.1016/j.memsci.2014.02.034

Liu, X.-M., Sheng, G.-P., Luo, H.-W., et al. (2010). Contribution of extracellular polymeric substances (EPS) to the sludge aggregation. *Environmental Science and Technology, 44*, 4355–4360. https://doi.org/10.1021/es9016766

Liu, X.-M., Sheng, G.-P., Wang, J., & Yu, H.-Q. (2008). Quantifying the surface characteristics and flocculability of Ralstonia eutropha. *Applied Microbiology and Biotechnology, 79*, 187–194. https://doi.org/10.1007/s00253-008-1426-1

Liu, X.-M., Sheng, G.-P., & Yu, H.-Q. (2007). DLVO approach to the flocculability of a photosynthetic H_2—producing bacterium, Rhodopseudomonas acidophila. *Environmental Science and Technology, 41*, 4620–4625. https://doi.org/10.1021/es070107n

Liu, Y., & Fang, H. H. P. (2003). Influences of extracellular polymeric substances (EPS) on flocculation, settling, and dewatering of activated sludge. *Critical Reviews in Environment Science and Technology, 33*, 237–273. https://doi.org/10.1080/10643380390814479

Lurie, M., & Rebhun, M. (1997). Effect of properties of polyelectrolytes on their interaction with particulates and soluble organics. *Water Science and Technology, 36*, 93–101. https://doi.org/10.1016/S0273-1223(97)00425-3

Mehrabani, J. V., Mousavi, S. M., & Noaparast, M. (2011). Evaluation of the replacement of NaCN with *Acidithiobacillus ferrooxidans* in the flotation of high-pyrite, low-grade lead-zinc ore. *Separation and Purification Technology, 80*, 202–208. https://doi.org/10.1016/j.seppur.2011.04.006

Mehrabani, J. V., Noaparast, M., Mousavi, S. M., et al. (2010). Depression of pyrite in the flotation of high pyrite low-grade lead-zinc ore using *Acidithiobacillus ferrooxidans*. *Minerals Engineering, 23*, 10–16. https://doi.org/10.1016/j.mineng.2009.08.008

Merma, A. G., Torem, M. L., Morán, J. J. V., & Monte, M. B. M. (2013). On the fundamental aspects of apatite and quartz flotation using a Gram positive strain as a bioreagent. *Minerals Engineering, 48*, 61–67. https://doi.org/10.1016/j.mineng.2012.10.018

Mietta, F. (2010). *Evolution of the floc size distribution of cohesive sediments Ontwikkeling van de vlokgrootteverdeling van cohesief sediment.*

Misra, M., Bukka, K., & Chen, S. (1996). The effect of growth medium of *Thiobacillus ferrooxidans* on pyrite flotation. *Minerals Engineering, 9*, 157–168.

More, T. T., Yadav, J. S. S., Yan, S., et al. (2014). Extracellular polymeric substances of bacteria and their potential environmental applications. *Journal of Environmental Management, 144*, 1–25. https://doi.org/10.1016/j.jenvman.2014.05.010

Muddanna, M. H., & Baral, S. S. (2019). A comparative study of the extraction of metals from the spent fluid catalytic cracking catalyst using chemical leaching and bioleaching by *Aspergillus niger*. *Journal of Environmental Chemical Engineering, 7*, 103335. https://doi.org/10.1016/J.JECE.2019.103335

Murphy, D. (1952). Structure of a levan produced by *Bacillus polymyxa*. *Canadian Journal of Chemistry, 30*, 872–878. https://doi.org/10.1139/v52-106

Nagaoka, T., Ohmura, N., & Saiki, H. (1999). A novel mineral flotation process using *Thiobacillus ferrooxidans*. *Applied and Environment Microbiology, 65*, 3588–3593.

Natarajan, K. A., & Das, A. (2003). Surface chemical studies on "*Acidithiobacillus*" group of bacteria with reference to mineral flocculation. *International Journal of Mineral Processing, 72*, 189–198. https://doi.org/10.1016/S0301-7516(03)00098-X

Nichols, C. M., Bowman, J. P., & Guezennec, J. (2005). Effects of incubation temperature on growth and production of exopolysaccharides by an Antarctic sea ice bacterium grown in batch culture. *Applied and Environment Microbiology, 71*, 3519–3523. https://doi.org/10.1128/AEM.71.7.3519-3523.2005

Olson, G. J., Brierley, J. A., & Brierley, C. L. (2003). Bioleaching review part B: Progress in bioleaching: Applications of microbial processes by the minerals industries. *Applied Microbiology and Biotechnology, 63*, 249–257. https://doi.org/10.1007/s00253-003-1404-6

Padukone, S. U., & Natarajan, K. A. (2011). Microbially induced separation of quartz from calcite using *Saccharomyces cerevisiae*. *Colloids Surfaces B Biointerfaces, 88*, 45–50. https://doi.org/10.1016/j.colsurfb.2011.05.054

Parikh, S. J., & Chorover, J. (2006). ATR-FTIR Spectroscopy Reveals Bond Formation During Bacterial Adhesion to Iron Oxide. *Langmuir, 22*, 8492–8500. https://doi.org/10.1021/la061359p

Patra, P., & Natarajan, K. A. (2003). Microbially-induced flocculation and flotation for pyrite separation from oxide gangue minerals. *Minerals Engineering, 16*, 965–973. https://doi.org/10.1016/S0892-6875(03)00268-1

Patra, P., & Natarajan, K. A. (2004a). Microbially induced flotation and flocculation of pyrite and sphalerite. *Colloids Surfaces B Biointerfaces, 36*, 91–99. https://doi.org/10.1016/j.colsurfb.2004.05.010

Patra, P., & Natarajan, K. A. (2004b). Microbially induced flocculation and flotation for separation of chalcopyrite from quartz and calcite. *International Journal of Mineral Processing, 74*, 143–155. https://doi.org/10.1016/j.minpro.2003.10.006

Patra, P., & Natarajan, K. A. (2006). Surface chemical studies on selective separation of pyrite and galena in the presence of bacterial cells and metabolic products of *Paenibacillus polymyxa*. *Journal of Colloid and Interface Science, 298*, 720–729. https://doi.org/10.1016/j.jcis.2006.01.017

Patra, P., & Natarajan, K. A. (2008). Role of mineral specific bacterial proteins in selective flocculation and flotation. *International Journal of Mineral Processing, 88*, 53–58. https://doi.org/10.1016/j.minpro.2008.06.005

Pecina, E. T., Rodríguez, M., Castillo, P., et al. (2009). Effect of *Leptospirillum ferrooxidans* on the flotation kinetics of sulphide ores. *Minerals Engineering, 22*, 462–468. https://doi.org/10.1016/j.mineng.2008.12.008

Pecina-Treviño, E. T., Ramos-Escobedo, G. T., Gallegos-Acevedo, P. M., et al. (2012). Bioflotation of sulfide minerals with *Acidithiobacillus ferrooxidans* in relation to copper activation and surface oxidation. *Canadian Journal of Microbiology, 58*, 1073–1083.

Poorni, S., & Natarajan, K. A. (2013a). Microbially induced selective flocculation of hematite from kaolinite. *International Journal of Mineral Processing, 125*, 92–100. https://doi.org/10.1016/j.minpro.2013.10.002

Poorni, S., & Natarajan, K. A. (2013b). Biomediated separation of kaolinite and hematite using *bacillus subtilis*. *Advances in Materials Research, 825*, 223–226. https://doi.org/10.4028/www.scientific.net/AMR.825.223

Poorni, S., & Natarajan, K. A. (2014). Flocculation behaviour of hematite-kaolinite suspensions in presence of extracellular bacterial proteins and polysaccharides. *Colloids Surfaces B Biointerfaces, 114*, 186–192. https://doi.org/10.1016/j.colsurfb.2013.09.049

Raichur, A. M., Misra, M., Bukka, K., & Smith, R. W. (1996). Flocculation and flotation of coal by adhesion of hydrophobic *Mycobacterium phlei*. *Colloids Surfaces B Biointerfaces, 8*, 13–24. https://doi.org/10.1016/S0927-7765(96)01312-4

Rao, M. K. Y., Natarajan, K. A., & Somasundaran, P. (1992). Effect of biotreatment with *Thiobacillus ferrooxidans* on the floatability of sphalerite and galena. *Mining, Metall Explor, 9*, 95–100.

Rijnaarts, H. H. M., Norde, W., Lyklema, J., & Zehnder, A. J. B. (1999). DLVO and steric contributions to bacterial deposition in media of different ionic strengths. *Colloids Surfaces B Biointerfaces, 14*, 179–195. https://doi.org/10.1016/S0927-7765(99)00035-1

Sadeghizadeh M, Hosseini MR, Ahmadi A (2017) Bioflocculation of Hematite and Goethite Using *Bacillus licheniformis* (Bioflocculation of Hematite and Goethite)

Salehizadeh, H., & Shojaosadati, S. A. (2001). Extracellular biopolymeric flocculants: Recent trends and biotechnological importance. *Biotechnology Advances, 19*, 371–385. https://doi.org/10.1016/S0734-9750(01)00071-4

Salehizadeh, H., & Yan, N. (2014). Recent advances in extracellular biopolymer flocculants. *Biotechnology Advances, 32*, 1506–1522. https://doi.org/10.1016/j.biotechadv.2014.10.004

San Martín, F., Kracht, W., & Vargas, T. (2018). Biodepression of pyrite using *Acidithiobacillus ferrooxidans* in seawater. *Minerals Engineering, 117*, 127–131. https://doi.org/10.1016/j.mineng.2017.11.005

San Martín, F., Kracht, W., Vargas, T., & Rudolph, M. (2020). Mechanisms of pyrite biodepression with *Acidithiobacillus ferrooxidans* in seawater flotation. *Minerals Engineering, 145*, 106067. https://doi.org/10.1016/j.mineng.2019.106067

Santhiya, D., Subramanian, S., & Natarajan, K. A. (2000). Surface chemical studies on galena and sphalerite in the presence of *Thiobacillus thiooxidans* with reference to mineral beneficiation. *Minerals Engineering, 13*, 747–763. https://doi.org/10.1016/S0892-6875(00)00059-5

Santhiya, D., Subramanian, S., & Natarajan, K. A. (2001a). Surface chemical studies on sphalerite and galena using *Bacillus polymyxa*: I. Microbially induced mineral separation. *Journal of Colloid and Interface Science, 235*, 289–297. https://doi.org/10.1006/jcis.2000.7374

Santhiya, D., Subramanian, S., Natarajan, K. A., et al. (2001b). Biomodulation of galena and sphalerite surfaces using *Thiobacillus thiooxidans*. *International Journal of Mineral Processing, 62*, 121–141. https://doi.org/10.1016/S0301-7516(00)00048-X

Santhiya, D., Subramanian, S., & Natarajan, K. A. (2002). Surface chemical studies on sphalerite and galena using extracellular polysaccharides isolated from *Bacillus polymyxa*. *Journal of Colloid and Interface Science, 256*, 237–248. https://doi.org/10.1006/jcis.2002.8681

Sarvamangala, H., Natarajan, K. A., & Girisha, S. T. (2013). Microbially-induced pyrite removal from galena using *Bacillus subtilis*. *International Journal of Mineral Processing, 120*, 15–21. https://doi.org/10.1016/j.minpro.2013.02.005

Schneider, I. A. H., Misra, M., & Smith, R. W. (1994). Bioflocculation of Hematite Suspensions with Products from Yeast Cell Rupture. *Developments in Chemical Engineering and Mineral Processing, 2*, 248–252. https://doi.org/10.1002/apj.5500020407

Selim, K. A., & Rostom, M. (2018). Bioflocculation of (Iron oxide – Silica) system using *Bacillus cereus* bacteria isolated from Egyptian iron ore surface. *Egyptian Journal of Petroleum, 27*, 235–240. https://doi.org/10.1016/j.ejpe.2017.07.002

Sheng, G.-P., Yu, H.-Q., & Li, X.-Y. (2010). Extracellular polymeric substances (EPS) of microbial aggregates in biological wastewater treatment systems: A review. *Biotechnology Advances, 28*, 882–894. https://doi.org/10.1016/j.biotechadv.2010.08.001

Simon, M., Grossart, H. P., Schweitzer, B., & Ploug, H. (2002). Microbial ecology of organic aggregates in aquatic ecosystems. *Aquatic Microbial Ecology, 28*, 175–211. https://doi.org/10.3354/ame028175

Smith, R. W., Misra, M., & Dubel, J. (1991). Mineral bioprocessing and the future. *Minerals Engineering, 4*, 1127–1141. https://doi.org/10.1016/0892-6875(91)90088-D

Sobeck, D. C., & Higgins, M. J. (2002). Examination of three theories for mechanisms of cation-induced bioflocculation. *Water Research, 36*, 527–538. https://doi.org/10.1016/S0043-1354(01)00254-8

Somasundaran, P., Deo, N., Deo, P., & Natarajan, K. A. (2005). Role of biopolymers on bacterial adhesion and mineral beneficiation. *Minerals & Metallurgical Processing, 22*, 1–11. https://doi.org/10.1007/bf03403189

Subramanian, S., Santhiya, D., & Natarajan, K. A. (2003). Surface modification studies on sulphide minerals using bioreagents. *International Journal of Mineral Processing, 72*, 175–188. https://doi.org/10.1016/S0301-7516(03)00097-8

Tang, F. H. M., Alonso-Marroquin, F., & Maggi, F. (2014). Stochastic collision and aggregation analysis of kaolinite in water through experiments and the spheropolygon theory. *Water Research, 53*, 180–190. https://doi.org/10.1016/j.watres.2014.01.026

Tang, F. H. M., & Maggi, F. (2016). A mesocosm experiment of suspended particulate matter dynamics in nutrient- and biomass-affected waters. *Water Research, 89*, 76–86. https://doi.org/10.1016/j.watres.2015.11.033

Tezuka, Y. (1969). Cation-dependent flocculation in a flavobacterium species predominant in activated sludge. *Applied Microbiology, 17*, 222–226. https://doi.org/10.1128/am.17.2.222-226.1969

Tsuneda, S., Aikawa, H., Hayashi, H., et al. (2003). Extracellular polymeric substances responsible for bacterial adhesion onto solid surface. *FEMS Microbiology Letters, 223*, 287–292. https://doi.org/10.1016/S0378-1097(03)00399-9

Vasanthakumar, B., Ravishankar, H., & Subramanian, S. (2013). Microbially induced selective flotation of sphalerite from galena using mineral-adapted strains of *Bacillus megaterium*. *Colloids Surfaces B Biointerfaces, 112*, 279–286. https://doi.org/10.1016/j.colsurfb.2013.07.034

Vasanthakumar, B., Ravishankar, H., & Subramanian, S. (2014). Basic studies on the role of components of *Bacillus megaterium* as flotation biocollectors in sulphide mineral separation. *Applied Microbiology and Biotechnology, 98*, 2719–2728. https://doi.org/10.1007/s00253-013-5251-9

Vasanthakumar, B., Ravishankar, H., & Subramanian, S. (2017). Selective bio-flotation of sphalerite from galena using mineral – adapted strains of *Bacillus subtilis*. *Minerals Engineering, 110*, 179–184. https://doi.org/10.1016/j.mineng.2017.05.002

Vijayalakshmi, S. P., & Raichur, A. M. (2002). Bioflocculation of high-ash Indian coals using *Paenibacillus polymyxa*. *International Journal of Mineral Processing, 67*, 199–210. https://doi.org/10.1016/S0301-7516(02)00044-3

Vijayalakshmi, S. P., & Raichur, A. M. (2003). The utility of *Bacillus subtilis* as a bioflocculant for fine coal. *Colloids Surfaces B Biointerfaces, 29*, 265–275. https://doi.org/10.1016/S0927-7765(03)00005-5

Vilinska, A., & Rao, K. H. (2008). *Leptosririllum ferrooxidans*-sulfide mineral interactions with reference to bioflotation nad bioflocculation. *Transactions of the Nonferrous Metals Society of China, 18*, 1403–1409. https://doi.org/10.1016/S1003-6326(09)60016-9

Vilinska, A., & Rao, K. H. (2011). Surface thermodynamics and extended DLVO theory of *Leptospirillum ferrooxidans* cells' adhesion on sulfide minerals. *Minerals & Metallurgical Processing, 28*, 151–158. https://doi.org/10.1007/bf03402248

Virk-Baker, M. K., Nagy, T. R., Barnes, S., & Groopman, J. (2014). Dietary acrylamide and human cancer: A systematic review of literature. *Nutrition and Cancer, 66*, 774–790. https://doi.org/10.1080/01635581.2014.916323

Waldeck, J., Daum, G., Bisping, B., & Meinhardt, F. (2006). Isolation and Molecular Characterization of Chitinase-Deficient *Bacillus licheniformis* Strains Capable of Deproteinization of Shrimp Shell Waste To Obtain Highly Viscous Chitin. *Applied and Environment Microbiology, 72*, 7879–7885. https://doi.org/10.1128/AEM.00938-06

Wilén, B.-M., Jin, B., & Lant, P. (2003). The influence of key chemical constituents in activated sludge on surface and flocculating properties. *Water Research, 37*, 2127–2139. https://doi.org/10.1016/S0043-1354(02)00629-2

Wilén, B.-M., Lumley, D., Mattsson, A., & Mino, T. (2008). Relationship between floc composition and flocculation and settling properties studied at a full scale activated sludge plant. *Water Research, 42*, 4404–4418. https://doi.org/10.1016/j.watres.2008.07.033

Wingender, J., Neu, T. R., & Flemming, H.-C. (1999). Microbial extracellular polymeric substances. In *Microbial extracellular polymeric substances* (pp. 49–72). Springer Berlin Heidelberg.

Ye, F., Ye, Y., & Li, Y. (2011). Effect of C/N ratio on extracellular polymeric substances (EPS) and physicochemical properties of activated sludge flocs. *Journal of Hazardous Materials, 188*, 37–43. https://doi.org/10.1016/j.jhazmat.2011.01.043

Zhuang, X., Wang, Y., Li, Q., et al. (2012). The production of bioflocculants by *bacillus licheniformis* using molasses and its application in the sugarcane industry. *Biotechnology and Bioprocess Engineering, 17*, 1041–1047. https://doi.org/10.1007/s12257-012-0213-0

Phosphate Minerals and Applications of Phosphate Solubilizing Microorganisms for Extraction of Critical Minerals and Rare Earth Elements

Homayoun Fathollahzadeh

Abstract

The key energy transition industries are reliant on critical minerals including but not limited to rare earth elements (REEs), copper (Cu), and lithium (Li). Furthermore, the ongoing COVID-19 pandemic, geopolitical conflicts, and climate change have exposed major bottlenecks in the security of global critical minerals and food supply chains. On the one hand, the major host minerals for REEs are primary and secondary phosphate minerals such as monazite and xenotime. On the other hand, phosphate and potash used in fertilizers are essential to improve the resilience of agricultural soils in a changing climate. Phosphate solubilizing microorganisms (PSMs) are often reported to be a major contributing factor for plant and soil nutrition by making insoluble phosphate into more soluble forms. Recent progress in understanding the molecular mechanisms of REEs solubilization from REEs-containing phosphate mineral, monazite, has suggested that in addition to protonation and complexation mechanisms either through organic acids production in heterotrophs or biogenic sulfuric acid in acidophiles, bacterial attachment has a pivotal role in bioleaching systems. This chapter highlights the recent application of PSMs for the bioleaching of critical minerals and emphasizes the significance of microbial function to narrow down potential microorganisms and associated dissolution pathways of these metals.

H. Fathollahzadeh (✉)
Curtin Medical School, Curtin University, Bentley, WA 6102, Australia
e-mail: homayoun.fathollahzadeh@curtin.edu.au

H. Fathollahzadeh
The Institute for Geoscience Research, School of Earth and Planetary Sciences, Curtin University, Bentley, WA 6102, Australia

Keywords

Phosphate solubilizing microorganisms · Phosphate minerals · Critical minerals · Rare earth elements · Bioleaching

1 Introduction

Mining and minerals are critical for development of green technologies to drive successful delivery of sustainable development goals (SDGs). To address risks associated with processing and supply of critical minerals as the building blocks for renewable energy and electric vehicles, technological innovation, metallurgical, and extractive advances are required. Demand for critical minerals needed for this transition is expected to increase. Critical minerals, including rare earth elements (REE), are the foundation of low-carbon circular economy (Lima et al., 2022).

The most important REEs-bearing mineral for REEs downstream processing is monazite. Currently, conventional extraction of critical minerals requires significant processing where downstream process relies on either an alkaline or an acidic lixiviants (e.g., concentrated sodium hydroxide and sulfuric acid, respectively) and high pressure and/or temperatures. These processes normally generate large amounts of toxic waste and neglect low-grade ore in mineral processing flowsheet as their processing result in higher costs and lower economical profits. Furthermore, improving the resilience of critical materials supply chains in a COVID-19 pandemic context and ongoing high-intensity military conflicts between producers and consumers has become even more crucial.

While more critical minerals projects around the world are in the pipeline, to close the supply chain gap, developing a new integrated approach toward sustainable mining during mine life cycle has become more urgent. Recent progress in the application of biomining tools reflects enhanced biologically catalyzed solubilization of metals from primary

S. Panda et al. (eds.), *Biotechnological Innovations in the Mineral-Metal Industry*,
Advances in Science, Technology & Innovation, https://doi.org/10.1007/978-3-031-43625-3_7

and secondary resources (Watling, 2016). The metabolic flexibility of biomining microbes for metals dissolution is abundant, allowing necessary survival mechanisms in polly-metallic and extreme environment (Watling et al., 2010). Furthermore, members of the genus *Acidihalobacter* have revealed multiple strategies that can be used to tolerate high levels of metals and oxidative stress found in bioleaching environments (Khaleque et al., 2020). A variety of mechanisms (Fig. 1) such as physical penetration (first attachment and then bio-corrosion) and the production of acidic and/or metal complexing metabolites (i.e., organic acids, inorganic acids, and siderophores) have been demonstrated to provide a means of metal biorecovery for solubilization of metals from solid matrices (Liang & Gadd, 2017; Fathollahzadeh et al., 2019).

Furthermore, by 2050, the world population is expected to grow by two billion people and so will increasing demand for phosphorus (P), an essential nutrient for all living organisms (Peterson et al., 2022). A recent system dynamics (SD) modeling has identified that global supply of phosphate will fall below global requirement in 2040 (Nedelciu et al., 2020). It has also been demonstrated that despite the fact some improvements have been achieved, the transition toward a circular economy of P is far from complete (Peterson et al., 2022). To match regional needs for phosphorus, the phosphate rock (PR) production is required to double by 2050 (Nedelciu et al., 2020). Also, findings from global P supply chain dynamics emphasize the importance of implementing enhanced P management, particularly for Latin America, South Asia, the Caribbean, and Sub-Saharan Africa regions (Nedelciu et al., 2020). Therefore, alternative sustainable approaches such as biotech solutions offering sustainable benefits have received more attention.

Phosphate solubilizing microorganisms (PSMs) play a critical role in the biogeochemical cycling of P in both terrestrial (e.g., agricultural soil) and aquatic (e.g., lake sediments) ecosystems by mineralizing organic P, solubilizing inorganic P minerals, and mediating biomass P retention (Qian et al., 2010; Tian et al., 2021). To date, studies of PSMs have largely been performed for enhanced fertilizer efficiency in agricultural systems as well as their role in transformation of P during eutrophication retention (Qian et al., 2010; Tian et al., 2021). However, less is known

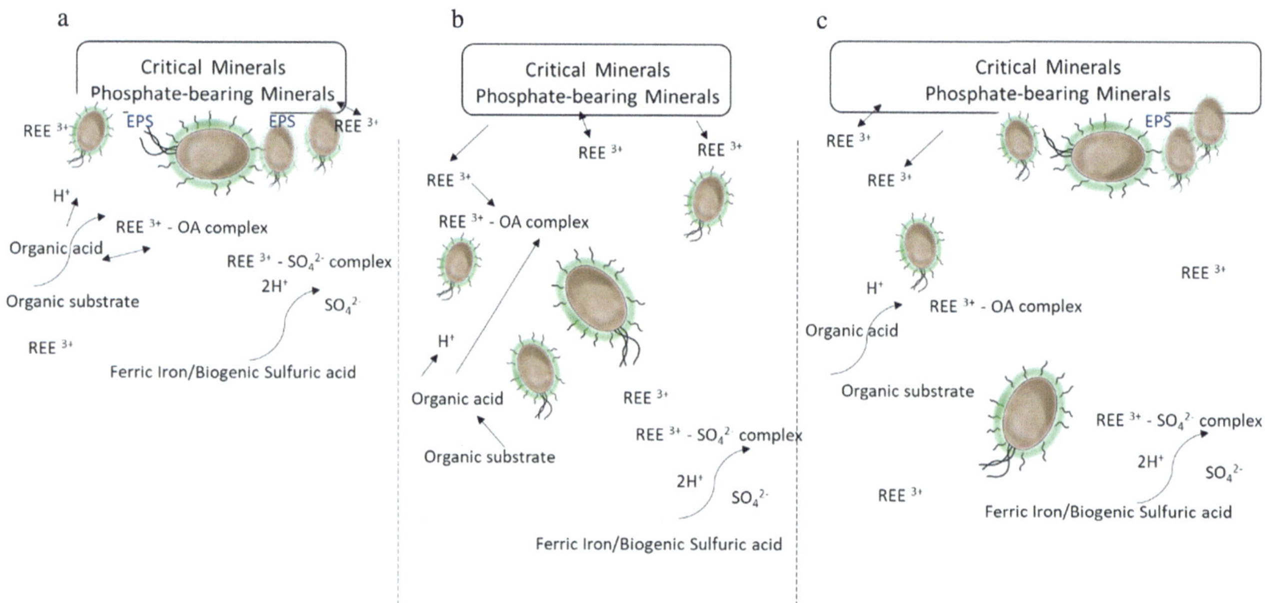

Fig. 1 Conceptual model showing of the proposed mechanisms of critical minerals phosphate-bearing minerals such as monazite. **a** During contact leaching, following attachment of microbes with the help of microbial tools such extracellular polymeric substances (EPS) and proton dissociation, phosphate (PO_4^{3-}), and REEs cations (REE^{3+}) is released into pregnant leachate (PL). In the presence of organic substrate, pH of PL decreases due to organic acids (OA) produced by the cells, and they may form complex with REE^{3+}. It is suggested that precipitation of PO_4^{3-} in cells enhances REE^{3+} solubility. In the presence of ferric iron and/or acidophilic bacteria capable of biogenic sulfuric acid generation, extra protons maintain acidic pH, followed by protons dissociation, free protons attacks to mineral surface and dissolution of REE^{3+}. **b** During non-contact leaching of phosphate-bearing minerals, floating cells indirectly with only help of proton dissociation contributing to REE^{3+} and PO_4^{3-} solubilization. In the presence of ferric iron and/or acidophilic bacteria capable of biogenic sulfuric acid generation, extra protons maintain acidic pH, follows by protons dissociation, free protons attacks to mineral surface and dissolution of REE^{3+}. **c** During a cooperative leaching, as result of teamwork of attached cells and suspended cells, the concentration and solubility of PO_4^{3-} and REE^{3+} in PL changes over time. In the presence of ferric iron and/or acidophilic bacteria capable of biogenic sulfuric acid, extra protons maintain acidic pH, follows by protons dissociation, free protons attacks to mineral surface and dissolution of REE^{3+}

about the impact of introducing PSMs in the bioleaching systems, including their effects on the dissolution of critical minerals (Fathollahzadeh et al., 2019).

This chapter will highlight an overview of the status of microbial bioleaching on phosphate-based critical minerals and will discuss the new insights and future developments for the integration of bio-based solutions into classic metallurgical system.

2 What Have We Learned from Soil Science and Agriculture Engineering?

Traditionally, studying genetics of phosphate solubilization and its potential applications to improve plant growth promoting rhizo-bacteria (PGPR) played an indispensable role in the development of novel agricultural solutions globally (Rodriguez & Fraga, 1999; Rodríguez et al., 2006). In this sense, the term "PSMs" have been used to simplify a type of microorganisms capable of mediating refractory phosphate into more soluble forms which is available to plants and biomass (i.e., the production of organic acids associated with soluble P originated from orthophosphates and the production acid phosphatase and/or phytase associated with organic phosphate) (Goldstein, 2007; Goldstein & Krishnaraj, 2007). The above dissolution mechanisms have been suggested to reflect the metabolic needs and diversity of microbial communities followed by fulfilling phosphate requirement and its availability to be obtained from the mineral surface and the leachate as well as the solubility, mobility, and/or reactivity of divalent and trivalent cations in the systems (e.g., Ca^{2+}, REE^{3+}) (Fathollahzadeh et al., 2019).

When investigating the highly interesting P solubilization traits of phosphate solubilizing bacteria (PSB) in the rhizosphere of rapeseed, faba bean, and winter pea, it has been demonstrated that plant-specific P requirement shaped with associated microbiome where PSB isolated from rapeseed soil resulted in higher isolates and solubilization efficiency relative to winter pea or faba bean (Amy et al., 2022a). The authors suggested that further studies are required to better understand the induced shifts in microbial community and processes linked to the rapeseed growth enhancement and/or P status or even a combined effect of PGPR traits and PSM abilities (Amy et al., 2022b).

Despite the known benefits of biofertilizer PSMs as effective and important components in the establishment of sustainable soil management systems (Alori et al., 2017; Richardson, 2001), very little translation of this knowledge into bioleaching of phosphate-based critical minerals has been performed. The following sections will highlight the latest findings on bioleaching of critical elements from phosphate-based critical minerals.

3 From Mineral Weathering to Microbial Leaching

The process of mineral weathering is the combination of abiotic and biotic processes by which nutrients are liberated from minerals. Over the last ten years, understanding bio-chemical interactions between minerals and microbes (bacteria or fungi) has shown that microbes effectively contribute to mineral weathering and plant growth (Uroz et al., 2022). The studying and isolation of effective microbes within the interfaces of minerals suggest that microbial communities and molecular mechanisms related to acidification, chelation, and physiological adaptation contribute to the mobilization of base cations and phosphorus (Uroz et al., 2022). From phosphate minerals, apatite is the most studied as it is the primary source of phosphorus in many natural environments. Apatite, sediment rocks, and lithiophilite-enriched layers normally cumulate and occur with critical minerals such REEs, Cu, Li, Ti, and V (Emsbo et al., 2015; Mao et al., 2016; Samarakoon et al., 2021; Von Gruenewaldt, 1993; Williams & Cesbron, 1977).

The study of the effect of *Bacillus megaterium* on the dissolution behavior of the mineral apatite has shown that the bacteria can accelerate rates without being in physical contact with the dissolving mineral (Hutchens et al., 2006). When investigating microbial release of apatite- and goethite-bound phosphate in acidic forest soils, it has been found that phosphate release from hydroxyapatite (HAP) was higher than from goethite (Pastore et al., 2020). The authors suggested that phosphate release from HAP was associated with acidification by organic acid released by microbes. Therefore, since microbial metabolites control phosphate mineral dissolutions, the idea of using PSMs becomes a step toward using microbial bioleaching systems for extraction and recovery of critical minerals. The secreted organic acids produced a range of PSMs which are summarized somewhere else (Fathollahzadeh et al., 2019).

Till 2023, only a few studies focused on the bioleaching of natural phosphate-based critical minerals, such as monazite concentrate and monazite sand (Brisson et al., 2016, 2020; Fathollahzadeh et al., 2019; Fujita et al., 2022). Nevertheless, recent studies have demonstrated that monazite dissolution and REEs leaching increased in the presence of bacterial species, and microscopic imaging (scanning electron microscopy-SEM, atomic force microscopy-AFM, and confocal Raman microscopy) showed microbial attachment to the mineral surface confirming that cell contact improved the bioleaching (Fathollahzadeh et al., 2018a, 2018b; Van Alin et al., 2023). Various microbial mechanisms and list of genes have been demonstrated or proposed to play a direct or indirect and/or cooperative role in monazite bioleaching by bacteria (Fathollahzadeh et al., 2019). Based on these findings, a conceptual model for bioleaching mechanisms (Fig. 1) of monazite was proposed (Fathollahzadeh et al., 2019).

Bioleaching experiments with *Aspergillus niger* were conducted on different mineral phosphates at 1% pulp density, and the phosphorous leaching yield was aluminum phosphate > turquoise (Cu–Al phosphate) > monazite (Castro et al., 2020). For monazite bioleaching, higher concentration of REEs was observed when the fungus was cultivated on minimal medium with the mineral as the sole phosphorous source (Castro et al., 2020). It has been demonstrated that the concentration of dissolved P in PL cannot be reliable indication of solubilization as absorbing of inorganic P by microbes from the PL for use in their metabolic processes reduces the soluble forms of P concentration in PL (Brisson et al., 2016; Corbett et al., 2017; Reynier et al., 2021). With excess extracellular P in the leachate which could incorporate REEs during precipitation, it has been suggested that level of P in the leachate and diffusion of P into biomass may influence REEs dissolution rate during bioleaching (Fathollahzadeh et al., 2019). Given that, low residual dissolved phosphate in bioleaching of REEs from end-of-life NiMH batteries has improved REEs leaching 11.0-fold (Rasoulnia et al., 2022). These results emphasize the importance of a better understanding of the phosphate regulation and associated molecular mechanisms that affect bioleaching. Future research needs to address microbial P solubilization from different range of critical minerals using heterotrophic and autotrophic cultures. Extending beyond primary critical minerals, bioleaching of REEs from retorted phosphor powder (RPP), containing 85–95% apatite mineral and 5–15% ceramic mixture, demonstrated that microbial exudate (i.e., gluconic acid) is more effective than abiotic organic acid solutions (Reed et al., 2016). In addition, economic and operational viability of waste products/tailing as the secondary source for extraction of REEs, Cu, Li or as feedstock for leaching of critical minerals warrant further investigation.

4 New Avenues for Mine Closure and Nutrient Management: PSMs Capability

For the successful establishment of vegetation and compliance with the relevant completion criteria for mine closure, the establishment of a robust growth media (i.e., topsoil) is critical for sustainable rehabilitation of mining sites. Throughout the mineral resource development cycle (i.e., mine closure), particularly prior to use growth media in rehabilitation, topsoil management along restoration model plans is required. Given that heavily degraded ecosystems, such as closed mine sites have extremely low levels of nutrients, including P, characterization of soil microbial, chemical, and physical properties is indispensable for successful plant growth and environmental rehabilitation. Major changes in soil health determined by measuring properties change from natural ecosystem to mining substrate (iron waste piles in the Carajás Mineral Province, Brazil) suggested that iron mining operations resulted in significant alteration of the chemical soil properties (Ramos et al., 2022). Mining materials are reported to contain a lower soil organic matter (SOM) and nutrients than undisturbed areas of *Cangas* and forests (Ramos et al., 2022). Recovery of soil P and nutrient on former mines through PSMs offer new opportunities for the enhanced soil P cycling following ecological restoration (Liang et al., 2020). In the context of mine restoration and establishing native vegetation that blends well with the surrounding areas, the present chapter outlines bio-based solutions to fortify soil health and quality at the benefit of supporting plant growth and associated microbiome. With advanced bio-prospecting and next-generation sequencing technologies to identify the genetic traits of beneficial microbial partners in "mine life cycle," apart from what bio-solutions offer for extraction and recovery of critical minerals, further research also should aim to unearth beneficial role of microbial boxes for each potential stage of mine life cycle.

5 Conclusion

In a metal-demanding world for green transition, bioleaching of critical minerals such REEs from primary REEs-bearing minerals is considered an environmentally friendly solution compared to conventional techniques. So far, in the last couple of years, a series of works have studied the potential of PSMs for the bioleaching of phosphates and REEs from phosphate-bearing minerals such as monazite; however, these studies are not enough to highlight mechanisms responsible for the mobilization of REEs. More bioleaching studies on types of phosphate minerals containing critical minerals, type of microbes, and their microbial-mineral interactions are needed to evaluate the potential of bioleaching systems for a more sustainable supply chain of these metals.

References

Alori, E. T., Glick, B. R., & Babalola, O. O. (2017). Microbial phosphorus solubilization and its potential for use in sustainable agriculture. *Frontiers in Microbiology, 8*, 971.

Amy, C., Avice, J. C., Laval, K., & Bressan, M. (2022a). Are native phosphate solubilizing bacteria a relevant alternative to mineral fertilizations for crops? Part I. When rhizobacteria meet plant P requirements. *Rhizosphere*, 100476.

Amy, C., Avice, J. C., Laval, K., & Bressan, M. (2022b). Are native phosphate-solubilizing bacteria a relevant alternative to mineral fertilizations for crops? Part II: PSB inoculation enables a halving of P input and improves the microbial community in the rapeseed rhizosphere. *Rhizosphere, 21*, 100480.

Brisson, V. L., Zhuang, W. Q., & Alvarez-Cohen, L. (2016). Bioleaching of rare earth elements from monazite sand. *Biotechnology and Bioengineering, 113*(2), 339–348.

Brisson, V. L., Zhuang, W. Q., & Alvarez-Cohen, L. (2020). Metabolomic analysis reveals contributions of citric and citramalic acids to rare earth bioleaching by a *Paecilomyces* fungus. *Frontiers in Microbiology, 10*, 3008.

Castro, L., Blázquez, M. L., González, F., & Muñoz, J. A. (2020). Bioleaching of phosphate minerals using *Aspergillus niger*: Recovery of copper and rare earth elements. *Metals, 10*(7), 978.

Corbett, M. K., Eksteen, J. J., Niu, X. Z., Croue, J. P., & Watkin, E. L. (2017). Interactions of phosphate solubilising microorganisms with natural rare-earth phosphate minerals: A study utilizing Western Australian monazite. *Bioprocess and Biosystems Engineering, 40*(6), 929–942.

Emsbo, P., McLaughlin, P. I., Breit, G. N., du Bray, E. A., & Koenig, A. E. (2015). Rare earth elements in sedimentary phosphate deposits: Solution to the global REE crisis? *Gondwana Research, 27*(2), 776–785.

Fathollahzadeh, H., Becker, T., Eksteen, J. J., Kaksonen, A. H., & Watkin, E. L. (2018a). Microbial contact enhances bioleaching of rare earth elements. *Bioresource Technology Reports, 3*, 102–108.

Fathollahzadeh, H., Hackett, M. J., Khaleque, H. N., Eksteen, J. J., Kaksonen, A. H., & Watkin, E. L. (2018b). Better together: Potential of co-culture microorganisms to enhance bioleaching of rare earth elements from monazite. *Bioresource Technology Reports, 3*, 109–118.

Fathollahzadeh, H., Eksteen, J. J., Kaksonen, A. H., & Watkin, E. L. (2019). Role of microorganisms in bioleaching of rare earth elements from primary and secondary resources. *Applied Microbiology and Biotechnology, 103*(3), 1043–1057.

Fujita, Y., Park, D., Lencka, M., Anderko, A., Reed, D., Thompson, V., Das, G., Eslamimanesh, A., & Jiao, Y. (2022). Beneficiation of REE: Prospects for biotechnology deployment. Earth and Space Science Open Archive.

Goldstein, A., & Krishnaraj, P. (2007). Phosphate solubilizing microorganisms vs. phosphate mobilizing microorganisms: What separates a phenotype from a trait? In *First International Meeting on Microbial Phosphate Solubilization* (pp. 203–213). Springer.

Goldstein, A. H. (2007). Future trends in research on microbial phosphate solubilization: One hundred years of insolubility. *First International Meeting on Microbial Phosphate Solubilization* (pp. 91–96). Springer.

Hutchens, E., Valsami-Jones, E., Harouiya, N., Chaïrat, C., Oelkers, E. H., & McEldoney, S. (2006). An experimental investigation of the effect of *Bacillus megaterium* on apatite dissolution. *Geomicrobiology Journal, 23*(3–4), 177–182.

Khaleque, H. N., Fathollazadeh, H., González, C., Shafique, R., Kaksonen, A. H., Holmes, D. S., & Watkin, E. L. (2020). Unlocking survival mechanisms for metal and oxidative stress in the extremely acidophilic, halotolerant *Acidihalobacter* genus. *Genes, 11*(12), 1392.

Li, Q., Liu, F., Li, M., Chen, C., & Gadd, G. M. (2021). Nanoparticle and nanomineral production by fungi. *Fungal Biology Reviews*.

Liang, X., & Gadd, G. M. (2017). Metal and metalloid biorecovery using fungi. *Microbial biotechnology, 10*(5), 1199–1205.

Liang, J. L., Liu, J., Jia, P., Yang, T. T., Zeng, Q. W., Zhang, S. C., Liao, B., Shu, W. C., & Li, J. T. (2020). Novel phosphate-solubilizing bacteria enhance soil phosphorus cycling following ecological restoration of land degraded by mining. *The ISME Journal, 14*(6), 1600–1613.

Lima, A. T., Kirkelund, G. M., Ntuli, F., & Ottosen, L. M. (2022). Screening dilute sources of rare earth elements for their circular recovery. *Journal of Geochemical Exploration, 238*, 107000.

Mao, M., Rukhlov, A. S., Rowins, S. M., Spence, J., & Coogan, L. A. (2016). Apatite trace element compositions: A robust new tool for mineral exploration. *Economic Geology, 111*(5), 1187–1222.

Nedelciu, C. E., Ragnarsdottir, K. V., Schlyter, P., & Stjernquist, I. (2020). Global phosphorus supply chain dynamics: Assessing regional impact to 2050. *Global Food Security, 26*, 100426.

Pastore, G., Kaiser, K., Kernchen, S., & Spohn, M. (2020). Microbial release of apatite-and goethite-bound phosphate in acidic forest soils. *Geoderma, 370*, 114360.

Peterson, H. M., Baker, L. A., Aggarwal, R. M., Boyer, T. H., & Chan, N. I. (2022). A transition management framework to stimulate a circular phosphorus system. *Environment, Development and Sustainability, 24*(2), 1713–1737.

Qian, Y., Shi, J., Chen, Y., Lou, L., Cui, X., Cao, R., Li, P., & Tang, J. (2010). Characterization of phosphate solubilizing bacteria in sediments from a shallow eutrophic lake and a wetland: Isolation, molecular identification and phosphorus release ability determination. *Molecules, 15*(11), 8518–8533.

Ramos, S. J., Gastauer, M., Martins, G. C., Guedes, R. S., Caldeira, C. F., Souza-Filho, P. W., & Siqueira, J. O. (2022). Changes in soil properties during iron mining and in rehabilitating minelands in the Eastern Amazon. *Environmental Monitoring and Assessment, 194*(4), 1–17.

Rasoulnia, P., Barthen, R., Lakaniemi, A. M., Ali-Löytty, H., & Puhakka, J. A. (2022). Low residual dissolved phosphate in spent medium bioleaching enables rapid and enhanced solubilization of rare earth elements from end-of-life NiMH batteries. *Minerals Engineering, 176*, 107361.

Reed, D. W., Fujita, Y., Daubaras, D. L., Jiao, Y., & Thompson, V. S. (2016). Bioleaching of rare earth elements from waste phosphors and cracking catalysts. *Hydrometallurgy, 166*, 34–40.

Reynier, N., Gagné-Turcotte, R., Coudert, L., Costis, S., Cameron, R., & Blais, J. F. (2021). Bioleaching of uranium tailings as secondary sources for rare earth elements production. *Minerals, 11*(3), 302.

Richardson, A. E. (2001). Prospects for using soil microorganisms to improve the acquisition of phosphorus by plants. *Functional Plant Biology, 28*(9), 897–906.

Rodríguez, H., & Fraga, R. (1999). Phosphate solubilizing bacteria and their role in plant growth promotion. *Biotechnology Advances, 17*(4–5), 319–339.

Rodríguez, H., Fraga, R., Gonzalez, T., & Bashan, Y. (2006). Genetics of phosphate solubilization and its potential applications for improving plant growth-promoting bacteria. *Plant and Soil, 287*(1–2), 15–21.

Samarakoon, N., Malaviarachchi, S. P., Senaratne, A., & Wijayasinghe, A. (2021). Gold occurrences in copper-magnetite-apatite deposit at Seruwila, Sri Lanka. *Ore and Energy Resource Geology, 8*, 100014.

Tian, J., Ge, F., Zhang, D., Deng, S., & Liu, X. (2021). Roles of phosphate solubilizing microorganisms from managing soil phosphorus deficiency to mediating biogeochemical P cycle. *Biology, 10*(2), 158.

Uroz, S., Picard, L., & Turpault, M. P. (2022). Recent progress in understanding the ecology and molecular genetics of soil mineral weathering bacteria. *Trends in Microbiology*.

Van Alin, A., Corbett, M. K., Fathollahzadeh, H., Tjiam, M. C., Putnis, A., Eksteen, J., Kaksonen, A. H., & Watkin, E. (2023). *Klebsiella aerogenes* adhesion behaviour during biofilm formation on monazite. *Microorganisms, 2023*(11), 1331.

Von Gruenewaldt, G. (1993). Ilmenite-apatite enrichments in the Upper Zone of the Bushveld Complex: A major titanium-rock phosphate resource. *International Geology Review, 35*(11), 987–1000.

Watling, H. (2016). Microbiological advances in biohydrometallurgy. *Minerals, 6*(2), 49.

Watling, H., Watkin, E., & Ralph, D. (2010). The resilience and versatility of acidophiles that contribute to the bio-assisted extraction of metals from mineral sulphides. *Environmental Technology, 31*(8–9), 915–933.

Williams, S. A., & Cesbron, F. P. (1977). Rutile and apatite: Useful prospecting guides for porphyry copper deposits. *Mineralogical Magazine, 41*(318), 288–292.

Role of Biosurfactants in Heavy Metal Removal and Mineral Flotation

Manisha Mahapatra, Snehalata Pradhan, Subham Preetam and Arun Kumar Pradhan

Abstract

Surfactants are chemical compounds produced from petroleum feedstock, agro-based waste materials and microbial fermentation having wide variety of use in industries, pharmaceutical, agriculture, cosmetics, etc. These are amphiphilic moieties and chemically synthesised. These chemical compounds are toxic and are responsible for various harmful environmental problems. Recently, biosurfactants have gained lots of interest worldwide, because they are green-alternatives for surfactants. Biosurfactants are produced naturally from microorganisms like yeast, fungi and bacteria. These have both hydrophobic and hydrophilic groups which makes its unique and important in different industries. These organisms produce surface active metabolites or secondary metabolites and grow on water immiscible or oily surface. The surface active molecules help them to absorb, emulsify, wetting, solubilise and disperse the water immiscible substances. Biosurfactants are in demand and commercially promising due to their properties, i.e., low toxicity, higher biodegradability, environmental compatibility, foaming properties, shows stable activity at extreme pH, temperature and salinity, etc. Biosurfactants play very crucial role in mineral flotation. Heavy metal removal and mineral flotation is a very crucial process for industries (which commercially separates metals from ores by collecting them on the surface/froth layer—so the metals can be used commercially) and also for the environment. Biosurfactant mediated mineral floatation and heavy metal removal involves the metal ion sorption to sorbent material followed by floatation and floatation product collection. Using biosurfactants in replacement of surfactants for heavy metal removal and mineral floatation are actually effective, low cost, recyclable, reusable and environmental friendly. This chapter emphasises on removal of some metals from their respective ores using different biosurfactants. A probable mechanism of flotation by biosurfactant is also discussed.

Keywords

Surfactants · Biosurfactants · Mineral floatation · Environment · Surface active metabolites · Heavy metals

1 Introduction

The constant requirement of adequate minerals for mankind from exhaustible resources needs the use of advanced technology. The advance technology benefits in increasing mineral products and reducing waste material during the flotation of minerals ores. Froth flotation is comprehensively used worldwide by many industries for commercial preparation of minerals due to it affordability, cheap or low cost (Tolley et al., 1996).

Natural resources have been always explored and have become a source of experimentation in the science field by utilising sophisticated technologies in this global industrialisation age, resulting in development of items with high aggregate value in the global market (e.g., biosurfactants). Amphiphilic microbial compounds having hydrophobic and hydrophilic moieties that separates at liquid/liquid, liquid/solid and liquid/gas interfaces are known as biosurfactants (Santos et al., 2016).

M. Mahapatra · S. Preetam · A. K. Pradhan (✉)
Centre for Biotechnology, Siksha 'O' Anushandhan (Deemed to Be University), Bhubaneswar 751003, India
e-mail: arunpradhan@soa.ac.in

S. Pradhan
Kabi Samrat Upendra Bhanja College of Teacher Education, Ganjam, Odisha 761126, India

S. Panda et al. (eds.), *Biotechnological Innovations in the Mineral-Metal Industry*,
Advances in Science, Technology & Innovation, https://doi.org/10.1007/978-3-031-43625-3_8

The properties like emulsification, foam generation, detergency and dispersion enable these particular biomolecules to play an important role and are very much desired attributes in many industrial sectors. Biosurfactant production/manufacturing is marked as one of the most important technologies for advancement of twenty-first century.

Apart from having a significant beneficial influence on the major global issues, biosurfactant manufacturing is critical to the implementation of sustainable industrial processes that involves use of renewable resources and "green" goods. Low toxicity and biodegradability of these biomolecules have led to an increase in scientific research involving wide variety of commercial uses for biosurfactants in the fields of bioremediation and others.

The froth flotation or mineral flotation used by many industries for extracting minerals/metals resulting in various toxic substances during the production process. Most of the industries contain toxic metals in their wastewater streams or their waste water treatment containers which leads to elevate the level of water pollution. These polluted waters consist of heavy metals and toxic elements as well as different mixture of hazardous substances of chemicals which is a big concern or threat for aquatic life as well as human life. This not only affects the aquatic life but it is also a threat for animal as well as humans which reaches us through food chain. Therefore, the ongoing research deals with different novel approaches or treatment technology for elimination of these toxic heavy metals from wastewater (Zamboulis et al., 2004). Similarly, for biomineral and metals processes like hydrometallurgical process and bioleaching are widely used by industries. Biomineral processing is used to remove/recover minerals from their ores by using microorganisms like fungi and bacteria. *Rhodococcus ruber* 9C strain degrades around 80% dibenzothiophene (DBT) while removing heavy sulphur content in Indian lignite, CPC and coal by 15.87%, 14.83% and 33.44%, respectively (Mishra et al., 2017). Another study shows, *Acidophilic ferroxidans*, an acidophilic chemolithotrophic bacteria used for copper recovery by using bioleaching and hydrometallurgical process (Panda et al., 2012, 2013, 2014). Similarly nickel cobalt recovery is also done by bio-reduction of chromite overburden by using DIRB (dissimilarity iron reducing bacterial Consortium) (Esther et al., 2013). These all methods show used of microorganisms as the major part. Similarly for minerals, sorptive floatation is used by industries, in which surfactants play a major role.

Sorptive flotation is widely used treatment technology for mineral which includes a sorbent and a surfactant. Sorptive floatation is a two staged process, and this method first involves adsorption or co-precipitation or occlusion of toxic metal ions (carried out in situ) followed by flotation by adding a suitable surfactant. Because both the process absorption and floatation can be carried out in same treatment unit this is referred as sorptive floatation (Matis et al., 2003).

Most of the toxic substances present are water soluble due to their hydrophilic nature. In order to separate the minerals or flotation, hydrophobicity and hydrophilicity is needed.

Flotation process involves various frothing agents which helps in air dispersion and increases strength of the bubble in flotation unit. Surfactants are important agents which are surface active molecules. These are basically amphiphilic in nature, i.e., hydrophilic as well as hydrophobic moieties are present. There are several types of surfactants available (Fig. 2). This chapter basically deals with the use of biosurfactant for mineral flotation. The use of rhamnolipids in coal and mineral ion floatation, use of surfactin in ion floatation and some green/biosurfactants is briefly discussed in this chapter. It also explains about the future aspects of biosurfactant or green surfactants in mineral ion floatation as well its use for environment and the potential value it holds for the future generation (Zouboulis et al., 2003).

2 What Are Surfactants and Biosurfactants?

Surfactants are amphiphilic, surface active and surface tension reducing material. Due to amphiphilicity, these have the ability to replace bulky molecules of higher energy resulting in reduction of free energy in the system. The hydrophilic part has very less affinity towards bulky medium whereas hydrophobic part of surfactant has higher affinity to bulky medium.

They readily disperse as immersion in any liquid or water by maintaining a reducing interfacial and surface tension between gas, liquid and solid.

A hydrocarbon chain is frequently used as a polar moiety, whereas considering the polar moiety can be ionic, i.e., cationic or anionic, non-ionic or amphoteric (Mao et al., 2015; Silva et al., 2014).

Surfactants make hydrophilic molecules more soluble, lowering the interfacial tension as well as surface tension between the oil/water contact (Banat et al., 2010a; Campos et al., 2013). The majority of surfactants on the market today are produced chemically. Synthetic tensioactive substances, on the other hand, are often poisonous and tough to degrade by the action of microbes. Over the past few years, such issues have forewarned the scientific community to accept the ecological/environment friendly surfactants (Vijayakumar & Saravanan, 2015).

This popularity of surfactants is the reason why surfactants are widely famous in different industry due to their adhesive nature, works as foaming agent, flocculation property, emulsifying property, etc., leading to their high demand for use in different purpose (Khoshdast et al., 2012; Vijayakumar & Saravanan, 2015).

These are chemical-based compounds mainly produced by petroleum feedstock, and some are generated synthetically in the laboratory. Even though these have very useful for different processes like floatation, but are toxic in nature, harmful for the environment as well as for mankind. The growing concern for environment protection has led everyone to find an alternative for these chemical surfactants that is green surfactant or biosurfactant or naturally occurring surfactant which are mainly produced by microbes, fungi or plants known as green surfactant or biosurfactants (Fig. 1). Furthermore, consumer worries about the environment, as well as new recent environmental control regulations, have led to achieve development of natural surfactants as a substitute to existing goods. Biosurfactant research began in the 1960s, and its application has grown in subsequent decades (Cerqueira et al., 2011; Silva et al., 2014). The perks of biosurfactants are low toxicity, structural diversity, greater biodegradability, the ability to function in a wide range of temperature, pH and salinity, lower CMC (critical micelle concentration) as well as greater selectivity, and most importantly its production involves industrial waste/renewable sources and industrial by-products, i.e., mainly natural resources; therefore, these biosurfactants have piqued the attention of various industries (Makkar et al., 2011; Pacwa-Plociniczak et al., 2011a; Preetam et al., 2022; Rosa et al., 2015). But there are some factors which affects the biosurfactant production in industries such as environment factors (like pH, salinity and temperature), presence of carbon substrate (diesel and crude oil are good source of carbon), its activity measurement like change in stabalisation/destablisation, change in surface and interfacial tensions and hydrophilic-lipophilic balance. Devices like tensiometer are widely used for these checking purpose. At the end, produced biosurfactant is analysed on the basis of their nature like—type, solvent, bacteria, etc. These surfactant works on an interface, i.e., liquid–liquid, solid–liquid and vapour-liquid, and this happens due to the immiscible phases present in solution or sample (Mulligan, 2005).

The hydrophilic part always attached towards the solution, i.e., liquid–liquid interface where the hydrophobic part sticks to the surface that is air–liquid interface. This process is efficient and reduces the work load to separate a molecule and bring it to surface. While reducing the surface tension of water micelle formation takes place, micelle formation has correlation with surface tension. Hence, good and efficient surfactant have low critical micelle concentration (CMC). CMC is defined as the surfactant's concentration in bulky phase of solution, upon which the micelle forms, i.e., from surfactants molecules and foaming starts. Biosurfactants are produced naturally as co-metabolites or secondary metabolites from particular type of microorganism. Therefore, these are produced extracellularly or as a part of cell membrane from bacteria, fungi or yeast and from few plants (Mulligan et al., 1993) (Table 1).

A surfactant generally has the property like solubility enhancement, surface tension reduction and critical micelle concentration and mainly used for the application like low surface tension, foaming capacity, detergency power, increasing solubility and wetting ability.

The basic criteria for industries to choose surfactants is their energy consumption or energy cost, their solubility

Fig. 1 Different types of surfactants

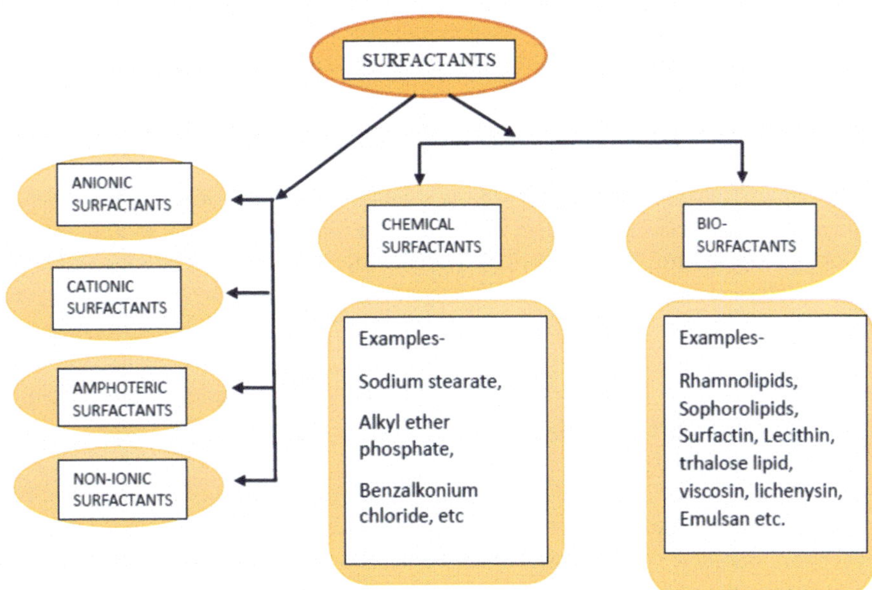

Table 1 Biosurfactants and the microorganisms they are produced from and its application

S. No.	Origin	Biosurfactant	Common applications	References
1	*Pseudomonas aeruginosa*	Rhamnolipids	Cd, Zn, Pb	Khoshdast et al. (2012), Fazaelipoor et al. (2010)
2	*Candida bombicola*	Sophorose lipids	Cd, Cu, Zn cosmetics and deodorant industries	Mulligan (2005)
3	*Bacillus subtilis*	Surfactin	Antibiotics, Cu, Zn	AytarÇelik et al. (2021)
4	*Pseudomonas fluorescens*	Viscosin	n-hexadecane mineralisation, etc.	Mulligan (2005), Mulligan et al. (1993)
5	*Arthrobacter parafineus*	Trehalose lipid	Used as antibacterial and antiadhesive agents	Mulligan et al. (1993)
6	*Anthrobacter* spp.	Glycolipids	Cu, Zn, bioleaching	Mulligan et al. (1993)
7	*Pseudomonas* spp.	Ornithine lipids	–	Desai and Banat (1997)
8	*Lactobacillus fermentum*	Diglycosyl diglycerides	Food additives/Food products	Mulligan et al. (2001a)
9	*Serratia marcescens*	Serrawettin	–	Okoliegbe and Agarry (2012), Zhang et al. (2009)
10	*Acinetobacter* spp.	Polymeric surfactant	Cosmetics, inks, drystuffs, paper coatings, agrochemicals, etc.	Nakar and Gutnick (2003), Sarma et al. (2019)

nature, charge type, adsorptive nature and physiochemical nature.

For metal removal along with surfactants some acids, bases and organic solvents are also used. Hence, the chemical surfactants are toxic as compared to biosurfactants which are fully biodegradable (Mulligan et al., 1999).

The biosurfactants are basically glycolipids, alkane, oil, sugars, lipopeptide, phospholipids, fatty acid, polymer, etc. The hydrophilic part of the surfactant attaches with amino acids, carbohydrates, cyclic peptide, phosphates, carboxylic acid or alcohol whereas the hydrophobic part attaches with long chain of fatty acid or Alpha alkyl beta hydroxy fatty acid.

Advantages of biosurfactants over chemical surfactants or synthetic surfactants are high specificity, biodegradability, environment compatibility less toxic or negligible toxic and have many environmental applications. The major industries using biosurfactants are soil washing of flushing, petroleum industries for oil removal application, mineral flotation in pharma industries, bioremediation of the contaminated landslides, etc. (Mulligan et al., 2001a) (Fig. 2).

3 Removal of Heavy Metals by Biosurfactants

Biosurfactants are widely used for heavy metal recovery, decontamination of organic substances and also for different scientific studies and processes. The organic substance decontamination or pollutants removal of hydrocarbons is carried out by the method of emulsification and pseudo-solubilisation (during cleaning process) for improving their bioavailability. Whereas for recovery of inorganic compounds, chelating agents are used during ion removal/cleaning aided with chemical interactions between metal ions and the amphipathic molecules (Banat et al., 2010b; Bodek et al., 1998; Pacwa-Plociniczak et al., 2011b).

In the restoration of heavy metal-contaminated soil, there are two primary techniques. The first method is used ex situ, which involves excavating contaminated soil, placing it in a glass column and washing it with a biotensioactive solution. The second method includes washing the soil at the original site, which involves the use of trenches and drainage tubes for collecting the biosurfactant solution (Mulligan et al., 2001b). Even a small amount of polluted soil can be treated with biosurfactants.

The biosurfactant metal complex is extracted from the soil in a massive cement mixer. After that, the soil is returned to its original site, and the biosurfactant metal complex undergoes treatment process so that the biosurfactant separates and precipitates while separating the metal behind (Sarubbo et al., 2015). In this way, the recovery is completed and metals are recovered. A probable mechanism for removal of heavy metal from contaminated soil with the help of biosurfactant is given in Fig. 3.

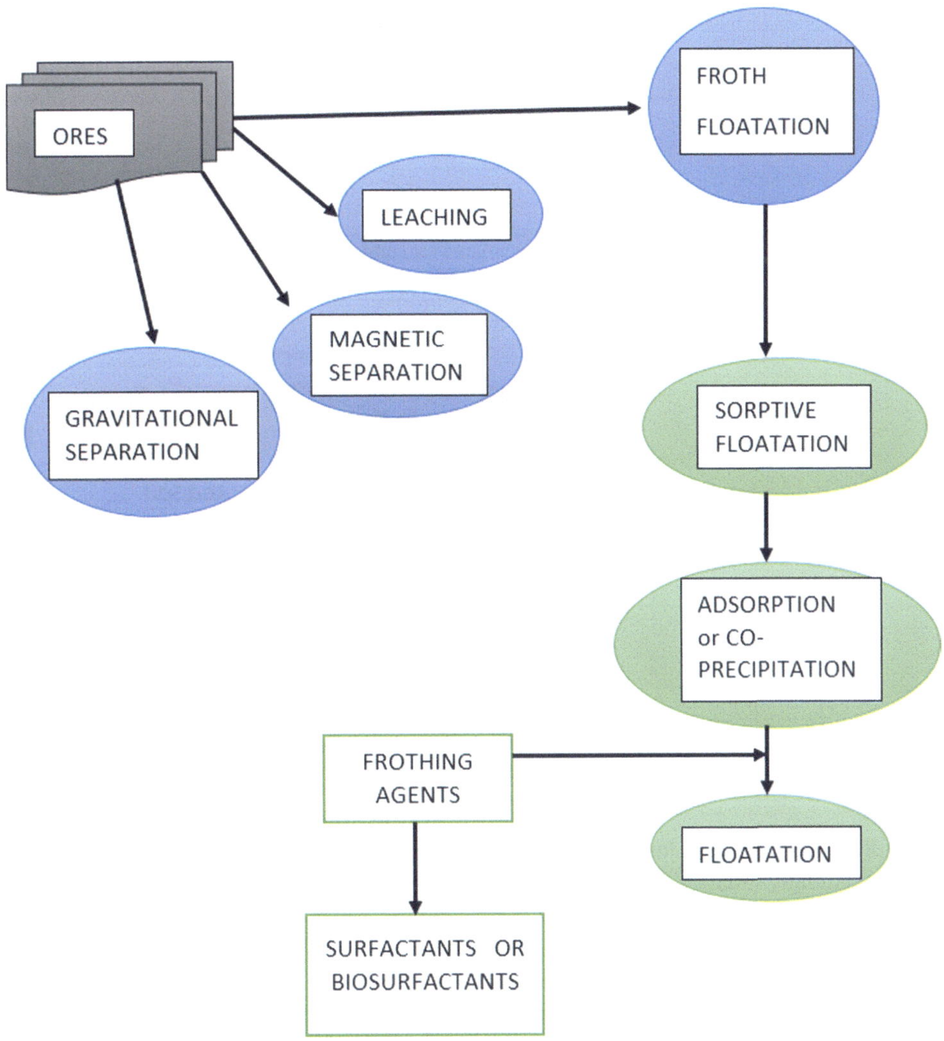

Fig. 2 Brief process of floatation and application of biosurfactant. Adapted from Refs. Matis et al. (2003), Zouboulis et al. (2003)

3.1 Heavy Metal Binding Mechanisms of Biosurfactants

The major interactions involved in biosurfactants and heavy metal binding are ionic interaction, electrostatic interactions, precipitation–dissolution and counter ion binding. Biosurfactants can directly bind to the sorbed metals and then collection of metals is done at solid–liquid interface having low interfacial tension. In the sorptive floatation process, metal–surfactant combination is seen while co-precipitating the toxic complexes. Anionic surfactants promote the interaction of metal with surfaces while the cationic surfactants reduce metal association by competing for certain but not all negatively charged surfaces. The concentrations and different kinds of biosurfactants have a different impact on heavy metal removal (Christofi & Ivshina, 2002; Frazer, 2000; Singh & Cameotra, 2004).

The most popular example is rhamnolipid which forms different micelle of ≫5 nm in diameter, vesicular structures of less than 50 nm diameter depending on pH of solution and at concentration above CMC. Increase in pH above 6.8 leads to spontaneously aggregate the surfactant molecules into complex structures called micelles. Ionic bond formation is seen when rhamnolipid, i.e., anionic biosurfactant carrying negative charge interacts with cationic metal (e.g., Cd(II) or Zn(II)) carrying positive charge (Açıkel, 2011; Zhang & Miller, 1992).

Micelles consists of polar head groups which binds to metal and makes them soluble in water.

Adsorbed metals are also dissolved by surfactant monomers, which create dissolved complexes. On the other hand, certain metals may bind to the anionic exterior of rhamnolipid micelles. Metal ions are coupled with oppositely charged ions, or they can be replaced with the same charged ions or they can form chelates on the micelle surface by complexing with agents. Hence, various binding mechanisms are seen between different biosurfactant and heavy metals. This method of using surfactant and micelle

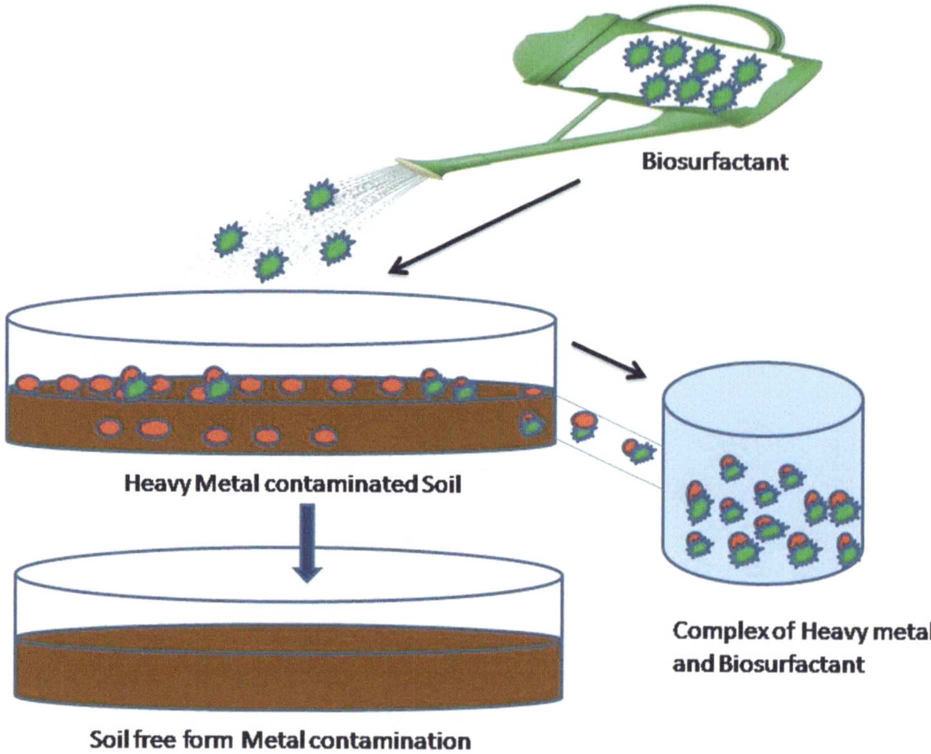

Fig. 3 Removal of heavy metal from contaminated soil by biosurfactant

formation not only helps making the process simpler and also user friendly (Açıkel, 2011; Zhang & Miller, 1992).

4 Biosurfactant in Mineral Floatation

4.1 Rhamnolipids in Mineral Ion Floatation

Rhamnolipids biosurfactants are the most popular, extensively studying or exploited by the biosurfactant among other biosurfactants. These are produced from bacteria *Pseudomonas aeruginosa*. About seven different types of homologues of these rhamnolipids are also found (Abalos et al., 2001). Two types of rhamnolipid are seen whose structure involves two rhamnose linked with beta hydroxy decanoic acid and one rhamnose attached to similar or identical fatty acid. *P. aeruginosa* takes up C12 and C11—glycerol, Glucose, fructose, mannitol, pyruvate, succinate, citrate, alkanes, etc., as raw materials to produce and form C8–C12. Some of them are also unsaturated with double and triple bonds (Robert et al., 1989). This *P. aeruginosa* are ubiquitous environmental bacterium which are isolated from water or soil and plants. Rhamnolipids are the most cultivated biosurfactants in the laboratory in large scale for different substrates and their applications.

Two major kinds of rhamnolipids, having molecular mass of 504 g mol^{-1} and 650 g mol^{-1}, respectively are RLL (R1) and RRLL (R2). RLL (R1), i.e., ($C_{26}H_{48}O_9$) is l-rhamnosyl-ß-hydroxydecanoyl-ß-hydroxydecanoate. RRLL (R2), i.e., ($C_{32}H_{58}O_{13}$) isl-rhamnosyl-ß-l-rhamnosyl-ß-hydroxydecanoyl-ß-hydroxydecanoate. The mixture of RLL, RRLL and mono- and di-rhamnolipid forms are specifically needed for removal of hydrocarbons and heavy metals from soil (Açıkel, 2011; Haba et al., 2003; Hamme et al., 2006).

4.2 Use of Rhamnolipid in Coal and Mineral Floatation

The experiment for coal and mineral floatation was initially carried out by fazaelipoor et al. They studied the rhamnolipid as frother as well as isolated, cultivated them in large scale. Here, they used diesel oil as solvents and heavy metal residues were seen after solvent dispersion. Upon adding rhamnolipid they did not see much frothability but the product recovery rate was high about 72–80%, with about 10–15% of ash content (Fazaelipoor et al., 2010; Khoshdast & Sam, 2012).

Khoshdast et al. briefly explain Rhamnolipid use, for which they isolated strains, cultivated them, produced them which followed up by measuring there, surface tension, pH, solubility and structural analysis as well as physical characterisation. Through this analysis, they concluded that the Rhamnolipid of 97.5% can reduce water from 72 to 30 mN $m^{-1}$with critical micelle concentration of 10 mg l^{-1} (Fazaelipoor et al., 2010; Khoshdast et al., 2011).

Another paper of Fazaelipoor et al. explains very beautifully the process of coal flotation test by desirable by surfactant (in this case Rhamnolipid). In this they added rhamnolipids as frother. Then surfactant concentration, particle size, oil concentration and solid present in concentration, where observed at intervals. The raw material of the solution was placed in 1 L capacity container along with the water of PH 7 and rotated in the speed of 1000 RPM in the impeller. The reaction was given about 3.2 L min^{-1}, and temperature was set to 25–26 °C. Initially, the raw materials or the collectors were added to the container and incubated for 2–3 min and then biosurfactant were added and incubated for 1 min more. The rotation continues for 6 min and the sample collection happens in the interval of 30, 60, 90, 150, 210 and 360 s to measure the kinetics of the floatation (Fazaelipoor et al., 2010).

Finally, the samples were filtered, dried, weighted and analysed, and ash content was determined. Software's like MINTLAB 14.1 were used for analysis of the floatation result and then statistical analysis was done. Because these are non-ionic molecules they also act as frother and their polar part links with the hydrogen bond of the water where is the non-polar and links with the surface or their air water interface. By changing the surface tension, we can also measure the surface activity. Rhamnolipid shows higher surface activity and biodegradability than the chemical surfactant (Fazaelipoor et al., 2010).

Having good ability and biodegradability make biosurfactant a promising frother and environmental friendly in protection of our nature. The result of these experiments was amazing and showing successful application of coal flotation. About 72 to 80% of solid combustible matter a recovery was done and about 10–15% ash content and 55–57% of efficient separation. This shows that rhamnolipids or biosurfactants can easily be used in mineral surfactants which are far better and more friendly than chemical or synthetic surfactants.

5 Surfactin in Ion Floatation

Bacillus subtilis, *Bacillus pumilus* and *Bacillus licheniformis* generate surfactin that is a cyclic peptide antibiotic composed of 7 amino acids bound to carboxyl and hydroxyl groups of a 14-carbon acid. Surfactin is made up of aspartic and glutamic amino acids. The glutamate residues have shown its binding affinity to metals that include Ca, Mg, Mn, Li, Ba and Rubidium. Because of the two charges on the surfactin molecule, the theoretical metal-surfactin ratio is 1 mol metal:1 mol surfactin (Açıkel, 2011). Surfactin has a potential benefit of having two charges owing to the aspartic and glutamic amino acids in its peptide structure, which makes it particularly potent and successful for ion flotation. Another method is precipitate floatation of metal ions for metal separation because as we know surfactants generates thick foam even at low concentrations. A low-flow foam fractionation technique is used that includes metal precipitation in aqueous solution involving constant binding of precipitates and its clusters to the rising bubbling foam. The metal content and solution's pH are two most important factors that influence metal removal through foam separation (Gurjar & Sengupta, 2015).

6 Other Biosurfactants in the Removal of Heavy Metals

LPS or Lipopolysaccharides are a form of biosurfactant that is made up of hydrophobic phospholipids and hydrophilic polysaccharide moiety. Langley and Beveridge were the first to test them for heavy metal extraction, demonstrating that lipopolysaccharides improve the outer cell walls hydrophilicity and allowing bacterial cells to absorb metallic cations. Kim and Vipulanandan investigated the elimination of lead from polluted soil and water (kaolinite) (Kim & Vipulanandan, 2006; Langley & Beveridge, 1999). *Flavobacterium* sp. grew on used vegetable oil to create the biotensioactive substance which is used for removal of lead from polluted water, hence, is quite effective.

Other forms of biosurfactants (mainly sophorolipids in nature) generated by *Candida* species have also been effectively used in heavy metal flotation. These can remove up to 90% of cations in the column and air-dissolved floatation process. Soil barrier is a unique method for separating heavy metals and petroleum by a biosurfactant produced from *Candida lipolytica*, i.e., a yeast. Biosurfactants also helps in removing Zn, Cu, Pb and Cd in groundwater by reducing soil permeability (Sarubbo et al., 2015). Sophorolipids are more effective than chemical surfactants at removing heavy metals from polluted environments. Heavy metals such as Cu and Zn have been shown to be removed from metal-polluted sediments using soil washing with sophorolipids. Water soluble anionic sophorolipids are negatively charged having COO– as head groups which makes strong ionic bonds with cationic elements hence proving that anionic sophorolipids are more efficient in metal removal than the non-ionic sophorolipids (Mishra et al., 2021).

Lipids derived from mannosylerythritol (MELs). Mannosylerythritol are glycolipid biosurfactants produced from microbial cells that are highly effective. Due to their outstanding biocompatibility, surface activity and healing capabilities, they have been widely employed in environmental sectors. Mannosylerythritol (MEL-AuNP) from the Ustilago maydis strain CGMCC5.203 is used for the synthesis of gold nanoparticles. These can also be used as reducing stabilising agents in the biosynthesis of nanoparticles (Mishra et al., 2021).

Plant-derived biosurfactants are also discovered and been used for the removal of heavy metals. Saponin is plant-based biosurfactant was investigates in Japan which removes heavy metal from soil from lakes construction sites (Mulligan et al., 2007). The soil contained zinc, copper, nickel and petroleum hydrocarbons in a bulky amount and the successive extraction revealed that the biosurfactant significantly decreased the zinc oxide, copper and organic fractions (Sarubbo et al., 2015).

Flotation is the only processing procedure that involves three phases of air, liquid and solid at the same time. The reagents involved in floatation are pH regulators, frothers and modifiers than can easily alter the chemical and physical properties of solid, liquid and air phases. These frothers are defined as surface active heteropolar molecules having hydrocarbon tail and polar head group. Hence, some biosurfactants can replace frothers as well playing their role very well. A biosurfactant produces from A *P. aeruginosa* act as frother in coal floatation giving around 72–79% of combustible matter recovery and having very little ash content, i.e., 10–15% with overall separating efficiency of 55–57%. Hence, these biosurfactants are very much efficient over chemical surfactants (Inès & Dhouha, 2015).

7 Conclusion and Future Aspects

This chapter shows that with the help of microorganisms like bacteria, fungi, yeast and plants we can cultivate biosurfactant in mineral medium along with gas oil. With the foam separation technique, it can be separated from fermentation medium, dried and used as per the requirement. This briefly explains about removal of heavy metal from soil in both the ex situ and in situ process, as well as their proper mechanism with some proved experimental methods, from literature. This indicates its benefit and is much less harmful because of the use of biological surfactants. Apart from the heavy metal recovery, mineral floatation is also focused, the use of rhamnolipid is very popular cause of its nature. Rahmnolipids are the most used biosurfactants in coal floatation and are still very user friendly, effective and environment friendly process.

But there are many different biosurfactants which are still needed to be study extensively to know their property and there use in different industries. These biosurfactant shows good profitability, good-frothability, good froth height as well as the froth stability, most importantly these are biodegradable, environmentally friendly, easily producible and less energy consuming. Rhamnolipids and surfactin can also be further investigated for its potential in different areas. This chapter also explain about the biosurfactants having low critical micelle concentration and high sorptive nature have strong potential. Therefore, these biomaterials not only used in mineral flotation but also in oil washing, biodegradation, in medicine industry, in petroleum industries and many more.

The future aspect of biosurfactant basically revolves around finding, determining, observing and analysing of different biosurfactants produced by different organisms. Placing them in an order or rank or criteria with the higher efficiency to their lower efficiency. This will help in finding many different kinds of biosurfactant and their quality as well as their ability in different fields. Therefore, these bio substances/biomaterials are much more efficient than the chemical ones.

References

Abalos, A., Pinaso, A., Infante, M. R., Casals, M., Garcia, F., & Manresa, A. (2001). Physicochemical and antimicrobial properties of new rhamnolipids by *Pseudomonas aeruginosa* AT10 from soybean oil refinery wastes. *Langmuir, 17*, 1367–1371.

Açıkel, Y. S. (2011). Use of biosurfactants in the removal of heavy metal ions from soils. In *Biomanagement of metal-contaminated soils* (pp. 183–223). Springer.

AytarÇelik, P., Çakmak, H., & ÖzAksoy, D. (2021). Green bioflotation of calcite using surfactin as a collector. *Journal of Dispersion Science and Technology*, 1–11.

Banat, I. M., Franzetti, A., Gandolfi, I., et al. (2010b). Microbial biosurfactants production, applications and future potential. *Applied Microbiology and Biotechnology, 87*, 427–444.

Banat, I. M., Franzetti, A., Gandolfi, I., Bestetti, G., Martinotti, M. G., Fracchia, L., Smyth, T. J., & Marchant, R. (2010a). Microbial biosurfactants production, applications. *Applied Microbiology and Biotechnology, 87*, 427–444.

Bodek, I., Lyman, W. J., Reehl, W. F., et al. (1998). *Environmental inorganic chemistry: Properties, processes and estimation methods*. Pergamon Press.

Campos, J. M., Stamford, T. L. M., Sarubbo, L. A., Luna, J. M., Rufino, R. D., & Banat, I. M. (2013). Microbial biosurfactants as additives for food industries. *Biotechnology Progress, 29*, 1097–1108.

Cerqueira, V. S., Hollenbach, E. B., Maboni, F., Vainstein, M. H., Camargo, F. A., do Carmo, M., Peralba, R., & Bento, F. M. (2011). Biodegradation potential of oily sludge by pure and mixed bacterial cultures. *Bioresource Technolology, 102*, 11003–11010.

Christofi, N., & Ivshina, I. B. (2002). Microbial surfactants and their use in field studies of soil remediation. *Journal of Applied Microbiology, 93*, 915–929.

Desai, J. D., & Banat, I. M. (1997). Microbial production of surfactants and their commercial potential. *Microbiology and Molecular Biology Reviews, 61*(1), 47–64.

Esther, J., Panda, S., Behera, S. K., Sukla, L. B., Pradhan, N., & Mishra, B. K. (2013). Effect of dissimilatory Fe (III) reducers on bio-reduction and nickel–cobalt recovery from Sukinda chromite-overburden. *Bioresource Technology, 146*, 762–766.

Fazaelipoor, M. H., Khoshdast, H., & Ranjbar, M. (2010). Coal flotation using a biosurfactant from *Pseudomonas aeruginosa* as a frother. *Korean Journal of Chemical Engineering, 27*(5), 1527–1531.

Frazer, L. (2000). Innovations: Lipid lather removes metals. *Environmental Health Perspectives, 108*, A320–A323.

Gurjar, J., & Sengupta, B. (2015). Production of surfactin from rice mill polishing residue by submerged fermentation using *Bacillus subtilis* MTCC 2423. *Bioresource Technology, 189*, 243–249.

Haba, E., Pinazo, A., Jauregui, O., Espuny, M. J., Infante, M. R., & Manresa, A. (2003). Physicochemical characterization and antimicrobial properties of rhamnolipids produced by *Pseudomonas aeruginosa* 47 T2 NCBIM 40044. *Biotechnology and Bioengineering, 81*, 316–322.

Inès, M., & Dhouha, G. (2015). Glycolipid biosurfactants: Potential related biomedical and biotechnological applications. *Carbohydrate Research, 416*, 59–69.

Khoshdast, H., Abbasi, H., Sam, A., & Noghabi, K. A. (2012). Frothability and surface behavior of a rhamnolipid biosurfactant produced by *Pseudomonas aeruginosa* MA01. *Biochemical Engineering Journal, 60*, 127–134.

Khoshdast, H., & Sam, A. (2012). An efficiency evaluation of iron concentrates flotation using rhamnolipid biosurfactant as a frothing reagent. *Environmental Engineering Research, 17*(1), 9–15.

Khoshdast, H., Sam, A., Vali, H., & Noghabi, K. A. (2011). Effect of rhamnolipid biosurfactants on performance of coal and mineral flotation. *International Biodeterioration & Biodegradation, 65*(8), 1238–1243.

Kim, J., & Vipulanandan, C. (2006). Removal of lead from contaminated water and clay soil using a biosurfactant. *Journal of Environmental Engineering, 132*, 777–786.

Langley, S., & Beveridge, T. J. (1999). Effect of O-side chain-lipopolysaccharide chemistry on metal binding. *Applied and Environment Microbiology, 65*, 489–498.

Makkar, R. S., Cameotra, S. S., & Banat, I. M. (2011). Advances in utilization of renewable substrates for biosurfactant production. *Applied Microbiology and Biotechnology, 1*, 1–19.

Mao, X., Jiang, R., Xiao, W., & Yu, J. (2015). Use of surfactants for the remediation of contaminated soils: A review. *Journal of Hazardous Materials, 285*, 419–435.

Matis, K. A., Zouboulis, A. I., Lazaridis, N. K., & Hancock, I. C. (2003). Sorptive flotation for metal ions recovery. *International Journal of Mineral Processing, 70*(1–4), 99–108.

Mishra, S., Lin, Z., Pang, S., Zhang, Y., Bhatt, P., & Chen, S. (2021). Biosurfactant is a powerful tool for the bioremediation of heavy metals from contaminated soils. *Journal of Hazardous Materials*, 126253.

Mishra, S., Panda, S., Pradhan, N., Satapathy, D., Biswal, S. K., & Mishra, B. K. (2017). Insights into DBT biodegradation by a native Rhodococcus strain and its sulphur removal efficacy for two Indian coals and calcined pet coke. *International Biodeterioration & Biodegradation, 120*, 124–134.

Mulligan, C. N., Oghenekevwe, C., Fukue, M., et al. (2007). Biosurfactant enhanced remediation of a mixed contaminated soil and metal contaminated sediment. Paper presented at Geoenvironmental Engineering Seminar. 7th Meeting; 2007 May 19–24; Japan–Korea–France

Mulligan, C. N. (2005). Environmental applications for biosurfactants. *Environmental Pollution, 133*(2), 183–198.

Mulligan, C. N., & Gibbs, B. F. (1993). Factors influencing the economics of biosurfactants. In N. Kosaric (Ed.), *Biosurfactants, production, properties, applications* (pp. 329–371). Marcel Dekker.

Mulligan, C. N., Yong, R. N., & Gibbs, B. F. (1999). Metal removal from contaminated soil and sediments by the biosurfactant surfactin. *Environmental Science and Technology, 33*, 3812–3820.

Mulligan, C. N., Yong, R. N., & Gibbs, B. F. (2001a). Heavy metal removal from sediments by biosurfactants. *Journal of Hazardous Materials, 85*, 111–125.

Mulligan, C. N., Yong, R. N., & Gibbs, B. F. (2001b). Remediation technologies for metal contaminated soils and groundwater: An evaluation. *Engineering Geology, 60*, 193–207.

Nakar, D., & Gutnick, D. L. (2003). Involvement of a protein tyrosine kinase in production of the polymeric bioemulsifier emulsan from the oil-degrading strain *Acinetobacter lwoffii* RAG-1. *Journal of Bacteriology, 185*(3), 1001–1009.

Okoliegbe, I. N., & Agarry, O. O. (2012). Application of microbial surfactant (a review). Department of Biological Sciences University of Abuja, Federal Capital Territory, Abuja, Nigeria. Accepted 28 February 2012

Pacwa-Plociniczak, M., Plaza, G. A., Piotrowska-Seget, Z., et al. (2011b). Environmental applications of biosurfactants: Recent advances. *International Journal of Molecular Sciences, 12*, 633–654.

Pacwa-Plociniczak, M., Plaza, G. A., Piotrowska-Seget, Z., & Cameotra, S. S. (2011a). Environmental applications of biosurfactants: Recent advances. *International Journal of Molecular Sciences, 13*, 633–654.

Panda, S., Pradhan, N., Mohapatra, U., Panda, S. K., Rath, S. S., Rao, D. S., Nayak, B., Sukla, L. B., & Mishra, B. K. (2013). Bioleaching of copper from pre and post thermally activated low grade chalcopyrite contained ball mill spillage. *Frontiers of Environmental Science & Engineering, 7*, 281–293.

Panda, S., Rout, P. C., Sarangi, C. K., Mishra, S., Pradhan, N., Mohapatra, U., Subbaiah, T., Sukla, L. B., & Mishra, B. K. (2014). Recovery of copper from a surface altered chalcopyrite contained ball mill spillage through bio-hydrometallurgical route. *Korean Journal of Chemical Engineering, 31*, 452–460.

Panda, S., Sarangi, C. K., Pradhan, N., Subbaiah, T., Sukla, L. B., Mishra, B. K., Bhatoa, G. L., Prasad, M., & Ray, S. K. (2012). Bio-hydrometallurgical processing of low-grade chalcopyrite for the recovery of copper metal. *Korean Journal of Chemical Engineering, 29*, 781–785.

Preetam, S., Dash, L., Sarangi, S. S., Sahoo, M. M., & Pradhan, A. K. (2022). Application of nanobiosensor in health care sector. *Bio-Nano Interface*, 251–270.

Robert, M., Mercade, M. E., Bosch, M. P., Parra, J. L., Espiny, M. J., Manresa, M. A., & Guinea, J. (1989). Effect of the carbon source on biosurfactant production by *Pseudomonas aeruginosa* 44T1. *Biotechnology Letters, 11*, 871–887.

Rosa, C. F. C., Freire, D. M. G., & Ferraz, E. C. (2015). Biosurfactant microfoam: Application in the removal of pollutants from soil. *Journal of Environmental Chemical Engineering, 3*, 89–94.

Santos, D. K. F., Rufino, R. D., Luna, J. M., Santos, V. A., & Sarubbo, L. A. (2016). Biosurfactants: Multifunctional biomolecules of the 21st century. *International Journal of Molecular Sciences, 17*(3), 401.

Sarma, H., Bustamante, K. L. T., & Prasad, M. N. V. (2019). Biosurfactants for oil recovery from refinery sludge: Magnetic nanoparticles assisted purification. In *Industrial and municipal sludge* (pp. 107–132). Butterworth-Heinemann.

Sarubbo, L. A., Rocha, R. B., Jr., Luna, J. M., Rufino, R. D., Santos, V. A., & Banat, I. M. (2015). Some aspects of heavy metals contamination remediation and role of biosurfactants. *Chemistry and Ecology, 31*(8), 707–723.

Silva, R. C. F. S., Almeida, D. G., Luna, J. M., Rufino, R. D., Santos, V. A., & Sarubbo, L. A. (2014). Applications of biosurfactants in the petroleum industry and the remediation of oil spills. *International Journal of Molecular Sciences, 15*, 12523–12542.

Singh, P., & Cameotra, S. S. (2004). Enhancement of metal bioremediation by use of microbial surfactants. *Biochemical and Biophysical Research Communications, 319*, 291–297.

Tolley, W., Kotlyar, D., & Van Wagoner, R. (1996). Fundamental electrochemical studies of sulfide mineral flotation. *Minerals Engineering, 9*(6), 603–637.

Van Hamme, J. D., Singh, A., & Ward, O. P. (2006). Physiological aspects Part 1 in a series of papers devoted to surfactants in microbiology and biotechnology. *Biotechnology Advances, 24*, 604–620.

Vijayakumar, S., & Saravanan, V. (2015). Biosurfactants-types, sources and applications. *Research Journal of Microbiology, 10*, 181–192.

Zamboulis, D., Pataroudi, S. I., Zouboulis, A. I., & Matis, K. A. (2004). The application of sorptive flotation for the removal of metal ions. *Desalination, 162*, 159–168.

Zhang, C. X., Yang, S. Y., Xu, M. X., Sun, J., Liu, H., Liu, J. R., Liu, H., Kan, F., Sun, J., Lai, R., & Zhang, K. Y. (2009). *Serratia nematodiphila* sp. nov., associated symbiotically with the entomopathogenic nematode Heterorhabditidoideschongmingensis (Rhabditida: Rhabditidae). *International journal of systematic and evolutionary microbiology, 59*(7), 1603–1608.

Zhang, Y., & Miller, R. M. (1992). Enhanced octadecane dispersion and biodegradation by a *Pseudomonas rhamnolipid* surfactant (biosurfactant). *Applied and Environment Microbiology, 58*, 3276–3282.

Zouboulis, A. I., Matis, K. A., Lazaridis, N. K., & Golyshin, P. N. (2003). The use of biosurfactants in flotation: Application for the removal of metal ions. *Minerals Engineering, 16*(11), 1231–1236.

Recovery of Metals from Leach Liquors: Biosorption *versus* Metal Sulfide Precipitation

Mehmet Ali Kucuker

Abstract

Extraction of metals (leaching) is chemical or biochemical processes that utilize acids or microorganisms to enhance the suspension of metals from the primary and secondary sources by making them more amenable to dissolution in aqueous solutions (leachate). Recovery of metals from the leachates is an essential stage supported by additional purification processes such as precipitation of impurities, electrowinning, solvent extraction, chemical or biological adsorption, and ion exchange. In this study, especially biosorption and metal sulfide precipitation are overviewed and discussed. Biosorption is a process by which particular biomass such as bacteria, fungi, yeast, agricultural wastes, algae, and biowastes can able to bind with specific ions or other molecules from aqueous solutions. Metal sulfide precipitation can be highly effective in obtaining a high degree of separation of metal cations from complex leachates. Each of these techniques has advantages and drawbacks. Sometimes, a technique may not be effective in attaining higher metal recovery. Therefore, different recovery techniques are needed to recover the target elements from the complex leachates. Maybe a combination of two or three recovery techniques is required to recover metals from complex leachates. Additionally, the research activity highlighted that metal sulfide precipitation and biosorption processes have to limit factors that could hinder the process scale-up. Thus, more research is needed to evaluate the environmental impacts of metal recovery from leach liquors.

Keywords

Leaching · Leachate · Metal recovery · Biosorption · Metal sulfide precipitation

1 Introduction

Metals are an essential part of today's life with numerous applications in medicine, mobility, communication, households, ornaments and accessories, security, and environmental protection. Therefore, the demand for metals has increased year by year (Işıldar et al., 2019; Kucuker & Kuchta, 2018). Metals are usually manufactured from the primary (ores) and secondary sources (mining waste, by-products of industrial activities, and so on) making substances more soluble in leachate which makes them easier to break down (Kucuker, 2018; Sahan et al., 2019; Tuncuk et al., 2012). The extraction of metals, which frequently provides widely to value of the primary and secondary sources, is therefore crucially important (Kucuker & Kuchta, 2018). Studies have shown that there are several processes to extract valuable metals from sources, using extraction techniques such as mechanical, hydrometallurgical, and biometallurgical (Işıldar et al., 2019; Kucuker, 2018; Marra, 2017; Panda et al., 2021; Tuncuk et al., 2012). Recently, researchers have received growing attention in the recovery of metals using hydrometallurgical and biometallurgical.

In the metallurgical processes, hydrometallurgical processes seem to cause a relatively low investment, no airborne emissions formation, functional selectivity and seem even suitable for small-scale applications (Tuncuk et al., 2012). Research on hydrometallurgy focuses primarily on the leaching of valuable metals (Sethurajan et al., 2019; Tuncuk et al., 2012). Obviously, it is essential to search for inexpensive and ecological processes as well as there are a plethora of cutting-edge technologies that can be utilized for the efficient leaching and recovery of primary metals.

M. A. Kucuker (✉)
Engineering Faculty, Department of Environmental Engineering, İzmir Institute of Technology, Gülbahçe Campus, Urla, 35430 İzmir, Turkey
e-mail: mehmetalikucuker@iyte.edu.tr

S. Panda et al. (eds.), *Biotechnological Innovations in the Mineral-Metal Industry*,
Advances in Science, Technology & Innovation, https://doi.org/10.1007/978-3-031-43625-3_9

Bio/hydrometallurgy is a highly effective method that utilizes various biological processes such as bioleaching and biosorption to extract valuable metals from different sources including ore, industrial waste, mining waste, bottom ash, and end-of-life vehicles. This innovative approach provides an efficient solution for recovering precious metals in both primary and secondary forms. By implementing this technique, we can minimize the environmental impact of traditional extraction methods while also reducing costs associated with metal recovery (Işıldar et al., 2019; Kucuker & Kuchta, 2018). Both processes need leaching to extract the metals from primary and secondary sources, and then removing metals from leach liquors (leachates) is a crucial stage supported by additional purification processes such as solvent extraction, electrowinning, chemical or biological adsorption, precipitation of impurities, and ion exchange. In this chapter, especially biosorption and metal sulfide precipitation are overviewed and discussed.

It is important to find ways to recycle metals cheaply and without harming the environment.

2 Metal Recovery from Leach Liquors

In recent years, researchers have been fascinated with metal recovery through hydrometallurgical processes (Sethurajan et al., 2019). The steps of the hydrometallurgical methods are shown in Fig. 1. Extracting metals (leaching) from resources using chemical agents or bioleaching is the primary step (Kucuker & Kuchta, 2018; Panda et al., 2012; Sethurajan et al., 2019) that encompasses selective chemical leaching of solid residue. Applying selective leaching methods, metal ions can be solved into liquid solutions selectively and separately, leading to the specific element composition in the final leachate (Lee et al., 2013). In addition, as a multi-element complex, selective leaching is an effective and essential way to separate different elements for the other metal recovery process (Kucuker et al., 2017; Lee et al., 2013; Sethurajan et al., 2019; Zhu et al., 2017). After the leaching process, raw metal recovery from the leaching solution is significant as it is the ultimate and essential stage. Numerous approaches such as solvent extraction, precipitation, electrowinning, and (bio)sorption have been utilized and applied to recover metals from the leachates. They are the most commonly used methodologies (Sethurajan, 2016). The recovery processes as precipitation and biosorption are criticized.

2.1 Metals Recovery by Sulfide Precipitation

Precipitation is a traditional procedure used for recovering metals from multimetallic solutions. Precipitation of metals from leachates can be reached by forming their respective sulfide, hydroxide, and carbonate salts (Manahan, 1990; Sethurajan, 2016). The solution pH and the metal concentrations in the leachate are the main parameters for the recovery competencies of the metals by precipitation (Hedrich et al., 2018; Sethurajan, 2016). If the ions' concentration in the solution surpasses the solubility product, metal precipitation is observed (Wang et al., 2005). Changes in the system's ionic equilibrium can be caused by adding the reaction products (either metal or sulfide/hydroxide) (Sethurajan, 2016). Certain chemicals such as sulfide can induce precipitation (precipitating agents) (Wang et al., 2005). Sulfide sources such as solid (FeS, CaS), aqueous (Na_2S, NaHS, NH_4S), and gaseous (H_2S) can impact the precipitation rate (Lewis, 2010). In addition, the reaction of sodium thiosulfate ($Na_2S_2O_3$), thiourea, thioacetamide, polyphenylene sulfide, and carbon disulfide can be used to remove sulfides from water by precipitation (Lewis, 2010). The sulfidation process of metal is similar to the precipitation of metal hydroxides (Pohl, 2020). For instance, the metal removal mechanism by sodium sulfide precipitation is shown in Eq. (1).

$$Na_2S + MSO_4 \rightarrow MS(\text{precipitate}) + Na_2SO_4. \quad (1)$$

In addition, the application of a biosource of sulfide (H_2S) has been used to remove the metals (Kumar et al., 2021) and the optimization of solid–liquid separation of precipitates (Uçar, 2017). In these studies, H_2S was generated using the bioreactors, and the configuration in the biogenic sulfide precipitation process should be under control due to the sulfide content that is critical in the final metal removal efficiency (Estay et al., 2021; Kumar et al., 2021; Pohl, 2020). Significantly, the biogenic sulfide precipitation process can be used to get metals such as Cu, As, Zn, and Pb from acid mine drainages (Estay et al., 2021), and the system can reach metal removal by almost 100% (Macingova & Luptakova, 2012).

There are several affecting factors such as reaction time, temperature, pH, and sulfide dosage for metal sulfide precipitation. The optimum conditions of metals recovery by sulfide precipitation are summarized in Table 1. According to the results in the literature, different metals have been removed from leachate under various conditions such as pH, sulfide dosage, retention time, and temperature. For all cases, the maximum removal rates were more than 90%. There is a piece of limited information on kinetic aspects; thus, the relation among the affecting factors could not be explained precisely (Estay et al., 2021).

In the sulfidogenic system, metal precipitation is obtained by adjusting the pH value (Migdisov et al., 2002; Sethurajan, 2016). The sulfide species concentration (S_2^-) in the solution is based on pH. This means that different metal sulfides have different solubility products and form at different pH values (Hedrich et al., 2018; Lewis, 2010; Sethurajan, 2016). Alternatively, metal sulfide precipitation

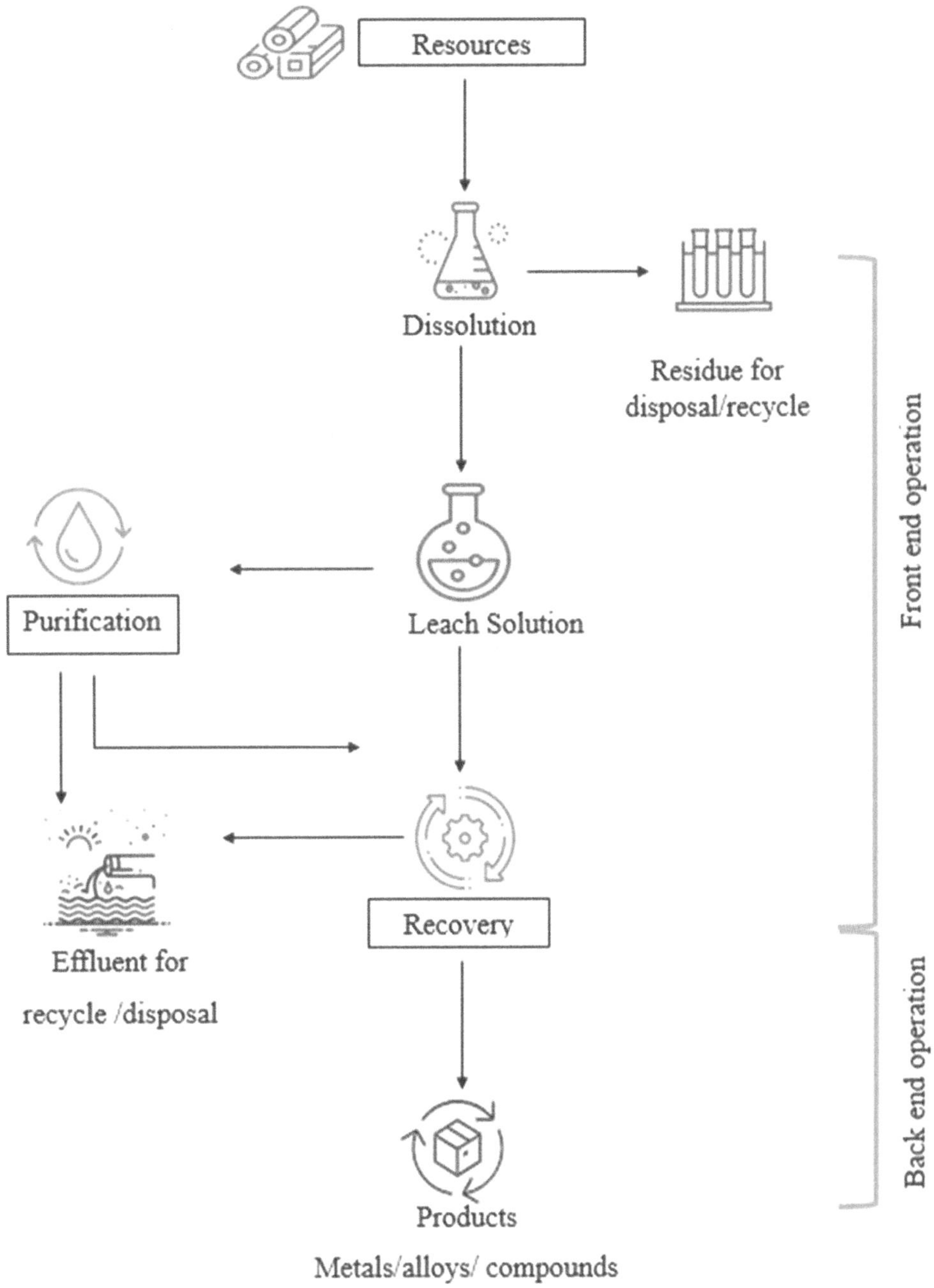

Fig. 1 Metal recovery from resources through hydrometallurgy. Adapted and redrawn from Gupta (2006)

has been carried out to accurately categorize the various types of reduced sulfur compounds present in solutions through pH optimization (Lewis, 2010). Therefore, the metal recovery efficiency can be increased by minimizing polysulfide formation (Mokone et al., 2009) (Fig. 2).

On the other hand, precipitation processes come with a multitude of drawbacks, for instance, the process demands an excessive amount of chemicals for pH regulation; as well as, dealing with sludge containing hazardous substances which poses significant challenges in terms of settling and dewatering issues (Ahalya et al., 2003; Lewis, 2010; Sethurajan, 2016). Moreover, the metal sulfides' precipitation that is susceptible to dosing the precipitation agent is a problematic method due to the fact that metal sulfides have a very poor solubility rate (Pohl, 2020). It is imperative to note that the precipitation of metal sulfides may lead to a detrimental issue—poor separation between solid and liquid components due to the formation of fine particles, including colloidal precipitates (Lewis & Van Hille, 2006; Sethurajan, 2016). Another risk is related to toxic hydrogen sulfide emissions during sulfide precipitation (Fu & Wang, 2011; Lewis, 2010; Pohl, 2020). In some cases, such as the recovery of Zn, Fe, Cd, and Cu, sulfide precipitation methods can be attractive since they

Table 1 Summary of the affecting factors of metal sulfide precipitation for selected elements

Element	Sulfide dosage, molar ratio (S^2/M)	Reaction time to reach maximum conversion (min)	pH	Temperature (°C)	Reference
Cu	0.5–0.6	< 1	3.5–5.0	15	Estay et al. (2020)
	1.4	5	10.0	25	Deng et al. (2019)
	1.0–1.2	5	10.6	25	Li et al. (2021)
Co	NR*	2.5	2.5	20–25**	Uçar (2017)
	10	75	1.0	20–25	Vemic et al. (2016)
	2	30	3	30	Choubey et al. (2021)
Zn	2	75	> 12.0	70	Zhang et al. (2020)
	NR*	45	5.5	45–85	Gharabaghi et al. (2012)
As	NR*	120	< 1	70	Hong et al. (2020)
	3	60	4.0	25	Bin et al. (2019)
	2.5	60	< 1	20–25**	Liu et al. (2016)
Ni	10	75	1.0	20–25	Vemic et al. (2016)
	NR*	45	7.5	45–85	Gharabaghi et al. (2012)
Re	NR*	60	< 1	70	Hong et al. (2020)
	NR*	360	< 1	25	Cai et al. (2021)
Fe	NR*	5	7.4	20–25**	Uçar (2017)
Mo	10	75	1.0	20–25	Vemic et al. (2016)

Adapted and modified from Estay et al. (2021)
* *NR* Not reported; ** Room temperature

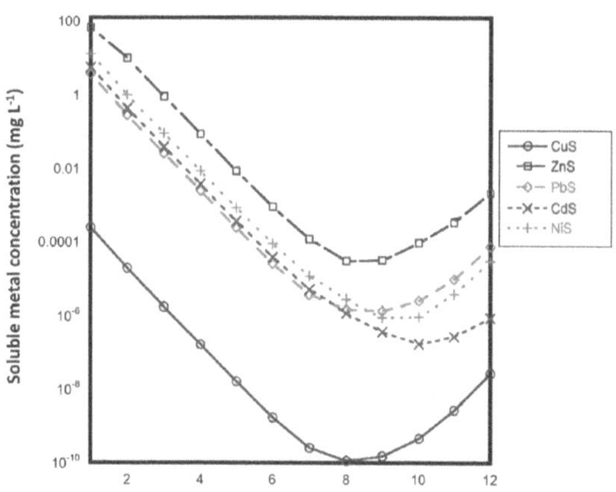

Fig. 2 pH dependence of metal sulfide for soluble metal concentration. Adapted and redrawn from Lewis (2010)

are comparatively uncomplicated to manage (Lewis, 2010; Sethurajan, 2016). All in all, defined limitations of sulfide precipitation have compelled the researcher to find innovative and efficient precipitation mixtures to expand active metal precipitation methods (Fu et al., 2007; Lewis, 2010; Thomas et al., 2014).

2.2 Metals Recovery by Biosorption

Biosorption is an incredibly promising approach to extracting metals from water-based solutions. It is the second crucial step in bio/hydrometallurgical processes and has proven highly effective thus far. Researchers have recently focused on developing metal recovery techniques from aqueous solutions such as leachates (Isıldar et al., 2019; Kucuker et al., 2017; Sethurajan et al., 2019; Panda et al., 2021). In this context, the most active investigation on the valuable metal recovery from leachates was related to bio/hydrometallurgical methods, which are reported to be more ecologically friendly (Kucuker et al., 2017; Panda et al., 2021).

Decades ago, biosorption technology has emerged as a promising solution for recovery (Volesky, 1987). It combines the benefits of adsorption such as chemisorption and physisorption with ion exchange and microprecipitation onto functional groups found in inactive cell walls of biomass. This innovative method has proven to be highly effective in recovering valuable resources while also being environmentally friendly (Kucuker et al., 2017; Robalds et al., 2016; Volesky, 2003). Both absorption and adsorption are the general terms of sorption. Absorption can be specified as a physical or a chemical phenomenon or a process in which a substance enters the gas, liquid, or solid material, i.e., absorption of carbon dioxide by sodium hydroxide (Fomina & Gadd, 2014; Gadd, 2009). Adsorption is the physisorption or chemisorption for binding ions onto the surface of another molecule, such as heavy metals recovery from industrial effluents using sorbent (Fomina & Gadd, 2014; Robalds et al., 2016). If the sorbent is a biological material, biosorption can be evaluated as a subcategory of adsorption (Michalak et al., 2013). Biosorption can be described as the observation of the experimental observation when the amount of sorbate (metal ion) in the solution absorbs after a certain period by the biosorbent (Robalds et al., 2016). Biosorption can be explained as the uptake of metals, mixtures, and particulate matter from the liquid using a biosorbent (biomass) (Gadd, 1993; Wang & Chen, 2009). Numerous biological raw materials has been tested for metal removals. Biosorption, an independent metabolic process, occurs on the cell walls of various microorganisms such as fungi, bacteria, and biowaste (including garden waste and wood); as well as, this natural phenomenon is known to be highly effective in removing contaminants from water (Mao et al., 2011). Because biosorption is often conducted with dead biomass, this results in ignoring any nutrient requirement (Das, 2010). Biosorption offers a remarkable benefit as it can be effectively utilized in situ without the need for any industrial process operations, provided that an appropriate design is implemented (Das, 2010; Tewari et al., 2005). Biosorbents can be classified as follows: bacteria (e.g., *Bacillus subtilis*), fungi (e.g., *Rhizopus arrhizus*), yeast (e.g., *Saccharomyces cerevisiae*), algae (e.g., *Chlorella vulgaris, A. gracilis, Sargassum sp.* and *A. densus*), industrial wastes (e.g., by-product of the food industry), agricultural wastes (e.g., leaves) (Kucuker, 2018; Kucuker et al., 2016; Vijayaraghavan & Yun, 2008; Wang & Chen, 2009). There is limited research on defining the compatibility of the biosorbent for accurate industrial scale (Wang & Chen, 2009). While biosorption has yet to be implemented on a full-scale basis, its benefits are undeniable and make it an ideal choice for large environmental metal removal projects.

The metal biosorption capacity is based on the formation of the biosorbent (Volesky, 2007), as the cell wall of the biosorbent is the main responsible area for metal uptake through biosorption (Kucuker et al., 2016; Volesky, 2007; Wang & Chen, 2009). The presence of certain biopolymers in cell walls, particularly those containing carboxylic groups, has been found to be highly effective in capturing and retaining metals (Volesky, 2007). The relevant fundamental functional groups are shown in Table 2. The O-, N-, S-, or P-containing groups have been proven to contribute to certain binding metals (Wang & Chen, 2009).

Table 2 Classes of organic compounds as a functional group in biomass (Kucuker, 2018; Talaro & Talaro, 2002; Volesky, 2007; Wang & Chen, 2009)

The formula of functional group	Name	Class of compounds
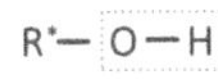	Hydroxyl	Alcohols, carbohydrates
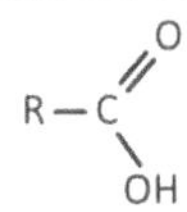	Carboxyl	Fatty acids, proteins, organic acids
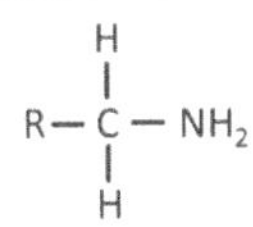	Amino	Proteins, nucleic acids
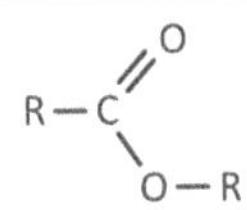	Ester	Lipids
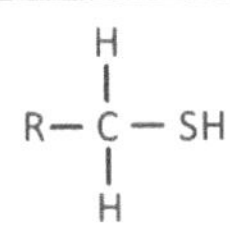	Sulfhydryl	Cysteine (amino acid), protein
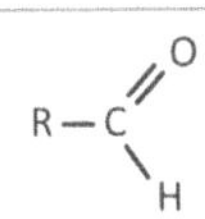	Carbonyl, the terminal end	Aldehydes, polysaccharides
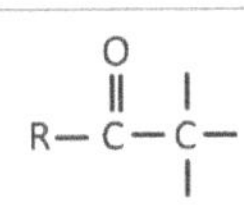	Carbonyl, internal	Ketones, polysaccharides
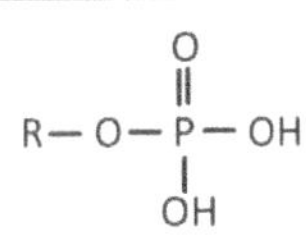	Phosphate	DNA, RNA, ATP

Moreover, researches have demonstrated that enhancing non-living biomaterials with chelating agents or implementing chemical and physical modifications such as altering functional groups, crosslinking agents, heat treatment, or ultrasonic treatment can lead to a significant increase in sorption capacity. This evidence highlights the effectiveness of these methods for improving material performance and should encourage their utilization in various applications. Such modifications can also be used to increase the durability of biomaterials in challenging environments (e.g., a very low or high pH), improve the sorption kinetics, and increase their regeneration potential. It is important to note that these additions frequently result in a significant rise in both the financial burden and ecological footprint of the technology (Fomina & Gadd, 2014; Gadd, 2009). In addition, a low price is one of the main advantages of biosorption media. However, pre-treatment of the biomass results in higher production costs and pre-treated biosorbents equivalent prices as the ones of artificial ion exchange resins. By implementing certain pre-treatment methods, the loading of Ru in *Rhodopseudomonas palustris* cells can be significantly improved. Mineral acids have proven to increase this loading from 86 mg/g dry cells to an impressive 145 mg/g dry cells while also enhancing selectivity over Ni and Zn (Kwak et al., 2013). Additionally, blending with chitosan and surface modification using polyethyleneimine (PEI) has resulted in a remarkable enhancement of biomass's ability to uptake Ru by up to 6.9 times its original capacity (Kwak et al., 2013). These innovative techniques offer promising solutions for maximizing the potential benefits that these microorganisms possess when it comes to metal recovery processes. Table 3 presents a comprehensive compilation of

Table 3 Recovery of REEs and PMs from aqueous sources through biosorption

Cation	Biosorbent	Biosorption capacity (mg/g)	References
Ag^{+}	*Chemically modified chitosan resin*	413.62	Donia et al. (2007)
	Bacillus cereus	91.75	Li et al. (2011)
	Saccharomyces cerevisiae	135.91	Chen et al. (2014)
	Klebsiella sp.3S1	141.10	Muñoz et al. (2017)
	Magnetospirillum gryphiswaldense	13.50	Wang et al. (2011)
Au^{3+}	*Fucus vesiculosus*	68.94	Mata et al. (2009)
	Chemically modified chitosan	669.66	Donia et al. (2007)
	Crosslinked chitosan resin	70.34	Fujiwara et al. (2007)
	Thiourea modified alginate	1668.25	Gao et al. (2017)
Pd^{2+}	*Racomitrium lanuginosum*	37.20	Sari et al. (2009)
	Collagen fiber (BTICF) membrane Chitosan derivatives	689.36	Chassary et al. (2005)
	Bayberry tannin	33.40	Ma et al. (2006)
Y^{3+}	NaOH-modified *Pleurotus ostreatus*	45.45	Hassanien et al. (2014)
La^{3+}	Prawn carapace	200.00	Varshini et al. (2014)
	Fish scales	250.00	Varshini et al. (2014)
	Grapefruit peel	171.20	Torab-Mostaedi et al. (2015)
	Pleurotus ostreatus basidiocarps	54.54	Hussien (2014)
	Chlamydomonas reinhardtii	142.86	Birungi and Chirwa (2014)
	Platanus orientalis	28.65	Sert et al. (2008)
	Sargassum sp.	91.68	Oliveira and Garcia (2009)
	Chlorella vulgaris	74.60	Birungi and Chirwa (2014)
Ce^{3+}	Grapefruit peel	159.30	Torab-Mostaedi et al. (2015)
	Prawn carapace	1000.00	Varshini et al. (2014)
	Fish scales	200.00	Varshini et al. (2014)
	Corn style	250.00	Varshini et al. (2014)
	Platanus orientalis	32.05	Sert et al. (2008)
Pr^{3+}	*Green seaweed (Ulva lactuca)*	69.75	Vijayaraghavan (2015)
	Free Turbinaria conoides (brown seaweed)	146.40	Vijayaraghavan and Jegan (2015)
	Polysulfone immobilized Turbinaria conoides	119.50	Vijayaraghavan and Jegan (2015)
	Crab shell	66.60	Varshini et al. (2015)
	Sargassum sp.	98.63	Oliveira et al. (2011)
Nd^{3+}	*Chlorella vulgaris*	157.21	Kucuker et al. (2016)
	Nostoc ellipsosporum	11.54	Heilmann et al. (2015)
	Physcomitrella patens	106.73	Heilmann et al. (2015)
	Calothrix brevissima	69.23	Heilmann et al. (2015)
	Tetraselmis chuii	51.92	Heilmann et al. (2015)
	Chlorella sorokiniana	23.08	Heilmann et al. (2015)
	Sargassum sp.	100.96	Oliveira and Garcia (2009)
Sm^{3+}	Activated biochars from cactus fibers (pH = 3.0)	90.00	Hadjittofi et al. (2016)
	Activated biochars from cactus fibers (pH = 6.5)	350.00	Hadjittofi et al. (2016)
	Sargassum sp.	105.25	Oliveira et al. (2011)
Eu^{3+}	*Sargassum sp.*	82.64	Oliveira and Garcia (2009)
	Raw cactus fibres	0.16	Prodromou and Pashalidis (2016)
	Modified cactus fibres (MnO2-coated))	0.46	Prodromou and Pashalidis (2016)
	Sargassum polycystum Ca-loaded biomass	62.30	Diniz and Volesky (2005)
Yb	*Sargassum polycystum* Ca-loaded biomass	48.45	Diniz and Volesky (2005)

Adapted from Kucuker (2018)

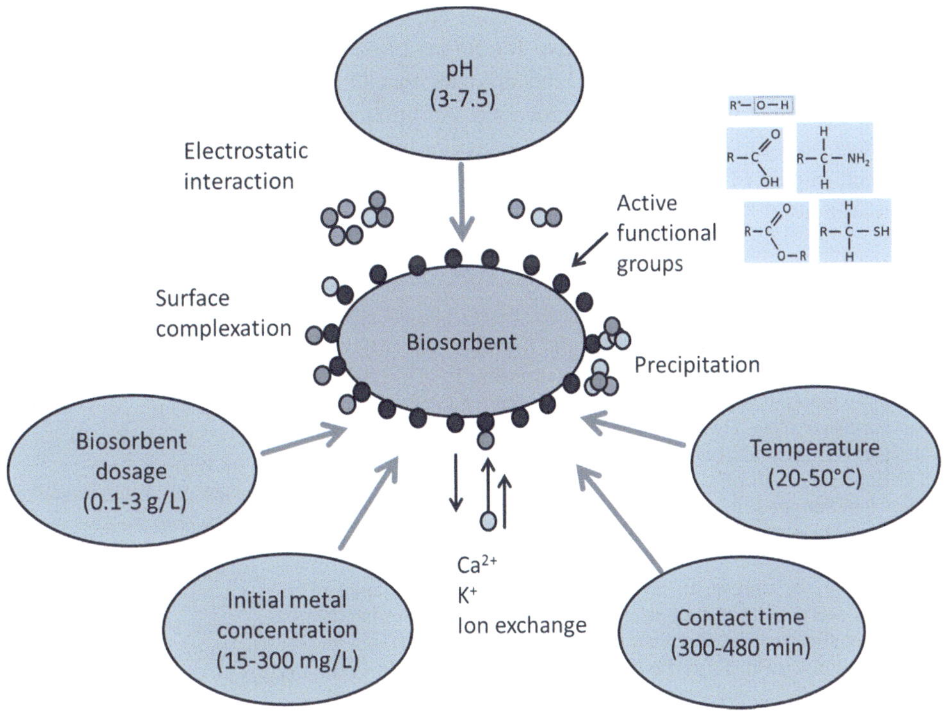

Fig. 3 Various mechanisms of biosorption and essential parameters affecting the biosorption of REEs. Adapted and modified from Das and Das (2013)

research studies conducted on the biosorption of rare earth elements (REEs) and precious metals (PMs).

Undoubtedly, the efficiency of biosorption processes can be influenced by various operational factors. These include pH levels, temperature variations, the dosage of biosorbent used, initial metal concentration in the solution under treatment as well as time and agitation rate applied during processing (Kucuker et al. 2017). The systems engaged in metal uptake on biosorption sites are various, involving complex processes (Fig. 3).

3 Future Perspective for Recovery of Metals from Leach Liquors Through Biosorption and Sulfide Precipitation

The metal sulfide precipitation and biosorption processes have been successfully tested in both pilot and industrial-scale applications, proving that they are highly effective alternatives for removing or recovering specific elements such as metals or metalloids from various sources (Estay et al., 2021). However, there are several gaps in the field of metal recovery through metal sulfide precipitation and biosorption, which are related to optimizing and expanding its application. Therefore, further studies in terms of selective recovery, kinetic reaction studies, reactor type, and supersaturation control should be conducted (Estay et al., 2021). While integrated methods like bioleaching, biosorption, and sulfide precipitation can be effective in extracting metals from leach liquors, they do come with certain limitations. These include the need for large quantities of chemicals, sorbents, or reagents as well as selective processes such as leaching, precipitation, and sorption. Moreover, the study revealed that biogenic sulfide precipitation and biosorption techniques necessitated an extended duration for treatment and yielded limited loading rates when compared to chemical methods. Further exploration is imperative in order to assess the ecological consequences of metal removal from leach liquors through a comprehensive approach.

4 Conclusions

Recovery of metals from the leachates is an essential stage supported by additional purification processes such as precipitation of impurities, solvent extraction, electrowinning, chemical or biological adsorption, and ion exchange. In this study, especially biosorption and metal sulfide precipitation are overviewed and discussed. Sulfide precipitates are not amphoteric, so a high degree of metal recovery can be obtained in a shorter time over a wide pH range compared with the biosorption process. However, colloidal precipitates are produced while metal sulfides precipitate, resulting in further separation issues. Moreover, a potential risk can be seen for the emission of toxic hydrogen sulfide during the sulfide precipitation. The studies have brought to light that biosorption techniques necessitate a lengthier duration for treatment and offer inferior loading rates when compared with chemical processes. It is crucial to address the limiting factors that may impede process scale-up. To overcome these obstacles, it is imperative for future research endeavors to concentrate on hybrid technologies that integrate both chemical and biological processes. Additionally, there must be a further exploration into assessing the environmental impact of metal recovery from primary and secondary sources.

References

Ahalya, N., Ramachandra, T. V., & Kanamadi, R. D. (2003). Biosorption of heavy metals. *Research Journal of Chemistry and Environment, 7*(4), 71–79.

Bin, H. U., Yang, T. Z., Liu, W. F., Zhang, D. C., & Lin, C. H. E. N. (2019). Removal of arsenic from acid wastewater via sulfide precipitation and its hydrothermal mineralization stabilization. *Transactions of Nonferrous Metals Society of China, 29*(11), 2411–2421.

Birungi, Z. S., & Chirwa, E. M. N. (2014). The kinetics of uptake and recovery of lanthanum using freshwater algae as biosorbents: Comparative analysis. *Bioresource Technology, 160*, 43–51.

Cai, X., Kong, L., Hu, X., & Peng, X. (2021). Recovery of Re (VII) from strongly acidic wastewater using sulphide: Acceleration by UV irradiation and the underlying mechanism. *Journal of Hazardous Materials, 416*, 126233.

Chassary, P., Vincent, T., Marcano, J. S., Macaskie, L. E., & Guibal, E. (2005). Palladium and platinum recovery from bicomponent mixtures using chitosan derivatives. *Hydrometallurgy, 76*(1), 131–147.

Chen, C., Wen, D., & Wang, J. (2014). Cellular surface characteristics of Saccharomyces cerevisiae before and after Ag (I) biosorption. *Bioresource Technology, 156*, 380–383.

Choubey, P. K., Dinkar, O. S., Panda, R., Kumari, A., Jha, M. K., & Pathak, D. D. (2021). Selective extraction and separation of Li, Co and Mn from leach liquor of discarded lithium ion batteries (LIBs). *Waste Management, 121*, 452–457.

Das, N. (2010). Recovery of precious metals through biosorption—a review. *Hydrometallurgy, 103*(1), 180–189. https://doi.org/10.1016/j.hydromet.2010.03.016

Das, N., & Das, D. (2013). Recovery of rare earth metals through biosorption: An overview. *Journal of Rare Earths, 31*(10), 933–943.

Deng, Z., Oraby, E. A., & Eksteen, J. J. (2019). The sulfide precipitation behaviour of Cu and Au from their aqueous alkaline glycinate and cyanide complexes. *Separation and Purification Technology, 218*, 181–190.

Diniz, V., & Volesky, B. (2005). Biosorption of La, Eu and Yb using Sargassum biomass. *Water Research, 39*(1), 239–247.

Donia, A. M., Atia, A. A., & Elwakeel, K. Z. (2007). Recovery of gold (III) and silver (I) on a chemically modified chitosan with magnetic properties. *Hydrometallurgy, 87*(3), 197–206.

Estay, H., Barros, L., & Troncoso, E. (2021). Metal sulfide precipitation: Recent breakthroughs and future outlooks. *Minerals, 11*(12), 1385.

Fomina, M., & Gadd, G. M. (2014). Biosorption: Current perspectives on concept, definition and application. *Bioresource Technology, 160*, 3–14.

Fu, F., & Wang, Q. (2011). Removal of heavy metal ions from wastewaters: A review. *Journal of Environmental Management, 92*, 407–418.

Fu, F., Zeng, H., Cai, Q., Qiu, R., Yu, J., & Xiong, Y. (2007). Effective removal of coordinated copper from wastewater using a new dithiocarbamate-type supramolecular heavy metal precipitant. *Chemosphere, 69*, 1783–1789.

Fujiwara, K., Ramesh, A., Maki, T., Hasegawa, H., & Ueda, K. (2007). Adsorption of platinum (IV), palladium (II) and gold (III) from aqueous solutions on l-lysine modified crosslinked chitosan resin. *Journal of Hazardous Materials, 146*(1), 39–50.

Gadd, G. M. (1993). Interactions of fungip with toxic metals. *New Phytologist, 124*(1), 25–60.

Gadd, G. (2009). Biosorption: Critical review of scientific rationale, environmental importance and significance for pollution treatment. *Journal of Chemical Technology and Biotechnology, 84*(1), 13–28.

Gao, X., Zhang, Y., & Zhao, Y. (2017). Biosorption and reduction of Au (III) to gold nanoparticles by thiourea modified alginate. *Carbohydrate Polymers, 159*, 108–115.

Gharabaghi, M., Irannajad, M., & Azadmehr, A. R. (2012). Selective sulphide precipitation of heavy metals from acidic polymetallic aqueous solution by thioacetamide. *Industrial and Engineering Chemistry Research, 51*(2), 954–963.

Gupta, C. K. (2006). *Chemical metallurgy: Principles and practice*. John Wiley & Sons.

Hassanien, W. A. G., Desouky, O. A. N., & Hussien, S. S. E.(2014). Bioleaching of some rare earth elements from Egyptian monazite using Aspergillus ficuum and Pseudomonas aeruginosa. *Walailak Journal of Science and Technology (WJST), 11*(9), 809–823.

Hadjittofi, L., Charalambous, S., & Pashalidis, I. (2016). Removal of trivalent samarium from aqueous solutions by activated biochar derived from cactus fibres. *Journal of Rare Earths, 34*(1), 99–104.

Hedrich, S., Joulian, C., Graupner, T., Schippers, A., & Guézennec, A.-G. (2018). Enhanced chalcopyrite dissolution in stirred tank reactors by temperature increase during bioleaching. *Hydrometallurgy, 179*, 125–131.

Heilmann, M., Jurkowski, W., Buchholz, R., Brueck, T., & Becker, A. M. (2015). Biosorption of neodymium by selected photoautotrophic and heterotrophic species. *Journal of Chemical Engineering and Process Technology, 6*(4), 1.

Hong, T., Zheng, T., Liu, M., Mumford, K. A., & Stevens, G. W. (2020). Investigation on the recovery of rhenium in the high arsenite wash acid solution from the copper smelting process using reducing sulfide precipitation method. *Hydrometallurgy, 195*, 105402.

Hussien, S. S. (2014). Biosorption lanthanum pleurotus ostreatus basidiocarp. *International Journal of Biomedical Research, 2*, 26–36.

Işıldar, A., van Hullebusch, E. D., Lenz, M., Du Laing, G., Marra, A., Cesaro, A., Panda, S., Akcil, A., Kucuker, M. A., & Kuchta, K. (2019). Biotechnological strategies for the recovery of valuable and critical raw materials from waste electrical and electronic equipment (WEEE)–A review. *Journal of Hazardous Materials, 362*, 467–481.

Kucuker, M. A. (2018). Biomining concept for recovery of rare earth elements (REEs) from secondary sources. Hamburger Berichte; Bd. 48; Verlag Abfall aktuell der Ingenieurgruppe RUK GmbH, Stuttgart, ISBN 978-3-9817572-8-6.

Kucuker, M. A., & Kuchta, K. (2012). Biosorption with Algae as a green technology for recovery of rare earth metals from E-waste (pp.4–6). Presented at the International conference on Recycling and Reuse.

Kucuker, M. A., & Kuchta, K. (2018). Biomining–biotechnological systems for the extraction and recovery of metals from secondary sources. *Global Nest Journal, 20*, 737–742.

Kucuker, M. A., Nadal, J. B., & Kuchta, K. (2016). Comparison between batch and continuous reactor systems for biosorption of neodymium (Nd) using microalgae. *International Journal of Plant, Animal and Environmental Sciences, 6*, 197–203.

Kucuker, M. A., Wieczorek, N., Kuchta, K., & Copty, N. K. (2017). Biosorption of neodymium on *Chlorella vulgaris* in aqueous solution obtained from hard disk drive magnets. *PLoS ONE, 12*(4), e0175255.

Kumar, M., Nandi, M., & Pakshirajan, K. (2021). Recent advances in heavy metal recovery from wastewater by biogenic sulfide precipitation. *Journal of environmental management, 278*, 111555.

Kwak, H. W., Yang, Y. S., Kim, M. K., Lee, J. Y., Yun, H., Kim, M. H., & Lee, K. H. (2013). Chromium (VI) adsorption behavior of silk sericin beads. *International Journal of Industrial Entomology, 26*(1), 47–53.

Lee, C. H., Chen, Y. J., Liao, C. H., Popuri, S. R., Tsai, S. L., & Hung, C. E. (2013). Selective leaching process for neodymium recovery from scrap Nd-Fe-B magnet. *Metallurgical and Materials Transactions A, 44*(13), 5825–5833.

Lewis, A. E. (2010). Review of metal sulfide precipitation. *Hydrometallurgy, 104*, 222–234.

Lewis, A., & van Hille, R. (2006). An exploration into the sulphide precipitation method and its effect on metal sulphide removal. *Hydrometallurgy, 81*(3), 197–204.

Li, L., Hu, Q., Zang, J. H., Qi, H. Y., & Zhuang, G. Q. (2011). Resistance and biosorption mechanism of silver ions by Bacillus cereus biomass. *Journal of Environmental Sciences, 23*(1), 108–111.

Li, H., Oraby, E., & Eksteen, J. (2021). Recovery of copper and the deportment of other base metals from alkaline glycine leachates derived from waste printed circuit boards (WPCBs). *Hydrometallurgy, 199*, 105540.

Liu, R., Yang, Z., He, Z., Wu, L., Hu, C., Wu, W., & Qu, J. (2016). Treatment of strongly acidic wastewater with high arsenic concentrations by ferrous sulfide (FeS): Inhibitive effects of S (0)-enriched surfaces. *Chemical Engineering Journal, 304*, 986–992.

Ma, H. W., Liao, X. P., Liu, X., & Shi, B. (2006). Recovery of platinum (IV) and palladium (II) by bayberry tannin immobilized collagen fibre membrane from water solution. *Journal of Membrane Science, 278*(1), 373–380.

Macingova, E., & Luptakova, A. (2012). Recovery of metals from acid mine drainage. *Chemical Engineering, 28*, 109–114.

Manahan, S. E. (1990). *Hazardous waste chemistry, toxicology, and treatment*. CRC Press.

Mao, Y., Hu, H., & Yan, Y. (2011). Biosorption of cesium (I) from aqueous solution by a novel exopolymers secreted from Pseudomonas fluorescens C-2: Equilibrium and kinetic studies. *Journal of Environmental Sciences, 23*(7), 1104–1112.

Marra, A. (2017). Innovative treatments for resource recovery from waste electrical and electronic equipment (PhD thesis), Università degli Studi di Salerno, Italy.

Mata, Y. N., Torres, E., Blazquez, M. L., Ballester, A., González, F. M. J. A., & Munoz, J. A. (2009). Gold (III) biosorption and bioreduction with the brown alga Fucus vesiculosus. *Journal of Hazardous Materials, 166*(2), 612–618.

Michalak, I., Chojnacka, C., & Witek-Krowiak, A. (2013). State of the art for the biosorption process—A review. *Applied Biochemistry and Biotechnology, 170*(6), 1389–1416.

Migdisov, A. A., Williams-Jones, A. E., Lakshtanov, L. Z., & Alekhin, Y. V. (2002). Estimates of the second dissociation constant of H2S from the surface sulfidation of crystalline sulfur. *Geochimica Et Cosmochimica Acta, 66*(10), 1713–1725.

Mokone, T. P., Van Hille, R. P., & Lewis, A. E. (2009). Mechanisms responsible for particle formation during metal sulphide precipitation processes. Water institute of Southern Africa and international mine water association. Pretoria.

Muñoz, A. J., Espínola, F., & Ruiz, E. (2017). Biosorption of Ag (I) from aqueous solutions by Klebsiella sp. 3S1. *Journal of Hazardous Materials, 329*, 166–177.

Oliveira, R. C., & Garcia, O. (2009). Study of biosorption of rare earth metals (La, Nd, Eu, Gd) by Sargassum sp. biomass in batch systems: Physicochemical evaluation of kinetics and adsorption models. In *Advanced materials research* (Vol. 71, pp. 605–608). Trans Tech Publications.

Oliveira, R. C., Jouannin, C., Guibal, E., & Garcia, O. (2011). Samarium (III) and praseodymium (III) biosorption on Sargassum sp.: batch study. *Process biochemistry, 46*(3), 736–744.

Panda, S., Costa, R. B., Shah, S. S., Mishra, S., Bevilaqua, D., & Akcil, A. (2021). Biotechnological trends and market impact on the recovery of rare earth elements from bauxite residue (red mud)–A review. *Resources, Conservation and Recycling, 171*, 105645.

Panda, S., Parhi, P. K., Pradhan, N., Mohapatra, U. B., Sukla, L. B., & Park, K. H. (2012) Extraction of copper from bacterial leach liquor of a low grade chalcopyrite test heap using LIX 984N-C". Hydrometallurgy (Elsevier) (Vol. 121–124, pp. 116–119). https://doi.org/10.1016/j.hydromet.2012.03.008

Pohl, A. (2020). Removal of heavy metal ions from water and wastewaters by sulfur-containing precipitation agents. *Water, Air, and Soil Pollution, 231*(10), 1–17.

Prodromou, M., & Pashalidis, I. (2016). Europium adsorption by non-treated and chemically modified opuntia ficus indica cactus fibres in aqueous solutions. *Desalination and Water Treatment, 57*(11), 5079–5088.

Robalds, A., Naja, G. M., & Klavins, M. (2016). Highlighting inconsistencies regarding metal biosorption. *Journal of Hazardous Materials, 304*, 553–556.

Sahan, M., Kucuker, M. A., Demirel, B., Kuchta, K., & Hursthouse, A. (2019). Determination of metal content of waste mobile phones and estimation of their recovery potential in Turkey. *International Journal of Environmental Research and Public Health, 16*(5), 887.

Sari, A., Mendil, D., Tuzen, M., & Soylak, M. (2009). Biosorption of palladium (II) from aqueous solution by moss (Racomitrium lanuginosum) biomass: Equilibrium, kinetic and thermodynamic studies. *Journal of Hazardous Materials, 162*(2), 874–879.

Sert, S., Kütahyali, C., Inan, S., Talip, Z., Cetinkaya, B., & Eral, M. (2008). Biosorption of lanthanum and cerium from aqueous solutions by Platanus orientalis leaf powder. *Hydrometallurgy, 90*(1), 13–18.

Sethurajan, M. (2016). Metallurgical sludges, bio/leaching and heavy metals recovery (Zn, Cu) (Doctoral dissertation, Paris Est).

Sethurajan, M., van Hullebusch, E. D., Fontana, D., Akcil, A., Deveci, H., Batinic, B., Leal, J. P., Gasche, T. A., Kucuker, M. A., Kuchta, K., Neto, I. F. F., Soares, H. M. V. M., & Neto, A. I. F. (2019). Recent advances on hydrometallurgical recovery of critical and precious elements from end of life electronic wastes-a review. *Critical Reviews in Environmental Science and Technology*, 1–64.

Talaro, K. P., & Talaro, A. (2002). Physical and chemical control of microbes. In *Foundations in Microbiology* (pp. 325–326). U.S. Mineral Commodity Summaries (2015).

Tewari, N., Vasudevan, P., & Guha, B. K. (2005). Study on biosorption of Cr(VI) by Mucor hiemalis. *Biochemical Engineering Journal, 23*(2), 185–192.

Thomas, M., Białecka, B., & Zdebik, D. (2014). Sources of copper ions and selected methods of their removal from wastewater from the printed circuits board production. *Inzynieria Ekologiczna, 37*, 31–49.

Torab-Mostaedi, M., Asadollahzadeh, M., Hemmati, A., & Khosravi, A. (2015). Biosorption of lanthanum and cerium from aqueous solutions by grapefruit peel: Equilibrium, kinetic and thermodynamic studies. *Research on Chemical Intermediates, 41*(2), 559–573.

Tuncuk, A., Stazi, V., Akcil, A., Yazici, E. Y., & Deveci, H. (2012). Aqueous metal recovery techniques from e-scrap: Hydrometallurgy in recycling. *Minerals Engineering, 25*(1), 28–37. https://doi.org/10.1016/j.mineng.2011.09.019

Uçar, D. (2017). Sequential precipitation of heavy metals using sulfide-laden bioreactor effluent in a pH controlled system. *Mineral Processing and Extractive Metallurgy Review, 38*(3), 162–167.

Varsihini, C. J. S., Das, D., Das, N. (2014).Optimization of parameters for Cerium(III) biosorption onto biowaste materials of animal and plant origin using 5-level Box–Behnken design: equilibrium, kinetic, thermodynamic and regeneration studies. *Journal of Rare Earths,32*(8), 745–758.

Vemic, M., Bordas, F., Comte, S., Guibaud, G., Lens, P. N. L., & Van Hullebusch, E. D. (2016). Recovery of molybdenum, nickel and cobalt by precipitation from the acidic leachate of a mineral sludge. *Environmental Technology, 37*(17), 2231–2242.

Vijayaraghavan, K. (2015). Biosorption of lanthanide (praseodymium) using Ulva lactuca: Mechanistic study and application of two, three, four and five parameter isotherm models. *Journal of Environment and Biotechnology Ressearch, 1*(1), 10–17.

Vijayaraghavan, K., & Jegan, J. (2015). Entrapment of brown seaweeds (Turbinaria conoides and Sargassum wightii) in polysulfone matrices for the removal of praseodymium ions from aqueous solutions. *Journal of Rare Earths, 33*(11), 1196–1203.

Vijayaraghavan, K., & Yun, Y. S. (2008). Bacterial biosorbents and biosorption. *Biotechnology Advances, 26*(3), 266–291. https://doi.org/10.1016/j.biotechadv.2008.02.002

Volesky, B. (1987). Biosorbents for metal recovery. *Trends in Biotechnology, 5*(4), 96–101.

Volesky, B. (2003). Biosorption process simulation tools. *Hydrometallurgy, 71*(1), 179–190.

Volesky, B. (2007). Biosorption and me. *Water Research, 41*(18), 4017–4029.

Wang, J., & Chen, C. (2009). Biosorbents for heavy metals removal and their future. *Biotechnology Advances, 27*(2), 195–226.

Wang, L. K., Yung-tse, H., & Shammas, N. K. (Eds.). (2005). *Physicochemical treatment processes* (Vol. 3). Humana Press.

Wang, W., Pranolo, Y., & Cheng, C. Y. (2011). Metallurgical processes for scandium recovery from various resources: A review. *Hydrometallurgy, 108*(1), 100–108.

Zhang, Y., Feng, X., & Jin, B. (2020). An effective separation process of arsenic, lead, and zinc from high arsenic-containing copper smelting ashes by Alkali leaching followed by sulfide precipitation. *Waste Management and Research, 38*(11), 1214–1221.

Zhu, X., Li, W., Tang, S., Zeng, M., Bai, P., & Chen, L. (2017). Selective recovery of vanadium and scandium by ion exchange with D201 and solvent extraction using P507 from hydrochloric acid leaching solution of red mud. *Chemosphere, 175*, 365–372.

Anaerobic Bioreactor Technology (ABT) for the Treatment of Acid Mine Drainage (AMD)

Karumanchi Bhavya, Sameena Begum and Anupoju Gangagni Rao

Abstract

Acid mine drainage (AMD) is considered as a widely spread environmental problem that affects several countries involved in mining activities. Because of its high acidity as well as high metal(loid)s content generating environmental and health toxicity, AMD poses a threat to the surrounding ecosystems. Generally, when exposed to air and water, sulfide minerals undergo oxidative dissolution, which results in formation of AMD. Treatment of AMD at source is regarded to be an effective option; however, this might not be possible at all the sites. Technologies for treating AMD can be governed through the application of various physical, chemical, and biological processes to defuse acidity and remove metal(loid)s from the liquid streams. However, the physicochemical techniques are intended to achieve process viability and cost-effective when the treatment stream is of high volume and sulfate rich. In contrast to this, biological processes are economical to run and do not require a high concentrations of sulfate in the targeted stream. The present chapter critically reviews the state-of-the-art on available aerobic and anaerobic bioreactor technologies with an emphasis on anaerobic bioreactors for the treatment of AMD. In the remediation of AMD, the anaerobic process is a type of biological remediation that relies on neutralizing acidity and precipitating the metal contaminants by natural microbial consortia preferably the sulfate-reducing bacteria (SRB). However, as the AMD is associated with low organic matter, a supply of source of an external factor carbon that is required to complete the remediation process. Anaerobic bioreactors, such as membrane bioreactors, continuous stirred tank reactors, bioelectrochemical systems, up-flow sludge blanket reactors, are suitable bioreactor processes for the treatment of AMD wherein the syntrophic activity of both SRBs and other fermentative and few methane forming bacteria takes place. These anaerobic reactors through the application of SRBs are paving its path in the treatment of AMD because of its efficacy and cost-effectiveness. However, adding of external organic substances are required during the treatment of AMD with SRB which could play a pivotal role in determining the cost of the technology. This chapter describes briefly about the aerobic reactors and detailed information on the different types of anaerobic bioreactors available that can be made suitable for AMD treatment. Comparing the passive and active SRB-based alternatives, their substrate choice, and the recent advances in the anaerobic treatment of AMD along with future perspectives as an alternative to conventional techniques are discussed.

Keywords

AMD · Anaerobic digestion · Mining · Sulfide · SRB · Toxic metal

K. Bhavya · S. Begum · A. Gangagni Rao (✉)
Bioengineering and Environmental Sciences (BEES) Group, Department of Energy and Environmental Engineering (DEEE), CSIR-Indian Institute of Chemical Technology, Hyderabad 500007, India
e-mail: agrao@iict.res.in; gangagnirao@gmail.com

K. Bhavya · A. Gangagni Rao
Academy of Scientific and Innovative Research (AcSIR), Ghaziabad 201002, India

1 Acid Mine Drainage (AMD) and Its Sources of Generation

The commercial exploitation of naturally occurring mineral resources via mining activities has tremendously increased in the past few years because of the escalating need for metals and their allied products. The mining sector contributes

S. Panda et al. (eds.), *Biotechnological Innovations in the Mineral-Metal Industry*, Advances in Science, Technology & Innovation, https://doi.org/10.1007/978-3-031-43625-3_10

significantly to improve the global economy claims the Mining, Minerals, and Sustainable Development Project (MMSD). Around 3500 active mining sites across the world have been reported and the majority of them present waste rock dumps and tailing dams (Tayebi-Khorami et al., 2019). As per the recent estimates, the production of mineral and metal supplies has resulted in 100 billion tons of solid waste generation globally (Tayebi-Khorami et al., 2019).

AMD involves the discharge of acidic water and metal conjugates in and around the mining areas (Roy Chowdhury et al., 2015). The seepage and mixing of ungoverned release of the dissolved metals containing high concentrations of sulfate and acids from abandoned mines and tailing piles into the nearby water bodies and pollutes them. Adverse impacts have been witnessed due to its low pH and high sulfate and metal(loid)s concentration in AMD that are toxic to the aquatic flora and fauna in many of the mining sites across the world (Gontia & Janssen, 2016). Further, long-term exposure of reactive sulfide minerals like the pyrite and pyrrhotite ores to oxygen and water in the lack of adequate neutralizing minerals, results in the weathering of mines and contributes greatly to the formation of AMD (Neculita et al., 2007; Tsukamoto et al., 2004). Acid mine drainage is thus a metal-rich, acidic wastewater, and other toxic substances like sulfuric acid and dissolved iron, generated from a mining site. The process of AMD formation during the mining activities and subsequent natural weathering is shown in Fig. 1. Considering the negative impacts of AMD to the environment, the pollution control bodies have brought up policies for the treatment and storage of mine wastes before releasing into the surroundings.

Additionally, the formation of AMD is prominent in both active as well as abandoned mines in addition to open pit sites. The damage caused in open pits and active sites cannot be evident while they are in fully operational condition, however its long-term operation results in AMD formation. As the water table rises during the constant pumping limits, the atmospheric air enters the mass of the rock leading

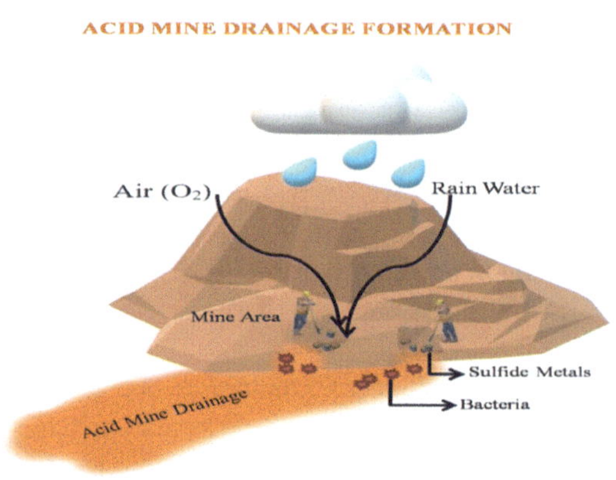

Fig. 1 Acid mine drainage formation due to natural weathering

to the oxidation of metals and other inorganic salts forming iron sulfate salts which can readily suspend in ground water, thus contaminate both the water as well as the ground. The dissolution of sulfate and iron salts in the water often results in the formation of free sulfuric acid thereby making the water more acidic and scarcely rich in organic materials. Due to the formation of a strong acid in the aqueous stream, the pH of the water drops to lower limits to as low as 2, which is extremely dangerous to any living being (Jong & Parry, 2003; Verburg et al., 2009). Due to its acidity, metal toxicity, sedimentation, and other unfavorable characteristics, AMD once developed can have a significant adverse effect on the environment in the vicinity. Thus, before being released into the environment, AMD must be collected and treated to remove metal ions and ensure that its pH is neutral (Neculita et al., 2007; Roy Chowdhury et al., 2015).

1.1 Characteristics of AMD: Sulfide Minerals and pH Profile

Generating metal ions and sulfate rich waste streams are not just limited to mineral and mining activities, many manufacturing processes such as scrubbing of flue gases, processing of paper and pulp, chemical manufacturing streams release effluents similar to that of AMD (Dhir, 2018; Gontia & Janssen, 2016; Rambabu et al., 2020). Sulfide minerals available in AMD are depicted in Fig. 2a. Sulfide mineral rocks such as marcasite, pyrite, and chalcopyrite weather to form AMD, when they come in contact with O_2 from air and H_2O from rain which is an aqueous geochemical process as shown in equation Eq. (1). The main redox reaction in the AMD formation is shown in equation Eq. (2). Among many sulfide minerals, pyrite (FeS_2) and marcasite (FeS_2) are the most common and, abundantly available in nature (Verburg et al., 2009). Under ideal conditions, neutrophilic and acidophilic sulfur-oxidizing bacteria can speed up this oxidation cycle resulting in the release of metal ions, sulfate ions, and sulfuric acid (Demersa et al., 2015; Kadnikov et al., 2019). The ensuing acidic water vigorously dissolves aluminum oxides, and carbonates of different minerals, contributing to pH buffering process leading to instant dissolution of Al, Ca, and with other substances in AMD (Ighalo et al., 2022; Kim et al., 2002).

$$2FeS_2 + 2H_2O + 7O_2 \rightarrow 2Fe^{2+} + 4SO_4{}^{2-} + 4H^+ \quad (1)$$

$$14Fe^{3+} + FeS_2 + 8H_2O \rightarrow 15Fe^{2+} + 2SO_4{}^{2-} + 16H^+ \quad (2)$$

AMD can be broadly classified into 3 categories based on its pH, it may be classified as saline drainage, acid mine drainage, or neutral mine drainage as shown in Fig. 2b. The typical drainage characteristics of acidic mine drainage is

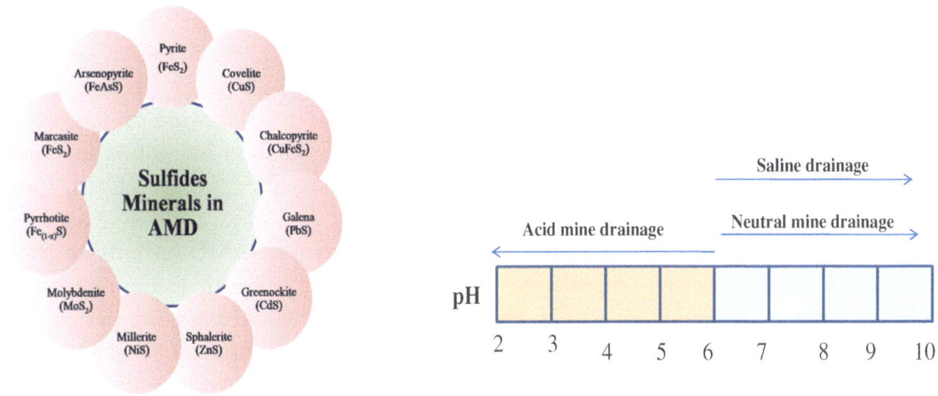

Fig. 2 The typical mineral characteristics of AMD and its classification based on pH

(**a**)Sulfide Minerals in AMD

(**b**)Types of Mine drainages produced by sulfide oxidation in relation with pH (modified from GARD Guide, 2009; http://www.gardguide.com/).

having an acidic pH ranging between 2 and 6 with moderate to elevated metal(loid)s and sulfate concentrations. On the other end, the neutral mine drainage has a near neutral to alkaline pH with low to medium concentration of metals, sulfate concentration. The third category of mine drainage is the saline drainage which has a neutral to alkaline pH with low metals and may have moderate levels of sulfate, iron, manganese, and calcium (Nordstrom et al., 2015).

1.2 Microbial Community in AMD

Despite the extreme acidity, toxicity, and high metal concentration, the AMD ecosystem does not limit the microbial diversity (Chen et al., 2016; Mendez-Garcia et al., 2015). Mineral–microbe interactions are critical in AMD ecosystems, as AMD is a prevalent environmental problem. Microbial activity accelerates acid production and may be accountable for the massive amount of AMD produced (Baker & Banfield, 2003).

The underlying mechanisms of microbial sulfide oxidation and the role of microbes in the amount of AMD formation are now well known (Edwards et al., 2000; Panda et al., 2016; Sheoran et al., 2010). Microorganisms are thought to be responsible for around 75% of the total AMD generated (Edwards et al., 2000). Advances in isolation, culturing, 16S rRNA gene sequencing, and molecular methods have enhanced our understanding of microbial diversity in connection to AMD ecosystems over several decades (Kuang et al., 2012). Microorganisms belonging to the phyla *Proteobacteria, Nitrospira, Actinobacteria, Firmicutes, and Acidobacteria, Aquificae, and Candidate division TM7*, to mention a few, are among the major bacterial lineages found in AMD. *Acidithiobacillus ferrooxidans, Leptospirillum ferrooxidans, Ferrovum spp., Acidiphillum, Acidocella, Acidicaldus, Acidomonas, Metallibacterium scheffleri, Acidithrix ferrooxidans, Ferrimicrobium acidiphilum, Alicyclobacillus spps.*, and other microorganisms have been found in AMD environments (Das et al., 2009). Microorganisms in AMD that live in such harsh environments are naturally evolved to greater potentially toxic metal concentration as well as having the unique capacity to decrease them to less toxic chemical forms.

1.3 Effects of AMD on the Environment

AMD has a vast array of dissolved minerals and metals due to the low pH which promotes the growth of acidophilic bacteria, which have been known of producing acidic waters as a catalyst from sulfide minerals (Gao et al., 2019; Sánchez Espana et al., 2005) by generating sulfuric acid (H_2SO_4) as it is a strong acid that dissociates into H^+ ions and SO_4^{2-}. The high concentration of metals in AMD easily makes their way into the surrounding soils, contaminating them (Gao et al., 2019). Enzymes and microorganisms are hampered by toxicity in the soil, resulting in a decrease in the biochemical characteristics and quality of the soil (Auld et al., 2013; Ferreira et al., 2021). AMD has the ability to contaminate groundwater, rendering it unsafe for use in agriculture, as well as for other applications and human intake. AMD runoff may seriously impact aquatic life; contaminate water sources and lower pH levels (Ighalo et al., 2022; Kaur et al., 2018). AMD also effects biodiversity, disturbs ecosystems, corrodes infrastructure, and water supplying systems are contaminated often in water scarce areas (Jong & Parry, 2003; Rambabu et al., 2020).

Because of the toxicity of AMD may severely affect the ecosystem, in recent decades, there has been an increase in the demand for efficient remediation methods for AMD and its afflicted environment (Gupta & Sar, 2020). AMD that has not been treated and is released into the environment has been shown to have a number of adverse effects on living things (Amanda & Moersidik, 2019; Ighalo et al., 2022). Rhizosphere functions also get inhibited when there is surge in absorbing metal ions like Al^{3+} when bound to cell membrane (Skousen et al., 2017). Plant root dysfunction and soil acidification is triggered by bacterial and fungal activity inside soils. Therefore, movement and absorption of AMD is affected by many factors like soil condition, presence of metallic ions, capacity of dissolved ions solubility, and related micro flora (Skousen et al., 2019). Increased suspended particles, potentially toxic metals mobilization, lower pH in water bodies, and groundwater pollution are all effects of AMD, as are potentially toxic metals penetration into the food chain and absorption by plants and animals, as well as the deterioration of water resources quality (Silva et al., 2013). Human and animal cells can be affected by toxic metals in water, which lowers the ability of cells to survive (Acharya & Kharel, 2020; Dutta et al., 2020). Acidic drainage has different effects depending on the location, past use of land, climate, the size of mining, geochemistry of excess material, and the composition of mine water. Accumulation of potentially toxic metals in soils and water bodies leads to increase in toxic amount of bio-concentration and bio-accumulation in flora, fauna, and humans through food chain and food web. Environmental risks from AMD are "second only to global warming and ozone depletion (Acharya & Kharel, 2020; Moodley et al., 2018)" according to the US Environmental Protection Agency (EPA).

2 Remediation Strategies

Different methods have been developed for the treatment of AMD through many years of research and broadly categorized into two types namely control at source and mitigation techniques (Kaksonen & Sahinkaya, 2012; Rambabu et al., 2020). In source control, the basis of the working principle is on preventing the seepage of oxygenated water by removing the O_2 and H_2O to cease the process of oxidation by sulfide minerals; however, these source control strategies are effective and demanding. Retrofitting is not always achievable as most mines ceased operations before the AMD hazards have been discovered (Ma et al., 2001). Mitigation control approach is based on reducing acidity by neutralizing pH of acid mine water and favoring metal precipitates formation (Yilmaz et al., 2019). They are categorized as active and passive systems (Garcia et al., 2001; Muyzer & Stams, 2008) (Table1). The above-mentioned two treatment techniques are effective in reducing acidity and lowering the concentration of potentially toxic metals (Johnson & Hallberg, 2005; Tsukamoto et al., 2004). However, the advancement in research have modified and re-classified treatment methods into abiotic and biotic methods which are additionally divided into active and passive treatment systems. Biological treatment methods provide many benefits that include the removal of sulfate ions and potentially toxic metals permanently from mine waters, while generating less hazardous water plus the recovery of valuable metals. The aim of these treatment technologies is to decrease the pollutants to permissible limit or to create conditions where they show near neutral or minimal impact on environment which is achieved through biological activity (Kaksonen & Puhakka, 2007; Mendez-Garcia et al., 2015). Sulfate-reducing bacteria (SRB) are mainly used in these biological systems. SRB's are considered as working agents and are therefore, further used in wetlands ecosystems or used as substrate barrier to neutralize pH and reduce metal leachate concentration.

Collection and treatment of waters contaminated by AMD is the major step in mitigating and control strategies used in AMD treatment. This approach mainly comprises of collecting all the mine wastes generated by AMD to be treated. The treatment process can be achieved by chemical or biological approach, by bringing the pH to neutral and eliminating metal precipitates and suspended solids. A conventional strategy for treating AMD comprises using alkali

Table 1 Treatment strategies for remediation of AMD

AMD treatment methods			
Active systems		Passive systems	
Abiotic	Biological (Biotic)	Abiotic	Biological (Biotic)
• Addition of lime for pH neutralization • Aeration for iron oxidation	• Sulfidogenic bioreactors or anaerobic reactors	• Anoxic limestone channels • Open lime stone channels	• Aerobic wetlands • Anaerobic wetlands • Permeable reactive barriers • Algal bioremediation

to decrease acidity, and neutralizing the pH of the water, and precipitate metals like hydroxides and carbonates. Other different treatment methods include ion exchange process, reverse osmosis and electro dialysis, but they are barely selected because of their high operational and maintenance costs. Biological treatment, which is also known as bioremediation involves SRB to treat waters polluted by AMD. SRB are proficient to generate biogenic H_2S. This is then further used to respond with potentially toxic metals, which results in metal sulfide precipitation (Jamil et al., 2013a, 2013b).

Chemical treatment methods are rapidly being replaced by biological treatment approaches for reducing sulfate. Both active and passive treatment technologies are efficient in treating ground and surface waters contaminated by AMD. But due to high operational costs and intensive manpower requirement for maintaining active treatment technologies, passive methods like constructed wetlands, anaerobic sulfate-reducing bioreactors, anoxic limestone channels, open limestone channels, limestone leach filter beds, and slag drain beds (Roy Chowdhury et al., 2015) are widely opted worldwide over active treatments.

2.1 Active Abiotic Technologies

Active abiotic treatment that involves adding a chemical-neutralizing agent to acidic effluents is the most common technique for treating acidic effluents (Coulton et al., 2003).

AMD's pH will increase if an alkaline substance is added to it.

AMD's pH will increase if an alkaline substance is added to it; in solution many metals will precipitate as hydroxides and carbonates, speeding up the rate of chemical oxidation of ferrous iron, which requires high levels of aeration or the addition of an oxidizing chemical like hydrogen peroxide. This results in a Fe-rich sludge that may also have other metals dissolved in it, such as lime and slaked lime, based on the chemistry of the processed mine water. This approach makes use of a variety of neutralizing substances, including lime (calcium oxide), calcium carbonate, sodium carbonate, sodium hydroxide, and magnesium oxide (Dhir, 2018).

The cost and effectiveness of these resources varies. When calcium-containing neutralizing chemicals are employed, sulfate may be partly removed as gypsum. Although active chemical treatment for AMD can be successful, it does come with certain drawbacks like regular maintenance needs for mechanical systems, use of high quantity chemical reagents, man power required for continuous operation and bulk sludge disposal problems (Dhir, 2018).

2.2 Passive Systems: Biotic and Abiotic

Passive biological treatment systems use natural geochemical processes and microbial activity to enhance the influent water condition, by neutralizing the acidity and reducing the potentially toxic elements loads from mine drainage. Although local abiotic variables like dissolved oxygen concentration and water quality may change over the application time potentially affecting rate of (bio)chemical reactions; these systems require minimum management and maintenance (Gazea & Kontopoulos, 1996; Kaksonen & Puhakka, 2007). pH, temperature, salinity, metal concentrations, and other factors all have a part in determining the efficacy of various AMD treatment methods (Ali et al., 2019a, 2019b).

Chemical, physical, and biological techniques are used to treat AMD in these passive treatment systems. The pH is maintained, sulfate and metal concentrations are reduced, and salinity is controlled using this treatment (Tsukamoto et al., 2004). In passive treatment system, microbes play a crucial role. These treatment methods are best for treating low-acidity mine streams. The benefit of a passive system is that it has a high rate of metal removal while requiring less maintenance and consuming less energy compared to active treatment systems (Neculita et al., 2007). The disadvantage is that it requires a lot of foot print area which may be limited in some cases. The following are some examples of passive bioremediation systems: Limestone ponds, Open limestone channels, Anoxic limestone drains (ALD), Aerobic wetlands, Anaerobic wetlands/compost reactors, Permeable reactive barriers (PRB), and Packed bed iron-oxidation bioreactors.

2.2.1 Injection of an Organic Substrate

Infusing rich organic material into mine shafts or boreholes that reach the depths of AMD sites is one of the techniques for in situ remediation. These organic substrates, which supply energy to SRB underground in AMD, might be ammonium phosphate added with organic substrate to stimulate the microbial activity or acetate-bearing compounds (Sahinkaya, 2009). According to the reported literature, mine water running through an area rich in organic substrate shows significant removal efficiency for Al, Cd, Co, and Zn, as well as a pH increase. During high flow rate seasons, however, this impact may be mitigated and precipitated metals could be resolubilized (Skinner & Schutte, 2006).

2.2.2 Permeable Reactive Barriers (PRB)

One more in situ remediation approach is the permeable reactive barrier (PRB), which involves installing a reactive medium perpendicular to the plume of polluted water, which frequently necessitates correct flow trajectory calculation. AMD will passively move across the reactive barrier due to the natural hydraulic gradient, undergoing neutralization and metal precipitation. Furthermore, reactive barriers frequently depend on natural flow to carry AMD through specified treatment zones, resulting in substantially longer processing times and less tractability. System efficiency can be harmed by diminishing the substrates and the mineral precipitation blockage (Jeen & Mattson, 2016).

2.2.3 Anoxic Ponds

To decrease dissolved oxygen and ferric ions (Fe^{3+}), anoxic ponds can be utilized upstream of more delicate treatment systems. Plastic liner put behind a gravel layer works as a gas barrier, preventing metal and acidic stream leakage. While collecting CO_2, apparatus is meant to reduce ambient oxygen intake. This method will generally improve the pH and high metal concentration in the effluent (Skousen et al., 2017).

2.2.4 Wetlands

Wetlands are the most prevalent treatment method, and they've been identified as a cost-effective AMD treatment option (Skinner & Schutte, 2006). Due to the combined impacts of physical, chemical, and biological processes that determine output water quality, wetlands are very complex ecosystems. Aerobic and anaerobic wetlands are the types of wetlands. Rich organic substrates, limestone, and SRB inoculum are submerged in the anaerobic wetland to improve the acidity of metalliferous waters and allow for the reduction of iron and sulfate compounds, whereas aerobic wetlands target on net alkaline waters. Planting vegetation on submerged substrate is a broad topic with a range of potential outcomes. For underlying microbial populations, surface vegetation is advised as a basis of cover and energy (Kaksonen & Sahinkaya, 2012). Surface plants have also been shown to negatively affect SRB performance (Gazea & Kontopoulos, 1996). Alongside these concerns, wetland remediation possibly is not appropriate in arid or semi-arid climatic conditions. Sediments with metal sulfides might be once again oxidized and dissolved, and re-acidified in the treatment area as water levels rise and fall over the seasons (Kaksonen & Puhakka, 2007).

3 Bioreactor Applications in AMD Treatment: Focus on Anaerobic Technologies

A number of reactor configurations have been described in the literature for the biological reduction of sulfate, includes batch reactors, biochemical reactors, sequential batch reactors, anaerobic membrane bioreactors, membrane bioreactors, up-flow anaerobic sludge blanket (UASB) reactors (Kaksonen & Puhakka, 2007). Few of the bioreactors along with their configuration, merits and demerits are described in Table 2.

3.1 Aerobic Bioreactor Technology for the Treatment of AMD

3.1.1 Algal-Based Bioreactors

Algal bioreactors are an attractive bioremediation technique because of its cost-effectiveness and high metal removal efficiency and sulfates. Algal bioremediation is a new and appealing biological method for AMD treatment. A variety of algal strains having been examined for the bioremediation of acidic streams, including *Anabaena*, *Chlamydomonas*, *Chlorella*, *Cladophora*, *Oscillatoria*, *Phaeodactylum*, *Scenedesmus*, *Spirulina* sp., and others (Dean et al., 2019). These algal strains behave as "hypersorbents" and "hyper-accumulators" for numerous metals and elements, exhibiting exceptional selectivity. In addition, the metabolisms of algal biomass generate high alkalinity, which helps to neutralize the acidic character of the drain stream and facilitates metal precipitation. However, because the efficiency of algal treatment method is highly influenced by the pH, oxygen level, and temperature of acidic streams. Therefore, the bioremediation using algae is constantly employed in combination with various treatment strategies. Recent studies reported that the use of macro algae as possible bioindicators for pollution detection and dissemination (Rambabu et al., 2020). Challenges and potential of algae-based bioreactors are algae can be easily grown from oxidation pond, high-rate algal ponds and mining lakes, reduces CO_2 in air and can be made into biofuels like ethanol, biohydrogen, biochar, and many value-added bio products like, antioxidants, vitamins, antimicrobial drugs. As it is an emerging technology more research must be done to know the reliability of fuels produced by algal bioreactors. Algae have some drawbacks, including their unpredictable responses to complex, changing environments, and crucial environmental factors like solar radiation, the availability of nutrients, temperature, and ecological succession are difficult to understand and must be continuously monitored for the process to be successful.

3.1.2 Biochemical Reactor (BCR) System

A BCR is an engineered treatment system comprises of three different reactive zones: oxidative zone, transitional zone, and sulfide zone, as well as a free water zone close to the media as shown in Fig. 3 that uses an organic substrate to promote microbial and chemical processes in

Table. 2 Summary of aerobic and anaerobic reactors used in AMD treatment along with their advantages and disadvantages

Reactor type	Advantage	Disadvantage
Continuous stirred tank reactor (CSTR)	Quick, dependable, and constant equilibrium conditions	Inadequate biomass retention
Anaerobic contact process (ACP)	Superior than CSTR in terms of biomass retention	Sludge and flocks are broken down by biomass circulation
Anaerobic filter reactor (AFR)	Minimal shear forces More time for sludge retention Down flow gravitational mechanism	Rise in pressure gradient
Fluidized-bed reactor (FBR)	Adequate surface area for SRB growth Substantial biomass retention Very slight pressure gradients Recycle flow results in lower influent concentrations	Carrier fluidization requires energy Shear force-induced biomass detaching Less biomass capacity is available than in a UASB reactor
Algal based bioreactors	Algae can be easily grown Valuable byproducts can be obtained	Algae is not stabile and is influenced by environmental factors
Membrane bioreactor (MBR)	Easy to operate and has a higher nitrogen removal rate	Fouling of membranes which may lead to membrane permeability loss
Biochemical reactor system (BCR)	Require low energy, and may have low maintenance	Space may restrict the effective design of a BCR
Anaerobic membrane bioreactor (AnMBR)	Highs solids retention, rejection of high molecular weight organics and less energy consumption	Membrane fouling and effluent nutrient control difficulty
Up-flow anaerobic sludge blanket reactor (UASB)	No flow channeling Sludge is not compacted Zero clogs formed Potential for high treatment rates	Flushing out biomass

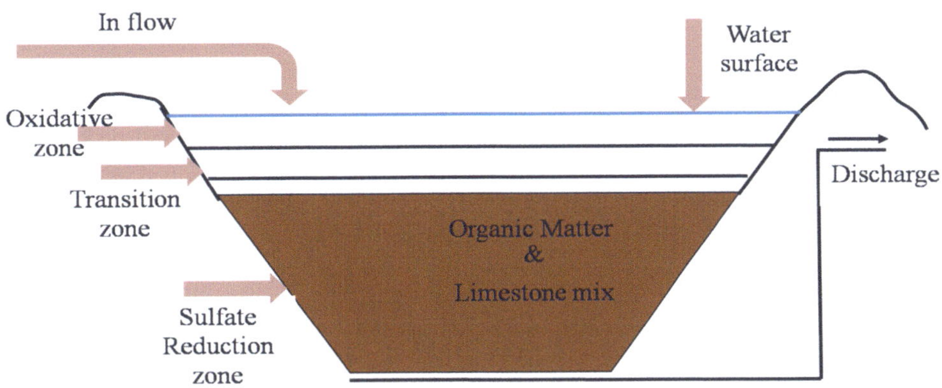

Fig. 3 Schematic diagram representing a biochemical reactor system (Kittrell, 2014)

acid mine water to reduce metal concentrations, acidity, and sulfate. BCRs can be designed in a variety of ways. Pre- and post-treatment units can be used to carry out each process (bioprocesses, chemical reactions, and solid separation) in a separate tank, or they can all be carried out in one unit by the supply of organic materials such as wood chips or manure. Limestone is frequently used with organic substances to provide buffering capacity and substrate permeability (ITRC, 2013). The pH causes the development of metal sulfide solids as alkalinity is added to or created in the BCR chemically. Many metals solubility is reduced when pH rises, and the metals solidify upon precipitation, which are confined in the solid substrate or caught in the downstream sedimentation cells. Sulfate must now be moved to the sulfide state under reducing circumstances. The BCR contains SRBs, cellulose degraders, and fermenters biologically. SRBs rely on cellulose degraders, such as Bacteroids and Clostridium, to breakdown the substrate, which is often a complex carbohydrate, into simpler carbon molecules (Neculita et al.,

2007). Cellulose degraders are able to thrive in both aerobic and anaerobic environments. Fermentative anaerobes will predominate in a BCR for sulfate reduction. The transitional zone is anoxic to slightly anaerobic, with iron oxidation and organic matter degradation due to increased microbial activity. The sulfide zone is anaerobic and extremely reduced, with a high level of microbial activity. Crushed gravel and perforated pipes are commonly used in drainage systems. BCR may use local materials for the substrate, lowering the initial material cost and simplifying the construction process. BCRs are simple to use and maintain, and they don't require any electricity. It has been demonstrated that a BCR can function for years without the need to replace or replenish the organic substrate, which is particularly advantageous given that they are frequently found in isolated locations with restricted access. BCRs require low energy, and may have low maintenance if properly designed. BCRs, on the other hand, might be troublesome for treating AMD since they frequently need pre- and post-treatment and hence are not stand-alone systems. Organics and nutrients may be discharged, and there may be an increase in biological oxygen demand and color in the effluent, causing it to fail to fulfill water quality criteria on a regular basis. Space may restrict the effective design of a BCR. Over time, it will be necessary to replace the organic substrate, and the BCRs permeability will alter (Kittrell, 2014). A BCR can be used at various mining sites like metals and coal mining and can also work in remote sites with limited infrastructure and extreme conditions. A BCR can be applied at variable pH, sulfate, and metals concentration.

3.1.3 Membrane Bioreactor (MBR)

MBRs, which combine biological treatment (Bioreactor) with a micro or ultra-filtration membrane, allow for process acceleration while also producing a consistent, high-quality effluent. To separate treated water from active biomass, MBR's biological reactor technology may be used with both aerobic and anaerobic suspended growth bioreactors. Membrane technology is regarded as the most modern AMD treatment option due to its low chemical requirement, lack of sludge formation, and small scale of operations (Al-Zoubi et al., 2010). pH, feed concentration, permeate flow, and temperature are other factors that affect how well metal and salt are rejected by the membrane-based AMD treatment (Rambabu et al., 2020). The benefits of MBR could be potential reuse of effluent water and it has smaller bioprocess footprint and low sludge yield. A MBR, as shown in Fig. 4, is easy to operate and has a higher nitrogen removal rate than any other treatment methods. Potential limitation could be fouling of membranes which may lead to membrane permeability loss. MBR is replacing traditional clarifiers and can be used for AMD treatment. It possesses high biological oxygen demand (BOD), chemical oxygen demand (COD), and ammoniacal nitrogen removal efficiencies.

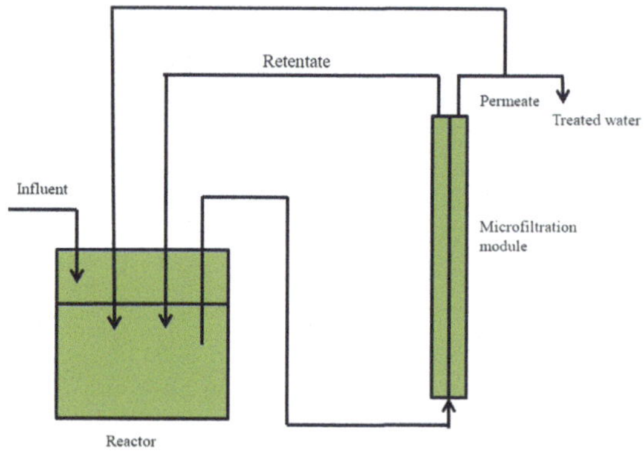

Fig. 4 Schematic diagram of MBR (Barreiros et al., 1998a, 1998b)

3.2 Anaerobic Bioreactor Technology for the Treatment of AMD

Treatment of AMD through anaerobic process for the removal of contaminants is a potential approach as it has a potential to combine SO_4^{2-}, metals, acidity removal in a single reactor with significantly low production of waste sludge in addition to the bioenergy recovery when an external source of organic material is added in excess. Sulfate reduction and potentially toxic metal removal in passive and active systems have been studied over the years, numerous bioreactor types and reactor designs have been used. There are several examples, including the anaerobic contact process (ACP), anaerobic filter (AFR), hybrid reactors, continuous stirred tank reactors (CSTRs), up flow anaerobic sludge blanket reactor (UASB), off-line sulfidogenic bioreactors, and fixed bed reactors (FBRs) (Bartzas et al., 2006; Cruz Viggi et al., 2010; Ňancucheo & Johnson, 2012). Sulfidogenic bioreactors are the active biological systems specially designed for the treatment of sulfate rich wastewaters that have a benefit over passive biological remediation in terms of performance and control, absorption of potentially toxic metals, and reduction in sulfate contents in the treated waters (Bai et al., 2013; Becerra, 2010). Pretreatment of AMD using chemical neutralization, precipitation, and permeable reactive barrier prior to anaerobic treatment enhances the overall process performance.

3.2.1 AMD Treatment in Anaerobic Bioreactors: Mechanism

The solid organic substrate matrix comes into contact with the AMD water moving horizontally or vertically through the reactor (Dhir, 2018; Nordwick et al., 2006) where the complex organic carbon compounds in the AMD are metabolized by the SRB. (Lu et al., 2011).

Microorganisms help in shifting the alkalinity generating processes and in the reduction of complexity of AMD by decreasing the metals and sulfate dissolved concentrations. An acidophilic heterotrophic bacterium which is present in acid mine water plays a key role in AMD treatment. By oxidizing ferrous ions, acidophilic heterotrophic bacteria catalyze the dissimilatory reduction of sulfate to sulfide. When a strong acid is transformed into hydrogen sulfide, alkalinity forms. Under anaerobic circumstances, heterotrophic bacteria such as *Pseudomonas, Clostridium*, and *Desulfovibrio* reduce Mn and Fe by using them as final electron acceptors. Ammonification and denitrification are biologically mediated processes that can help neutralize the AMD. Bacterial species such as *Pseudomonas, Paracoccus, Flavobacterium, Alcaligenes, and Bacillus spp.* support this process. SRBs such as *Desulfovibrio spp.*, use acidic mine water as an electron donor to produce bicarbonate and convert sulfate to sulfide when organic carbon nutrition sources are present. Reduced sulfate forms sulfides, which increase the quantity of bicarbonate that causes alkalinity, produce insoluble metal complexes (Sand et al., 2001) as shown in Eq. (3)

$$2CH_2O + SO_4^{2-} \rightarrow 2HCO_3^- + H_2S \quad (3)$$

Metals in high concentrations are eliminated as hydroxides as a result of precipitation or co-precipitation (Jong & Parry, 2003).

The overall AMD treatment in anaerobic sulfate-reducing bioreactors is shown in Fig. 5. Metals are effectively removed by precipitation when the pH is increased. Metals such as Cu, Zn, Cd, Pb, Ag, and Fe mostly precipitate as metal sulfides as a result of hydrogen sulfide generated during sulfate reduction.

3.2.2 Anaerobic Sequencing Batch Reactor (ASBR)

Anaerobic sequencing batch reactors (ASBRs) are high-throughput anaerobic treatment systems that follow a cyclic process that includes feed, reaction, settling, and decantation (Fig. 6). The initial stage is to introduce the wastewater into the reactor, which is continually mixed with the contents. The amount of substrate supplied is determined by several criteria, such as the target hydraulic retention time (HRT), organic loading rate, and predicted settling characteristics. Because of its improved biological solids retention and process control, ASBRs can be utilized as an alternative to continuous stirred tank reactors for wastewater treatment, resulting in better effluent quality. Furthermore, by properly controlling the cycle duration and discharge operation of batch reactions, effluent regulations may be more readily met when the influencing elements are at adequate levels (Akil & Jayanthi, 2012). In the bioreactor with a sequential design, SRB is the biological agent and the potential pollutants that can be reduced by ASBR are manganese, calcium, magnesium, and other potentially toxic metals. In a contact time of 172 days at a pH 4.5 the removal efficiency of sulfate can be reached to 84.7, 80% of manganese removal, and calcium by 50% and magnesium by 38%. These elements could be precipitated in the form of carbonate or hydroxide in the bioreactor leading to the increased pH of the reactor contents. The system's ability to remove potentially harmful metals is enhanced by the addition of tailing leachate. An ASBR designed for treatment of AMD was evaluated by measuring its capacity to hold the metals concentrations by studying COD and sulfate removal kinetics as reported by Martins Costa et al. (2019) where biological sludge was used for treatment of AMD with high metals concentration.

The microbial population was dominated by *Desulfovibrionaceae sp.* (Gomez et al., 2021; Ighalo et al.,

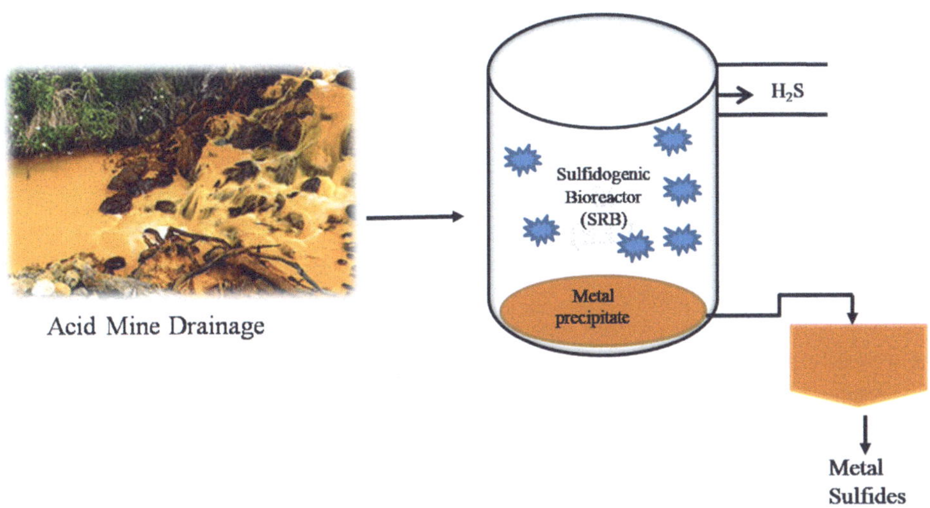

Fig. 5 Anaerobic sulfate-reducing bioreactors

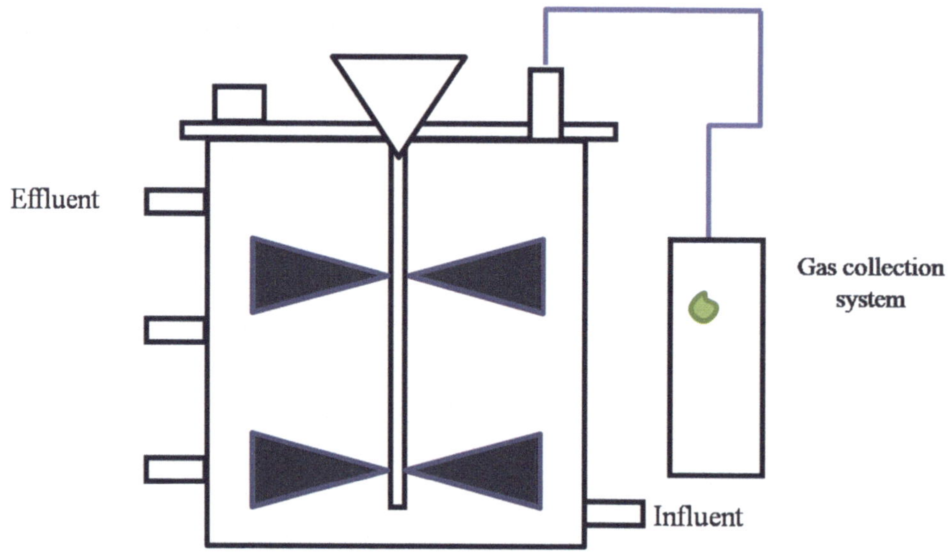

Fig. 6 Schematic diagram of ASBR (Park et al., 2012)

2022). An ASBR requires less space and this system has minimal footprint with high nutrient removal capabilities but one of the potential limitations is the continuous monitoring and maintenance required for the system's steady operation.

3.2.3 Up-Flow Anaerobic Sludge Blanket Reactor (UASB)

Up-flow anaerobic sludge blanket reactors (UASB) are often used to treat domestic sewage and industrial wastewater (Fig. 7). Three-phase separation in the vertical tank mechanism distinguishes this reactor by its simplicity. The basic idea behind the UASB concept is to create circumstances that allow a substantial volume of biological sludge to be retained in the reactor's interior without the requirement for an inert support. These conditions may be obtained by using a three-phase separator, which is linked to various system operating factors such as slow outflow rates and the generation and maintenance of good sedimentation characteristic granules. As a result, the HRT of the reactor may be separated from the sludge biomass retention time (Rodriguezet al., 2012). UASB reactor includes 2 zones, a reactor zone and a settling zone. Granular sludge, which has great mechanical strength and superior settling qualities and is resistant to toxic shocks, is what sets UASB apart from other anaerobic reactors and with high methanogenic properties. UASB is the most commonly used in industrial wastewater treatment and is highly efficient with less space requirement and less energy consumption with less sludge production. It is associate with less operating costs as well as efficient in achieving 65–75% of COD removal. Challenges with UASB would be low pathogen and nutrient removal, odor problems and long start up. Application of UASB for AMD treatment was reported by Leal-Gutierrez et al. (2021) for converting sulfate into sulfide and to determine the effect of pH and COD: SO_4^{2-} ratio. It was concluded that UASB system achieved 69% of sulfate to sulfide bioconversion.

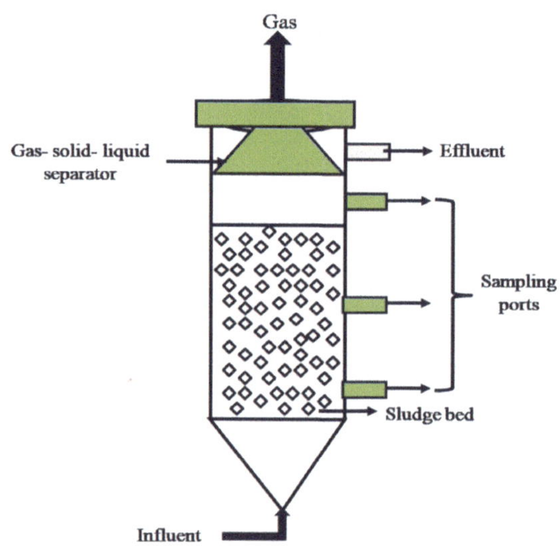

Fig. 7 Schematic diagram of UASBR

3.2.4 Anaerobic Membrane Bioreactor (AnMBR)

An anaerobic membrane bioreactor (AnMBR) is the application of membrane filtration process for the treatment of wastewater without exposing it to air/oxygen (Fig. 8). AnMBR consists of 2 parts, a sludge bed and the supernatant in which a hollow fiber membrane will be submerged as they work on a similar principle of aerobic and membrane reactors, but they leverage the advantage of the benefits of anaerobic degradation. When compared to their aerobic equivalents, AnMBR can treat wastewater without aeration, produce biogas for energy purposes and substantially less biosolids (Uman et al., 2021). Many of the

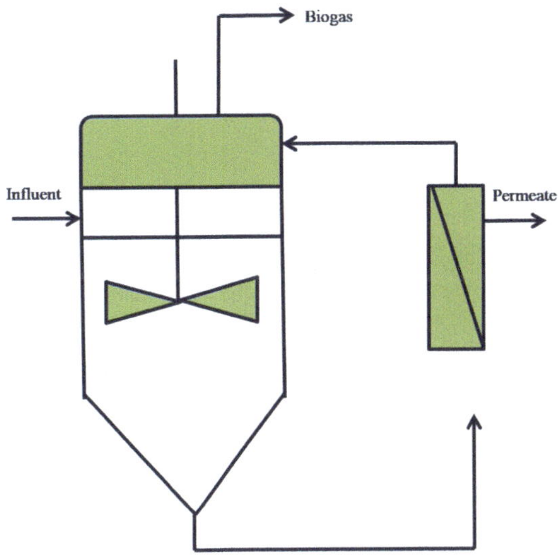

Fig. 8 Schematic diagram of AnMBR (Chang, 2014)

standards may be satisfied by AnMBR. However, issues including membrane fouling, dissolved methane recovery, and management of effluent nutrients must be addressed. Membrane fouling is a primary issue among them, since research suggests that present fouling mitigation strategies constitute a major energy demand for AnMBR (Gong et al., 2019). Low energy consumption, high solids retention, and rejection of high molecular weight organics are the few benefits of AnMBR. The drawbacks of this system include membrane fouling and the challenge of achieving efficient membrane scouring. AnMBR is appropriate for the treatment of both industrial and municipal wastewaters. AnMBR for the treatment of AMD was evaluated by Sahinkaya et al. (2019) where sulfate and COD concentrations of 1500 and 1000 mg L^{-1} at pH4 were maintained. High COD and sulfate removal efficiency of 95% was reported even at low COD/sulfate ratio. Over 99% of iron, copper, zinc, nickel was removed because of metal sulfide precipitation.

3.2.5 Bioelectrochemical Treatment System (BES)

Bioelectrochemical systems (BES) have emerged as an intriguing technology in terms of wastewater treatment and energy consumption in recent years. BES is based on the metabolic processes of exoelectrogenic microorganisms that can catalyze electrochemical reactions on electrode surfaces of electrochemical cells (Ren, 2013). Bioelectrochemical systems have been built in a variety of configurations which has a cathode and anode chambers with anion exchange membrane separating the two chambers. Exoelectrogenic bacteria oxidize the substrate in the anode, and the electrons are released to the electrode. The electrons are used in the cathode to carry out a reduction process, such as converting O_2 to H_2O, protons to hydrogen (H_2) gas, or reducing other chemical compounds to less refractory forms. In this regard, the possibility of recovering dispersed metal ions in their elemental form by reducing them cathodically in their oxidized elemental form that can be retrieved is quite interesting (Kim et al., 2015). Electrical energy may be acquired from the electrical circuit in a microbial fuel cell (MFC) based on the thermodynamic energy balance, however electrical energy must be provided by a power source in a microbial electrolysis cell (MEC) (Ghangrekar & Chatterjee, 2017). BES has many advantages similar to that of microbial fuel cells (MFC); microbial electrical cells (MEC) as they are primarily meant for waste water treatment for pollutants removal with simultaneous power generation. AMD can also be treated using BES; however, an external organic carbon source is required as the AMD is deprived of organic material and rich in inorganic compounds. Example of metal recovery from AMD using BES was reported by Lefebvre et al. (2022) where iron was removed from AMD by increasing the pH. For the treatment of AMD with high ferrous iron content, a proton exchange membrane MFC was effective (Fe^{3+} is reduced to Fe^{2+}).

3.2.6 Anaerobic Sulfate-Reducing Bioreactors: Active Biotic Systems

Anaerobic digestion (AD) is one of the promising biological processes for the treatment and stabilization of solid and liquid wastes. The wastewaters rich in organic material are amenable for AD, however, treatment of AMD via AD process could also prove beneficial as the anaerobic reactors contain mixed microbial consortia which also include SRBs, methanogens, acidogens, and so on. As the environmental conditions required for the growth of SRBs and methanogens are same, SRBs and methanogens compete for organic matter but SRBs convert sulfate to sulfide whereas methanogens convert the organic matter to biogas which is a mixture of methane and carbon dioxide. The predominance of SRBs in anaerobic reactors is high when the wastewaters rich in sulfates are treated. Therefore, anaerobic sulfates reducing bioreactors is a promising approach for AMD remediation with a potential to combine SO_4^{2-}, metals in a single reactor with significantly low waste sludge generation in addition to the bioenergy recovery. SRBs are used in the biological process of anaerobic treatment for sulfate-rich effluents like AMD. Because they are heterotrophic bacteria, SRB need organic matter to serve as electron donors for sulfate reduction. Electron donors may be from complex carbon molecules (Skinner & Schutte, 2006). Potentially toxic metals create insoluble compounds with biogenic sulfides, causing them to precipitate and to be removed as sulfides (Panda et al., 2016). SRB can thrive in pH ranges of 5–9 and shows high activity in this range therefore the AMD pH shall be corrected before being treated in anaerobic processes. In these reactors, a thick layer of organic-rich materials combined with limestone forms the foundation of anaerobic sulfate-reducing bioreactors. Under the organic layer, a thin coating of limestone is used to provide extra alkalinity while also supporting the underlying drainage channels. The AMD is released into the drainage system after passing through the organic layer and limestone bed vertically. SRBs feed on the organic layer and convert SO_4^{2-} to H_2S and oxidize organic matter (CH_2O) to bicarbonate ions (HCO_3^-) (Li et al., 2018) as shown in Eq. (4). The energy generated in this process is used by sulfate-reducing bacteria to grow and develop.

$$2CH_2O(aq) + SO_4^{2-} + H^+ \rightarrow H_2S + 2HCO_3^- \quad (4)$$

The bicarbonates (HCO_3^-) generated subsequently react with hydrogen (H^+) ions to form CO_2 and water (H_2O). As a result of the consumption of H^+ ions, the pH of AMD water rises. Metal sulfides, oxides, hydroxides, and carbonates begin to precipitate at high pH levels. Metal sulfide precipitation is the most prevalent type in anaerobic sulfate-reducing system (Waybrant et al., 2002). Sulfate-reducing bioreactors therefore aid in the reduction of acidity, metal toxicity, and sulfate content in AMD water, and also enhance overall water quality.

Maintenance of suitable biochemical environment favors the remediation process by the SRBs resulting in the precipitation of dissolved metals and their immobilization as sulfides. Sulfate, anaerobic conditions, and the availability of organic carbon all contribute to this type of environment. Most metals may be successfully removed from mine waters if such conditions are created within a reactive barrier or field-bioreactor (Santos et al., 2015). Sulfate-reducing passive bioreactors have recently gained a lot of interest as a viable technology for AMD treatment. They have several benefits, including high metal removal at low pH, stable sludge, cheap operating costs, and low energy usage. The intended method of pollutant removal is sulfide precipitation; however, in passive bioreactors, several other processes, such as sulfate-reducing passive bioreactors depend on the activity of an anaerobic micro flora, including SRB, which is primarily controlled by the reactive mixture composition, their efficiency is occasionally limited by the adsorption and precipitation of metal carbonates and hydroxides. The source of organic carbon is the most important component in the mixture. (Nordwick et al., 2006).

SRB use organic carbon to reduce sulfate while also producing biogenic hydrogen sulfide (H_2S) and alkalinity.

This causes heavy metal accumulation in AMD, as well as a rise in pH and alkalinity. As described in the equations below, biogenic H_2S reacts with metallic ions (Me^{2+}) present in AMD to create metal sulfides (MeS), whereas hydroxide ions (HCO_3^-) combine with protons (H^+) to neutralize acidic waters (Kaksonen & Puhakka, 2007) Eqs. (5, 6)

$$Me^{2+} + HS^- \rightarrow MeS + H^+ \quad (5)$$

$$HCO_3^- + H^+ \rightarrow CO_2 + H_2O \quad (6)$$

A reduction in sulfate content and potentially toxic metals, as well as an improvement in pH and alkalinity, are predicted in the resultant sulfate-reducing environment.

Metabolism of SRB for Sulfate Reduction

Dissimilatory Sulfate Reduction Pathway

There are two types of biological sulfur reduction: assimilatory and dissimilatory. The bacteria, algae, fungi, and plants all take up the SO_4^{2-} ion in the assimilatory reduction process, where it is lowered and fixed in sulfur-containing amino acids like cysteine and methionine inside cells. The dissimilatory reduction happens in two steps. The first involves using sulfate as the final acceptor in the electron transport system under anaerobic circumstances, converting sulfate to sulfide. The second stage is the oxidation of sulfide to elemental sulfur with the help of sulfur-oxidizing bacteria (SOB), which may be used as a fertilizer or a substrate in bioleaching processes (Janssen et al., 1999). SRB treatment for AMD has a low operating cost. The conversion of sulfate ions to sulfide by SRB under anaerobic conditions is used in this approach (Costa et al., 2020).

The most prominent way of sulfate reduction is via the dissimilatory sulfate reduction pathway. This is a metabolic pathway occurring in sulfur reducing bacteria (SRB). The most stable form of sulfur, which is the sulfate, is first reduced to sulfite followed by reduction to sulfide. The microbial cell initially uptakes the sulfate containing compounds. The ATP present inside the cell activates the sulfate and converts it into an intermediate product called Adenosine-5'-phosphosulphate (APS) along with release of two inorganic phosphates. This first step is catalyzed by sulfate adenylyl transferase (Sat) enzyme. The second step is the conversion of APS into sulfite with adenosine monophosphate (AMP) as the byproduct with APS reductase enzyme (AprBA) being the catalyst. The third step is the most crucial step of the pathway where the sulfur atom present in sulfite forms a complex trisulfide bond with reduced DsrC protein. DsrC is a protein with a highly conserved C-terminal arm containing two cysteines which are separated by 11 amino acid residues. This DsrC acts as a substrate for the reaction catalyzing heterodimer protein complex, DsrAB. The C-terminal arm of DsrC inserts itself into a cleft between DsrA and DsrB proteins which is near to the substrate binding site of DsrAB complex. Cysteine of DsrC in their reduced sulphydryl (R-SH) form will get

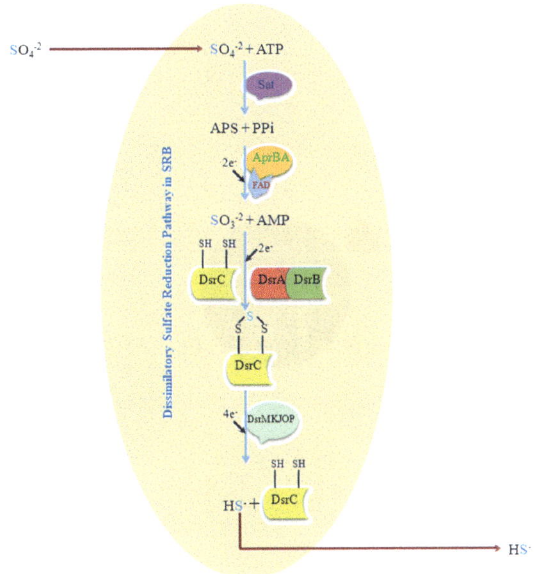

Fig. 9 Dissimilatory sulfate reduction pathway

oxidized and binds with the S atom of sulfite forming a trisulfide-protein complex (Fig. 9). In the fourth and final step, the trisulfide is reduced to sulfide and released from the cell with the DsrC protein being restored. The final step is catalyzed by a membrane protein complex called DsrMKJOP. The whole reduction process requires eight electrons out of which two are needed during APS reduction, another two required for trisulfide-protein complex formation, and last four in the final step of sulfite to sulfide conversion (Santos et al., 2015). The reactions required for sulfate reduction by SRB metabolism are listed below (Eqs. 7, 8, 9) (Xingyu et al., 2013).

$$SO_4^{2-} + AMP^{4-} + H^+ \rightarrow APS^{2-} + HP_2O_7^{3-} \quad (7)$$

$$APS^{2-} + H^+ + 2e^{2-} \rightarrow HSO_3^- + AMP^{2-} \quad (8)$$

$$HSO_3^- + 6H^+ + 6e^{2-} \rightarrow HS^- + H_2O \left(M^{2+} - \text{Metal cation}\right) \quad (9)$$

3.3 Critical Performance Indicators of Anaerobic Technologies

3.3.1 PH

pH is one of the important process performance indicators that shift the reaction pathway from one to another with slight changes. The anaerobic treatment of wastewaters is carried out at neutral or weak acidic environments. The optimal pH range of 5 to 6 is required for sulfate reduction and at this range optimum H_2S generation occurs (Broco et al., 2005).

3.3.2 Organic Substrates for Treatment of AMD: Direct versus Indirect Substrate

The major limitation of the biochemical reaction by the SRB is the accessibility of carbon sources. The carbon supply in AMD water is limited, requiring extra or external carbon sources for treatment to be successful (Kolmert et al., 2000). Sulfate reduction is a high-energy intensive process that necessitates a large volume of high-energy reductant (Martins et al., 2009). As a result, the effectiveness and cost viability of bioremediation technology will be influenced by the carbon sources used. The composition of organic material needs to be studied since it influences the efficacy of SRB eco-technology. While functioning as a readily available carbon source, a substrate must be able to establish a proper low redox environment.

Microbial communities are more resilient and sustainable when made up of a variety of readily biodegradable materials and organic carbon sources (Neculita et al., 2007; Sheoranet al., 2010). A crucial component in the development of the substrate for sulfate-reducing bioreactors is organic material. Such products may be purchased for a lower price or for no price at all as they are frequently regarded as waste items. The only expense may be incurred during the transportation to treatment site (Gusek, 2004).

SRB prefers simple organic substrates as a food source, which can be provided directly or indirectly. Maple wood chips, sphagnum peat moss, leaf compost, conifer compost, chicken manure, and conifer sawdust are all examples of indirect organic substrates that can be used (Jamil et al., 2013a, 2013b).

Direct organic substrate sources that do not need to be degraded before being consumed by SRB are alcohols, organic acids, and sugars. In the meantime, indirect organic sources such as organic compost, wood or paper waste, and food manufacturing byproducts must be further degraded in order to generate the required output. Indirect substrate will be more suitable because mining operations are located far from urban areas. In long-term conditions, indirect substrate will be more suited than direct organic substrate since mining sites are located far from metropolitan areas. Even while basic substrates have the benefit of allowing SRB to utilize energy sources rapidly and directly, they are quickly depleted. Indirectly, substrates must be supplied into the system on a continual basis, raising operating, and maintenance costs (Hiibel et al., 2011; Jamil et al., 2013a, 2013b). Effective reactive mixes have an organic carbon source (different organic/cellulosic wastes), a bacterial source or SRB inoculum (river sediment/animal manure), a solid porous medium (gravel/sand), a nitrogen source (urea), and a neutralizer (limestone) (Dhir, 2018). The reactive mixture's composition predominantly controls the activity of SRB, determines the effectiveness of passive bioreactors. Microbial communities are more likely to be long-lived and sustainable if they are made up of a combination of rapidly degradable materials and different organic carbon sources (Jamil et al., 2013a, 2013b; Nordwick et al., 2006).

3.3.3 Hydraulic Retention Time

For direct organic substrate, a hydraulic retention time (HRT) of 3–5 days is necessary for the precipitation of sulfide metals, whereas indirect organic substrate requires an HRT of 7–10 days for adequate microbial development (Gonçalves et al., 2007).

3.3.4 Temperature

The effect of temperature on SRB growth and sulfate reduction kinetics is significant. SRB can live in both mesophilic and thermophilic environments (Tassé et al., 2003). The ability of sulfate-reducing bacteria to degrade organic complex substrate to a simple form is also influenced by the activity of other anaerobic bacteria. Methanogens, on the other hand, are sensitive to low temperatures, requiring mesophilic environments to develop.

SRBs are found to thrive at temperatures ranging from 0 to 80 ºC. Despite the fact that SRB are active in arctic habitats (at temperatures below 5 °C). Low temperatures reduce the effectiveness of passive AMD treatment by lowering the biogeochemical activity. A study reported by Ben Ali et al. (2019a, 2019b) for the treatment of AMD in arctic conditions in their review, for example preliminary laboratory testing of a synthetic AMD in PBR columns at 4 °C versus 25 °C revealed a direct and substantial influence on how a temperature has an impact on microbial activity drop, as well as Cd, Zn, and sulfate removal (Kawaja et al., 2006).

3.3.5 Solid Support

Sand, gravel, and glass beds (Choudhary & Sheoran, 2011) are used as solid supports for SRB and may have beneficial impacts on bacterial growth because of their large pore size, low surface area, and big volume, as well as enhancing metal precipitation. To prevent clogging in the bioreactor, it is desirable to use a solid support with a high pore size, low surface area, and a large volume of solids (Jamil et al., 2013a, 2013b).

3.3.6 Inhibitory Effect

In AMD, high amounts of metallic ions including iron, zinc, copper, and manganese can prevent SRB from growing. 2–50 mg Cu/L, 13–40 mg Zn/L, 75–125 mg Pb/L, 4–54 mg Cd/L, 10–20 mg Ni/L, 60 mg Cr/L, 74 mg Hg/L are the lethal ranges for SRB populations (Tang et al., 2009). These ranges may change depending on the species of SRB that are available (Fig. 10).

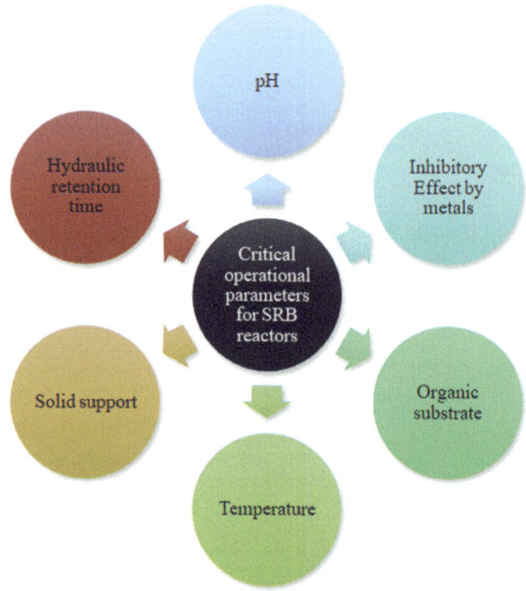

Fig. 10 Sulfidogenic bioreactor parameters

4 Present State of Art and Future Perspective

Due to variety of federal and state laws, commercial and government entities are required to develop various AMD treatments or control technologies (Skinner & Schutte, 2006). AMD pollutes the ecosystem, thus avoiding AMD development or migration from its source is generally thought to be the best solution. According to research, bioremediation of AMD using sulfate-reducing bacteria has caught the attention of numerous researchers. Previously, research topics on sulfidogenic bioreactors were focused on substrates with liquids, such as lactate and ethanol, solid substrate materials, on the other hand, having the capacity to be an effective supply of substrate for systems using sulfate-reducing bacteria. The way that SRB activity occurs when a solid substrate material is used, on the other hand, is poorly understood. The factor that restricts the rate of sulfate reduction by SRB is the breakdown of complex organic matter. The system design and location of the biological treatment plant, the profitability of metal recovery, the choice of substrate, and the discharge criteria are only a few of the variables that affect a biological treatment plant's overall operating costs. Finding suitable low-cost substrate substitutes, such as organic solid waste and food waste byproducts, may boost the implementation of SRB technology (Jamil et al., 2013a, 2013b). As biological treatment of AMD can be divided it into two separate components—treatment of AMD and prevention of leaching and remediating the AMD contaminated environment. The soil and water bodies are the most commonly polluted by AMD. Bioreactors and other types of wetlands being reported as having succeeded to treat AMD prior to being released it into the surroundings or to minimize leaching. Algal remediation, microbiological remediation, and wetland remediation are the three most commonly reported strategies for effectively treating AMD polluted aquatic environments (Ighalo et al., 2022).

Because there is no single dependable approach for treating AMD, researchers from all over the world have been working to develop effective and beneficial strategies for dealing with acidic mine effluents. According to recent AMD research, biological treatment technologies are particularly promising since they avoid the issues of high operational costs and sludge disposal that chemical treatment systems have. For a number of acidic effluents with various concentrations and other aqueous characteristics, bioremediation procedures are very simple and viable to execute. According to recent research along with anaerobic bioreactors permeable reactive barriers, microbial and algae-based bioremediation and wetlands has uncovered new avenues and possibilities for AMD treatment (Ighalo et al., 2022). Developing prediction models for mapping algorithms that correlate microbiological characteristics with the chemical composition of mining sites, as well as the creation of mineral-specific AMD treatment methods, are two areas of future research scope. Bio-sorbed or precipitated metal sulfides can be used to successfully recover metal, is also an area of research that can be explored using contemporary biological techniques (Rambabu et al., 2020).

5 Conclusion

AMD, as highlighted in this chapter, is a global problem that harms the state of environment and subsequently, health of humans. Conventional methods of treating AMD may not always provide the desired degree of sulfate reduction, and waste disposal needs extra landfill area. As a result, more efficient and sustainable methods must be developed in order to recycle and utilize the trash created. Among the available methods, this evaluation emphasized the treatment of AMD with SRB, a low-cost, highly efficient option showing potential for resource recovery. The SRB converts sulfate ions to hydrogen sulfide by dissimilatory metabolism, which combines with metallic ions to produce metal sulfide precipitation. Because of their efficacy and cost-effectiveness, anaerobic bioreactors are a potential approach for treating AMD contaminated water. There are a number of bioreactors that uses Anaerobic Bioreactor Technology that can be employed to treat AMD which are discussed in this chapter. The novel reactor configurations in this chapter would be a combination of membrane technology and anaerobic process—Anaerobic membrane bioreactor (AnMBR); Bioelectrochemical systems (BES). However, the effectiveness of the system is determined by the SRB's activity, which in turn determined by the reactive mixture and organic carbon supply available. These anaerobic bioreactors are capable of operating at pH 5.0 and temperatures ranging from 2 to 68 °C. These characteristics make this a method that can be used on a large scale, even at pilot scale. In addition to the reactive mixture's composition and the presence of SRB, the anaerobic bioreactor's overall productivity and long-term operation are also influenced by the reactor's design, Eh, hydraulic retention time, and COD/sulfate ratios. The use of Anaerobic Bioreactor Technology to control AMD has been widely developed to limit the negative consequences. Transportation expenses for liming materials and; the size and terrain of the accessible area; sludge disposal or waste stream generation if poorly managed; as well as labor and maintenance costs, are all important considerations. Importantly, the elements described above should be examined as a function of one another rather than being analyzed separately. As can be observed from current trends, optimizing highly efficient bioreactors uses significantly less space, allowing for a reduction in land requirements. The emphasis should be on improving overall process design that incorporates life cycle evaluation. Furthermore, AMD remediation may also be viewed as a way to capitalize on the extraction of renewable raw materials, including metal recovery via bio-treatment techniques, may yield a strong economic benefit along with waste treatment.

Acknowledgements The authors are thankful to the Director, CSIR-Indian Institute of Chemical Technology (IICT) for the support to carry out this research work (IICT/Pubs./2021/241). The authors are grateful to the Academy of Scientific and Innovative Research (AcSIR). Also, sincere thanks to the Department of Science and Technology (DST) for providing the DST-INSPIRE fellowship to Ms. Bhavya Chowdary Karumanchi (IF190579).

References

Acharya, B. S., & Kharel, G. (2020). Acid mine drainage from coal mining in the United States—An overview. *Journal of Hydrology*, *588*. https://doi.org/10.1016/j.jhydrol.2020.125061, PubMed: 125061.

Akil, K., & Jayanthi, S. (2012). Anaerobic sequencing batch reactors and its influencing factors: An overview. *Journal of Environmental Science and Engineering*, *54*(2), (317–322).

Al-Zoubi, H., Rieger, A., Steinberger, P., Pelz, W., Haseneder, R., & Härtel, G. (2010). Optimization study for treatment of acid mine drainage using membrane technology. *Separation Science and Technology*, *45*(14), 2004–2016. https://doi.org/10.1080/01496395.2010.480963

Amanda, N., & Moersidik, S. S. (2019). Characterization of sludge generated from acid mine drainage treatment plants. *Journal of Physics: Conference Series, 1351*(1), 012113. https://doi.org/10.1088/1742-6596/1351/1/012113

Auld, R. R., Myre, M., Mykytczuk, N. C., Leduc, L. G., & Merritt, T. J. (2013). Characterization of the microbial acid mine drainage microbial community using culturing and direct sequencing techniques. *Journal of Microbiological Methods, 93*(2), 108–115. https://doi.org/10.1016/j.mimet.2013.01.023

Bai, H., Kang, Y., Quan, H., Han, Y., Sun, J., & Feng, Y. (2013). Treatment of AMD by SRB with iron in *bench scale runs. Bioresource Technology, 128*, 818–822. https://doi.org/10.1016/j.biortech.2012.10.070

Baker, B. J., & Banfield, J. F. (2003). Microbial communities in acid mine drainage. *FEMS Microbiology Ecology, 44*(2), 139–152. https://doi.org/10.1016/S0168-6496(03)00028-X

Barreiros, A. M., et al. (1998a). Membrane bioreactor for drinking water denitrification. *Bioprocess Engineering, 18*(4), 297. Available at: https://doi.org/10.1007/s004490050445

Barreiros, A. M., Rodrigues, C. M., Crespo, J. P., & Reis, M. A. (1998b). Membrane bioreactor for drinking water denitrification. *Bioprocess Engineering, 18*(4), 297–302. https://doi.org/10.1007/s004490050445

Bartzas, G., Komnitsas, K., & Paspaliaris, I. (2006). Laboratory evaluation of Fe0 barrier*s to treat acidic leachates. Minerals Engineering, 19*(5), 505–514. https://doi.org/10.1016/j.mineng.2005.09.032

Becerra, C. A. (2010). *Role of sulfate-reducing bacteria in the attenuation of AMD through sulfate and iron reduction*. University of Massachusetts—Amherst.

Ben Ali, H. E. B., Neculita, C. M., Molson, J. W., Maqsoud, A., & Zagury, G. J. (2019a). Efficiency of batch biochemical reactors for mine drainage treatment at low temperature and high salinity. *Applied Geochemistry, 103*, 40–49. https://doi.org/10.1016/j.apgeochem.2019.01.014

Ben Ali, H. E., Neculita, C. M., Molson, J. W., Maqsoud, A., & Zagury, G. J. (2019b). Performance of passive systems for mine drainage treatment at low temperature and high salinity: A review. *Minerals Engineering, 134*, 325–344. https://doi.org/10.1016/j.mineng.2019.02.010

Broco, M., Rousset, M., Oliveira, S., & Rodrigues-Pousada, C. (2005). Deletion of flavoredoxin gene in *Desulfovibriogigas* reveals its participation in thiosulfate reduction. *FEBS Letters, 579*(21), 4803–4807. https://doi.org/10.1016/j.febslet.2005.07.044

Chang, S. (2014). Anaerobic membrane bioreactors (anmbr) for wastewater treatment. *Advances in Chemical Engineering and Science, 04*(01), 56–61. https://doi.org/10.4236/aces.2014.41008

Chen, L. X., Huang, L. N., Méndez-García, C., Kuang, J. L., Hua, Z. S., Liu, J., Shu, W. S., Xing, L., et al. (2016). Microbial communities, processes and functions in acid mine drainage ecosystems. *Current Opinion in Biotechnology, 38*, 150–158. https://doi.org/10.1016/j.copbio.2016.01.013

Choudhary, R. P., & Sheoran, A. S. (2011). Comparative study of cellulose waste versus organic waste as substrate in a sulfate reducing bioreactor. *Bioresource Technology, 102*(6), 4319–4324. https://doi.org/10.1016/j.biortech.2010.08.126

Costa, J. M., Castro, K. C., Rodriguez, R. P., & Sancinetti, G. P. (2020). Anaerobic reactors for the treatment of sulphate and metal-rich wastewater: A review. *International Journal of Environmental Analytical Chemistry*, 1–12. https://doi.org/10.1080/03067319.2020.1728261

Coulton, R., Bullen, C., & Hallett, C. (2003). The design and optimisation of active mine water treatment plants. *Land Contamination and Reclamation, 11*(2), 273–279. https://doi.org/10.2462/09670513.825

Cruz Viggi, C. C. C., Pagnanelli, F., Cibati, A., Uccelletti, D., Palleschi, C., & Toro, L. (2010). Biotreatment and bioassessment of heavy metal removal by sulphate reducing bacteria in fixed bed reactors. *Water Research, 44*(1), 151–158. https://doi.org/10.1016/j.watres.2009.09.013

Das, B. K., Roy, A., Koschorreck, M., Mandal, S. M., Wendt-Potthoff, K., & Bhattacharya, J. (2009). Occurrence and role of algae and fungi in AMD environment with special reference to metals and sulfate immobilization. *Water Research, 43*(4), 883–894. https://doi.org/10.1016/j.watres.2008.11.046

Dean, A. P., Hartley, A., McIntosh, O. A., Smith, A., Feord, H. K., Holmberg, N. H., & Pittman, J. K. (2019). Metabolic adaptation of a *Chlamydomonasacidophila* strain isolated from acid mine drainage ponds with low eukaryotic diversity. *Science of the Total Environment, 647*, 75–87. https://doi.org/10.1016/j.scitotenv.2018.07.445

Demers, I., Benzaazoua, M., Mbonimpa, M., Bouda, M., Bois, D., & Gagnon, M. (2015). Valorisation of acid mine drainage treatment sludge as remediation component to control acid generation from mine wastes, part 1: Material characterization and laboratory kinetic testing. *Minerals Engineering, 76*, 109–116. https://doi.org/10.1016/j.mineng.2014.10.015

Dhir, B. (2018). Biotechnological tools for remediation of AMD (removal of metals from wastewater and leachate). In *Biogeotechnologies for minesiterehabilitation* (pp. 67–82). Elsevier.

Dutta, M., Islam, N., Rabha, S., Narzary, B., Bordoloi, M., Saikia, D., & Saikia, B. K. (2020). Acid mine drainage in an Indian high-sulfur coal mining area: Cytotoxicity assay and remediation study. *Journal of Hazardous Materials, 389*, 121851. https://doi.org/10.1016/j.jhazmat.2019.121851

Edwards, K. J., Bond, P. L., & Banfield, J. F. (2000). Characteristics of attachment and growth of *Thiobacillus caldus* on sulphide minerals: A chemotactic response to sulphur minerals? *Environmental Microbiology, 2*(3), 324–332. https://doi.org/10.1046/j.1462-2920.2000.00111.x

Ferreira, R. A., Pereira, M. F., Magalhães, J. P., Maurício, A. M., Caçador, I., & Martins-Dias, S. (2021). Assessing local acid mine drainage impacts on natural regeneration-revegetation of São Domingos mine (Portugal) using a mineralogical, biochemical and textural approach. *Science of the Total Environment, 755*(1), 142825. https://doi.org/10.1016/j.scitotenv.2020.142825

Gao, P., Sun, X., Xiao, E., Xu, Z., Li, B., & Sun, W. (2019). Characterization of iron-metabolizing communities in soils contaminated by acid mine drainage from an abandoned coal mine in Southwest China. *Environmental Science and Pollution Research International, 26*(10), 9585–9598. https://doi.org/10.1007/s11356-019-04336-6

García, C., Moreno, D. A., Ballester, A., Blázquez, M. L., & González, F. (2001). Bioremediation of an industrial acid mine water by metal-tolerant sulphate-reducing bacteria. *Minerals Engineering, 14*(9), 997–1008. https://doi.org/10.1016/S0892-6875(01)00107-8

Gazea, K. A., & Kontopoulos, A. (1996). A review of passive systems for the treatment of AMD, *Miner. Engineering*. https://doi.org/10.1016/0892-6875(95)00129-8

Ghangrekar, M. M., & Chatterjee, P. (2017). A systematic review on bioelectrochemical systems research. *Current Pollution Reports, 3*(4), 281–288. https://doi.org/10.1007/s40726-017-0071-7

Gomez, D. V., Serrano, A., Peces, M., Ryan, B., Hofmann, H., & Southam, G. (2021). A sequential bioreactor adaption strategy enhanced the precipitation of metals from tailings' leachates. *Mining Engineering, 170*, 107051. https://doi.org/10.1016/j.mineng.2021.107051

Gonçalves, M. M. M., de Mello, L. A. O., & da Costa, A. C. A. (2007). The use of seaweed and sugarcane bagasse for the biological treatment of metal-contaminated waters under sulfate-reducing conditions. In *Biotechnology for Fuels and Chemicals* (pp. 465–473). Humana Press.

Gong, H., Jin, Z., Xu, H., Yuan, Q., Zuo, J., Wu, J., & Wang, K. (2019). Enhanced membrane-based pre-concentration improves wastewater organic matter recovery: Pilot-scale performance and membrane fouling. *Journal of Cleaner Production, 206*, 307–314. https://doi.org/10.1016/j.jclepro.2018.09.209

Gontia, P., & Janssen, M. (2016). Life cycle assessment of bio-based sodium polyacrylate production from pulp mill side streams: Case study of thermo-mechanical and sulfite pulp mills. *Journal of Cleaner Production, 131*, 475–484. https://doi.org/10.1016/j.jclepro.2016.04.155

Gupta, A., & Sar, P. (2020). Characterization and application of an anaerobic, iron and sulfate reducing bacterial culture in enhanced bioremediation of acid mine drainage impacted soil. *Journal of Environmental Science and Health. Part A, Toxic/Hazardous Substances and Environmental Engineering, 55*(4), 464–482. https://doi.org/10.1080/10934529.2019.1709362

Gusek, J. J. (2004). Scaling up design challenges for large scale sulfate reducing bioreactors. *Journal American Society of Mining and Reclamation, 2004*(1), 752–765. https://doi.org/10.21000/JASMR04010752

Hiibel, S. R., Pereyra, L. P., Breazeal, M. V. R., Reisman, D. J., Reardon, K. F., & Pruden, A. (2011). Effect of organic substrate on the microbial community structure in pilot-scale sulfate-reducing biochemical reactors treating mine drainage. *Environmental Engineering Science, 28*(8), 563–572. https://doi.org/10.1089/ees.2010.0237

Ighalo, J. O., Kurniawan, S. B., Iwuozor, K. O., Aniagor, C. O., Ajala, O. J., Oba, S. N., & Igwegbe, C. A. (2022). A review of treatment technologies for the mitigation of the toxic environmental effects of acid mine drainage (AMD). *Process Safety and Environmental Protection, 157*, 37–58. https://doi.org/10.1016/j.psep.2021.11.008

ITRC. (2013). *Biochemical reactors for mining-influenced water*. Interstate Technology & Regulatory Council.

Jamil, I. N., & Clarke, W. P. (2013a). Bioremediation for acid mine drainage: Organic solid waste as carbon sources for sulfate-reducingbacteria: A review. *Journal of Mechanical Engineering and Sciences, 5*, 569–581. https://doi.org/10.15282/jmes.5.2013.3.0054

Jamil, I. N., & Clarke, W. P. (2013b). Bioremediation for AMD: Organic solid waste as carbon sources for sulfate-reducing bacteria: A review. In Third Malaysian Postgraduate Conference (MPC) (p. 203).

Janssen, A. J. H., Lettinga, G., & de Keizer, A. (1999). Removal of hydrogen sulphide from wastewater and waste gases by biological conversion to elemental sulphur. *Colloids and Surfaces a: Physicochemical and Engineering Aspects, 151*(1–2), 389–397. https://doi.org/10.1016/S0927-7757(98)00507-X

Jeen, S. W., & Mattson, B. (2016). Evaluation of layered and mixed passive treatment systems for AMD. *Environnement et Technique (UK)*. https://doi.org/10.1080/09593330.2016.1167249

Johnson, D. B., & Hallberg, K. B. (2005). AMD remediation options: A review. *Science of the Total Environment, 338*(1–2), 3–14. https://doi.org/10.1016/j.scitotenv.2004.09.002

Jong, T., & Parry, D. L. (2003). Removal of sulfate and heavy metals by SRB in short-term bench scale up flow anaerobic packed bed reactor runs. *Water Research, 37*(14), 3379–3389. https://doi.org/10.1016/S0043-1354(03)00165-9

Kadnikov, V. V., Gruzdev, E. V., Ivasenko, D. A., Beletsky, A. V., Mardanov, A. V., Danilova, E. V., & Ravin, N. V. (2019). Selection of a microbial community in the course of formation of acidmine-drainage. *Microbiology, 88*(3), 292–299. https://doi.org/10.1134/S0026261719030056

Kaksonen, A. H., & Puhakka, J. A. (2007). Sulfate reduction based bioprocesses for the treatment of AMD and the recovery of metals. *Engineering in Life Sciences, 7*(6), 541–564. https://doi.org/10.1002/elsc.200720216

Kaksonen, A. H., & Sahinkaya, E. (2012). Precipitation of Zn(II), Cu(II) and Pb(II) at bench-scale using biogenic hydrogen sulfide from the utilization of *volatile fatty acids*. Review of sulfate reduction based bioprocesses for AMD treatment and metals recovery. In M. T. Alvarez, C. Crespo & B. Mattiasson (Eds.), *International Minewater Association Symposium, Chemosphere, 66*(9), 1677–1683.B (pp. 207–214). Edith Cowan University.

Kaur, G., Couperthwaite, S. J., Hatton-Jones, B. W., & Millar, G. J. (2018). Alternative neutralisation materials for acid mine drainage treatment. *Journal of Water Process Engineering, 22*, 46–58. https://doi.org/10.1016/j.jwpe.2018.01.004

Kawaja, J. D. E., et al. (2006). *A performance comparison of arsenic, cadmium and zinc treatment in a duplicate column study of an anaerobic bioreactor operated at 4 degrees C and 25 degrees C.*

Kim, J. R., Song, Y. E., Munussami, G., Kim, C., & Jeon, B. H. (2015). Recent applications of bioelectrochemical system for useful resource recovery: Retrieval of nutrient and metal from wastewater. *Geosystem Engineering, 18*(4), 173–180. https://doi.org/10.1080/12269328.2015.1037966

Kim, J., Yang, J., Ha, C., & Whang, K. (2002). A modified ramp reset waveform for high contrast ratio in AC PDPs. *Journal of Information Display, 3*(4), 13–18. https://doi.org/10.1080/15980316.2002.9651903

Kittrell, E. F. (2014). Potential acid mine drainage treatment utilizing acidophilic sulfate reducing bacteria in an upflow bioreactor. *Masters Theses, 7245.*

Kolmert, A., Wikström, P., & Hallberg, K. B. (2000). A fast and simple turbidimetric method for the determination of sulfate in sulfate-reducing bacterial cultures. *Journal of Microbiological Methods, 41*(3), 179–184. https://doi.org/10.1016/s0167-7012(00)00154-8

Kuang, J. L., Huang, L. N., Chen, L. X., Hua, Z. S., Li, S. J., Hu, M., Shu, W. S., et al. (2013). Contemporary environmental variation determines microbial diversity patterns in AMD. *ISME Journal, 7*(5), 1038–1050. https://doi.org/10.1038/ismej.2012.139

Leal-Gutiérrez, M. J., Cuéllar-Briseño, R., Castillo-Garduño, A. M., Bernal-González, M., Chávez-Castellanos, Á. E., Solís-Fuentes, J. A., Bazúa-Rueda, E. R., et al. (2021). Precipitation of heavy metalions (Cu, Fe, Zn, and Pb) from mining flotation effluents using a laboratory-scale Upflow anaerobic sludge blanket reactor. *Water, Air, and Soil Pollution, 232*(5). https://doi.org/10.1007/s11270-021-05042-1

Lefebvre, O., Neculita, C. M., Yue, X., & Ng, H. Y. (2022, 2012). Bioelectrochemical treatment of acid mine drainage dominated with iron. *Journal of Hazardous Materials, 241–242*, 411–417. https://doi.org/10.1016/j.jhazmat.2012.09.062

Li, X., Lan, S. M., Zhu, Z. P., Zhang, C., Zeng, G. M., Liu, Y. G., & Wu, S. H. (2018). The bioenergetics mechanisms and applications of sulfate-reducing bacteria in remediation of pollutants in drainage: A review. *Ecotoxicology and Environmental Safety, 158*, 162–170. https://doi.org/10.1016/j.ecoenv.2018.04.025

Lu, J., Chen, T., Wu, J., Wilson, P. C., Hao, X., & Qian, J. (2011). Acid tolerance of an AMD bioremediation system based on biological sulfate reduction. *Bioresource Technology, 102*(22), 10401–10406. https://doi.org/10.1016/j.biortech.2011.09.046

Ma, J. F., Ryan, P. R., & Delhaize, E. (2001). Aluminium tolerance in plants and the complexing role of organic acids. *Trends in Plant Science, 6*(6), 273–278. https://doi.org/10.1016/s1360-1385(01)01961-6

Martins Costa, J., Piacentini Rodriguez, R., & Sancinetti, G. P. (2019). Anaerobic batch reactor treating acid mine drainage: Kinetic stability on sulfate and COD removal. *Journal of Water Process Engineering, 31*, 100825. https://doi.org/10.1016/j.jwpe.2019.100825

Martins, M., Faleiro, M. L., Barros, R. J., Veríssimo, A. R., & Costa, M. C. (2009). Biological sulphate reduction using food industry wastes as carbon sources. *Biodegradation, 20*(4), 559–567. https://doi.org/10.1007/s10532-008-9245-8

Méndez-García, C., Peláez, A. I., Mesa, V., Sánchez, J., Golyshina, O. V., & Ferrer, M. (2015). Microbial diversity and metabolic networks in acid mine drainage habitats. *Frontiers in Microbiology, 6*, 475. https://doi.org/10.3389/fmicb.2015.00475

Moodley, I., Sheridan, C. M., Kappelmeyer, U., & Akcil, A. (2018). Environmentally sustainable acid mine drainage remediation: Research developments with a focus on waste/by-products. *Minerals Engineering, 126*, 207–220. https://doi.org/10.1016/j.mineng.2017.08.008

Muyzer, G., & Stams, A. J. (2008). The ecology and biotechnology of sulphate-reducing bacteria. *Nature Reviews Microbiology, 6*(6), 441–454. https://doi.org/10.1038/nrmicro1892

Ňancucheo, I., & Johnson, D. B. (2012). Selective removal of transition metals from acidic mine waters by novel consortia of acidophilic sulfidogenic bacteria. *Microbial Biotechnology, 5*(1), 34–44. https://doi.org/10.1111/j.1751-7915.2011.00285.x

Neculita, C. M., Zagury, G. J., & Bussière, B. (2007). Passive treatment of acid minedrainagein bioreactors using sulfate-reducingbacteria: Critical review and research needs. *Journal of Environmental Quality, 36*(1), 1–16. https://doi.org/10.2134/jeq2006.0066

Nordstrom, D. K., Blowes, D. W., & Ptacek, C. J. (2015). Hydro geochemistry and microbiology of mine drainage: An update. *Applied Geochemistry, 57*, 3–16. https://doi.org/10.1016/j.apgeochem.2015.02.008

Nordwick, S., Zaluski, M., Park, B., & Bless, D. (2006). Advances in development of bioreactors applicable to the treatment of ARD. *Journal American Society of Mining and Reclamation, 2006*(2), 1410–1420. https://doi.org/10.21000/JASMR06021410

Panda, S., Mishra, S., & Akcil, A. (2016). Bioremediation of acidic mine effluents and the role of sulfidogenic bio systems: A mini-review. *Euro-Mediterranean Journal for Environmental Integration, 1*(1).https://doi.org/10.1007/s41207-016-0008-3

Park, J.-K., Kang, J. H., Chong, Y., & Lee, N. (2012). A study on mass balance of carbon in a solid waste landfill. *Journal of Korea Society of Waste Management, 29*, 348–355.

Rambabu, K., Banat, F., Pham, Q. M., Ho, S. H., Ren, N. Q., & Show, P. L. (2020). Biological remediation of AMD: Review of past trends and current outlook. *Environmental Science and Ecotechnology, 2*, 100024. https://doi.org/10.1016/j.ese.2020.100024

Ren, Z. J. (2013). The principle and applications of bioelectrochemical systems. *Biofuel Technologies*, 501–527.

Rodriguez, R. P., Oliveira, G. H. D., Raimundi, I. M., & Zaiat, M. (2012). Assessment of a UASB reactor for the removal of sulfate from acid mine water. *International Biodeterioration and Biodegradation, 74*, 48–53. https://doi.org/10.1016/j.ibiod.2012.07.012

Roy chowdhury, A., Sarkar, D., & Datta, R. (2015). Remediation of AMD-impacted.

Sahinkaya, E. (2009). Biotreatment of zinc-containing wastewater in a sulfidogenic CSTR: Performance and artificial neural network (ANN) modelling studies. *Journal of Hazardous Materials, 164*(1), 105–113. https://doi.org/10.1016/j.jhazmat.2008.07.130

Sahinkaya, E., Isler, E., Yurtsever, A., & Coban, I. (2019). Sulfidogenic treatment of acid mine drainage using anaerobic membrane bioreactor. *Journal of Water Process Engineering, 31*, 100816. https://doi.org/10.1016/j.jwpe.2019.100816

Sánchez Espana, J., López Pamo, E., Santofimia, E., Aduvire, O., Reyes, J., & Barettino, D. (2005). Acid mine drainage in the Iberian pyrite Belt (Odiel river watershed, Huelva, SW Spain): Geochemistry, mineralogy and environmental implications. *Applied Geochemistry, 20*(7), 1320–1356. https://doi.org/10.1016/j.apgeochem.2005.01.011

Sand, W., Gehrke, T., Jozsa, P. G., & Schippers, A. (2001). Biochemistry of bacterial leaching—Direct versus indirect bioleaching. *Hydrometallurgy, 159–175*, 386x(00)00180-8. https://doi.org/10.1016/S0304

Santos, A. A., Venceslau, S. S., Grein, F., Leavitt, W. D., Dahl, C., Johnston, D. T., & Pereira, I. A. (2015). A protein trisulfide couples dissimilatory sulfate reduction to energy conservation. *Science, 350*(6267), 1541–1545. https://doi.org/10.1126/science.aad3558

Sheoran, A. S., Sheoran, V., & Choudhary, R. P. (2010). Bioremediation of acid-rock drainage by sulphate-reducing prokaryotes: A review. *Minerals Engineering, 23*(14), 1073–1100. https://doi.org/10.1016/j.mineng.2010.07.001

Silva, L. F., Fdez-Ortiz de Vallejuelo, S., Martinez-Arkarazo, I., Castro, K., Oliveira, M. L., Sampaio, C. H., Madariaga, J. M., et al. (2013). Study of environmental pollution and mineralogical characterization of sediment rivers from Brazilian coal mining acid drainage. *Science of the Total Environment, 447*, 169–178. https://doi.org/10.1016/j.scitotenv.2012.12.013

Skinner, S. J. W., & Schutte, C. F. (2006). The feasibility of a permeable reactive barrier to treat acidic sulphate-and nitrate-contaminated ground water. *Water SA, 32*(2). https://doi.org/10.4314/wsa.v32i2.5253

Skousen, J. G., Ziemkiewicz, P. F., & McDonald, L. M. (2019). AMD formation, control and treatment: Approaches and strategies. *Extractive Industries and Society, 6*(1), 241–249. https://doi.org/10.1016/j.exis.2018.09.008

Skousen, J., Zipper, C. E., Rose, A., Ziemkiewicz, P. F., Nairn, R., McDonald, L. M., & Kleinmann, R. L. (2017). Review of passive systems for AMD treatment. *Mine Water and the Environment, 36*(1), 133–153. https://doi.org/10.1007/s10230-016-0417-1

Tang, K., Baskaran, V., & Nemati, M. (2009). Bacteria of the sulphur cycle: An overview of microbiology, biokinetics and their role in petroleum and mining industries. *Biochemical Engineering Journal, 44*(1), 73–94. https://doi.org/10.1016/j.bej.2008.12.011

Tassé, N., Isabel, D., & Fontaine, R. (2003). Wood Cadillac tailings: Designing a biofilter for arsenic control. In *Proceedings of the Sudbury, Mining and the Environment conference*. Sudbury, UK, ON, Canada.

Tayebi-Khorami, M., Edraki, M., Corder, G., & Golev, A. (2019). Re-thinking mining waste through an integrative approachled by circular economy aspirations. *Minerals, 9*(5), 286. https://doi.org/10.3390/min9050286

Tsukamoto, T. K., Killion, H. A., & Miller, G. C. (2004). Column experiments for microbiological treatment of AMD: Low- temperature, low-pH and matrix investigations. *Water Research, 38*(6), 1405–1418. https://doi.org/10.1016/j.watres.2003.12.012

Uman, A. E., Bair, R. A., & Yeh, D. H. (2021). Assessment of an anaerobic membrane bioreactor (AnMBR) treating medium-strength synthetic wastewater under cyclical membrane operation. *Membranes, 11*(6), 415. https://doi.org/10.3390/membranes11060415

Verburg, R., Bezuidenhout, N., Chatwin, T., & Ferguson, K. (2009) The global acid rock drainage guide[GARD guide]. *Mine Water and the Environment, 28*(4). https://doi.org/10.1007/s10230-009-0078-4

Waybrant, K. R., Ptacek, C. J., & Blowes, D. W. (2002). Treatment of mine drainage using permeable reactive barriers: Column experiments. *Environmental Science and Technology, 36*(6), 1349–1356. https://doi.org/10.1021/es010751g

Xingyu, L., Gang, Z., Xiaoqiang, W., Laichang, Z., Jiankang, W., & Renman, R. (2013). A novel low pH sulfidogenic bioreactor using activated sludge as carbon source to treat AMD (AMD) and recovery metal sulfides: Pilot scale study. *Minerals Engineering, 48*, 51–55. https://doi.org/10.1016/j.mineng.2012.11.004

Yilmaz, T., Yucel, A., Cakmak, Y., Uyanik, S., Yurtsever, A., & Ucar, D. (2019). Treatment of acidic mine drainage in up-flow sulfidogenic reactor: Metal recovery and the pH neutralization. *Journal of Water Process Engineering, 32*. https://doi.org/10.1016/j.jwpe.2019.100916

Integration of Bioleaching and Biorefinery Technologies for the Recovery of Base and Critical Elements from Electronic Waste

Alexandre H. Kompalitch and Eric D. van Hullebusch

Abstract

Several profound societal changes such as the shift towards renewable energies have created an ever-increasing demand for base and critical metals. Electronic wastes constitute a significant secondary source of such elements and a potential environmental hazard if disposed of improperly. In contrast to traditional methods of recycling e-waste, biohydrometallurgy is an environmentally friendly, low-cost, and energy-efficient alternative. Although processes developed in laboratories display promising yields, it is still premature to implement these biotechnological strategies on a larger scale as the bioleaching and biorefinery mechanisms are still poorly understood. Moreover, very few studies focus on fully biological processes, and most opt for more efficient hybrid approaches. Thus, this book chapter compiles the optimal parameters reported in recent studies, from waste pre-treatment to metal biorecovery, along with insights to complete and close the biohydrometallurgical recycling loop.

Keywords

Biohydrometallurgy · Bioleaching · Biorecovery · E-waste · Pregnant leaching solution · Process optimization · Critical metals

1 Introduction

The advent of advanced technologies, increasing urbanization, and global economic development are the main causes for the exponential production and consumption of electrical and electronic equipment (EEE). It ranges from small day-to-day products, such as monitors, personal computers, cell phones, microwaves, and lamps, to large household products, namely refrigerators, electric stoves, washing machines, and air conditioners. Programmed obsolescence and constant technological innovation, combined with increasing consumer demand for affordable equipment and high living standards, drastically reduce these products' lifespan. These phenomena lead to an unprecedented production of waste (Tansel, 2017), more commonly known as waste electrical and electronic equipment (WEEE or E-waste). E-waste management and recycling are a real challenge but also an opportunity for our modern societies. E-waste recycling represents a non-negligible source of metals. The average e-waste contains around 60% of metals and metalloids (Gopikrishnan et al., 2020), ranging from rare earth elements (REE) to platinum group metals (PGM), that are essential for the production of EEE and the development of low-carbon energies.

E-waste is the most prevalent and fastest-growing source of waste (Darby & Obara, 2005). In 2019, 53.6 Mt of e-waste was globally generated, but only 17.4% (9.3 Mt) is documented to be collected and recycled. E-waste production is estimated to reach 74.7 Mt by 2030, an increase of 39% (Forti et al., 2020). The huge amount of e-waste generated each year represents a serious threat to the environment when improperly managed. It is usually discarded in landfills or transferred to developing countries. E-waste contains toxic substances such as arsenic, lead, cadmium, mercury, and plastics that can leach into the soil, contaminate the environment and therefore the human habitat (Widmer et al., 2005). Although 78 countries have adopted legislation on e-waste treatment, most have insufficient policies or poor enforcement. In 2019, Europe was the continent with the most functional e-waste policy, achieving a collection and recycling rate of 42.5%. It was followed by Asia (11.7%), Americas (9.4%), Oceania (8.8%), and Africa (0.9%) (Forti et al., 2020).

A. H. Kompalitch · E. D. van Hullebusch (✉)
Université Paris Cité, Institut de Physique du Globe de Paris, CNRS, 75005 Paris, France
e-mail: vanhullebusch@ipgp.fr

S. Panda et al. (eds.), *Biotechnological Innovations in the Mineral-Metal Industry*,
Advances in Science, Technology & Innovation, https://doi.org/10.1007/978-3-031-43625-3_11

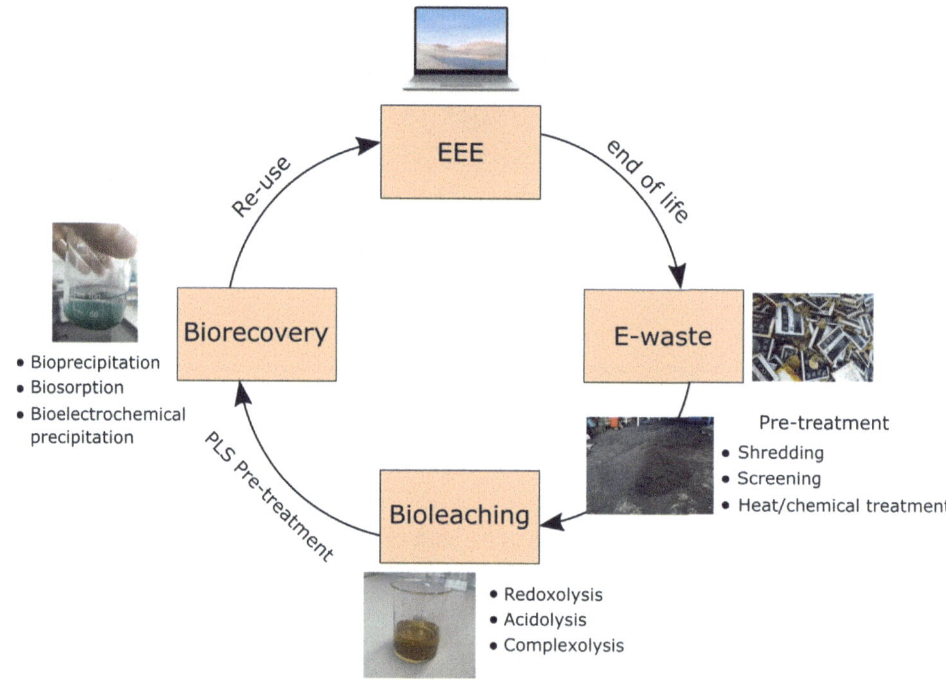

Fig. 1 Schematic representation of a biohydrometallurgical process cycle as conceptualized in this review

Currently, e-waste is industrially recycled through pyrometallurgical and hydrometallurgical processes. Both routes have major drawbacks. The former approach consists of smelting waste at high temperatures of up to 900 °C, is energy-intensive, lacks selectivity, and involves loss of some metals (Becci et al., 2020; Sethurajan & Gaydardzhiev, 2021). The main issues related to the hydrometallurgical route are its intensive use of toxic chemicals causing environmental problems, the high costs, and the production of hazardous substances (Mishra et al., 2021). Biohydrometallurgy uses microorganisms in an aqueous medium to recover metals from various sources (Kaksonen et al., 2018). It is considered as an environmentally friendly and sustainable alternative to established e-waste treatment and metal recovery processes. It allows for simpler, less energy-intensive, and cost-effective processes (Ilyas et al., 2014). These techniques are already being used successfully for mining primary and secondary ores (e.g. biomining, heap bioleaching) (Jerez, 2017; Johnson & Du Plessis, 2015; Thenepalli et al., 2019) and are extremely promising for e-waste recycling. Research on the bioprocessing of e-waste for the recovery of critical metals is at an early stage but is growing rapidly. It generally involves two steps in which microorganisms have central roles in the transformation processes. The first is bioleaching, which is the conversion of metals from their solid to their soluble form through the bioproduction of leaching compounds or metabolites. Biorecovery is the second step and allows the bioprecipitation of dissolved metals that could be selectively recovered in some specific cases.

Very few studies related to biohydrometallurgy fully combine bioleaching and bioprecipitation steps (e.g. Sethurajan et al., 2017). These two processes are often coupled with hydrometallurgy, i.e. either biorecovery preceded by chemical leaching (or the use of a synthetic pregnant leaching solution) (Calvert et al., 2019; Creamer et al., 2006; Kim et al., 2018) or bioleaching followed by chemical precipitation (Ijadi Bajestani et al., 2014). Moreover, most biorecovery studies are focused on acid mine drainage and wastewater treatment, as well as primary ore biomining. Therefore, there is a clear need to integrate real pregnant bioleaching solutions (PLSs) coming from e-waste with biorecovery setups to complete the biohydrometallurgy recycling loop (Fig. 1).

This book chapter attempts to reflect on the means and critical parameters to be considered to achieve this. To this end, various up-to-date studies on the bioprocessing of e-waste and the biorecovery of base and critical metals have been reviewed. First of all, pre-treatment of the three types of e-waste that have been most frequently found in recent biometallurgical studies are documented: printed circuit boards, liquid–crystal displays, and lithium-ion batteries. Several bioleaching and biorecovery parameters are then examined to determine which are most critical for the successful integration of bioleaching for biorecovery technologies.

2 Electronic Waste Pre-treatment

Pre-treatment of e-waste is a crucial step that must be taken prior to metal recovery in order to optimize conditions for biohydrometallurgical processes. Depending on the type of waste treated, the structure of electronic scrap can be extremely complex due to the manufacturing process and the diversity of the components used.

The pre-treatment of lithium-ion batteries, printed circuit boards, and liquid–crystal displays are presented, as these are the WEEEs that are the most commonly treated in recent bioleaching and bioprecipitation studies.

2.1 Printed Circuit Boards

Printed circuit boards (PCBs) are the basic building block of all electrical and electronic equipment since they support and connect all other components. PCBs represent 4–7 wt% of the total EEE generated and 2–3% of the global amount of E-waste generated (Dalrymple et al., 2007; Zhou & Qiu, 2010).

PCBs' most common base material is FR-4, which is a non-conductive composite material made of epoxy resin reinforced with woven fibreglass cloth. Because of its conductive properties, copper foil laminate is either printed or engraved on and inside the non-conductive substrate. Electric components such as diodes, capacitors, and transistors are then mounted and soldered on the substrate.

Generally, PCBs can contain up to 40% metals, 30% plastics, and 30% glass and ceramics (Kang et al., 2021; Ribeiro et al., 2019). The glass substrate is mainly composed of SiO_2, CaO, Al_2O_3, MgO, and BeO (Ghosh et al., 2015). Copper constitutes about 10–20% of PCBs as it is the main conductive material, but other metals are like Au, Ag, Pd, Pb, Sn, Sb, Ni, Al, Be, and Cd are also found. It has been reported that in PCB, Au and Ag concentrations can be as high as 700 g/t and 800 g/t, respectively (Kasper & Veit, 2018; Ribeiro et al., 2019). The concentration of Cu in waste PCBs is 13–26 times higher than in primary ores and that of Au is up to 50 times higher (Akcil et al., 2015; Cui & Zhang, 2008; Zhang & Forssberg, 1998). PCBs are commonly treated with brominated flame retardants (BFRs) such as polybrominated diphenyl ethers (PBDEs), tetrabromobisphenol A (TBBP-A), and hexabromocyclododecane (HBCD) (Alaee & Wenning, 2002). When incinerated, these compounds can react with the oxygen present in the atmosphere to form dioxins and furans (e.g. polybrominated dibenzo-p-dioxins and dibenzofurans (PBDD/Fs), polybromochlorodibenzo-p-dioxins and dibenzofurans (PXDD/Fs)) which are highly toxic for the human body (Wong et al., 2007).

Dismantling is the first step in waste PCB pre-treatment. Electrical components are dismounted and treated separately. This process can be done manually or automatically by applying pressure and destroying the solders, composed of a 63–37% Sn–Pb alloy. Although melting the solders has also been reported as an efficient way of dismounting ECs (Yokoyama & Iji, 1997), it can also produce enough heat to release dioxins. The laminated structure of PCB waste requires a reduction in particle size to increase the contact surface between the metals and the leachate. Industries employ several hammer crushers, rotary crushers, and disc crushers, but shredding and cutting remain the most effective methods (Ghosh et al., 2015).

2.2 Lithium-Ion Batteries

Lithium-ion batteries (LIBs) are nowadays the most common energy storage device because they are manufactured in products such as smartphones and laptops that have become part of our everyday life. They also play a decisive role in the energy transition as they are a central element in the construction of electric vehicles, due to their lightweight, high-energy capacity per unit of weight, and relatively long lifespan of up to 5 years (Mylarappa et al., 2017). Hence, spent LIBs potentially represent a very large waste stream. More than one million electric vehicles were sold globally in 2017, which could represent ultimately almost 250,000 tonnes of battery waste (Lie et al., 2020; Sethurajan & Gaydardzhiev, 2021).

Spent LIBs are a significant source of critical metals. LIBs function by exchanging lithium ions between a cathode and an anode via an electrolyte to generate electricity. Generally, the anode is composed of graphite, a polyvinylidene fluoride (PVDF) binder, and the cathode of lithium cobalt dioxide ($LiCoO_2$). Other lithium transition metal oxides such as LiV_2O_3, $LiMn_2O_4$, $LiNiO_2$, $LiFePO_4$, and $Li[Ni_xMn_yCo_z]O_2$ are less popular than $LiCoO_2$ because of their lower energy density and lifespan. Each anode and cathode are surrounded by current collectors, made of aluminium and copper foil, and the surrounding casing is made of iron.

The first step in the pre-treatment of spent LIBs is discharging. This step is essential because dismantling can cause short circuits or spontaneous combustion (Zheng et al., 2018). This is usually done by shorting the battery under a rolling press and then immersing it in a salt solution. One of the issues encountered in this process is the leakage of some of the internal components of the battery into the solution. Salts such as NaCl and KCl contain chloride ions that can corrode the iron casing. It has been reported that the use of $FeSO_4$ can provide a good compromise between discharge performance and leakage limitation (Kim et al., 2021; Xiao et al., 2020). The next step is to separate the electrodes from the other parts of the battery. Different methods are used such as heat treatment, organic solvent, and alkaline dissolution, but mechanical methods such as sieving and wet or dry crushing seem to prevail for bioleaching studies, as particle size reduction is a crucial parameter affecting leaching efficiency. Studies generally aim to achieve a fine particle size of around 75 μm (Heydarian et al., 2018; Horeh et al., 2016), but coarser particles are also used (Biswal et al., 2018; Roy et al., 2021).

2.3 Liquid–Crystal Displays

Liquid–crystal displays (LCDs) are flat panel displays that use the optical properties of liquid crystals combined with a polarizer. LCD production has regularly exceeded 200 million units since 2010 (Zhang et al., 2015). In addition, the lifetime of an LCD panel is typically 3–5 years. End-of-life LCDs can therefore be considered as a non-negligible secondary metal resource.

LCD monitors are constituted of an LCD panel, printed circuit boards (PCBs), and a metal frame. The LCD panel features a glass substrate and a backlight unit, which can account for 15–40 wt% each (Li et al., 2009). Spent LCDs as a secondary resource are mostly interesting for the recovery of indium, a post-transition metal used in the manufacture of optoelectronic and semiconductor compounds. About 70% of the indium is found in indium tin oxide (ITO) film, an In–Sn alloy composed of indium oxide (In_2O_3) and tin oxide (SnO_2) (Zhang et al., 2015). Indium is generally reported to be present in ITO at a concentration of 102 mg/kg (Li et al., 2009).

To recover indium from LCDs, the ITO film contained in the glass panel needs to be separated from other components. The ITO glass is a multi-layered surface, composed of glass panels, polarizing films, the ITO film, and the liquid–crystal panel which is glued together. Used LCDs must be dismantled to break the plastic case, remove the backlight, and retrieve the LCD panels. Manual dismantling is reported to be the most efficient and cost-effective way to recover metals compared to the mechanical way (Kopacek, 2010).

A second step would be the separation of the ITO glass from the liquid–crystal layer and the polarizing films, which are made of organic polymers such as polyethylene terephthalate, cellulose triacetate, polyvinyl alcohol, triphenyl phosphate, and benzene. Pyrolysis is a viable method of removing organic components from the glass substrate. Some studies (Lu et al., 2012; Wang & Xu, 2014) were able to remove the polarizing film and the liquid–crystal layer by heating and processing it into oils (e.g. acetic acid) and gases that can be recycled. A more environmentally friendly approach would be a combination of both physical and chemical treatments (Zhang et al., 2015). The organic layers could be softened by thermal shock (Li et al., 2009) or using acetone (Zhao et al., 2013) and then removed manually by hand. Once the ITO glass is recovered, crushing and milling are required to reduce the size of the particles, increase surface area, and enhance the leaching process. Zheng et al. (2023) have recently reviewed the state of the art regarding the recovery approaches of indium from spent LCD screens.

3 Bioleaching of Electronic Waste

The bioleaching of e-waste is the first step in the metal recovery process. It refers to the mobilization of base and critical metals from the solid state to their soluble form using various direct or indirect biological mechanisms. Several recent bioleaching studies and their optimal and initial conditions have been reported in Table 1.

3.1 Bioleaching Mechanisms

3.1.1 Redoxolysis

Redoxolysis is a process that involves the microbial production of compounds that can modulate the redox potential of the leaching solution. This process can occur by microbial attachment, mobilizing metals via electron transfer from the mineral to the microorganisms, but can also occur indirectly with the bioproduction of leachates oxidizing the targeted metals. The main leachate used for the bioleaching of e-waste is Fe^{3+}, which is produced by the microbially driven oxidation of Fe^{2+} (Eq. 1). Fe^{3+} can thus in turn oxidize the metals and solubilize them (Lee & Pandey, 2012) (Eq. 2). The reduced iron can then be bio-oxidized again, resulting in a cyclic process.

$$2Fe^{2+} + 0.5O_2 + 2H^+ \xrightarrow{At.\,ferrooxidans} 2Fe^{3+} + H_2O, \quad (1)$$

$$2Fe^{3+} + M^0 \rightarrow 2Fe^{2+} + M^{2+}. \quad (2)$$

Several microbes can oxidize iron, while the most commonly used microbe for mineral bioleaching is *Acidithiobacillus ferrooxidans*. It is an acidophilic chemolithoautotrophic proteobacterium that uses the energy derived from sulfur and iron oxidation for its growth (Valdès et al., 2008). Hence, the Fe^{3+}/Fe^{2+} ratio and the redox potential of the solution are key factors to monitor and understand the chemistry of the leaching solution. Among the sulfur/iron-oxidizing bacteria, *At. ferrooxidans* have been used in many recent studies involving e-waste bioleaching with very high efficiency (Arshadi & Mousavi, 2015; Becci et al., 2020; Heydarian et al., 2018; Hubau et al., 2020; Işıldar et al., 2016). Although bioleaching is mainly driven by redoxolysis, the acidic medium in which the bacterium thrives can improve the leaching efficiency.

Table 1 Literature overview of bioleaching conditions

Leaching mechanism	Microorganism(s)	E-waste	Culture medium(s)	Leaching time	Initial pH	Pulp density	T (°C)	Leaching efficiency	Reference
Redoxolysis	*Acidiphilium acidophilum*	PCB	$(NH_4)_2SO_4$ 3 g/L $MgSO_4 \cdot 7H_2O$ 1 g/L KH_2PO_4 0.5 g/L KCl 0.1 g/L $Ca(NO_3)_2$ 0.018 g/L $FeSO_4 \cdot 7H_2O$ 0.01 g/L S^0 25 g/L	60 d	2.5	1 g/L	30	Cu 79% Ni 39% Zn 29% Pb 10%	Priya and Hait (2018)
Redoxolysis	*Acidithiobacillus ferrooxidans*	LCD	$(NH_4)_2SO_4$ 0.4 g/L K_2HPO_4 3 g/L $MgSO_4 \cdot 7H_2O$ 0.5 g/L $CaCl_2 \cdot 2H_2O$ 0.25 g/L	8 d	2	15 g/L	30	In 100%	Xie et al. (2019)
Redoxolysis	*Acidithiobacillus caldus* *Leptospirillum ferriphilum* *Sulfobacillus thermosulphidooxans* *Ferroplasma*	PCB	$MgSO_4$ 0.5 g/L $(NH_4)_2SO_4$ 0.4 g/L K_2SO_4 0.4 g/L $CaH_4(PO_4)_2$ 0.29 g/L	26 d	1–1.5	10 g/L	45	Cu 96.8% Zn 68% Al 45%	Akbari and Ahmadi (2019)
Redoxolysis	*Acidithiobacillus ferrooxidans*	LED	$(NH_4)_2SO_4$ 3 g/L K_2HPO_4 0.5 g/L $MgSO_4 \cdot 7H_2O$ 0.5 g/L KCl 0.1 g/L $Ca(NO_3)_2$ 0.01 g/L $FeSO_4 \cdot 7H_2O$ 44.22 g/L	29 d	2–2.5	20 g/L	30	Cu 84% Ni 96% Ga 60%	Pourhossein and Mousavi (2018)
Redoxolysis	*Acidithiobacillus ferrooxidans*	PCB	$(NH_4)_2SO_4$ 3 g/L K_2HPO_4 0.5 g/L KCl 0.1 g/L $MgSO_4 \cdot 7H_2O$ 0.5 g/L $Ca(NO_3)$2 0.01 g/L $FeSO_4 \cdot 7H_2O$ 0.23 g/L	7 d	3	36 g/L	30	Cu 87.5%	Li et al. (2015)
Redoxolysis	*Acidithiobacillus ferrooxidans*	PCB	$(NH_4)_2SO_4$ 2 g/L K_2HPO_4 0.1 g/L $MgSO_4 \cdot 7H_2O$ 0.25 g/L KCl 0.1 g/L $FeSO_4 \cdot 7H_2O$ 0.25 g/L S^0 10 g/L	7 d	1.5	50 g/L	30	Cu>95%	Wang et al. (2018)
Redoxolysis	*Acidithiobacillus ferrooxidans* (DSMZ 1927)	LiB	$(NH_4)_2SO_4$ 0.50 g/L KCl 0.05 g/L $K_2HPO_4 \cdot 3H_2O$ 0.20 g/L $MgSO_4 \cdot 7H_2O$ 0.50 g/L $Ca(NO_3)_2$ 0.01 g/L $FeSO_4 \cdot 7H_2O$ 45.0 g/L	3 d	2	100 g/L	30	Co 94% Li 60%	Roy et al. (2021)

(continued)

Table 1 (continued)

Leaching mechanism	Microorganism(s)	E-waste	Culture medium(s)	Leaching time	Initial pH	Pulp density	T (°C)	Leaching efficiency	Reference
Redoxolysis	*Acidithiobacillus ferrooxidans* (DSMZ 583) *Leptospirillum ferrooxidans* (DSM 2705) *Acidithiobacillus thiooxidans* (DSMZ 9463)	PCB	$FeSO_4{\cdot}7H_2O$ 9 g/L $(NH_4)_2SO_3$ 3 g/L $MgSO_4{\cdot}7H_2O$ 0.5 g/L KH_2PO_4 0.5 g/L Ca $(NO_3)_2{\cdot}4H_2O$ 0.1 g/L S^0 1 g/L	8 d	1.8	10 g/L	30	Al 55.8% Cu 97.3% Ni 79.3% Zn 66.8%	Erust et al. (2020)
Redoxolysis Acidolysis	*Acidithiobacillus ferrooxidans* (PTCC 1647) *Acidithiobacillus thiooxidans* (PTCC 1717)	LiB	– $(NH_4)_2SO_4$ 3.0 g/L KCl 0.1 g/L K_2HPO_4 0.5 g/L $MgSO_4{\cdot}7H_2O$ 0.5 g/L $Ca(NO_3)_2$ 0.01 g/L $FeSO_4{\cdot}7H_2O$ 44.2 g/L – $(NH_4)_2SO_4$ 2 g/L $K_2HPO_4{\cdot}3H_2O$ 0.25 g/L $MgSO_4{\cdot}7H_2O$ 0.25 g/L KCl 0.1 g/L S^0 10 g/L	16 d	1.5	4–40 g/L	32	Li 99.2% Ni 89.4% Co 50.4%	Heydarian et al. (2018)
Redoxolysis	*Acidithiobacillus ferrooxidans* (PTCC1626)	PCB	$(NH_4)_2SO_4$ 3 g/L K_2HPO_4 0.5 g/L $MgSO_4{\cdot}7H_2O$ 0.5 g/L KCl 0.1 g/L $Ca(NO_3)_2$ 0.01 g/L $FeSO_4{\cdot}7H_2O$ 44.22 g/L	20 d	1	8.5 g/L	30	Cu 100% Ni 100%	Arshadi and Mousavi (2015)
Redoxolysis Acidolysis	*Acidithiobacillus thiooxidans* (DSM 9463) *Acidithiobacillus ferrooxidans* (DSM 17,398)	PCB	$(NH_4)_2SO_4$ 2 g/L $MgSO_4{\cdot}7H_2O$ 0.25 g/L KH_2PO_4 0.1 g/L KCl 0.1 g/L $FeSO_4{\cdot}7H_2O$ 8 g/L S^0 10 g/L	20 d	1	10 g/L	27	Cu 98%	Işıldar et al. (2016)
Acidolysis	*Acidithiobacillus thiooxidans* (PTCC 1717)	LCD	$(NH_4)_2SO_4$ 2 g/L K_2HPO_4 0.25 g/L $MgSO_4{\cdot}7H_2O$ 0.25 g/L KCl 0.1 g/L S^0 5 g/L	15 d	2	16 g/L	30	In 100% Sr 10%	Jowkar et al. (2018)
Complexolysis	*Aspergillus fumigatus* (KUC1520) *Aspergillus flavipes* (KUC5033) *Aspergillus japonicus* (KUC5035) *Aspergillus tubingensis* (KUC5037) *Aspergillus versicolor* (KUC5201) *Aspergillus niger* (KUC5254)	Zn-Mn/Ni–Cd batteries	– ME medium: Malt extract 20 g/L – Sucrose medium: Sucrose 100 g/L $NaNO_3$ 1.5 g/L KH_2PO_4 0.5 g/L $MgSO_4{\cdot}7H_2O$ 0.025 g/L KCl 0.025 g/L Yeast extract 1.6 g/L	4–8 d	2.47–7.47	1 g/L	27	Zn-Mn: Zn 100% Mn 92% Ni–Cd: Ni, Zn, Cd>90%	Kim et al. (2018)

(continued)

Table 1 (continued)

Leaching mechanism	Microorganism(s)	E-waste	Culture medium(s)	Leaching time	Initial pH	Pulp density	T (°C)	Leaching efficiency	Reference
Complexolysis	*Aspergillus niger*	LCD	Sucrose 50 g/L $NaNO_3$ 1.5 g/L KH_2PO_4 0.5 g/L $MgSO_4{\cdot}7H_2O$ 0.025 g/L KCl 0.025 g/L Yeast extract 1.6/L	90 min	4	10 g/L	70	In 100%	Cui et al. (2020)
Complexolysis	*Aspergillus niger*	LCD	Sucrose 100 g/L $NaNO_3$ 1.5 g/L KH_2PO_4 0.5 g/L $MgSO_4{\cdot}7H_2O$ 0.025 g/L KCl 0.025 g/L Yeast extract 1.6 g/L	15 d	2.5–3	10 g/L	30	In 100%	Cui et al. (2021)
Complexolysis	*Aspergillus niger*	PCB	Sucrose 100 g/L $NaNO_3$ 1.5 g/L KH_2PO_4 0.5 g/L $MgSO_4{\cdot}7H_2O$ 0.025 g/L KCl 0.025 g/L Yeast extract 1.6 g/L	2 h	1.97	10 g/L	80	All metals 100%	Jadhav et al. (2016)
Complexolysis	*Aspergillus niger* (PTCC 5210)	LiB	Sucrose 100 g/L $NaNO_3$ 1.5 g/L KH_2PO_4 0.5 g/L $MgSO_4{\cdot}7H_2O$ 0.025 g/L KCl 0.025 g/L Yeast extract 1.6 g/L	14 d	6	10 g/L	30	Cu 100% Li 95% Mn 70% Al 65% Co 45% Ni 38%	Bahaloo-Horeh et al. (2016)
Complexolysis	*Aspergillus niger* (PTCC 5210)	LiB	Sucrose 116.90 g/L $NaNO_3$ 1.5 g/L KH_2PO_4 0.5 g/L $MgSO_4{\cdot}7H_2O$ 0.025 g/L KCl 0.025 g/L Yeast extract 1.6 g/L	8 d	5.44	10/20 g/L	30	Cu 100% Li 100% Mn 77% Al 75% Co 64% Ni 54%	Bahaloo-Horeh and Mousavi (2017)
Complexolysis	*Chromobacterium violaceum* (ATCC12471)	PCB	Peptone 10 g/L Yeast extract 5 g/L Glycine 5 g/L	7 d	3	36 g/L	30	Au 70.6%	Li et al. (2015)
Redoxolysis	*Leptospirillum ferriphilum* *Sulfobacillus thermosulfidooxidans*	PCB	$(NH_4)_2SO_4$ 3 g/L K_2HPO_4 0.5 g/L $MgSO_4{\cdot}7H_2O$ 0.5 g/L KCl 0.1 g/L $Ca(NO_3)_2$ 0.01 g/L $FeSO_4{\cdot}7H_2O$ 74.5 g/L	9 d	1.5	100 g/L	55	Cu 93.4%	Wu et al. (2018)

(continued)

Table 1 (continued)

Leaching mechanism	Microorganism(s)	E-waste	Culture medium(s)	Leaching time	Initial pH	Pulp density	T (°C)	Leaching efficiency	Reference
Redoxolysis	*Leptospirillum ferrooxidans* *Acidithiobacillus ferrooxidans* *Sulfobacillus thermosulfidooxidans*	PCB	$(NH_4)_2SO_4$ 0.4 g/L H_3PO_4 85% 0.81 g/L $MgSO_4{\cdot}7H_2O$ 0.52 g/L KOH 0.48 g/L $FeSO_4{\cdot}7H_2O$ 0.52 g/L	2 d	1.3	50 g/L	36	Cu 96% Zn 85% Ni 73% Co 93% Fe 93%	Hubau et al. (2020)
Redoxolysis	*Leptospirillum ferrooxidans* (DSMZ 2705T) *Acidithiobacillus ferrooxidans* (DSMZ 14882T)	PCB	DSMZ 882. *Leptospirillum* (HH) medium	7 d	1.6	50 g/L	30	Cu 91% Zn 73%	Becci et al. (2020)
Complexolysis	*Pseudomonas balearica* SAE1	PCB	LB medium 5 g/L glycine	7 d	8	10 g/L	30	Au 68.5% Ag 33.8%	Kumar et al. (2018)
Complexolysis	*Pseudomonas putida* (WSC361) *Pseudomonas fluorescens* (E11.3)	PCB	Meat extract 1 g/L Yeast extract 2 g/L Peptone 2 g/L NaCl 5 g/L Glycine	5 d	7	5 g/L	30	Au 44%	Işıldar et al. (2016)
Complexolysis	*Sphingomonas sp.* MXB8 *Candida orthopsilosis* MXL20 *Aspergillus niger* MX7, MX5, MXPE6	PCB	– Potato Dextrose Agar (DB Bioxon®)—25 mL containing 20 g Dextrose 1.5 g $(NH_4)_2SO_4$ 0.5 g KH_2PO_4 0.025 g $MgCl_2$ 0.1 g $CaCl_2$ – 25 mL containing 20 g Dextrose 1.5 g $(NH_4)_2SO_4$ 0.5 g NaH_2PO_4 0.1 g $MgCl_2$ 0.1 g $CaCl_2$	35 d	<7	–	14–24	Ag 54% Cu 2.08% Au 0.53%	Díaz-Martínez et al. (2019)
–	*Streptomyces albidoflavus* (TN10)	PCB	Yeast extract malt extract broth	3 d	6	6 g/L	28	Al 66% Ca 74% Cu 68% Cd 65% Fe 42% Ni 81% Zn 82% Ag 56% Pb 46%	Kaliyaraj et al. (2019)

3.1.2 Complexolysis

Complexolysis involves a chelating agent produced (secondary product or by-product) by a selected microorganism and metal ions to form a metal–organic complex. The chelating agent also called a ligand creates a coordinate covalent bond with the metal ions leading to their solubilization. The ligands can be organic compounds such as siderophores. These molecules are usually produced in the environment with the aim of solubilizing iron to make it available to plants and bacteria (Ahmed & Holmström, 2014). Siderophores' structure variability makes them able to chelate other metals such as In, Ga, and REE (Hofmann et al., 2020; Johnstone & Nolan, 2015).

In biohydrometallurgical processes, complexolysis is mainly carried out by cyanogenic bacteria and fungi in alkaline conditions. Cyanide (CN^-) is generally used to mobilize gold and other noble metals like Ag, Pd, or Pt. In fact, since 1898, cyanide has been widely used to extract gold from primary and secondary ores (Mudder & Botz, 2006). Gold dissolution by cyanide consists of anodic (Eq. 3) and cathodic (Eq. 4) equations and can be summarized by Eq. 5 below:

$$4Au + 8CN^- \rightarrow 4Au(CN)_2^- + 4e^-, \quad (3)$$

$$O_2 + 2H_2O + 4e^- \rightarrow 4OH^-, \quad (4)$$

$$4Au + 8CN^- + O_2 + 2H_2O \rightarrow 4Au(CN)_2^- + 4OH^-. \quad (5)$$

This reaction can be performed by many heterotrophic bacteria such as *Chromobacterium violaceum*, *Pseudomonas fluorescens*, and *Pseudomonas putida* (Işıldar et al., 2016; Li et al., 2015). Some cyanogenic bacteria can detoxify cyanide and transform it into nontoxic β-cyanoalanine (Knowles, 1976), which renders biogenic cyanide production a safer alternative than chemical cyanide treatment. Fungi such as *Aspergillus niger* have also been used for metal recovery from WEEE (Bahaloo-Horeh & Mousavi, 2017; Cui et al., 2021; Jadhav et al., 2016). They produce organic acids (e.g. malic, oxalic, citric acids) that can chelate metals and thus mobilize them.

3.1.3 Acidolysis

Acids are widely used for metal leaching in hydrometallurgy. The protons can weaken the bond between the metal ions, allowing their mobility. In case of bioleaching, the acids are produced by microorganisms, but the principle remains the same. For instance, chemolithotrophic bacteria such as *Acidithiobacillus thiooxidans* use elemental sulfur as its primary energy source and oxidize it to sulfuric acid (Eq. 6). Sulfuric acid then reacts with the metal and solubilizes it from the host material (Eq. 7).

$$2S^0 + 3O_2 + 2H_2O \rightarrow 2H_2SO_4, \quad (6)$$

$$2M^0 + O_2 + 2H_2SO_4 \rightarrow 2M^{2+} + 2SO_4^{2-} + 2H_2O. \quad (7)$$

In recent bioleaching studies, acidolysis is often coupled with complexolysis and redoxolysis to enhance the leaching efficiency (Baniasadi et al., 2019; Işıldar et al., 2016). Biogenic acids are also used to control the pH of the leaching process (Heydarian et al., 2018). Still, acidolysis remains a powerful mechanism to utilize in bioleaching processes. Jowkar et al. (2018) achieved a 100% and 10% leaching rate for In and Sr, respectively, from LCD panels using only *At. thiooxidans*.

3.2 Bioleaching Optimization and Limiting Factors

3.2.1 Microorganisms

Bioleaching relies on the metabolic capacity of microorganisms to mobilize metal ions from the original substrate. The microorganisms involved produce leaching agents such as oxidizing agents (Fe^{3+}), complexing agents (CN^-), siderophores as well as organic (oxalic, malic, citric) or inorganic acids (H_2SO_4). The bioleaching efficiency lies in optimizing the production of these leaching agents. Whereas bioleaching is limited by abiotic factors such as pH, temperature, particle size, substrate concentration or nutrient, the bacterial concentration and metal tolerance are biotic factors that are critical for one-step processes (Jagannath et al., 2017; Sun et al., 2020).

Researchers have recommended the use of consortia for the leaching of e-waste. Microbial consortia can be found in a variety of habitats, where they work cooperatively to carry out metabolic activities, resulting in improved productivity over monocultures (Frey-Klett et al., 2011; Liu et al., 2011; Valix, 2017). Furthermore, the paring of suitable strains can result in reinforced metal resistance which can be attributed to physiological changes and a higher metal efflux (Ilyas et al., 2014).

3.2.2 pH

Microorganisms require optimum pH conditions for their growth and activity. Optimum pH conditions result in better leaching agents' production and bioleaching capability. Although most research on cyanogenic bacteria focuses less on optimum pH rather than initial pH, a few studies have shown that *C. violaceum* requires a slightly alkaline pH of 8 for growth, while cyanide production is maximized around pH 10 (Faraji et al., 2021; Li et al., 2015; Pourhossein et al., 2021; Shin et al., 2013). For acidophilic bioleaching involving iron-oxidizing bacteria, a pH above 2.5 causes jarosite to precipitate and trap Fe^{3+} in a thin layer that is deposited on the waste particle surface. This results in a decreased bioleaching efficiency. Moreover, an increase in pulp density can cause an increase in pH since most waste material is alkaline in nature (Pourhossein & Mousavi, 2018; Roy et al., 2021). Thus, most research tries to maintain a pH lower than 2. Initial pH is generally adjusted with the addition of sulfuric acid and can be controlled by the production of biogenic H_2SO_4 by sulfur-oxidizing bacteria such as *At. thiooxidans* (Işıldar et al., 2016). Arshadi and Mousavi (2015) showed that the mobilization of Ni is the most impacted for PCB bioleaching, achieving the maximum recovery for Ni and Cu at pH 1. With the same substrate, Erust et al. (2020) found the optimum pH for Cu, Ni, Al, and Zn to be around 1.8. A recent study on LIB bioleaching using a combination of *At. thiooxidans* and *At. ferrooxidans* indicated the optimum pH for Co, Ni, and Li to be 1.5 (Heydarian et al., 2018).

3.2.3 Temperature

Microbial growth activity is influenced by temperature, which is reflected in bacterial bioleaching capability. Microbes are classed as psychrophiles, mesophiles (e.g. *At. ferrooxidans*), or thermophiles (e.g. *Sulfobacillus thermosulfidooxidans*) according to their temperature dependency. Extremophilic bacteria such as psychrophiles and thermophiles grow at extremely low (−20 °C to 10 °C) or high temperatures (40–85 °C), while mesophiles thrive at temperatures ranging from 10 °C to 40 °C (Babu et al., 2015; Sethurajan & Gaydardzhiev, 2021). The temperature of spent medium can be increased to facilitate the dissolution of metals without changing the optimal growth temperature. This was performed by Cui et al. (2020) and Jadhav et al. (2016), where they performed the bioleaching of LCD and PCB, respectively, using *A. niger*. Temperatures of 70 °C and 80 °C were achieved, resulting in improved metal mobilization. Most bioleaching studies use mesophiles, so the optimum temperature usually lies around 30 °C. Studies that use consortia of mesophile and thermophile require temperatures around 45 °C (Akbari & Ahmadi, 2019).

3.2.4 Growth Medium

The composition of growth medium has a significant impact on the efficiency of bioleaching. A medium with an optimized substrate concentration allows the better growth of the cultivated microorganisms, as well as, improved leaching efficiency. The substrate used differs depending on the microorganism.

Chemolitotrophs such as *At. ferrooxidans* and *At. thiooxidans* utilize ferrous iron in the form of $FeSO_4$ and elemental sulfur, respectively. The primary energy source used by cyanide-producing heterotrophic bacteria is glycine and many fungi use organic substrates such as sucrose or glucose.

Since *At. ferrooxidans* is an autotrophic microorganism, the rate of iron and sulfur oxidation is partly dependent on the metabolic fixation of CO_2 (Silver, 1970; Valdés et al., 2008). Guezennec et al. (2018) performed bioleaching on a copper concentrate while increasing the partial pressure of CO_2 in the incoming gas. This study showed that the fraction of CO_2 in the air was insufficient for optimal kinetics and that enrichment along with a more active bacterial community was necessary for more efficient CO_2 uptake.

Many studies investigated the influence of Fe^{3+} concentration on bioleaching experiments. High Fe^{3+} concentration is a major constraint as it leads to the precipitation of jarosite (Eq. 8).

$$K^+ + 5OH^- + 3Fe^{3+} + SO_4^{2-} + H_2O \rightarrow KFe_3(SO_4)_2(OH)_6 + H^+. \quad (8)$$

Jarosite formation decreases initial Fe^{3+} content and thus leaching efficiency. This leachate loss is a major issue for metal recovery. To overcome this issue, Wang et al. (2018) have proposed an experimental model to mobilize copper from PCBs by minimizing the initial $FeSO_4$ input and using a combination of iron-oxidizing and sulfur-oxidizing bacteria. A copper leaching rate of over 95% was achieved using only 0.25 g/L of $FeSO_4 \cdot 7H_2O$ (and 10 g/L S^0), which is 239 times lower than in similar studies.

In recent studies, statistical models are often used to determine the optimal parameters for maximum mobilization of metals in solution. In particular, the response surface method (RSM) is widely used, and its purpose is to explore the relationships between different sets of variables, whether dependent or independent. This method is usually done in central composite design (CDD), which takes the form of a polynomial quadratic equation, and allows to predict for a given set of variables the number of experiments needed to determine the desired response (e.g. percentage of metal recovered). In e-waste bioleaching experiments, these variables are typically initial pH, inoculum size, substrate concentration, pulp density, and particle size.

Following this method, Arshadi and Mousavi (2015) optimized four critical parameters for a mobile phone PCBs leaching experiment which are pH, Fe^{3+} concentration, pulp density, and particle size. It was found that Fe^{3+} concentrations between 2 and 8 g/L were suitable for better metal recovery. Erust et al. (2020) showed that a Fe^{3+} concentration of 9 g/L displayed the highest leaching yield for all metals as well as the highest ORP, compared to lower concentrations. Heydarian et al. (2018) performed the bioleaching of mobile phone LIBs using a combination of *At. ferrooxidans* and *At. thiooxidans*. For this purpose, the initial concentration of the two substrates $FeSO_4$ and S^0 were investigated. Maximum recovery was achieved with a concentration of 36.7 g/L for $FeSO_4$ and 5.0 g/L for S^0, with an initial pH of 1.5.

RSM was also used by Jowkar et al. (2018) to determine the optimum S^0 concentration for LCD bioleaching using *At. thiooxidans* and resulted in 8.6 g/L. The same experimental design was used by Bahaloo-Horeh and Mousavi (2017) to investigate the parameters for LIB bioleaching using the fungus *A. niger*, which uses organic compounds such as sucrose or glucose as substrate. They reported that the optimum sucrose concentration was 116.90 g/L. On the other hand, for the bioleaching of In from LCD panels, Cui et al. (2021) reduced the sucrose concentration from 100 to 50 g/L to increase its utilization. This resulted in an enhancement of oxalic acid production for the bioleaching of In from LCD panels. Kim et al. (2016) investigated the leaching potential of six *Aspergillus* species for the recovery of metals from Zn–Mn and Ni–Cd batteries. They were able to compare the production of organic acids using either sucrose or maltose as substrate. Citric acid production was strongly favoured by the use of sucrose medium, while maltose medium led to a higher production of oxalic acid. Greater recovery of Zn, Co, Ni, and Mn was observed using the sucrose medium. It was suggested that high concentrations of oxalic acid produce precipitates of metal complexes and thus hinder the recovery of metals (Biswas et al., 2013).

Substrate optimization was also investigated in biogenic cyanide-assisted bioleaching. Cyanogenic bacteria depend on glycine to produce cyanide. Kumar et al. (2018) report that increasing the glycine concentration from 2.5 to 7.5 g/L enhanced the metal leaching capabilities of *Pseudomonas balearica* SAE1 bacterial strain, but a higher concentration decreased the leaching of the targeted metals. This phenomenon was also observed by Işıldar et al. (2016), where a 10 g/L glycine concentration had an inhibitory effect on the cyanide production and growth of *P. fluorescens*. Li et al. (2015) reported that the quantity of cyanide produced by *C. violaceum* can be increased by adding a small amount of nutritional salts to the growth medium. There was a noticeable improvement in gold leaching efficiency with the addition of NaCl and $MgSO_4{\cdot}7H_2O$.

Hubau et al. (2018) compared the effect of two distinct culture media on the bioleaching of PCBs. For this purpose, a consortium of bacteria belonging to the genus *Leptospirillum*, *Acidithiobacillus*, and *Sulfobacillus* was cultured in continuous mode. The two media, 0 km and 0Cm, had the same composition, except for the amount of $(NH_4)_2SO_4$ which was 3.70 g/L and 0.4 g/L, respectively. Reducing the NH_4^+ content allowed a better bio-oxidation rate with an influent Fe^{2+} concentration of 9 g/L. The decrease in NH_4^+ concentration limited the precipitation of ammonium jarosite, thereby restricting the coating of the activated charcoal used as solid support. The author suggests that modifying the concentration of monovalent ions such as potassium or ammonium in the medium could decrease the formation of jarosite, since potassium jarosite is more stable than ammonium jarosite (Kaksonen et al., 2014). This in turn, improves the bacterial attachment onto the charcoal surface.

3.2.5 Pulp Density

Pulp density is one of the most important parameters to consider in e-waste bioleaching. It refers to the ratio of solid waste per unit of liquid media. The higher the pulp density, the greater amount of metals is available for bioleaching, which increases the potential feasibility of the processes on an industrial scale. However, most laboratory-scale studies achieve a higher metal leaching rate when working with low pulp densities. E-waste is highly concentrated in metals and toxic compounds that can be harmful to microorganisms and reduce their activity (Ilyas et al., 2010; Liang et al., 2010). Moreover, high pulp densities lead to an increase in pH and thus a decrease of protons available to react with metals (Bahaloo-horeh & Mousavi, 2017). It can also result in reduced transfer rates of oxygen and carbon dioxide, and waste particles can potentially cause mechanical damage to the microorganisms (Haghshenas et al., 2009; Roy et al., 2021).

PCB bioleaching was performed by Kumar et al. (2018) using *Chromobacterium violeaceum* and *P. balearica* SAE1 at pulp densities ranging from 10 to 500 g/L. While the *Chromobacterium* strain was inhibited at 200 g/L of e-waste within 24 h, SAE1 was able to fully grow at a concentration of 500 g/L. Bacterial growth has a very important effect on the amount of leachate produced and is essential for efficient leaching. At 10 g/L waste, the amount of Ag and Au leached was 56.4% and 30.5%, respectively, while at 100 g/L, the recovery dropped to 8.3% and 1.8%. The same behaviour was observed in one of the most recent PCB bioleaching studies (Erust et al., 2020; Heydarian et al., 2018). Studies on bioleaching of LIBs show a similar effect of pulp density on metal recovery (Liu et al., 2020). Fungi bioleaching carried out by Bahaloo-horeh and Mousavi (2017) exhibits high recovery rates for low pulp density: 100% Cu, 100% Li, 77% Mn, and 75% Al at 20 g/L and 64% Co and 54% Ni at 10 g/L.

Several techniques are used to overcome the toxic effects of high pulp densities. The first is the development of an adaptation by subjecting the microorganisms to increasingly high concentrations of waste. This was done by Pouhossein and Mousavi (2018) using *At. ferrooxidans* at concentrations ranging from 5 to 25 g/L of PCB powder. The adapted bacteria culture displayed higher redox potential, bacterial concentration, and Fe^{3+} concentration than the non-adapted one at 20 g/L, which is the limit before the activity was greatly inhibited. The delay in Fe^{2+} oxidation caused by PCB toxicity can be greatly reduced by successive subcultures (Anaya-Garzon et al., 2021).

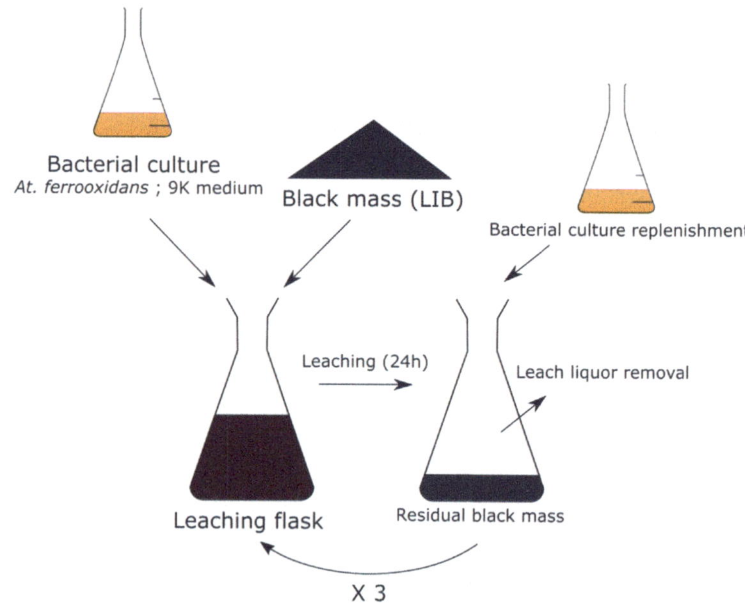

Fig. 2 Diagram illustrating the multiple-step leaching performed by Roy et al. (2021)

The multiple-step process has been highlighted as a successful way to diminish PCB toxicity on bacteria (Işıldar et al., 2016). Liang et al. (2010) have shown that three-step acidophilic bioleaching could improve the yield by 19% compared to a two-step approach. Moreover, it has been proven that iron-oxidative bacteria's major role is to regenerate Fe^{3+} and a non-contact mechanism is sufficient to leach copper from waste PCB (Wu et al., 2018). Fungal bioleaching of PCB has also been performed using the organic acids contained in the spent medium (Jadhav et al., 2016, Bahaloo-horeh et al., 2016). This method permits bypassing the toxicity and optimizing the production of the leachates. Roy et al. (2021) achieved a leaching rate of 94% of Co and 60% of Li at 100 g/L of LIB powder by removing the leach liquor and replenishing the bacterial culture every 24 h for a total of three cycles (Fig. 2).

4 Biorecovery of Critical Metal from the Pregnant Leaching Solution

Pre-treatment and bioleaching are two necessary steps in the recovery of metals from e-waste, but it is the bioprocessing of PLS that determines the economic feasibility of biotechnologies in the recovery of metals from e-waste. However, very few studies focus on the biorecovery of metals from e-waste PLS and research is at a very early stage. Promising techniques exist and have been applied successfully for the processing of PLS but mostly for the treatment of metal-rich waters such as acid mine drainage (AMD) and metallurgical leachates. A few recent biorecovery studies and key parameters are documented in Table 2.

4.1 Comparison Between AMD and E-Waste PLS and Required Pre-treatment

AMD is the result of contact between oxygen, water, and the geological subsurface during mining activities. This leads to the oxidation of minerals being in a reduced state, notably sulfides such as pyrite (FeS_2) which is the most common (Christensen et al., 1996). The oxidation of pyrite can be summarized with this reaction (Eq. 9):

$$2FeS_2 + 7O_2 + 2H_2O \rightarrow 2Fe^{2+} + 4SO_4^{2-} + 4H^+. \quad (9)$$

Oxidation of ferric iron is also occurring, catalyzed by iron-oxidizing bacteria such as *At. ferrooxidans* (Stumm & Morgan, 1981), following this reaction (Eq. 10):

$$Fe^{2+} + 1/4O_2 + H^+ \rightarrow Fe^{3+} + \frac{1}{2H_2O}. \quad (10)$$

Ferrous iron then precipitates as hydroxide or jarosite at pH values above 2 (Eq. 11), and the residual Fe^{3+} participates in the oxidation of pyrite (Eq. 12) (Akcil & Koldas, 2006):

$$Fe^{3+} + 3H_2O \rightarrow Fe(OH)_3 + 3H^+, \quad (11)$$

$$FeS_2 + 14Fe^{3+} + 8H_2O \rightarrow 15Fe^{2+} + 2SO_4^{2-} + 16H^+. \quad (12)$$

The combined reaction (Eq. 13) leads to the production of acid which facilitates the dissolution of pyrite:

$$4FeS_2 + 15O_2 + 14H_2O \rightarrow 4Fe(OH)_3 + 8SO_4^{2-} + 16H^+. \quad (13)$$

Besides the fact that other sulfides (ZnS, $CuFeS_2$) undergo similar oxidation processes, the acidic conditions of the

Table 2 Literature review of biorecovery of critical metals from pregnant leaching solutions

Mechanism	Microorganism(s)	E-waste	Culture medium	Recovery time	T (°C)	Initial pH	Metal recovered	Reference
Biosorption	*Aspergillus carbonarius* dead biomass	Synthetic solution ($K_2Cr_2O_7$)	Cornflour medium	12 h	37	2	Cr(VI) 92.43%	Lakshmi (2020)
Bioreduction Nanoparticles biosynthesis	*Bacillus licheniformis* (FZUL-63)	Synthetic polymetallic solution	LB medium	48 h	30	7	Au 90.4%	Cheng et al. (2019)
Biosorption	*Chlorella vulgaris*	Nd–Fe–B hard disk magnets	/	90 min	35	5	Nd 63.96%	Kucuker et al. (2017)
Bioreduction Nanoparticles biosynthesis	*Desulfovibrio desulfuricans* (ATCC 29577)	Electronic scrap	Post-gate medium C	Three-step process: 24 h, 24 h, 4 h	/	/	Cu > 99% Pd > 95%	Creamer et al. (2006)
Bioreduction Nanoparticles biosynthesis	*Magnetospirillum* MTB consortia	PCB						Sannigrahi and Suthindhiran (2019)
Bioreduction Carbonates	Metal reducing bacteria	PCB	$NaHCO_3$ 2.5 g/L $CaCl_2 \cdot 2H_2O$ 0.08 g/L NH_4Cl 1.0 g/L $MgCl_2 \cdot 6H_2O$ 0.2 g/L NaCl 10 g/L Yeast extract 0.5 g/L HEPES 7.2 g/L Trace mineral 10 mL Vitamin 1 mL	14 d	25	7.7	Mn	Yumi Kim et al. (2018)
Bioreduction Sulfides	Mixed bacterial culture (SRB)	Ni–Cd battery	KH_2PO_4, 0.5 K_2HPO_4, 0.1 $NaHCO_3$ 0.5–0.8 NH_4Cl 0.3 $CaCl_2 \cdot 2H_2O$ 0.05 $NaMo_4 \cdot 2H_2O$ 0.03 KCl 0.05 $CoCl_2 \cdot 6H_2O$ 0.02 $MgCl_2 \cdot 6H_2O$ 0.05 $NiSO_4 \cdot 6H_2O$ 0.01 Sodium ascorbate 0.05 Sodium thioglycollate 0.05	10 d	30	7		Paul et al. (2020)
Bioreduction Sulfides	SRB	Synthetic spent catalyst solution	NH_4Cl 1 g/L $Na_2SO_4 \cdot 7H_2O$ 4.15–5.5 g/L $CaCl_2 \cdot 6H_2O$ 0.06 g/L $MgSO_4 \cdot 7H_2O$ 0.06 g/L Yeast extract 0.1 g/L KH_2PO_4 0.5 g/L Ascorbic acid 0.05 g/L Nathioglycollate 0.05 g/L $FeSO_4 \cdot 7H_2O$ 0.1 g/L Sodium lactate	90 d	20–25	2	Mo 72% Co 20%	Cibati et al. (2013)
Bioreduction Sulfides	SRB	Synthetic AMD	Glycerol 5 mM Yeast extract 0.1 g/L	100 d	30	2.2	Cu	Colipai et al. (2018)
Bioreduction Sulfides	SRB	Metallurgical leach residues	K_2HPO_4 0.5 g/L NH_4Cl 1.0 g/L $CaSO_4$ 1.0 g/L $FeSO_4 \cdot 7H_2O$ 0.5 g/L Sodium lactate 3.5 g/L $MgSO_4 \cdot 7H_2O$ 2.0 g/L Yeast extract 1.0 g/L Ascorbic acid 0.1 g/L	15 d			Zn 97%	Sethurajan et al. (2017)

(continued)

Table 2 (continued)

Mechanism	Microorganism(s)	E-waste	Culture medium	Recovery time	*T* (°C)	Initial pH	Metal recovered	Reference
Bioreduction Sulfides	SRB community	LiB	K_2HPO_4 0.5 g/L NH_4Cl 1.0 g/L Na_2SO_4 1.0 g/L $MgSO_4{\cdot}7H_2O$ 2.0 g/L $FeSO_4{\cdot}7H_2O$ 0.5 g/L $CaCl_2{\cdot}2H_2O$ 0.1 g/L Sodium thioglycolate 0.1 g/L Ascorbic acid 0.1 g/L Yeast extract 1.0 g/L Sodium-DL-lactate 2.0 g/L	181 d	30–40	7.5	Al 99% Ni 99% Co 99% Fe 99.5% Mg 49.1% Mn 98.9% Cd 98.6% Zn 98.4% Cu 99.9%	Calvert et al. (2019)

environment lead to the dissolution of other metals associated with mining activities such as cadmium, zinc, copper, lead, and arsenic (Stumm & Morgan, 1996). Thus, AMDs can be characterized as aqueous, highly acidic matrices that contain high concentrations of dissolved metal and sulfate ions in which microorganisms are key contributors to mineral leaching. In this regard, AMDs are extremely similar to acidic e-waste PLS resulting from bioleaching processes such as redoxolysis and acidolysis. Since very few studies focus on real e-waste bioleachate, AMD biotreatment methods for the recovery of metal could be a reliable source of insight for future e-waste downstream bioprocesses.

As previously mentioned, most of the bioleaching processes are taking place in acidic conditions and iron is either added in the leaching solution as oxidant (Yazici & Deveci, 2014; Sethurajan & van Hullebusch, 2019; Wang et al., 2018) being regenerated by iron-oxidizing bacteria or originating from the leaching of WEEE material processed (Işildar, 2019). In view of the heterogeneity in metal composition of WEEE materials and selective nature of bioleaching systems, many metals dissolve in the leaching process, leading to the undesired contamination of PLS. PLS mostly originating from waste of printed circuit boards (WPCBs) leaching contains many impurity metals (Fe in particular, mostly present in its ferrous form) at significant levels (Işildar, 2019). These impurities should be reduced to acceptable levels to avoid problems in downstream recovery processes, e.g. low current efficiencies in electrowinning of copper from sulfate solutions in the presence of ferric/ferrous ions (Wang et al., 2018) or base metal sulfide sequential precipitation (Sethurajan & van Hullebusch, 2019). For instance, the PLS can be purified by goethite (FeOOH) and jarosite ($NaFe(III)_3(OH)_6(SO_4)_2$) precipitation before being processed for recovery step (Wang et al., 2018; Yazici & Deveci, 2014). Jarosite precipitation is widely applied in hydrometallurgy to remove iron from PLS (Miettinen et al., 2019). The precipitation as jarosite also allows the removal of sulfate ion which usually contained in the PLS as sulfuric acid is often used for performing the acidic (bio) leaching. Wang et al. (2018) showed that iron can be readily removed (in the form of goethite) from synthetic/real leach solutions of WPCBs to a large extent ($\geq$93%) at the expense of high copper losses particularly at pH 4.0. Such performance has been possible by implementing the biological oxidation of ferrous to ferric ion contained in the PLS by *At. ferrooxidans* leading to the formation of iron precipitates like FeOOH and Fe_3O_4 (magnetite). Hydrogen peroxide (H_2O_2) addition, before the precipitation with NaOH, allows overcoming the poor chemical oxidation of Fe^{2+} with yields higher than 85% as reported by Amato et al. (2020) allowing also the formation of iron oxides' precipitates.

4.2 Bioprecipitation

Precipitation is a rather straightforward approach to the recovery of base metals. It consists of shifting the chemical equilibrium of a system by causing a reaction in which one of the products is poorly soluble in the reaction solvent. This is most commonly achieved with the addition of precipitants such as hydroxides (NaOH, KOH, $Ca(OH)_2$), carbonates ($CaCO_3$), and sulfides (Na_2S, NaHS, H_2S) (Huang & Yang, 2021; Michalkova et al., 2013). In bioprecipitation processes, the formation of minerals is the result of an interaction between metabolic by-products and the elements in the matrix. Precipitation is dependent on the solubility of the solid in a given solvent, with PLSs generally in the form of an aqueous matrix. The solubility of metal compounds depends on several factors: pH, the concentration of ions in solution, and temperature. In particular, pH adjustment plays an important role in the selective recovery of metals. The use of bases and acids to regulate the pH of the solution is necessary to achieve the desired precipitation pH.

4.2.1 Sulfide Bioprecipitation

Precipitation and bioprecipitation are widely used techniques for the treatment of AMD. The traditional method is to use an alkali to raise the pH and precipitate the metals (Tabak et al., 2003). Bhattacharya et al. (1979, 1981) showed that the use of sulfide for the treatment of AMD displays significantly superior removal rates compared to lime and sodium hydroxide. Moreover, sulfide precipitation presents several advantages compared to hydroxide precipitation: metal sulfides have low solubility products which is practical for acidic solutions; selectivity is achieved much more efficiently; reaction kinetics are faster; less sludge is produced due to the high density of sulfides (Cibati et al., 2013). Bioprocesses based on sulfate reduction have recently emerged as promising alternatives to traditional chemical metal-containing water treatment and metal recovery. Biogenic sulfides are produced by anaerobic bacteria called "sulfate-reducing bacteria" (SRB) which play a crucial part in the carbon and sulfur cycle (Muyzer & Stams, 2008). SRB respire using organic carbon sources to reduce the sulfate ions to hydrogen sulfide (Eq. 14). Biogenic hydrogen sulfide gas then reacts with metal cations to precipitate insoluble metal sulfide (Eq. 15).

$$2CH_2O + SO_4^{2-} \rightarrow H_2S + 2HCO_3^-, \quad (14)$$

$$H_2S + M^{2+} \rightarrow MS\,(s) + 2H^+. \quad (15)$$

($2CH_2O$: organic carbon source; M^{2+}: metal cations).

One of the main advantages of biogenic sulfur precipitation is the selectivity of the process, which is highly dependent on pH. Sethurajan et al. (2017) studied the selective recovery of Zn from metallurgical leach residues. By removing first more than 85% of Fe through the neutralization of the bioleachate with the addition of NaOH, a removal rate of 97% of Zn was achieved. In fact, Fe is one of the major factors that affects the selective precipitation of sulfides as it can precipitate at low pH (2–3), forming jarosite or goethite (Hu et al., 2012). Calvert et al. (2019) performed a non-selective biogenic sulfide bioprecipitation of metals from a LIB leachate in a fluidised bed reactor (FBR). This study displayed excellent recovery rates (>98%) for most metals. The authors stated that selectivity could be achieved by optimizing precipitation pH and reagent concentrations. Cibati et al. (2013) achieved selectivity and high purity indexes with a less complex leachate which contained only four metals (Mo^{6+}, Ni^{2+}, Co^{2+}, V^{4+}). However, vanadium was recovered as hydroxide and the separation of Co and Ni sulfide was unfeasible due to their solubility product being very similar (Pauling, 1970). The pH of each precipitation vessel was adjusted using NaOH. Colipai et al. (2018) investigated the selective synthesis of covellite (CuS) nanoparticles from a polymetallic synthetic AMD using acidophilic sulfate-reducing bacteria (aSRB) in a two-step process without any pH adjustment. Also, Janyasuthiwong et al. (2016) achieved 90% Cu recovery by using continuous SRB bioreactor that could precipitate Cu from PCB leachates (1.0 M HNO_3). Calvert et al. (2019) reported that over 99% precipitation efficiency was achieved for Al, Ni, Co, and Cu using biogenic sulfide from lithium-ion battery (LIB) leachates. Paul et al. (2020) reported that Cd and Ni were microbially recovered from digested Ni–Cd battery liquor as Ni and Cd sulfides' nanoparticles that have potential to be used as semi-conducting material.

4.2.2 Reductive Bioprecipitation

Reductive bioprecipitation describes the biologically assisted reduction of metals or metalloids (generally, precious metals) from their ionic form to a less soluble state. It occurs either through direct contact with the cell surface or via extracellular electron transfer (EET) and metal chelators (Manzella et al., 2013). Although the nature of the metal that can be reduced is entirely dependent on the microorganism and its enzymatic activity, some bacteria such as *D. desulfuricans* are active for a range of metals (Fe(III), U(IV), Cr(VI)) through the use of the hydrogenase enzyme. Similarly, many bacteria and archaea that can reduce Fe have been shown to reduce Au(III) to insoluble Au(0) (Kashefi et al., 2001). Thus, selectivity in reductive bioprecipitation processes ought to be achieved through different pathways.

The biological reduction of precious metals to metallic nanoparticles has attracted some attention in recent years (Pat-Espadas & Cervantes, 2018; Wong-Pinto et al., 2020). The production of nanomaterials from waste streams using microbes is particularly interesting because: (i) the nanomaterials can be directly reuse in catalytic/industrial processes, (ii) the microbial biomass is considered as economical catalyst and the bioconversion can be conducted at ambient temperature and pressure. The treatment of electronic waste leachate by Creamer et al. (2006) with *D. desulfuricans* is a key reference in this matter. In a three-step process, it was possible to precipitate Au^0, Pd^0, and Cu^{2+} separately. First of all, high content of Cu^{2+} inhibited the hydrogenase-driven removal of Pd^{2+} which enabled the extracellular precipitation of Au^0. Secondly, a fraction of *D. desulfuricans* biomass was pre-treated with the addition of Pd^{2+}, which accumulated on the biomass in the absence of Cu^{2+}. These "palladised cells" catalyzed the reduction of Pd^{2+} to Pd^0. The reduction of Au and Pd is enabled by the sparging of H_2. At last, the Pd-depleted leachate was treated with dimethyl disulfide produced by *Klebsiella pneumoniae* and *Escherichia coli* (Essa et al., 2006) and >99% of Cu^{2+} was recovered as a mix of sulfate and hydroxide.

Fungi are also able to accumulate Au from chloroauric acid solutions ($HAuCl_4$) both intra- and extracellularly. Au(III) reduction in Au(0) nanoparticles (AuNPs) can occur both within the cell wall but also inside cell compartments, such as the plasma membrane, the cytoplasm, and/or the nucleus (Bindschedler et al., 2017). In view of their metabolic capability, fungi should be able to first mobilize Au from e-waste through leaching processes and second to ideally immobilize it in a form that is easily recovered through biomineralization processes (AuNPs synthesis) or biosorption (Bindschedler et al., 2017).

4.2.3 Microbially Induced Carbonate Precipitation (MICP)

MICP is a biological process that uses the metabolic activity of microorganisms to precipitate carbonates. Bioprecipitation of calcium carbonate is a geochemical phenomenon that is already well known as it is common in seawater, sediments, freshwater, soil (Castro-Alonso et al., 2019) and involves several mechanisms including sulfate reduction, denitrification, ammonification, ureolysis, and photosynthesis (Zhu & Dittrich, 2016). In the last 20 years, MICP has received particular attention for its applications in solidification of biocements (Belie, 2010; Ramachandran et al., 2001), soil improvement (Ashraf et al., 2017; Wang et al., 2017) and for the remediation and immobilization of heavy metals (Achal et al., 2011; Li et al., 2013; Qiao et al., 2021). Hydrolysis of urea is the most employed metabolic pathway, generally involving the bacillus *Sporosarcina pasteurii*. Ureolysis-driven carbonate precipitation occurs in several steps: first, the urease enzyme hydrolyzes urea into ammonia and carbamic acid (Eq. 16) which spontaneously reacts with water to form ammonia and carbonic acid (Eq. 17). Ammonia is then transformed into ammonium (Eq. 18), a reaction that releases hydroxides ions which lowers the pH and induces the carbonate ions' formation (Eqs. 19 and 20).

$$CO(NH_2)_2 + H_2O \xrightarrow{\text{Uréase}} NH_2COOH + NH_3, \quad (16)$$

$$NH_2COOH + H_2O \leftrightarrow H_2CO_3 + NH_3, \quad (17)$$

$$H_2CO_3 \leftrightarrow HCO_3^- + NH_3, \quad (18)$$

$$HCO_3^- + H^+ + 2OH^- \leftrightarrow CO_3^{2-} + 2H_2O. \quad (19)$$

Finally, divalent cations (e.g. Ca^{2+} or divalent metal M^{2+}) react with carbonate ions to precipitate as carbonates (Eq. 20).

$$M^{2+} + CO_3^{2-} \leftrightarrow MCO_{3(s)}. \quad (20)$$

MICP has rarely been used for the downstream processes of e-waste leaching solutions. Rhodochrosite ($MnCO_3$) was precipitated by Kim et al. (2018) from a waste mobile phone leaching solution using a metal reducing bacteria isolated from soil sample. Although the initial purpose of the experiment was to form reduced metals, it was suggested that $MnCO_3$ precipitation was initiated by the production of bicarbonate as a metabolic by-product.

The immobilization of heavy metals via the ureolytic pathway has shown promising results. Li et al. (2013) achieved high removal rates (88–99%) of Cd, Ni, Cu, Pb, Co, and Zn from a synthetic solution of metal chloride. The experiments were performed using a bacterium isolated from soil, as well as, *Sporosarnica pasteurii* and *Terrabacter tumescens*. The precipitates were identified as carbonates and showed rhombohedral ($NiCO_3$, $CoCO_3$), spherical ($CuCO_3$, $CdCO_3$), and needle-shaped ($PbCO_3$, $ZnCO_3$) structures.

Qiao et al. (2021) tested the heavy metal removal effect of isolated MICP strains from mining soils. The selected ureolytic consortium showed excellent performance in immobilizing Cd, Cu, Pb, Ni, and Zn, with removal rates up to 98%. The author emphasizes that the volume of inoculation must be controlled for the precipitation of carbonate crystals. The NH_3 produced during the hydrolysis of urea can lead to the formation of ammonia liquor which may have caused the dissolution of the precipitates (Stephen, 1975). Thus, except for Cd which does not seem to be affected, the optimal volume of bacteria would be between 1.25 and 4%. Only Cd and Pb precipitates could be characterized as carbonates by X-ray diffraction, while copper precipitates were identified as hydroxides. This may indicate that copper removal is mainly driven by the increase in pH during urea hydrolysis and that hydroxide and carbonate precipitation both compete for copper removal.

4.3 Biosorption

Biosorption is a physicochemical and metabolically independent biological process based on a variety of mechanisms including sorption (absorption and adsorption), ion exchange, surface complexation, and precipitation that represents a biotechnological cost-effective innovative way for the recovery of base, precious, and critical metals from aqueous solutions.

Additionally, for this extracellular process, live cells are not required; therefore, the process can be performed under extreme temperature and pH conditions (high and low, respectively). This results in an advantage from an industrial point of view as the process conditions' parameters are less difficult to manage. However, if live cells are used, greater attention needs to be paid in order to control the process and achieve the highest efficiencies. Another advantage of the extracellular process is that recovered materials can be easily retrieved by centrifugation and filtration. The main disadvantages of this process are the low volume of metals/metalloids recovered and the long process duration.

Applying biosorption to metal recovery from e-waste is a viable option. Chemical processes including complexation, redox reactions, chelation, and precipitation take place, as well as physical interactions such as ion exchange between the biosorbent and the metals in solution. Each biosorbent has a particular set of properties depending on its functional groups which determine its biosorption capacity. The biosorbent and the operating conditions, such as pH, temperature, biosorbent concentration, and metal charge, determine the efficiency of the process.

Processes involving the use of bacteria for metal recovery from pregnant leachate solutions have been used in the hydrometallurgical industry for a long time, as well as for metal recovery from mine drainage. The practices have led the industry towards the search for environmentally friendlier and non-harmful processes. However, as it was previously mentioned, the biosorption process takes place in an aqueous phase, which poses a major setback for its application to e-waste recycling since metals have to be first leached out of the solid. In addition, in order to maximize the capacity of the biosorbent, it is necessary to chemically modify its surface. These factors make the process feasible only at laboratory scale, requiring further investigation for its scale-up and commercialization (Ramanayaka et al., 2020).

Several review papers have detailed the sorption mechanism as well as the physicochemical conditions influencing the sorption efficiency (Das & Ting, 2017; Dodson et al., 2015; Fomina & Gadd, 2014; Gadd, 2009; Gupta et al., 2019; Mack et al., 2007; Volesky, 2007; Zeraatkar et al., 2016). However, a limited number of reports are available with respect to the recovery of base, precious and critical metals from WEEE in the context of biosorption: living or inactive biomass from bacterial (Brewer et al., 2019), fungal (Sinha et al., 2018), microalgae (Čížková et al., 2019; Kucuker et al., 2017); macroalgae (e.g. *Ascophyllum nodosum*, Pennesi et al., 2019), or plant origin (e.g. persimmon tannin, Gurung et al., 2013; wood pulp, Varshney et al., 2017). Metal containing solutions such as PLS are often characterized by a pH value below 3 as well as complex composition including different competing elements that affect the biosorptive properties. Consequently, the major challenges of biosorption approaches are stability of the sorbing materials, selectivity, effectiveness, and cost efficiency.

A study of a biosorption–bioreduction process for the recovery of Te and Cd from used solar cells was published by Rajwade and Paknikar (2003) as reported in Nancharaiah et al., (2016). Using *Pseudomonas mendocina,* to remove Ag and Cd from the acid leachate of the solar cells, Te was recovered as TeO precipitate. A more recent study describes the use of *Pseudomonas* sp. strain EPR3 for directly releasing TeO from insoluble materials including tellurium dioxide (TeO_2), cadmium telluride (CdTe), and bismuth telluride (Bi_2Te_3) (Bonificio & Clarke, 2014).

4.3.1 Bacterially Mediated Surface Adsorption

Microbially mediated surface adsorption offers a potentially cost-effective and eco-friendly approach for metal recovery. Microorganisms (bacteria and fungi) have long been exploited as biosorbents for base, precious, and critical metals' extraction, most notably for wastewater treatment. Microbial surface adsorption is a potential method for recovering metals from aqueous solutions containing metals extracted from WEEE PLS. However, there is limited research on WEEE resource recovery using this method due to the limited binding specificity of the surface functional groups, which can bind many cations with high affinity. Pollmann et al. (2018) reviewed some recent biotechnological developments in the field of biosorption and its use for critical metals' recovery from different solid waste streams including e-waste. However, only few of these developments are commercialized due to high implementation costs in comparison to conventional methods as well as its low element selectivity. Several studies have focused on the addition of metal-binding peptide tags onto bacterial surface proteins, and they have been shown to sequester more metal such as critical elements than controls. Nevertheless, these peptide tags offer limited specificity to the metals that are sorbed (Pollmann et al., 2018; Zhu et al., 2019).

In view of the increasing demand for rare earth elements (REEs) which are indispensable components of many green technologies (Dev et al., 2020; Park et al., 2020; Yu et al., 2020), numerous studies dedicated to the use of engineered bacteria have been published. Several studies have shown significant REE adsorption capacity of various microbes, highlighting their potential for REE extraction (Dev et al., 2020). For instance, Arunraj et al. (2019) reported the use of yeast (*Saccharomyces cerevisiae*) immobilized in a cellulose matrix as an efficient sorbent for the recovery of Eu(III) from an aqueous medium containing spent fluorescent lamp powder (Arunraj et al., 2019). The biosorbent regeneration was accomplished by using EDTA as complexing agent for Eu(III). The biosorbent reached a maximum adsorption capacity of 25.91 mg g^{-1} as determined by Langmuir isotherm model (Arunraj et al., 2019). However, the poor metal-binding specificity of the cell wall functional groups (e.g. carboxyl and phosphoryl) offers challenges for selective enrichment of REEs. To improve specific binding of metal ions, bioengineering approaches have been used as a cost-effective means to achieve selective metal-binding peptides and proteins on the cell surface. Lanthanide-binding tags (LBTs) made of 15–20 amino acids can complex terbium (Tb(III)) ion with high affinity (Sculimbrene & Imperiali, 2006). The Tb(III) adsorption capacity of *E. coli* was reported to increase to 28.3 ± 1.2 mg g^{-1} dry cell weight after the expression of the fusion protein OmpA-LBT on the membrane surface, which is twice as high as that observed in the un-induced control (Park et al., 2017). By using an emulsion technique, a high cell density *E. coli* strain that had been previously modified to display lanthanide-binding tags on the cell surface was enclosed inside a permeable polyethylene glycol diacrylate (PEGDA) hydrogel. By functionalizing the OmpA protein with 16 copies of an LBT, Park et al. (2016) have previously characterized modified *E. coli* cells for improved cell surface-mediated extraction of REEs. Park et al. (2016) showed through biosorption tests using leachates from rare earth and metal-mine tailings that functionalizing the cell surface with LBT produced a number of distinguishable benefits over the non-engineered control. First, a 2–10-fold improvement in the distribution coefficients for various REEs demonstrated improved REE adsorption efficiency from all leachates. Second, with the exception of Cu, the relative affinity of the cell surface for REEs increased compared to non-REEs. Thirdly, LBT display provided a way to distinguish between valuable heavy REEs and less valuable light REEs by progressively increasing the affinity of the cell surface for REEs as a function of decreasing atomic radius. Later on, Chang et al. (2020) showed that this bioengineered *E. coli* displays an increased sorption capacity and affinity for rare earths. The engineered LBT sites have higher affinity for rare earths than wild-type surface sites and REE-cell surface binding mode depended on both pH and aqueous concentration as displayed in Fig. 3. Brewer et al. (2019) have discussed the creation of a unique flow-through biosorption-based technique for the selective recovery of REE from electronic waste (NdFeB magnet). According to Brewer et al. (2019), this microbe bead adsorbent had a uniform distribution of cells whose surface functional groups were nevertheless readily accessible and efficient for the selective adsorption of REEs. With the aid of groundbreaking research, it was shown that successful Nd extraction at flow rates of up to 3 m h^{-1} at pH 4–6 could be achieved using microbe beads packed into fixed-bed columns. After nine consecutive cycles of adsorption and desorption, the microbe bead columns were stable enough to be reused and still held 85% of their sorption capacity. When compared to non-REE impurities, REEs showed a two-bed volume increase in breakthrough points, and the adsorbed fraction of the REEs was 97% pure after breakthrough, according to a bench-scale breakthrough curve with a NdFeB magnet leachate. These findings pave the path for a biomass-based REE recovery system by showing that the microbe beads can repeatedly separate REEs from non-REE metals in a column system. The performance of incorporating a biosorption technique into a large-scale process to create commercially viable total rare earth oxides (TREOs) from diverse feedstocks was studied in Jin et al. (2017) techno-economic study of the aforementioned approach. This study established the economic viability of applying biosorption to low-grade feedstocks for REE recovery.

4.3.2 Microalgae Mediated Surface Adsorption

Among widely available biosorbents, microalgal biomass possesses relatively high binding capacities for several base, precious, and critical elements (Aksu, 2002; Monteiro et al., 2012; Wilde & Benemann, 1993). In view of their

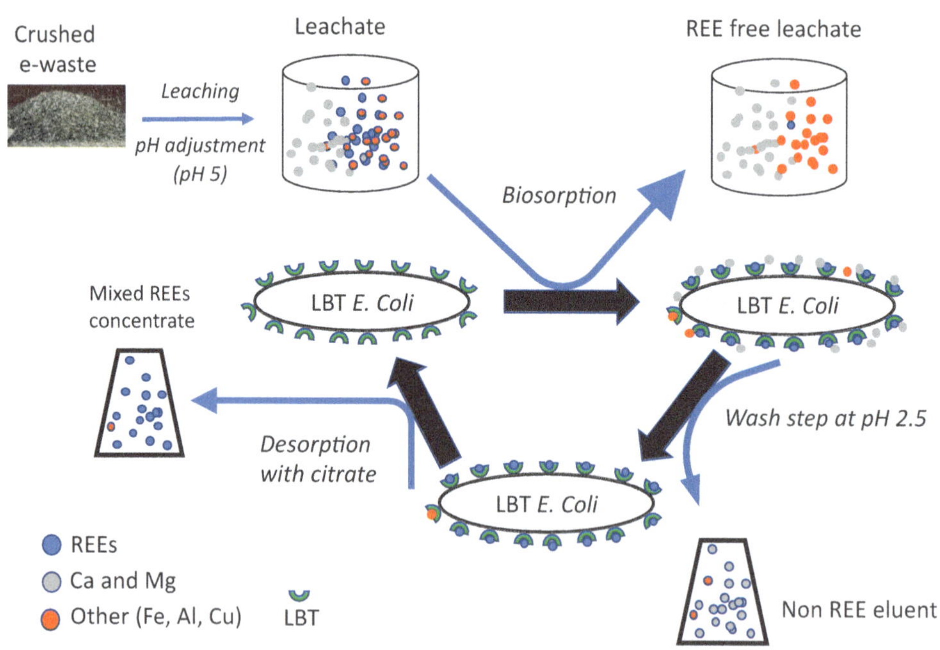

Fig. 3 Diagram of the biosorption-based REE recovery process using crushed e-waste. Crushed e-waste was acid leached after preprocessing, such as crushing and milling, which is generally non-selectively released REEs into solution. The REEs are subsequently separated from competing metal ions using a biosorption/desorption process once the pH of the REE-bearing leachate solutions has been adjusted. This procedure results in concentrated mixed rare earth solutions. The remarkable stability of LBT even after going through numerous adsorption/desorption cycles supports the reuse of biosorbents (adapted from Park et al., 2020)

inexpensive growth requirements (solar light and CO_2) and the advantage of being utilized simultaneously for multiple technologies (e.g. carbon mitigation, biofuel production, and wastewater treatment), microalgae are also seen as suitable candidates for eco-friendly biosorption technologies (Kumar et al., 2018). Microalgae generally display significant surface area-to-volume ratio available for contact with the surrounding environment (Monteiro et al., 2012), and this property provides an additional advantage for biosorption leading to the recovery of critical metals. This is explained by the presence of carboxylic, hydroxyl, amino, phosphate, and sulfhydryl functional groups in the microalgal cell wall that can act as metal-binding sites (Kumar et al., 2018). Microalgae have been used for the recovery of several critical elements from different wastewater streams; however, only a few have been applied so far for PLS. For instance, Kücüker et al. (2017) used dried green microalgae (*Chlorella vulgaris*) to study Nd removal from neodymium permanent magnet leachates in batch and continuous sorption systems. With a biosorbent dosage of 500 mg L^{-1} and an initial neodymium content in the mixed leachate solution of 250 mg L^{-1} at 35 °C, the maximum Nd absorption ($q = 157.21$ mg g^{-1} sorbent) was ascertained at pH 5 under these conditions. The sorption capabilities of microalgae have been reported for other critical elements such as indium and rare earth elements. Nicomel et al. (2020) investigated the In(III) adsorption by microalgal biomass in batch experiments. Adsorption isotherms have well-fitted with the Freundlich model. The estimated maximum In(III) adsorption capacity of microalgae was 0.14 mmol g^{-1}, which is higher than that of some chemically modified adsorbents reported in the literature (Nicomel et al., 2020). Selectivity for In(III) over other metals, such as Cu(II), Zn(II), and Al(III), was also observed. Furthermore, repeated cycles were performed, where the microalgae biosorbent was regenerated using 0.1 M HCl solution, along with up to 80% In(III) recovery. These results indicated the potential of microalgae for In(III) biosorption from aqueous solution. Čížková et al. (2019) investigated the growth of green microalgae on red mud and the intracellular accumulation of lanthanides. It was observed that *Desmodesmus quadricauda* was able to accumulate lanthanides to the highest level (27.3 mg/kg/day), in comparison to *Chlamydomonas reinhardtii* and *Parachlorella kessleri* (24.5 and 12.5 mg kg^{-1} per day, respectively). The accumulated lanthanide content was reported to increase in the *Desmodesmus quadricauda* within 2 days from zero in red-mud free culture to 12.4, 39.0, and 54.5 mg kg^{-1} of dry mass at red-mud concentrations of 0.03, 0.05, and 0.1%, respectively. Thus, green algae are prospective vehicles for biomining or bioleaching of lanthanides from red mud (Čížková et al., 2019). In general, if microalgae could be considered as effective biosorbent, substantial pH adaptation may be needed prior to biosorption, which may be a bottleneck especially when the aqueous solutions contain many different elements that may potentially lead to the loss of critical elements during pH correction or strongly act a competitor during critical elements' sorption. Due to the above-mentioned bottlenecks, most biosorption processes are still at the laboratory scale and poorly applied to WEEE leachates.

Several studies have been carried out for the development of different biosorbents, not only their structure but also their metabolic pathway. Even though these studies were mainly focused on removing toxic elements from wastewaters, the developed biotechnologies can also be applied to recycling processes including e-waste (Pollmann et al., 2018).

Despite these shortcomings, biological recovery of critical metals from several End-of-Life (EoL) products through a process known as "bioleaching and biosorption" has been reported by Kucuker and Kuchta (2018) and has attractive advantages compared to conventional metal processing technologies. According to a literature survey performed by Kucuker and Kuchta (2018), a schematic diagram for the recovery of critical metals from WEEE has been proposed (Fig. 4). The process includes physical separation (pre-treatment) and bio/hydrometallurgical (bioleaching and biosorption with microalgae) process for the recovery of critical metals from WEEE under a biomining concept.

4.4 Bioelectrochemical Metal Recovery Systems

Bioelectrochemical systems (BESs) have been utilized for metal removal and recovery from different waste streams including metallurgical, industrial, and wastewaters. BES consists of an anode and a cathode chamber often separated by an ion-selective membrane (Fig. 5). In the anode chamber microbial oxidation of biodegradable material (e.g. organic carbon) releases electrons to the electrode and protons to the electrolyte. The electrons flow through a circuit to the cathode and are used to reduce metallic ions in the cathode chamber. The circuit is completed by the migration of protons from the anode to the cathode through the membrane. Depending on the relative magnitude of the redox potentials at the anode and cathode, particular metals will be precipitated at the cathode (Nancharaiah et al., 2016).

The majority of studies have used synthetic solutions and have provided proof-of-principle for a few precious and rare metal(loid)s that are frequently recovered in WEEE PLS, including Ag, Au, Co, and Se (Dominguez-Benetton et al., 2018; Guo & Kim, 2020; Nancharaiah et al., 2015, 2016; Wang & Ren, 2014). Rodenas Motos et al. (2015) could achieve high recovery rates for soluble Cu(II) as Cu(0)

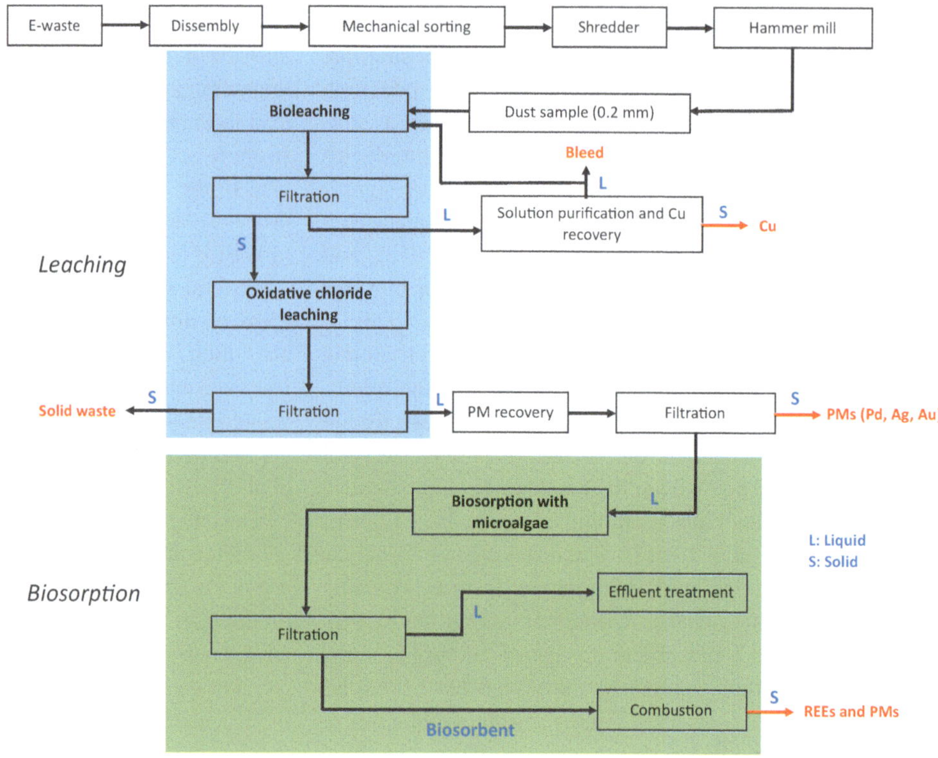

Fig. 4 Process flow sheet proposed for the recycling system linking bio/hydrometallurgical and biosorption process for recovery of REEs and precious metals from e-waste (modified from Kucuker & Kuchta, 2018)

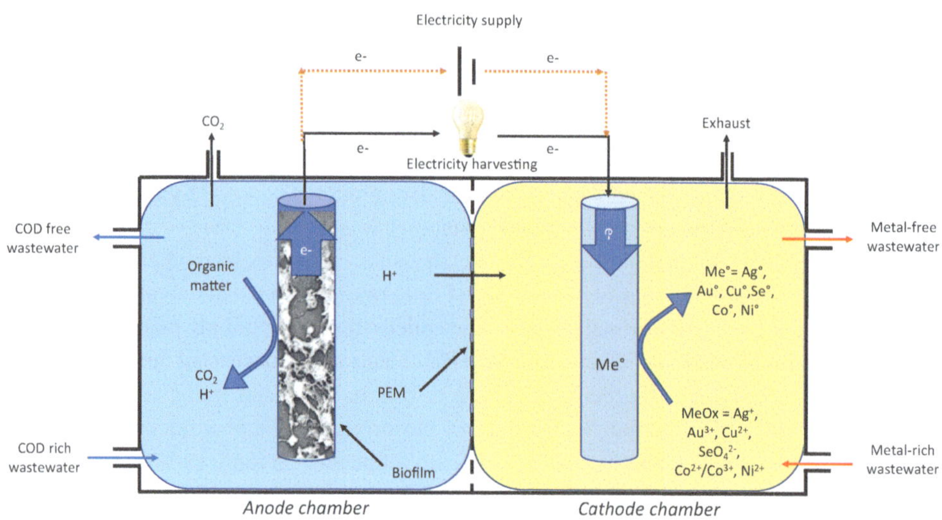

Fig. 5 Working principles of dual-chamber BES applied for metals' biorecovery. In the anode chamber, electroactive bacteria oxidize COD-rich wastewater and release reducing equivalents (electrons and protons). The electrons are used for reduction of metal ions at the cathode. Recovery of metal ions from WEEE PLS is studied through selective reduction of metal ions at the cathode. Abbreviations: *COD* chemical oxygen demand; *Me* metal; *PEM* proton exchange membrane (modified from Nancharaiah et al., 2015, 2016)

deposits. More than 90% removal efficiency was achieved in acidic solutions containing up to 2000 mg L^{-1} Cu(II). Numerous applications addressing the recovery of critical elements that are frequently found in WEEE PLS have been published. For instance, the bioelectrochemical recovery of Au^{3+}, Co^{2+}, and Fe^{3+} metal ions has been investigated by Varia et al. (2013) and the influence of aqueous gold speciation has been investigated by Ho and Babel (2019). Ho and Babel (2019) looked at the recovery of Ag and electricity production from different Ag(I)-containing synthetic wastewaters. This study found that different types of Ag(I)-containing solutions (i.e. Ag^+ solution, $[Ag(NH_3)_2]^+$, $[Ag(S_2O_3)_2]^{3-}$ complex, and mixed multi-metal solution containing Ag^+, Fe^{3+}, Cu^{2+}) employed in BES reactors resulted in different performances for Ag recovery and simultaneous electricity production. In all four reactors, Ag(I) was reduced electrochemically to form deposits on the cathode surface. The highest reduction rate (i.e. kinetic reaction constant, k of 1.55 h^{-1}) was found with Ag^+ solution, as compared to the other Ag(I) solutions (i.e. $[Ag(NH_3)_2]^+$ complex, k of 0.117 h^{-1}; $[Ag(S_2O_3)_2]^{3-}$ complex, k of 0.041 h^{-1}; and mixed multi-metal solution, k of 0.129 h^{-1}).

This study shows how the aqueous solution composition may affect the removal kinetic and efficiency. Zhang et al. (2020) investigated stacked bioelectrochemical systems (BESs), composed of microbial electrolysis cells (MECs) driven by microbial fuel cells (MFCs) which provide an alternative approach for the recovery and separation of mixed W(VI) and Mo(VI) without input of external energy. In another study, Huang et al. (2019) demonstrated the feasibility of larger-scale single-chamber MECs (40 L cylindrical single-chamber MEC fed acetate) for efficient treatment of W(VI) and Mo(VI), moving metallurgical MECs closer to commercialization for PLS treatment of these two metals. All the above-mentioned studies refer to critical elements that one can find in WEEE PLS. Dominguez-Benetton et al. (2018), in a recent paper, highlighted the challenges and opportunities to turn microbial electrometallurgy into a sustainable industrial technology in the near future. Figure 6 summarizes metals studied to date using BES technology and highlights those which are highly profitable for recovery (Dominguez-Benetton et al., 2018), especially for critical metals such as PGM and lanthanides (CRM Report, 2017).

To scale up the process and study the prospect of recovering metals from WEEE leachate streams using BES, more research is still required. In order to increase the effectiveness of bioelectrochemical processes, additional research is required to understand the interactions between microbes and metals, electron transport pathways, and electrode materials (Dominguez-Benetton et al., 2018). It is still possible to further explore the BES platform's potential use in the recovery of a variety of other priceless and technologically important metals, including metals found in WEEE.

5 Integration of Bioleaching for Biorecovery Processes

5.1 The Importance of PLS Characterization

The optimal conditions for a transition from a bioleaching process to a biorecovery process are an area that needs to be addressed. It should be considered as a crucial link to complete the biohydrometallurgical loop and to strengthen the overall technology readiness. For instance, thorough characterization of pregnant leaching solutions is lacking and should be further investigated, as the application of the appropriate biorecovery method is necessary. Most PLS is not only enriched in metals from the base substrate but also comprises ions contained in the culture growth medium or resulting from metabolic production during bioleaching. It has already been mentioned previously that the heterogeneous nature of WEEE and the non-selectivity of most bioleaching processes can lead to contamination of PLS. These contaminations take the form of solvents, organic compounds, as well as, unwanted toxic metals. Depending on the method used, this chemical complexity may represent a constraint in the processing of PLS (Zhuang et al., 2015). Iron, whether employed as a nutrient for *At. ferrooxidans* bacterium or contained in different types of WEEE, is an impurity commonly found in PLS, but the presence of other elements can also inhibit the selective biorecovery of metals. Chemical reactions such as coprecipitation or sorption are commonly occurring in biorecovery studies from e-waste. Thus, additional steps between bioleaching and biorecovery might be necessary to ensure the selective precipitation of the desired metal (i.e. Kumar et al., 2022). As PLS often has extreme pH values and ionic strength, it is important to define which processes favour the existence of precipitating agents in the given pH range. For instance, biogenic carbonate ions (CO_3^{2-}) required for the precipitation of metal carbonates prevail at pH above 12.3. Thus, it is extremely challenging to integrate acidic PLS with biorecovery process based on alkaline pH. Most biorecovery studies rely on synthetic solutions containing the target metals. The solutions are prepared from metallic salts, such as metal chloride and sulfate. The resulting pH of the solution ranges from neutral to slightly acidic which is often not representative of most PLS. It is therefore essential to consider the constraints of integrating real PLS into biorecovery processes in the scope of the future studies. This would provide useful information on the actual biorecovery yields, ways to mitigate these constraints during upstream processing or bioleaching, and thus close the biohydrometallurgical loop through the best possible method.

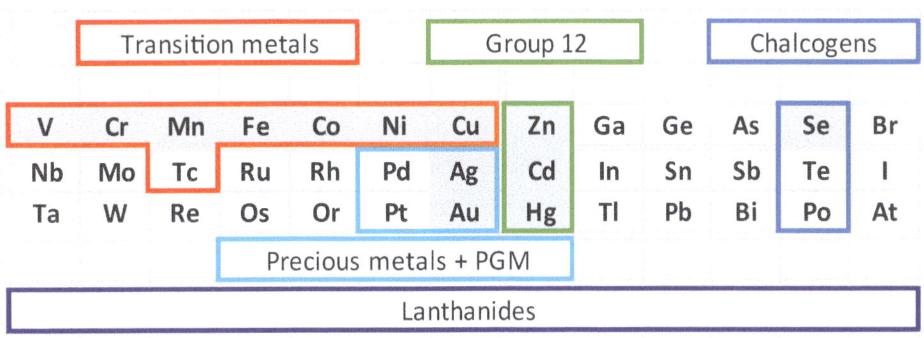

Fig. 6 Reported metals recovered using BES. Modified from Dominguez-Benetton et al. (2018)

5.2 Selective Biomineralization and Metal Speciation

Living organisms produce mineral crystals generally as part of a process for overcoming metal toxicity. A number of innovative critical metal biomineralization processes have been studied and reported as its application has potential for the production of nanomaterials; however, mechanisms need to be further investigated as they are not completely understood (Yu et al., 2020). Particularly in e-waste recycling, biomineralization is rising as a viable technology for the recovery of platinum group metals (PGMs). Furthermore, it is a selective PGMs recovery technique which offers advantages over conventional metal recovery methods in case of complex leachates such as e-waste (Rene et al., 2017). Some case studies have shown that the recovery yield of some critical elements is strongly affected by the chemical speciation in the bulk solution. However, this aspect is too often neglected and a better knowledge of the speciation of metals in solution could help to better understand the differences in performance observed in the literature. The speciation of metallic ions in solution directly influences the sorption and reduction capacity of microorganisms. In fact, metabolic pathways such as intracellular sequestration are controlled by the speciation of metals in the medium (Worms et al., 2006). Moreover, efflux mechanisms developed by some microorganisms may change the speciation of metals by expulsing uncharged metallic atoms (Worms et al., 2006). The contribution of exopolymeric substances (EPS) to changes in metal speciation in biomineralization processes is often ignored (Gadd & Pan, 2016). Maes et al. (2016) investigated the effect of varying high salt concentrations on the recovery of platinum from a halophilic microbial community. It has been reported that high salt concentration and chloride-rich matrices may influence Pt speciation thermodynamics. In addition, a study performed in 2017 by Maes et al. used five different platinum complexes (K_2PtCl_4, K_2PtCl_6, $Pt(NH_3)_4{\cdot}2HCO_3$, cisplatin, and carboplatin) to investigate the effect of Pt speciation on platinum recovery by axenic cultures (*S. oneidensis*, *C. metallidurans*, *B. toyonensis*, and *P. stutzeri*). Not only the kinetics of the platinum bioreduction were in general faster for $Pt(II)Cl_4^{2-}$, but each species cultured displayed different biomineralization behaviours depending on Pt speciation and Pt complexes. The recovery of europium via anaerobic reduction by a *Clostridium* bacterial strain has been performed by Maleke et al. (2019) and showed some potential for the selective reduction of Eu^{3+} to Eu^{2+}. Such bioreduction processes involve the transition of metal from an oxidized state to a reduced state and are therefore entirely controlled by the speciation of that metal. Electrochemical bioreduction (i.e. Ho & Babel, 2019) also has the same requirements for the respective oxidation state of the target metals. To tackle and address metal speciation influence on the recovery yield, a more selective bioleaching of target metals should be further pursued, primarily through the use of organic acids and siderophores for selective extraction of targeted metals.

6 Conclusions

Biohydrometallurgy paves the way for less energy-intensive, less costly, and more environmentally friendly methods of recycling electronic waste. Although research is still in its infancy, many building blocks have been laid through the optimization of several process units. The constraints of traditional bioleaching methods and their optimal parameters are well documented, but there are still several gap areas regarding selective forms of bioleaching such as those using organic acids or biogenic siderophores. Techniques have been developed to establish a balance between ideal growth parameters and increased pulp density, which would allow industrial applicability. In order to demonstrate the industrial applicability of a fully operational biotechnological recycling process, it is essential to integrate real pregnant leaching solutions with the downstream process studies. To do so, these solutions ought to be further characterized to facilitate this transition. Overall, although bioleaching and biorecovery strategies have proven to be effective at the laboratory scale, a better understanding of biorecovery mechanisms and associated constraints will allow a higher level of technological readiness.

Acknowledgements The authors thank the Horizon 2020 ERA-MIN2 Project Bio-assisted Closed Loop Recycling for E-mobility Metals from waste PCBs and Li-Ion Batteries—BaCLEM (Project ANR-20-MIN2-0002, 2021–2024) for financial support.

References

Achal, V., Pan, X., & Zhang, D. (2011). Remediation of copper-contaminated soil by *Kocuria flava* CR1, based on microbially induced calcite precipitation. *Ecological Engineering, 37*(10), 1601–1605. https://doi.org/10.1016/j.ecoleng.2011.06.008

Ahmed, E., & Holmström, S. J. M. (2014). Siderophores in environmental research: Roles and applications: Siderophores in environmental research. *Microbial Biotechnology, 7*(3), 196–208. https://doi.org/10.1111/1751-7915.12117

Akcil, A., & Koldas, S. (2006). Acid Mine Drainage (AMD): Causes, treatment and case studies. *Journal of Cleaner Production, 14*(12–13), 1139-1145. ISSN 0959-6526, https://doi.org/10.1016/j.jclepro.2004.09.006

Akbari, S., & Ahmadi, A. (2019). Recovery of copper from a mixture of printed circuit boards (PCBs) and sulphidic tailings using bioleaching and solvent extraction processes. *Chemical Engineering and Processing—Process Intensification, 142*, 107584. https://doi.org/10.1016/j.cep.2019.107584

Akcil, A., et al. (2015). Precious metal recovery from waste printed circuit boards using cyanide and non-cyanide lixiviants—A review. *Waste Management, 45*, 258–271. https://doi.org/10.1016/j.wasman.2015.01.017

Aksu, Z. (2002). Determination of the equilibrium, kinetic and thermodynamic parameters of the batch biosorption of nickel (II) ions onto *Chlorella vulgaris*. *Process Biochemistry, 38*(1), 89–99. https://doi.org/10.1016/S0032-9592(02)00051-1

Alaee, M., & Wenning, R. J. (2002). The significance of brominated flame retardants in the environment: Current understanding, issues and challenges. *Chemosphere, 46*(5), 579–582. https://doi.org/10.1016/S0045-6535(01)00224-7

Amato, A., Becci, A., & Beolchini, F. (2020) Sustainable recovery of Cu, Fe and Zn from end-of-life printed circuit boards. *Resources, Conservation and Recycling, 158*, 104792. https://doi.org/10.1016/j.resconrec.2020.104792

Anaya-Garzon, J., et al. (2021). Bioleaching of E-waste: Influence of printed circuit boards on the activity of acidophilic iron-oxidizing bacteria. *Frontiers in Microbiology, 12*, 669738. https://doi.org/10.3389/fmicb.2021.669738

Arshadi, M., & Mousavi, S. M. (2015). Multi-objective optimization of heavy metals bioleaching from discarded mobile phone PCBs: Simultaneous Cu and Ni recovery using *Acidithiobacillus ferrooxidans*. *Separation and Purification Technology, 147*, 210–219. https://doi.org/10.1016/j.seppur.2015.04.020

Arunraj, B., Sathvika, T., Rajesh, V., & Rajesh, N. (2019). Cellulose and *Saccharomyces cerevisiae* embark to recover europium from phosphor powder. *ACS Omega, 4*(1), 940–952. https://doi.org/10.1021/acsomega.8b02845

Ashraf, M., Azahar, S., & Yusof, N. Z. (2017). Soil improvement using MICP and biopolymers: A review. *IOP Conference Series: Materials Science and Engineering, 226*, 012058. https://doi.org/10.1088/1757-899X/226/1/012058

Babu, P., Chandel, A. K., & Singh, O. V. (2015). *Extremophiles and their applications in medical processes*. Springer International Publishing (SpringerBriefs in Microbiology). https://doi.org/10.1007/978-3-319-12808-5

Bahaloo-Horeh, N., & Mousavi, S. M. (2017). Enhanced recovery of valuable metals from spent lithium-ion batteries through optimization of organic acids produced by *Aspergillus niger*. *Waste Management, 60*, 666–679. https://doi.org/10.1016/j.wasman.2016.10.034

Baniasadi, M., et al. (2019). Advances in bioleaching as a sustainable method for metal recovery from e-waste: A review. *Journal of Industrial and Engineering Chemistry, 76*, 75–90. https://doi.org/10.1016/j.jiec.2019.03.047

Becci, A., et al. (2020). An innovative biotechnology for metal recovery from printed circuit boards. *Resources, Conservation and Recycling, 153*, 104549. https://doi.org/10.1016/j.resconrec.2019.104549

Bhattacharyya, D., Jumawan, A. B., & Grieves, R. B. (1979). Separation of toxic heavy metals by sulfide precipitation. *Separation Science and Technology, 14*(5), 441–452. https://doi.org/10.1080/01496397908058096

Bhattacharyya, D., Jumawan, A. B., Sun, G., Sund-Hagelberg, C., & Schwitzgebel, K. (1981). Precipitation of heavy metals with sodium sulfide: bench-scale and full-scale experimental results. In *AIChE Symposium Series* (Vol. 77, No. 209).

Bindschedler, S., et al. (2017). *Fungal biorecovery of gold from E-waste*. In *Advances in applied microbiology* (pp. 53–81). Elsevier. https://doi.org/10.1016/bs.aambs.2017.02.002

Biswal, B. K., et al. (2018). Biological leaching and chemical precipitation methods for recovery of Co and Li from spent lithium-ion batteries. *ACS Sustainable Chemistry & Engineering, 6*(9), 12343–12352. https://doi.org/10.1021/acssuschemeng.8b02810

Biswas, S., et al. (2013). Bioleaching of nickel and cobalt from lateritic chromite overburden using the culture filtrate of *Aspergillus niger*. *Applied Biochemistry and Biotechnology, 170*(7), 1547–1559. https://doi.org/10.1007/s12010-013-0289-9

Bonificio, W. D., & Clarke, D. R. (2014) Bacterial recovery and recycling of tellurium from tellurium-containing compounds by *Pseudoalteromonas* sp. EPR 3. *Journal of Applied Microbiology, 117*(5), 1293–1304. https://doi.org/10.1111/jam.12629

Brewer, A., et al. (2019). Microbe encapsulation for selective rare-earth recovery from electronic waste leachates. *Environmental Science & Technology, 53*(23), 13888–13897. https://doi.org/10.1021/acs.est.9b04608

Calvert, G., et al. (2019). Recovery of metals from waste lithium ion battery leachates using biogenic hydrogen sulfide. *Minerals, 9*(9), 563. https://doi.org/10.3390/min9090563

Castro-Alonso, M. J., et al. (2019). Microbially induced calcium carbonate precipitation (MICP) and its potential in bioconcrete: Microbiological and molecular concepts. *Frontiers in Materials, 6*, 126. https://doi.org/10.3389/fmats.2019.00126

Cheng, Y., et al. (2019). *Selective mineralization and recovery of Au(III) from multi-ionic aqueous systems by Bacillus licheniformis FZUL-63* (p. 10). https://doi.org/10.3390/min9070392

Chang, E., et al. (2020). Surface complexation model of rare earth element adsorption onto bacterial surfaces with lanthanide binding tags. *Applied Geochemistry, 112*, 104478. https://doi.org/10.1016/j.apgeochem.2019.104478

Christensen, B., Laake, M., & Lien, T. (1996). Treatment of acid mine water by sulfate-reducing bacteria: Results from a bench scale experiment. *Water Research, 30*, 1617–1624. https://doi.org/10.1016/0043-1354(96)00049-8

Cibati, A., et al. (2013). Selective precipitation of metals from synthetic spent refinery catalyst leach liquor with biogenic H2S produced in a lactate-fed anaerobic baffled reactor. *Hydrometallurgy, 139*, 154–161. https://doi.org/10.1016/j.hydromet.2013.01.022

Čížková, M., Mezricky, D., Rucki, M., Tóth, T., Náhlík, V., Lanta, V., Bišová, K., Zachleder, V., & Vítová, M. (2019). *Bio-mining of lanthanides from red mud by green microalgae*. *Molecules, 24*(7), 1356. https://doi.org/10.3390/molecules24071356

Colipai, C., et al. (2018). Synthesis of copper sulfide nanoparticles using biogenic H2S produced by a low-pH sulfidogenic bioreactor. *Minerals, 8*(2), 35. https://doi.org/10.3390/min8020035

Creamer, N. J., et al. (2006). Palladium and gold removal and recovery from precious metal solutions and electronic scrap leachates by *Desulfovibrio desulfuricans*. *Biotechnology Letters, 28*(18), 1475–1484. https://doi.org/10.1007/s10529-006-9120-9

CRM Report. (2017). European Commission, Directorate-General for Internal Market, Industry, Entrepreneurship and SMEs. Study on the review of the list of critical raw materials—Final report, Publications Office. https://data.europa.eu/doi/10.2873/876644

Cui, J., et al. (2020). Rapid and green process for valuable materials recovery from waste liquid crystal displays. *Resources, Conservation and Recycling, 153*, 104544. https://doi.org/10.1016/j.resconrec.2019.104544

Cui, J., Zhu, N., Mao, F., Wu, P., & Dang, Z. (2021). Bioleaching of indium from waste LCD panels by *Aspergillus niger*: Method optimization and mechanism analysis. *Science of the Total Environment, 790*, 148151. https://doi.org/10.1016/j.scitotenv.2021.148151

Cui, J., & Zhang, L. (2008). Metallurgical recovery of metals from electronic waste: A review. *Journal of Hazardous Materials, 158*(2–3), 228–256. https://doi.org/10.1016/j.jhazmat.2008.02.001

Dalrymple, I., et al. (2007). An integrated approach to electronic waste (WEEE) recycling. *Circuit World, 33*(2), 52–58. https://doi.org/10.1108/03056120710750256

Darby, L., & Obara, L. (2005). Household recycling behaviour and attitudes towards the disposal of small electrical and electronic equipment. *Resources, Conservation and Recycling, 44*(1), 17–35. https://doi.org/10.1016/j.resconrec.2004.09.002

Das, S., & Ting, Y.-P. (2017). Improving gold (bio)leaching efficiency through pretreatment using hydrogen peroxide assisted sulfuric acid: sustainable green solutions. *CLEAN—Soil, Air, Water, 45*(6), 1500945. https://doi.org/10.1002/clen.201500945

De Belie, N. (2010). Microorganisms versus stony materials: A love–hate relationship. *Materials and Structures, 43*(9), 1191–1202. https://doi.org/10.1617/s11527-010-9654-0

Dev, S., Sachan, A., Dehghani, F., Ghosh, T., Briggs, B. R., & Aggarwal, S. (2020). Mechanisms of biological recovery of rare-earth elements from industrial and electronic wastes: A review. *Chemical Engineering Journal, 397*, 124596. https://doi.org/10.1016/j.cej.2020.124596

Díaz-Martínez, M. E., et al. (2019). Microbial bioleaching of Ag, Au and Cu from printed circuit boards of mobile phones. *Current Microbiology, 76*(5), 536–544. https://doi.org/10.1007/s00284-019-01646-3

Dodson, J. R., Parker, H. L., García, A. M., Hicken, A., Asemave, K., Farmer, T. J., He, H., Clark, J. H., & Hunt, A. J. (2015). Bio-derived materials as a green route for precious & critical metal recovery and re-use. *Green Chemistry, 17*(4), 1951–1965. https://doi.org/10.1039/C4GC02483D

Dominguez-Benetton, X., Varia, J. C., Pozo, G., Modin, O., Ter Heijne, A., Fransaer, J., & Rabaey, K. (2018). Metal recovery by microbial electro-metallurgy. *Progress in Materials Science., 94*, 435–461. https://doi.org/10.1016/j.pmatsci.2018.01.007

Erust, C., et al. (2020). Intensified acidophilic bioleaching of multi-metals from waste printed circuit boards (WPCBs) of spent mobile phones. *Journal of Chemical Technology & Biotechnology, 95*(8), 2272–2285. https://doi.org/10.1002/jctb.6417

Essa, A. M. M., Creamer, N. J., Brown, N. L., & Macaskie, L. E. (2006). A new approach to the remediation of heavy metal liquid wastes via off-gases produced by Klebsiella pneumoniae M426. *Biotechnology and Bioengineering*.

Faraji, F., et al. (2021). A green and sustainable process for the recovery of gold from low-grade sources using biogenic cyanide generated by *Bacillus megaterium* : A comprehensive study. *ACS Sustainable Chemistry & Engineering, 9*(1), 236–245. https://doi.org/10.1021/acssuschemeng.0c06904

Fomina, M., & Gadd, G. M. (2014). Biosorption: Current perspectives on concept, definition and application. *Bioresource Technology, 160*, 3–14. https://doi.org/10.1016/j.biortech.2013.12.102

Forti, V., Baldé, C. P., Kuehr, R., & Bel, G. (2020). The Global E-waste Monitor 2020: Quantities, flows and the circular economy potential. United Nations University (UNU)/United Nations Institute for Training and Research (UNITAR)—co-hosted SCYCLE Programme, International Telecommunication Union (ITU) & International Solid Waste Association (ISWA), Bonn/Geneva/Rotterdam.

Frey-Klett, P., et al. (2011). Bacterial-fungal interactions: hyphens between agricultural, clinical, environmental, and food microbiologists. *Microbiology and Molecular Biology Reviews, 75*(4), 583–609. https://doi.org/10.1128/MMBR.00020-11

Gadd, G. M., & Pan, X. (2016). Biomineralization, bioremediation and biorecovery of toxic metals and radionuclides. *Geomicrobiology Journal, 33*(3–4), 175–178. https://doi.org/10.1080/01490451.2015.1087603

Gadd, G. M. (2009). Biosorption: Critical review of scientific rationale, environmental importance and significance for pollution treatment. *Journal of Chemical Technology & Biotechnology: International Research in Process, Environmental & Clean Technology, 84*(1), 13–28. https://doi.org/10.1002/jctb.1999

Gadd, G. M., & Griffiths, A. J. (1977). Microorganisms and heavy metal toxicity. *Microbial Ecology, 4*(4), 303–317. https://doi.org/10.1007/BF02013274

Ghosh, B., et al. (2015). Waste Printed Circuit Boards recycling: An extensive assessment of current status. *Journal of Cleaner Production, 94*, 5–19. https://doi.org/10.1016/j.jclepro.2015.02.024

Gopikrishnan, V., et al. (2020). Microbial leaching of heavy metals from e-waste. In *Biovalorisation of wastes to renewable chemicals and biofuels* (pp. 189–216). Elsevier. https://doi.org/10.1016/B978-0-12-817951-2.00010-9

Guezennec, A.-G., et al. (2018). CO_2 mass transfer in bioleaching reactors: CO_2 enrichment applied to a complex copper concentrate. *Hydrometallurgy, 180*, 277–286. https://doi.org/10.1016/j.hydromet.2018.08.006

Guo, H., & Younggy, K. (2020). Mechanisms of heavy metal separation in bioelectrochemical systems and relative significance of precipitation. In *Microbial electrochemical technologies* (pp. 128–144). CRC Press.

Gupta, N. K., Gupta, A., Ramteke, P., Sahoo, H., & Sengupta, A. (2019). Biosorption-a green method for the preconcentration of rare earth elements (REEs) from waste solutions: A review. *Journal of Molecular Liquids, 274*, 148–164. https://doi.org/10.1016/j.molliq.2018.10.134

Gurung, M., Adhikari, B. B., Kawakita, H., Ohto, K., Inoue, K., & Alam, S. (2013). Recovery of gold and silver from spent mobile phones by means of acidothiourea leaching followed by adsorption using biosorbent prepared from persimmon tannin. *Hydrometallurgy, 133*, 84–93. https://doi.org/10.1016/j.hydromet.2012.12.003

Haghshenas, D. F., et al. (2009). Adaptation of *Acidithiobacillus ferrooxidans* to high grade sphalerite concentrate. *Minerals Engineering, 22*(15), 1299–1306. https://doi.org/10.1016/j.mineng.2009.07.011

Heydarian, A., et al. (2018). Application of a mixed culture of adapted acidophilic bacteria in two-step bioleaching of spent lithium-ion laptop batteries. *Journal of Power Sources, 378*, 19–30. https://doi.org/10.1016/j.jpowsour.2017.12.009

Ho, N. A. D., & Babel, S. (2019). Electrochemical reduction of different Ag(I)-containing solutions in bioelectrochemical systems for recovery of silver and simultaneous power generation. *RSC Advances, 9*(52), 30259–30268. https://doi.org/10.1039/C9RA06369B

Hofmann, M., Retamal-Morales, G., & Tischler, D. (2020). Metal binding ability of microbial natural metal chelators and potential applications. *Natural Product Reports, 37*(9), 1262–1283. https://doi.org/10.1039/C9NP00058E

Horeh, N. B., Mousavi, S. M., & Shojaosadati, S. A. (2016). Bioleaching of valuable metals from spent lithium-ion mobile phone batteries using *Aspergillus niger*. *Journal of Power Sources, 320*, 257–266. https://doi.org/10.1016/j.jpowsour.2016.04.104

Hu, S.-H., Hu, S.-C., & Fu, Y.-P. (2012). Resource recovery of copper-contaminated sludge with jarosite process and selective precipitation. *Environmental Progress & Sustainable Energy, 31*(3), 379–385. https://doi.org/10.1002/ep.10559

Huang, L., Tian, F., Pan, Y., Shan, L., Shi, Y., & Logan, B. E. (2019). Mutual benefits of acetate and mixed tungsten and molybdenum for their efficient removal in 40 L microbial electrolysis cells. *Water Research, 162*, 358–368. ISSN 0043-1354. https://doi.org/10.1016/j.watres.2019.07.003

Huang, Z., & Yang, T. (2021). Comparative study of acid mine drainage using different neutralization methods. *Water Practice and Technology, 16*(3), 1026–1035. https://doi.org/10.2166/wpt.2021.039

Hubau, A., et al. (2018). Continuous production of a biogenic ferric iron lixiviant for the bioleaching of printed circuit boards (PCBs). *Hydrometallurgy, 180*, 180–191. https://doi.org/10.1016/j.hydromet.2018.07.001

Hubau, A., et al. (2020). Recovery of metals in a double-stage continuous bioreactor for acidic bioleaching of printed circuit boards (PCBs). *Separation and Purification Technology, 238*, 116481. https://doi.org/10.1016/j.seppur.2019.116481

Ijadi Bajestani, M., Mousavi, S. M., & Shojaosadati, S. A. (2014). Bioleaching of heavy metals from spent household batteries using *Acidithiobacillus ferrooxidans*: Statistical evaluation and optimization. *Separation and Purification Technology, 132*, 309–316. https://doi.org/10.1016/j.seppur.2014.05.023

Ilyas, S., et al. (2010). Column bioleaching of metals from electronic scrap. *Hydrometallurgy, 101*(3–4), 135–140. https://doi.org/10.1016/j.hydromet.2009.12.007

Ilyas, S., Lee, J., & Kim, B. (2014). Bioremoval of heavy metals from recycling industry electronic waste by a consortium of moderate thermophiles: Process development and optimization. *Journal of Cleaner Production, 70*, 194–202. https://doi.org/10.1016/j.jclepro.2014.02.019

Işıldar, A., et al. (2016). Two-step bioleaching of copper and gold from discarded printed circuit boards (PCB). *Waste Management, 57*, 149–157. https://doi.org/10.1016/j.wasman.2015.11.033

Işıldar, A., et al. (2019). Biotechnological strategies for the recovery of valuable and critical raw materials from waste electrical and electronic equipment (WEEE)—A review. *Journal of Hazardous Materials, 362*, 467–481. https://doi.org/10.1016/j.jhazmat.2018.08.050

Jadhav, U., Su, C., & Hocheng, H. (2016). Leaching of metals from printed circuit board powder by an *Aspergillus niger* culture supernatant and hydrogen peroxide. *RSC Advances, 6*(49), 43442–43452. https://doi.org/10.1039/C6RA04169H

Jagannath, A., Shetty K. V., & Saidutta, M. B. (2017). Bioleaching of copper from electronic waste using *Acinetobacter* sp. Cr B2 in a pulsed plate column operated in batch and sequential batch mode. *Journal of Environmental Chemical Engineering, 5*(2), 1599–1607. https://doi.org/10.1016/j.jece.2017.02.023

Janyasuthiwong, S., et al. (2016). Effect of operational parameters on the leaching efficiency and recovery of heavy metals from computer printed circuit boards: Recovery of heavy metals from computer printed circuit boards. *Journal of Chemical Technology & Biotechnology, 91*(7), 2038–2046. https://doi.org/10.1002/jctb.4798

Jerez, C. A. (2017). Biomining of metals: How to access and exploit natural resource sustainably. *Microbial Biotechnology, 10*(5), 1191–1193. https://doi.org/10.1111/1751-7915.12792

Jin, H., et al. (2017). Techno-economic assessment for integrating biosorption into rare earth recovery process. *ACS Sustainable Chemistry & Engineering, 5*(11), 10148–10155. https://doi.org/10.1021/acssuschemeng.7b02147

Johnson, D. B., & du Plessis, C. A. (2015). Biomining in reverse gear: Using bacteria to extract metals from oxidised ores. *Minerals Engineering, 75*, 2–5. https://doi.org/10.1016/j.mineng.2014.09.024

Johnstone, T. C., & Nolan, E. M. (2015). Beyond iron: Non-classical biological functions of bacterial siderophores. *Dalton Transactions, 44*(14), 6320–6339. https://doi.org/10.1039/C4DT03559C

Jowkar, M. J., et al. (2018). Bioleaching of indium from discarded liquid crystal displays. *Journal of Cleaner Production, 180*, 417–429. https://doi.org/10.1016/j.jclepro.2018.01.136

Kaksonen, A. H., et al. (2014). Biohydrometallurgical iron oxidation and precipitation: Part II—Jarosite precipitate characterisation and acid recovery by conversion to hematite. *Hydrometallurgy, 147–148*, 264–272. https://doi.org/10.1016/j.hydromet.2014.04.015

Kaksonen, A. H., et al. (2018). Recent progress in biohydrometallurgy and microbial characterisation. *Hydrometallurgy, 180*, 7–25. https://doi.org/10.1016/j.hydromet.2018.06.018

Kaliyaraj, D., et al. (2019). Bioleaching of heavy metals from printed circuit board (PCB) by Streptomyces albidoflavus TN10 isolated from insect nest. *Bioresources and Bioprocessing, 6*(1), 47. https://doi.org/10.1186/s40643-019-0283-3

Kang, K. D., et al. (2021). Assessment of pre-treatment techniques for coarse printed circuit boards (PCBs) recycling. *Minerals, 11*(10), 1134. https://doi.org/10.3390/min11101134

Kashefi, K., et al. (2001). Reductive precipitation of gold by dissimilatory Fe(III)-reducing *Bacteria* and *Archaea*. *Applied and Environmental Microbiology, 67*(7), 3275–3279. https://doi.org/10.1128/AEM.67.7.3275-3279.2001

Kasper, A. C., & Veit, H. M. (2018). Gold recovery from printed circuit boards of mobile phones scraps using a leaching solution alternative to cyanide. *Brazilian Journal of Chemical Engineering, 35*, 931–942. https://doi.org/10.1590/0104-6632.20180353s20170291

Kim, M.-J., et al. (2016). Bioleaching of spent Zn–Mn or Ni–Cd batteries by *Aspergillus* species. *Waste Management, 51*, 168–173. https://doi.org/10.1016/j.wasman.2015.11.001

Kim, Y., Seo, H., & Roh, Y. (2018). Metal recovery from the mobile phone waste by chemical and biological treatments. *Minerals, 8*(1), 8. https://doi.org/10.3390/min8010008

Kim, S., et al. (2021). A comprehensive review on the pretreatment process in lithium-ion battery recycling. *Journal of Cleaner Production, 294*, 126329. https://doi.org/10.1016/j.jclepro.2021.126329

Knowles, C. J. (1976). Microorganisms and cyanide. *Bacteriological reviews, 40*(3), 652–680. https://doi.org/10.1128/br.40.3.652-680.1976

Kopacek, B. (2010). ReLCD recycling and re-use of LCD panels. In *Proceedings of the 2010 IEEE international symposium on sustainable systems and technology (ISSST)* (pp. 1–3). IEEE. https://doi.org/10.1109/ISSST.2010.5507735

Kucuker, M. A., & Kuchta, K. (2018) Biomining—Biotechnological systems for the extraction and recovery of metals from secondary sources. *Global NEST Journal, 20*(4), 737–742. https://doi.org/10.30955/gnj.002692

Kucuker, M. A., Wieczorek, N., Kuchta, K., & Copty, N. K. (2017). Biosorption of neodymium on *Chlorella vulgaris* in aqueous solution obtained from hard disk drive magnets. *PLoS ONE, 12*(4), e0175255. https://doi.org/10.1371/journal.pone.0175255

Kumar, A., Saini, H. S., & Kumar, S. (2018). Bioleaching of gold and silver from waste printed circuit boards by *Pseudomonas balearica* SAE1 isolated from an e-waste recycling facility. *Current Microbiology, 75*(2), 194–201. https://doi.org/10.1007/s00284-017-1365-0

Kumar, R., et al. (2022). Downstream recovery of Li and value-added metals (Ni Co, and Mn) from leach liquor of spent lithium-ion batteries using a membrane-integrated hybrid system. *Chemical Engineering Journal, 447*, 137507. https://doi.org/10.1016/j.cej.2022.137507

Lakshmi, S., et al. (2020). Hexavalent chromium sequestration from electronic waste by biomass of Aspergillus carbonarius. *Bioengineered, 11*(1), 708–717. https://doi.org/10.1080/21655979.2020.1780828

Lee, J., & Pandey, B. D. (2012). Bio-processing of solid wastes and secondary resources for metal extraction—A review. *Waste Management, 32*(1), 3–18. https://doi.org/10.1016/j.wasman.2011.08.010

Li, J., et al. (2009). Recovery of valuable materials from waste liquid crystal display panel. *Waste Management, 29*(7), 2033–2039. https://doi.org/10.1016/j.wasman.2008.12.013

Li, J., Liang, C., & Ma, C. (2015). Bioleaching of gold from waste printed circuit boards by *Chromobacterium violaceum*. *Journal of Material Cycles and Waste Management, 17*(3), 529–539. https://doi.org/10.1007/s10163-014-0276-4

Li, M., Cheng, X., & Guo, H. (2013). Heavy metal removal by biomineralization of urease producing bacteria isolated from soil. *International Biodeterioration & Biodegradation, 76*, 81–85. https://doi.org/10.1016/j.ibiod.2012.06.016

Liang, G., Mo, Y., & Zhou, Q. (2010). Novel strategies of bioleaching metals from printed circuit boards (PCBs) in mixed cultivation of two acidophiles. *Enzyme and Microbial Technology, 47*(7), 322–326. https://doi.org/10.1016/j.enzmictec.2010.08.002

Lie, J., Tanda, S., & Liu, J.-C. (2020). Subcritical water extraction of valuable metals from spent lithium-ion batteries. *Molecules, 25*(9), 2166. https://doi.org/10.3390/molecules25092166

Liu, H., Gu, G., & Xu, Y. (2011). Surface properties of pyrite in the course of bioleaching by pure culture of *Acidithiobacillus ferrooxidans* and a mixed culture of *Acidithiobacillus ferrooxidans* and *Acidithiobacillus thiooxidans*. *Hydrometallurgy, 108*(1–2), 143–148. https://doi.org/10.1016/j.hydromet.2011.03.010

Liu, X., et al. (2020). Oxidative stress induced by metal ions in bioleaching of $LiCoO_2$ by an acidophilic microbial consortium. *Frontiers in Microbiology, 10*, 3058. https://doi.org/10.3389/fmicb.2019.03058

Lu, R., Ma, E., & Xu, Z. (2012). Application of pyrolysis process to remove and recover liquid crystal and films from waste liquid crystal display glass. *Journal of Hazardous Materials, 243*, 311–318. https://doi.org/10.1016/j.jhazmat.2012.10.035

Mack, C., Wilhelmi, B., Duncan, J. R., & Burgess, J. E. (2007). Biosorption of precious metals. *Biotechnology Advances., 25*(3), 264–271. https://doi.org/10.1016/j.biotechadv.2007.01.003

Maes, S., et al. (2016). Platinum recovery from industrial process streams by halophilic bacteria: Influence of salt species and platinum speciation. *Water Research, 105*, 436–443. https://doi.org/10.1016/j.watres.2016.09.023

Maleke, M., et al. (2019). Biomineralization and bioaccumulation of Europium by a thermophilic metal resistant bacterium. *Frontiers in Microbiology, 10*, 81. https://doi.org/10.3389/fmicb.2019.00081

Manzella, M. P., Reguera, G., & Kashefi, K. (2013). Extracellular electron transfer to Fe(III) Oxides by the hyperthermophilic archaeon *Geoglobus ahangari* via a direct contact mechanism. *Applied and Environmental Microbiology, 79*(15), 4694–4700. https://doi.org/10.1128/AEM.01566-13

Michalková, E., et al. (2013). Metals recovery from acid mine drainage and possibilities for their utilization. *Polish Journal of Environmental Studies, 22*(4), 1111–1118.

Miettinen, V., Mäkinen, J., Kolehmainen, E., Kravtsov, T., & Rintala, L. (2019). Iron control in atmospheric acid laterite leaching. *Minerals, 9*(7), 404. https://doi.org/10.3390/min9070404

Mishra, S., et al. (2021). A review on chemical versus microbial leaching of electronic wastes with emphasis on base metals dissolution. *Minerals, 11*(11), 1255. https://doi.org/10.3390/min11111255

Monteiro, C. M., Castro, P. M., & Malcata, F. X. (2012). Metal uptake by microalgae: Underlying mechanisms and practical applications. *Biotechnology Progress, 28*(2), 299–311. https://doi.org/10.1002/btpr.1504

Mudder, T. I., & Botz, M. M. (2006). Chemistry and treatment of cyanidation wastes (p. 391).

Muyzer, G., & Stams, A. J. M. (2008). The ecology and biotechnology of sulphate-reducing bacteria. *Nature Reviews Microbiology, 6*(6), 441–454. https://doi.org/10.1038/nrmicro1892

Mylarappa, M., et al. (2017). Resource recovery and material characterization of metals from waste li-ion batteries by an eco-friendly leaching agent. *Materials Today: Proceedings, 4*(11), 12215–12222. https://doi.org/10.1016/j.matpr.2017.09.152

Nancharaiah, Y. V., Mohan, S. V., & Lens, P. N. L. (2016). Biological and bioelectrochemical recovery of critical and scarce metals. *Trends in Biotechnology, 34*(2), 137–155. https://doi.org/10.1016/j.tibtech.2015.11.003

Nancharaiah, Y. V., Venkata Mohan, S., & Lens, P. N. L. (2015). Metals removal and recovery in bioelectrochemical systems: A review. *Bioresource Technology, 195*, 102–114. https://doi.org/10.1016/j.biortech.2015.06.058

Nicomel, N. R., Otero-Gonzalez, L., Arashiro, L., Garfí, M., Ferrer, I., Van Der Voort, P., Verbeken, K., Hennebel, T., & Du Laing, G. (2020). Microalgae: A sustainable adsorbent with high potential for upconcentration of indium(III) from liquid process and waste streams. *Green Chemistry, 22*(6), 1985–1995. https://doi.org/10.1039/C9GC03073E

Park, D. M., et al. (2017). Recovery of rare earth elements from low-grade feedstock leachates using engineered bacteria. *Environmental Science & Technology, 51*(22), 13471–13480. https://doi.org/10.1021/acs.est.7b02414

Park, D. M., Reed, D. W., Yung, M. C., Eslamimanesh, A., Lencka, M. M., Anderko, A., Fujita, Y., Riman, R. E., Navrotsky, A., & Jiao, Y. (2016). Bioadsorption of rare earth elements through cell surface display of lanthanide binding tags. *Environmental Science & Technology., 50*(5), 2735–2742. https://doi.org/10.1021/acs.est.5b06129

Park, D., Middleton, A., Smith, R., Deblonde, G., Laudal, D., Theaker, N., Hsu-Kim, H., & Jiao, Y. (2020). A biosorption-based approach for selective extraction of rare earth elements from coal by products. *Separation and Purification Technology, 241*, 116726. https://doi.org/10.1016/j.seppur.2020.116726

Pat-Espadas, A. M., & Cervantes, F. J. (2018). Microbial recovery of metallic nanoparticles from industrial wastes and their environmental applications. *Journal of Chemical Technology & Biotechnology, 93*(11), 3091–3112. https://doi.org/10.1002/jctb.5681

Paul, S., Shakya, A. K., & Ghosh, P. K. (2020). Bacterially-assisted recovery of cadmium and nickel as their metal sulfide nanoparticles from spent Ni–Cd battery via hydrometallurgical route. *Journal of Environmental Management, 261*, 110113. https://doi.org/10.1016/j.jenvman.2020.110113

Pauling, L. (1970). *General chemistry* (3rd ed.). W.H. Freema.

Pennesi, C., Amato, A., Occhialini, S., Critchley, A. T., Totti, C., Giorgini, E., Conti, C., & Beolchini, F. (2019). Adsorption of indium by waste biomass of brown alga *Ascophyllum nodosum*. *Scientific Reports, 9*(1), 16763. https://doi.org/10.1038/s41598-019-53172-8

Pollmann, K., Kutschke, S., Matys, S., Raff, J., Hlawacek, G., & Lederer, F. L. (2018). Bio-recycling of metals: Recycling of technical products using biological applications. *Biotechnology Advances, 36*(4), 1048–1062. https://doi.org/10.1016/j.biotechadv.2018.03.006

Pourhossein, F., et al. (2021). Novel green hybrid acidic-cyanide bioleaching applied for high recovery of precious and critical metals from spent light emitting diode lamps. *Journal of Cleaner Production, 298*, 126714. https://doi.org/10.1016/j.jclepro.2021.126714

Pourhossein, F., & Mousavi, S. M. (2018). Enhancement of copper, nickel, and gallium recovery from LED waste by adaptation of *Acidithiobacillus ferrooxidans*. *Waste Management, 79*, 98–108. https://doi.org/10.1016/j.wasman.2018.07.010

Priya, A., & Hait, S. (2018). Feasibility of bioleaching of selected metals from electronic waste by *Acidiphilium acidophilum*. *Waste and Biomass Valorization, 9*(5), 871–877. https://doi.org/10.1007/s12649-017-9833-0

Qiao, S., et al. (2021). Multiple heavy metals immobilization based on microbially induced carbonate precipitation by ureolytic bacteria and the precipitation patterns exploration. *Chemosphere, 274*, 129661. https://doi.org/10.1016/j.chemosphere.2021.129661

Rajwade, J. M., & Paknikar, K. M. (2003). Bioreduction of tellurite to elemental tellurium by *Pseudomonas mendocina* MCM B-180 and its practical application. *Hydrometallurgy, 71*(1–2), 243–248. https://doi.org/10.1016/S0304-386X(03)00162-2

Ramachandran, S. K., Ramakrishnan, V., & Bang, S. S. (2001). Remediation of concrete using microorganisms. *ACI Materials Journal, 98*(1). https://doi.org/10.14359/10154

Ramanayaka, S., Keerthanan, S., & Vithanage, M. (2020). 2—Urban mining of E-waste: Treasure hunting for precious nanometals. In: M. N. V. Prasad, M. Vithanage, & A. Borthakur (Eds.), *Handbook of electronic waste management* (pp. 19–54). Butterworth-Heinemann. https://doi.org/10.1016/B978-0-12-817030-4.00023-1

Rasoulnia, P., et al. (no date). Impacts of phosphorous source on organic acid production and heterotrophic bioleaching of rare earth elements and base metals from spent nickel-metal-hydride batteries. *Waste and Biomass Valorization*, 16. https://doi.org/10.1007/s12649-021-01398-x

Rene, E. R., et al. (Eds.). (2017). *Sustainable heavy metal remediation*. Environmental chemistry for a sustainable world. Springer International Publishing. https://doi.org/10.1007/978-3-319-58622-9

Ribeiro, P. P. M., dos Santos, I. D., & Dutra, A. J. B. (2019). Copper and metals concentration from printed circuit boards using a zig-zag classifier. *Journal of Materials Research and Technology, 8*(1), 513–520. https://doi.org/10.1016/j.jmrt.2018.05.003

Rodenas Motos, P., ter Heijne, A., van der Weijden, R., Saakes, M., Buisman, C. J., & Sleutels, T. H. (2015). High rate copper and energy recovery in microbial fuel cells. *Frontiers in Microbiology, 6*, 527. https://doi.org/10.3389/fmicb.2015.00527

Roy, J. J., Madhavi, S., & Cao, B. (2021). Metal extraction from spent lithium-ion batteries (LIBs) at high pulp density by environmentally friendly bioleaching process. *Journal of Cleaner Production, 280*, 124242. https://doi.org/10.1016/j.jclepro.2020.124242

Sannigrahi, S., & Suthindhiran, K. (2019). Metal recovery from printed circuit boards by magnetotactic bacteria. *Hydrometallurgy, 187*, 113–124. https://doi.org/10.1016/j.hydromet.2019.05.007

Sculimbrene, B. R., & Imperiali, B. (2006). Lanthanide-binding tags as luminescent probes for studying protein interactions. *Journal of the American Chemical Society, 128*(22), 7346–7352. https://doi.org/10.1021/ja061188a

Sethurajan, M., et al. (2017). Leaching and selective zinc recovery from acidic leachates of zinc metallurgical leach residues. *Journal of Hazardous Materials, 324*, 71–82. https://doi.org/10.1016/j.jhazmat.2016.01.028

Sethurajan, M., & Gaydardzhiev, S. (2021). Bioprocessing of spent lithium ion batteries for critical metals recovery—A review. *Resources, Conservation and Recycling, 165*, 105225. https://doi.org/10.1016/j.resconrec.2020.105225

Sethurajan, M., & van Hullebusch, E. D. (2019). Leaching and selective recovery of Cu from printed circuit boards. *Metals, 9*(10), 1034. https://doi.org/10.3390/met9101034

Shin, D., et al. (2013). Evaluation of bioleaching factors on gold recovery from ore by cyanide-producing bacteria. *Minerals Engineering, 48*, 20–24. https://doi.org/10.1016/j.mineng.2013.03.019

Silver, M. (1970). Oxidation of elemental sulfur and sulfur compounds and CO_2 fixation by *Ferrobacillus ferrooxidans* (*Thiobacillus ferrooxidans*). *Canadian Journal of Microbiology, 16*(9), 845–849. https://doi.org/10.1139/m70-142

Sinha, R., Chauhan, G., Singh, A., Kumar, A., & Acharya, S. (2018). A novel eco-friendly hybrid approach for recovery and reuse of copper from electronic waste. *Journal of Environmental Chemical Engineering, 6*(1), 1053–1061. https://doi.org/10.1016/j.jece.2018.01.030

Stephen, W. I. (1975). *L. Sucha* and *S. Kotrly*. Solution equilibria in analytical chemistry. *Analytica Chimica Acta, 75*, 372–490

Stumm, W., & Morgan, J. J. (1996). *Aquatic chemistry, chemical equilibria and rates in natural waters*. (3rd ed.). John Wiley & Sons, Inc.

Sun, J., et al. (2020). Impacts of operational parameters on the morphological structure and uranium bioleaching performance of bio-ore pellets in one-step bioleaching by *Aspergillus niger*. *Hydrometallurgy, 195*, 105378. https://doi.org/10.1016/j.hydromet.2020.105378

Tabak, H. H., Scharp, R., Burckle, J., Kawahara, F. K., & Govind, R. (2003). Advances in biotreatment of acid mine drainage and biorecovery of metals: 1. Metal precipitation for recovery and recycle. *Biodegradation, 14*, 423–436. https://doi.org/10.1023/A:1027332902740

Tansel, B. (2017). From electronic consumer products to e-wastes: Global outlook, waste quantities, recycling challenges. *Environment International, 98*, 35–45. https://doi.org/10.1016/j.envint.2016.10.002

Thenepalli, T., et al. (2019). A brief note on the heap leaching technologies for the recovery of valuable metals. *Sustainability, 11*(12), 3347. https://doi.org/10.3390/su11123347

Valdés, J., et al. (2008). *Acidithiobacillus ferrooxidans* metabolism: From genome sequence to industrial applications. *BMC Genomics, 9*(1), 597. https://doi.org/10.1186/1471-2164-9-597

Valix, M. (2017). Bioleaching of electronic waste: Milestones and challenges. In *Current developments in biotechnology and bioengineering* (pp. 407–442). Elsevier. https://doi.org/10.1016/B978-0-444-63664-5.00018-6

Varia, J., Martínez, S. S., Orta, S. V., Bull, S., & Roy, S. (2013). Bioelectrochemical metal remediation and recovery of Au^{3+}, Co^{2+} and Fe^{3+} metal ions. *Electrochimica Acta, 95*, 125–131. https://doi.org/10.1016/j.electacta.2013.02.051

Varshney, S., Jain, P., & Srivastava, S. (2017). Application of ameliorated wood pulp to recover Cd (II), Pb (II), and Ni (II) from e-waste. *Journal of Material Cycles and Waste Management, 19*(4), 1446–1456. https://doi.org/10.1007/s10163-016-0539-3

Volesky, B. (2007). Biosorption and me. *Water Research, 41*(18), 4017–4029. https://doi.org/10.1016/j.watres.2007.05.062

Wang, H., & Ren, Z. J. (2014). Bioelectrochemical metal recovery from wastewater: A review. *Water Research, 66*, 219–232. https://doi.org/10.1016/j.watres.2014.08.013

Wang, L., et al. (2018). A novel approach for recovery of metals from waste printed circuit boards and simultaneous removal of iron from steel pickling waste liquor by two-step hydrometallurgical method. *Waste Management, 71*, 411–419. https://doi.org/10.1016/j.wasman.2017.10.002

Wang, Z., Zhang, N., Cai, G., Jin, Y., Ding, N., & Shen, D. (2017). Review of ground improvement using microbial induced carbonate precipitation (MICP). *Marine Georesources & Geotechnology, 35*. https://doi.org/10.1080/1064119X.2017.1297877

Wang, R., & Xu, Z. (2014). Pyrolysis mechanism for recycle renewable resource from polarizing film of waste liquid crystal display panels. *Journal of Hazardous Materials, 278*, 311–319. https://doi.org/10.1016/j.jhazmat.2014.05.095

Widmer, R., et al. (2005). Global perspectives on e-waste. *Environmental Impact Assessment Review, 25*(5), 436–458. https://doi.org/10.1016/j.eiar.2005.04.001

Wilde, E. W., & Benemann, J. R. (1993). Bioremoval of heavy metals by the use of microalgae. *Biotechnology Advances, 11*(4), 781–812. https://doi.org/10.1016/0734-9750(93)90003-6

Wong, M. H., et al. (2007). Export of toxic chemicals—A review of the case of uncontrolled electronic-waste recycling. *Environmental Pollution, 149*(2), 131–140. https://doi.org/10.1016/j.envpol.2007.01.044

Wong-Pinto, L., Menzies, A., & Ordóñez, J. I. (2020). Bionanomining: Biotechnological synthesis of metal nanoparticles from mining waste—Opportunity for sustainable management of mining environmental liabilities. *Applied Microbiology and Biotechnology., 104*(5), 1859–1869. https://doi.org/10.1007/s00253-020-10353-0

Worms, I., Simon, D. F., Hassler, C. S., & Wilkinson, K. J. (2006). Bioavailability of trace metals to aquatic microorganisms: Importance of chemical, biological and physical processes on biouptake. *Biochimie, 88*. 1721–1731. https://doi.org/10.1016/j.biochi.2006.09.008

Wu, W., et al. (2018). Bioleaching of copper from waste printed circuit boards by bacteria-free cultural supernatant of iron–sulfur-oxidizing bacteria. *Bioresources and Bioprocessing, 5*(1), 10. https://doi.org/10.1186/s40643-018-0196-6

Xiao, J., et al. (2020). A cleaner approach to the discharge process of spent lithium ion batteries in different solutions. *Journal of Cleaner Production, 255*, 120064. https://doi.org/10.1016/j.jclepro.2020.120064

Xie, Y., et al. (2019). Leaching of indium from end-of-life LCD panels via catalysis by synergistic microbial communities. *Science of the Total Environment, 655*, 781–786. https://doi.org/10.1016/j.scitotenv.2018.11.141

Yokoyama, S., & Iji, M. (1997). Recycling of printed wiring boards with mounted electronic parts. In *Proceedings of the 1997 IEEE International Symposium on Electronics and the Environment. ISEE-1997* (pp. 109–114). IEEE. https://doi.org/10.1109/ISEE.1997.605282

Yu, Z., et al. (2020). Recent advances in the recovery of metals from waste through biological processes. *Bioresource Technology, 297*, 122416. https://doi.org/10.1016/j.biortech.2019.122416

Yazici, E. Y., & Deveci, H. (2014). Ferric sulfate leaching of metals from waste printed circuit boards. *International Journal of Mineral Processing, 133*, 39–45. https://doi.org/10.1016/j.minpro.2014.09.015

Zeraatkar, A. K., Ahmadzadeh, H., Talebi, A. F., Moheimani, N. R., & McHenry, M. P. (2016). Potential use of algae for heavy metal bioremediation, a critical review. *Journal of Environmental Management, 181*, 817–831. https://doi.org/10.1016/j.jenvman.2016.06.059

Zhang, S., & Forssberg, E. (1998). Mechanical recycling of electronics scrap—The current status and prospects. *Waste Management & Research: THe Journal for a Sustainable Circular Economy, 16*(2), 119–128. https://doi.org/10.1177/0734242X9801600204

Zhang, Y., Xu, Z., Huang, L., Kong, W., & Li, M. (2020). Preferable individual rather than sequential feedings under air exposure conditions for deposition of W and Mo in stacked bioelectrochemical systems. *Environmental Engineering Science, 37*(6), 439–449. https://doi.org/10.1089/ees.2019.0465

Zhang, K., et al. (2015). Recycling indium from waste LCDs: A review. *Resources, Conservation and Recycling, 104*, 276–290. https://doi.org/10.1016/j.resconrec.2015.07.015

Zhao, K., et al. (2013). Study on recycling process for EOL liquid crystal display panel. *International Journal of Precision Engineering and Manufacturing, 14*(6), 1043–1047. https://doi.org/10.1007/s12541-013-0140-9

Zheng, X., et al. (2018). A mini-review on metal recycling from spent lithium ion batteries. *Engineering, 4*(3), 361–370. https://doi.org/10.1016/j.eng.2018.05.018

Zheng, K., Benedetti, M. F., & van Hullebusch, E. D. (2023). Recovery technologies for indium, gallium, and germanium from end-of-life products (electronic waste)–A review. *Journal of Environmental Management, 347*, 119043. https://doi.org/10.1016/j.jenvman.2023.119043

Zhou, Y., & Qiu, K. (2010). A new technology for recycling materials from waste printed circuit boards. *Journal of Hazardous Materials, 175*(1–3), 823–828. https://doi.org/10.1016/j.jhazmat.2009.10.083

Zhu, B., Chen, Y., & Wei, N. (2019). Engineering biocatalytic and biosorptive materials for environmental applications. *Trends in Biotechnology, 37*(6), 661–676. https://doi.org/10.1016/j.tibtech.2018.11.005

Zhu, T., & Dittrich, M. (2016). Carbonate precipitation through microbial activities in natural environment, and their potential in biotechnology: A review. *Frontiers in Bioengineering and Biotechnology, 4*. https://doi.org/10.3389/fbioe.2016.00004

Zhuang, W. Q., Fitts, J. P., Ajo-Franklin, C. M., Maes, S., Alvarez-Cohen, L., & Hennebel, T. (2015). Recovery of critical metals using biometallurgy. *Current Opinion in Biotechnology, 33*, 327–335. https://doi.org/10.1016/j.copbio.2015.03.01

Mineral Processing in Bioreactors: Fundamentals and Design Aspects

W. Ashane M. Fernando, P. C. P. De Silva,
Adriana Guatame-Garcia, Bas Vriens and C. A. N. Fernando

Abstract

Bioreactors have proven capabilities to facilitate the biometallurgical extraction of important major and trace metal commodities, even from low-grade mineral ores and mine wastes. Yet, effective mineral processing with bioreactors requires a detailed and quantitative understanding of the underlying biogeochemical and mineralogical processes, the prevailing mass and energy transport limitations, as well as process control and monitoring concepts. In this chapter, we aim to introduce critical aspects of bioreactor design and operation, ranging from the pre-conditioning and properties of bioreactor material feeds, relevant geochemical reactions, kinetics, microorganisms and extraction conditions to process and performance control aspects. We also discuss select industrial applications, case studies and emerging technologies. We conclude that mineral processing with bioreactors is a challenging task that requires the input of concepts from various scientific and engineering fields and will play a critical role in satisfying the growing demand for metals and sustainable resource extraction in future.

W. A. M. Fernando (✉) · C. A. N. Fernando
Department of Nano Science Technology, Wayamba University of Sri Lanka, Kuliyapitiya 60200, Sri Lanka
e-mail: ashanef@wyb.ac.lk

P. C. P. De Silva
Ipanera Labs, Negombo 11500, Sri Lanka

A. Guatame-Garcia · B. Vriens (✉)
Department of Geological Sciences and Engineering, Queen's University, 36 Union Street, Kingston, ON K7L 1N6, Canada
e-mail: bas.vriens@queensu.ca

1 Introduction

Our current quality of life is highly dependent on the various functional properties of major and trace metals, whether used as fundamental building materials or for high-tech applications in space travel, nanomedicine or the clean energy transition (Akcil et al., 2020). To meet growing global demand, the prospecting, exploration, mining and processing of metal ore has been—and continues to be—expanding (Lee et al., 2020; Northey et al., 2017). However, declining ore grades near the Earth's surface bring challenges to the continued metal supply. In addition, increasing ore complexities and growing societal and environmental consciousness require more advanced extraction technologies and energy-efficient, environmentally sustainable production (Lottermoser, 2007). Therefore, the last decades have seen major innovations in science and engineering fields across the mining cycle, and novel approaches to metal recovery are no exception (Aznar-Sánchez et al., 2019; Litvinenko & Sergeev, 2019; Lusty & Gunn, 2014).

Conventional metal recovery techniques are primarily classified as pyrometallurgical, hydrometallurgical and electrometallurgical. Pyrometallurgical processes include ore calcining, roasting and smelting to convert target metals to their elemental form at elevated temperatures (Habashi, 2009). Hydrometallurgical techniques extract metals through solid–liquid interaction (i.e., leaching), which can be cost-effective for metal recovery from specific mineral hosts (Bautista, 2013; Peters, 1992). In electrometallurgy, metals are mobilised, deposited or plated in an electrolytic cell that may contain an aqueous or non-aqueous solution (Popov et al., 2007). Each of these techniques has intrinsic merits and demerits, and in industrial operations, one may encounter a combination of them. The focus of this chapter is on aspects of (bio)hydrometallurgy; reviews on other recovery techniques can be found elsewhere (e.g., Habashi, 1993; Mambote et al., 2000; Mooiman et al., 2005).

S. Panda et al. (eds.), *Biotechnological Innovations in the Mineral-Metal Industry*,
Advances in Science, Technology & Innovation, https://doi.org/10.1007/978-3-031-43625-3_12

Hydrometallurgical approaches require relatively low capital investment and operational costs (Schlesinger et al., 2011; Stanković et al., 2020) and they are increasingly feasible for low-grade ore deposits or even below-grade materials otherwise considered waste (e.g., heap or dump leaching; Ilankoon et al., 2018). However, increasing ore complexities challenge successful hydrometallurgy and many oxidised ores have been extensively mined and exhausted, leaving more reduced and gangue-loaded (i.e., refractory) ores for future mining (Kaksonen et al., 2018; Northey et al., 2014; Prior et al., 2012). Adaptation to improve the efficacy of conventional hydrometallurgical routes is therefore required.

The use of microorganisms to promote metal recovery in hydrometallurgy has long been considered in metal processing (Ehrlich, 2004), and there are now industrial-scale facilities dedicated to metal extraction using microorganisms. This field is known as biohydrometallurgy or biomining (herein referred to as biometallurgy), encompassing two main domains: bioleaching and biooxidation (Schippers et al., 2014). Bioleaching promotes the leaching of metals into an aqueous solution, whereas biooxidation enhances the removal of gangue rock matrix while leaving targeted ore minerals behind (Acevedo, 2000; Kundu & Kumar, 2014). Bioleaching is particularly popular in the copper (Cu) industry, where it is used for ~ 20% of global production (Yin et al., 2018). Bioleaching of chalcopyrite ($CuFeS_2$), the most abundant Cu resource at present (Vardanyan & Vardanyan, 2021), may even outcompete the recovery efficiencies obtained in conventional hydrometallurgical processing (Vakylabad et al., 2022). Biooxidation, in turn, is heavily deployed in the gold (Au) industry, specifically for refractory ore deposits that are not favourable for cyanidation but where gangue removal may be facilitated by bacterial oxidation (Khaing et al., 2019; Sugai & Sasaki, 2019). Although extraction using biometallurgy currently constitutes only ~ 5% of annual global Au processing, a shift towards bioextraction is noticeable (Kaksonen et al., 2020a, 2020b). Successful biometallurgical operations are also reported for various other base and precious metals, e.g., Zn, Ni, Mn, Co, Mo and U (Ghassa et al., 2014; Sugai & Sasaki, 2019; Watling, 2015; Watling et al., 2015), emerging technology critical elements (e.g., Swain, 2017) as well as Rare Earth Elements (REE) (Rasoulnia et al., 2021).

Industrial applications of biometallurgy are plentiful and can be performed in heaps, dumps, in situ, as well as in bioreactors. Conventionally, heap and dump leaching are performed for low-grade ore deposits (e.g., Cu grade < 1% (Saldaña et al., 2021)) with nutrient-rich solutions allowed to percolate large pads of material. Because the geochemical reactions that mobilise metals from low-grade ore deposits can be slow under the ambient conditions that prevail in heaps and dumps, the economic recovery of a target metal may take years (e.g., Gericke, 2012; Kaksonen et al., 2018; Wu et al., 2007). In situ leaching follows a similar strategy, where nutrient and/or microorganism-enriched solutions are introduced to an intact ore deposit underground. This technique has been particularly successful in uranium mining (Yin et al., 2018). However, overall, metal recovery rates obtained in heap, dump and in situ leaching operations may be limited by a reduced packing permeability and complex pore networks that cause long retention times, low O_2/CO_2 transfer rates and mineralogical heterogeneity (Fernando et al., 2020a, 2020b; Petersen, 2016; Vriens et al., 2019, 2020a). In addition, microorganism proliferation and activity depend on prevailing temperatures, redox potential, physicochemical material heterogeneity (Rawlings, 2005) and infiltration rates of the solution. Poor control or suboptimal conditions of these parameters hinder uniform bacterial colonisation and reduce metal recovery (Fagan et al., 2014). Also, because of the relatively large spatial footprints and prolonged operations required for economic metal recovery, heap and in situ leaching require careful and long-term management and environmental monitoring to prevent deleterious effects on the ground and surface waters (Fernando et al., 2018; Lottermoser, 2007).

In this regard, performing similar biogeochemical metallurgical reactions under fully enclosed bioreactor conditions often increases control and operational flexibility and monitoring capabilities, ultimately improving efficiency. For example, reaction kinetics can be expedited for enhanced recovery rates, i.e., days compared to years in heap leaching operations. Bioreactors may also be more readily adapted to changes in feed properties, offer processes to be automated at various scales and generally allow for better management of waste streams.

In this chapter, the following aspects of bioreactor design are discussed:

i. processes and parameters in bioreactors used for mineral processing,
ii. process control and (novel) monitoring techniques (e.g., Artificial Intelligence),
iii. selected industrial applications, and
iv. opportunities and emerging directions.

A selection of previous works on various aspects of bioreactor design and operation for biometallurgy is presented in Table 1.

Table 1 Overview of selected reviews relevant to bioreactor design and operation, covering a range of reactor types and methods

Article/chapter title	References
The use of reactors in biomining processes	Acevedo (2000)
Bioleaching: A microbial process of metal recovery; A review	Mishra et al. (2005)
Bioreactor design fundamentals and their application to gold mining	Acevedo and Gentina (2007)
Biomining	Rawlings and Johnson (2007)
How will biomining be applied in future?	Brierley (2008)
Biodiversity and interactions of acidophiles: understanding and optimising microbial processing of ores and concentrates	Johnson (2008)
Biohydrometallurgical prospects	Brierley (2010)
Progress in bioleaching: Fundamentals and mechanisms of bacterial metal sulphide oxidation	Vera et al. (2013)
The role of microorganisms in gold processing and recovery—A review	Kaksonen et al. (2014)
Review of biohydrometallurgical metals extraction from polymetallic mineral resources	Watling (2015)
Microbiological advances in biohydrometallurgy	Watling (2016)
A review of sulphide minerals microbially assisted leaching in stirred tank reactors	Mahmoud et al. (2017)
The evolution, current status and future prospects of using biotechnologies in the mineral extraction and metal recovery sectors	Johnson (2018)
Recent progress in biohydrometallurgy and microbial characterisation	Kaksonen et al. (2018)
Biotechnology for environmentally benign gold production	Natarajan (2019)
Prospective directions for biohydrometallurgy	Kaksonen et al., (2020a, 2020b)

2 Processes and Parameters in Mineral Processing Bioreactors

2.1 Metallurgical Unit Operations

Bioreactor design for hydrometallurgy and mineral processing requires consideration of the entire mineral processing system. Different unit operations may be involved before and after bioreactor operation (Fuerstenau & Han, 2009), although bioreactors in mineral processing generally occur in the final stages of the overall extractive process. This means that preceding unit operations are critical to the success of a biometallurgical reactor: ore or ore concentrate particles need to be carefully characterised and their properties translated into engineering design parameters. For instance, pre-conditioning through size reduction and separation based on mineralogy is required to obtain optimal mass transfer through precise control of, e.g., temperature, aeration and mixing (Brierley, 2008).

a. Particle characterisation

Geometallurgical variables at the processing plant largely dictate the engineering parameters required for designing bioreactors. Such variables can be classified into (1) primary variables that refer to the rock attributes (e.g., mineralogy, lithology, grain size, metal grade, texture, deleterious elements, among others) and (2) response variables, which are properties derived from the material's response to the process (e.g., grindability, flotability, acid-soluble metal content, acid consumption, metallurgical recovery, intact rock strength) (Coward et al., 2009). Primary variables can be retrieved from core logging and mineralogical, geochemical and physical assays. Primary variables integrated with information derived from the deposit block model allow one to constrain these properties spatially, enabling the development of processing plants and bioreactors that can account for temporal variability (Coward et al., 2009; Dominy et al., 2018). Response variables are derived from geometallurgical test work conducted at laboratory and pilot-scale facilities. The integration of data retrieved from primary and response variables enables an improved prediction of the metallurgical performance of the bioreactor (e.g., metallurgical efficiency, product quality, energy, and reagent consumption during processing) and supports the definition of operational parameters (Dominy et al., 2018; Lund & Lamberg, 2014; Pell et al., 2021). In addition, the mineral characterisation unit can also incorporate environmental variables (e.g., carbon footprint, particle emissions) to proactively design processes with better environmental performance (Pell et al., 2021).

b. Size reduction and liberation

Size reduction is achieved by comminution, which is the process of blasting, crushing and grinding material to the required particle size. Comminution aims to liberate valuable minerals from gangue to the coarsest permissible particle size. The material's breakage increases the reactive surface area to facilitate the geochemical reactions of microorganisms in a bioreactor (Fuerstenau & Han, 2009). A critical aspect of size reduction is the energy required for comminution, which depends on the hardness of the rock, the feed size and the targeted size for liberation (Gupta & Yan, 2016). Size reduction is an energy-intensive process that accounts for ~ 50% of the operating costs of a mineral processing plant, with grinding accounting for 50% of the energy consumption of the unit (Radziszewski, 2013). As a result, a significant challenge lies in achieving a particle size that is large enough to ensure energy savings but small enough to allow sufficient interaction between microorganisms and the ore or gangue minerals (Herbst et al., 2003). The optimum particle size is determined partly by the grain size and mineral texture (i.e., relationships of the minerals at the grain scale). In particular, bioreactors and flotation circuits require exposure of at least a fraction of the mineral's surface. Generally, target sizes for metal sulphides range between 100 μm for coarse-grain ores and 10 μm for fine-grain ores (Evans et al., 2015; Wills & Finch, 2016).

c. Concentration

Before introducing feeds to a bioreactor, further separation of metals from gangue particles may be required. The concentration methods depend on material properties such as density, magnetic susceptibility and electric conductivity (Fuerstenau & Han, 2009; Wills & Finch, 2016). Common techniques include (optical) sensor-based sorting, gravity concentration, froth flotation, and electrostatic and magnetic separation, all of which may be deployed in combination with a bioreactor unit. Biometallurgy can be used as a pre-treatment step to enhance the flotation performance or after flotation to complete the extraction of the metals.

2.2 Biometallurgical Mechanisms

Achieving successful metal extraction in bioreactors requires optimum leaching conditions that allow controlled microbial activity and growth (Sajjad et al., 2019). Such optimal conditions will vary widely depending on the type of ore, particle size and the involved microbial consortia, which may require different temperatures and pH, nutrient levels and oxygen and CO_2 content (Acevedo & Gentina, 1989; Sajjad et al., 2019). Regardless of the targeted process, establishing and maintaining optimal bioleaching conditions requires one to quantitatively understand variability in material feeds (Sect. 2.1), the major biogeochemical reactions (including stoichiometry, thermodynamics, kinetics and the biocatalytic effects of microorganisms), the compatibility between microorganisms and the ore-bearing mineral slurry, and the controls of the prevailing conditions on microbial growth and activity, among others. Bioleaching in engineered systems may be performed by diverse microorganisms, including bacteria, archaea and fungi. There have been many advances in metallurgical microbiology over the past decades, and the reader is referred to existing reviews on the diverse microbiology of bioleaching systems (Bosecker, 1997; Kundu & Kumar, 2014; Schippers et al., 2010; Watling, 2016).

Metal recovery from sulphide minerals is typically achieved through the activity of chemolithotrophic bacteria, well-known species of which include *Acidithiobacillus ferrooxidans* and *thiooxidans*, *Leptospirillum ferriphilum*, as well as thermophile archaea such as *Sulfolobus metallicus* or *Acidianus* sp. (Blackmore et al., 2018; Edwards et al., 2000; Jia et al., 2019; Sajjad et al., 2019; Schippers et al., 2010; Schrenk et al., 1998). Genetic engineering of these and other microorganisms to improve biometallurgical extraction is an active field of research (Kaksonen et al., 2020a, 2020b). The mechanisms by which microbial species convert poorly soluble sulphides into metal-sulphates may vary. Two types of pathways are typically distinguished based on the acid-solubility of the targeted sulphide: the thiosulfate pathway for acid-insoluble sulphides (e.g., pyrite [FeS_2]) and the polysulfide pathway for acid-soluble sulphides (e.g., sphalerite [ZnS] or galena [PbS]). Acid-insoluble pyrite oxidation by Fe^{3+} generates acids, dissolved thiosulfate ($S_2O_3^{2-}$) and Fe^{2+} that can be re-oxidised to Fe^{3+} Eqs. (1 and 2), after which dissolved $S_2O_3^{2-}$ may be further oxidised to sulphate (SO_4^{2-}), either directly (Eq. 3, 4) or via (poly)thionate intermediary species (Eq. 5–6), according to the following reactions (Schippers, 2004):

$$FeS_2 + 6Fe^{3+} + 3H_2O \rightarrow 7Fe^{2+} + S_2O_3^{2-} + 6H^+ \quad (1)$$

$$4Fe^{2+} + O_2 + 4H^+ \rightarrow 4Fe^{3+} + 2H_2O \quad (2)$$

$$S_2O_3^{2-} + 2O_2 + H_2O \rightarrow 2SO_4^{2-} + 2H^+ \quad (3)$$

$$S_2O_3^{2-} + 8Fe^{3+} + 5H_2O \rightarrow 8Fe^{2+} + 2SO_4^{2-} + 10H^+ \quad (4)$$

$$5S_2O_3^{2-} + 6H^+ \rightarrow 2/8S_8 + 2S_4O_6^{2-} + 3H_2O \quad (5)$$

$$S_4O_6^{2-} + 3Fe^{3+} + 2.75O_2 + 4.5\,H_2O \rightarrow 4SO_4^{2-} + 3Fe^{3+} + 9H^+ \quad (6)$$

The oxidation of acid-soluble sulphides also occurs through Fe^{3+} and proton attack, but instead releases divalent metal cations and sulphide radicals (Eq. 7), the latter which are further oxidised via elemental S to SO_4^{2-} (Eq. 8, 9) (Vera et al., 2013):

$$MS + Fe^{3+} + H^+ \rightarrow M^{2+} + 0.5H_2S_n + Fe^{2+}(n \geq 2) \quad (7)$$

$$0.5\,H_2S_n + Fe^{3+} \rightarrow 1/8\,S_8 + 2Fe^{2+} + H^+ \quad (8)$$

$$1/8S_8 + 1.5O_2 + H_2O \rightarrow SO_4^{2-} + 2H^+ \quad (9)$$

The catalysis of the above reactions by autotrophic bacteria increases reaction kinetics by orders of magnitude (Nordstrom, 1982), whereby the oxidation of Fe^{2+} to Fe^{3+} by oxygen is considered rate-limiting (Singer & Stumm, 1970), particularly under higher pH values where Fe-oxides are poorly soluble. Overall, microbially mediated oxidation of sulphides to sulphate involves the transfer of 7 or 8 electrons for each S atom, depending on whether the mineral is a mono-sulphide or di-sulphide. Thus, it constitutes a complex, multi-step process that involves biochemical and electrochemical reactions in which the kinetics of each step are affected by factors such as pH, pO_2, and the specific surface area and morphology of the sulphide (Blowes et al., 2014; Evangelou, 1998; Evangelou & Zhang, 1995; Keith et al., 2018). Different microbial species may be active at different steps in both pathways (e.g., performing Fe- versus S-oxidation), but species such as *Acidithiobacillus* and *Leptospirillum* are involved in steps in both pathways (Johnson, 2018; Schippers & Sand, 1999). These bacteria are acidophilic and have optimal growth conditions around pH 2–3 but can survive up to pH 6–7 (Nordstrom, 1982). Neutrophilic S-oxidising bacteria may coexist with acidophilic bacteria and contribute to oxidative sulphide dissolution under neutral conditions (Skłodowska & Matlakowska, 2007). In addition to the above mechanisms where bacteria oxidise dissolved Fe that can then attack metal sulphides, certain microbial species can oxidise mineral surfaces directly or produce dissolved organic acids that act as chelating agents that promote dissolution. Depending on the mechanism involved, microorganism activity is thus desired closer to or further from the mineral interface (Fig. 1). For instance, thiobacilli may secrete extracellular polymeric substances (EPS) that may enhance leaching and facilitate attachment of the bacteria to the mineral surface, affecting solute-surface exchange. Detailed information on mineralogy-dependent S-oxidation pathways, including the kinetics of microbial catalysis, is available in Vera et al., 2013; Rohwerder et al., 2007; Schippers et al., 2010; and Tu et al., 2017, among others (Table 1).

In addition to oxidative bioprocessing, hydrometallurgical processes using microorganisms can also be reductive, e.g., the dissolution of oxide minerals in anaerobic environments. Examples include a variety of commodities (e.g., Co, rare earth elements) that can be mobilised from materials such as laterites under reducing conditions (Johnson et al., 2013; Nancucheo et al., 2019; Smith et al., 2017; Yang et al., 2019).

2.3 Metallurgical Bioreactor Design

Bioreactors for mineral processing have various designs: batch reactors, stirred tanks, column reactors, including fixed or fluidised beds, membrane, bubbling or airlift columns, as well as solid substrate- and heap reactors (Gonzalez-Sanchez et al., 2008; Midha et al., 2012; Sahinkaya et al., 2011). From a chemical engineering perspective, bioleaching in agitated aerated tanks is typically considered a continuous-flow and steady-state process (Acevedo & Gentina, 2007). Regardless of the reactor mode and configuration (e.g., in serial versus parallel arrangement), the design and optimisation of a hydrometallurgical bioreactor process require (i) a time-dependent characterisation of material feeds, which might not be of constant quantity and quality in mining operations (Sect. 2.1), (ii) a quantitative understanding of the occurring reactions including stoichiometry, thermodynamics, and (bio)kinetics (Sect. 2.2) and (iii) balance models for mass and heat at a variety of spatiotemporal scales. Ultimately, the interplay between reaction- and transport processes poses a critical control on product formation and selectivity, efficiency (e.g., mass transfer versus kinetic rate limitations, power consumption), yield, reactor stability, etc. (Acevedo & Gentina, 2007; Morin, 2007). While not intended to review reactor engineering principles, this section addresses some key bioreactor design aspects.

Firstly, a kinetics study will reveal reaction rates, including substrate consumption and product synthesis rates, controlling kinetic factors, reaction orders and by-product formation. Well-characterised reactor kinetics may help avoid mineral passivation with biofilm or precipitation of secondary mineral phases (e.g., jarosite) that can reduce mass transfer (Kaksonen et al., 2014; Kartal et al., 2020), and identify potentially toxic constituents that can inhibit microorganism growth (e.g., chlorides, cyanides; Natarajan, 2019). A quantitative understanding of intermediary product formation and their thermodynamic versus kinetic stability in a bioreactor is critical. Still, when potentially hundreds of reactions are involved, and intermediary species are not rate-limiting on the scales of operation, the overall stoichiometry may be considered. The kinetics of

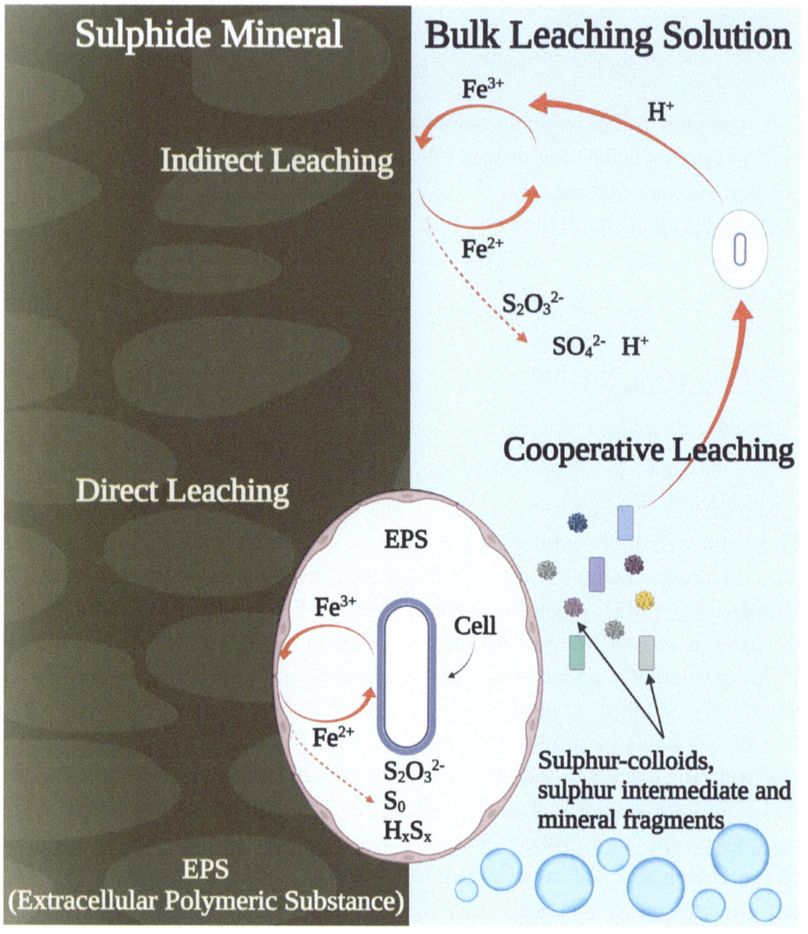

Fig. 1 Schematic illustration of various bio-assisted leaching and oxidation mechanisms in the presence of sulphide minerals (modified after Mahmoud et al. (2017) and Rawlings (2002))

oxidative dissolution of sulphide minerals r [units of mass or moles per volume or surface per time] may be generically expressed as

$$r = k[A]^{\alpha}[B]^{\beta}\ldots \quad (10)$$

where k is the reaction rate constant and [A] and [B] are the aqueous activities of solutes involved in the reaction kinetics. In the case of metal sulphide oxidation, these may include $[Fe^{3+}]$, $[O_2]$, or pH ($[H^+]$). The exponents, α, β, etc., are the reaction orders concerning components A, B, etc. Typically, reaction orders depend highly on the targeted mineral assemblage and leaching conditions tested. For instance, pyrite oxidation may depend on the aqueous proton activity to an order of −0.5 to −0.11 and the dissolved Fe^{3+} activity to an order of 0.3 to 0.58 (Holmes & Crundwell, 2000; Huminicki & Rimstidt, 2009; Williamson & Rimstidt, 1994). Certain design parameters such as temperature may be readily accounted for by adjusting k, e.g., using Arrhenius-type relationships. However, across widely variable conditions, the entire microbial community and bioreaction pathway (and thus kinetics) may change depending on the targeted process. Parameters such as liberated surface area and mineral morphology will affect bulk reaction rates and may be explicitly expressed in a rate equation as well. However, due to challenging quantification, microscale parameters often remain unresolved in the reaction rate constant k, complicating the extrapolation of kinetic studies across different applications (Vriens et al., 2020b). Overall, large variability may occur between bioreactor conditions, and kinetic reaction descriptions often have non-negligible uncertainties (Crundwell, 2000). To facilitate pragmatic reactor design, certain kinetic parameters can be reasonably assumed constant (e.g., temperature or bulk slurry pH may be adequately controlled). The resulting rate equations subsequently relate process kinetics directly to design parameters, such as total substrate or biomass M, average substrate particle size d_{50}, temperature T, or pH:

$$r = f(M, d_{50}, T, \ldots) \quad (11)$$

In addition to relevant reaction kinetics, a series of mass and energy balance equations form the foundation of bioreactor design. These will help define reactor and flow dimensions, ensure residence and mixing times allow for reactions to proceed as desired, avoid needless washout, and that the relevant mass and heat transfer processes ensure the reactor system is controlled and stable (Belyi et al., 2018; Miller, 1997). These balance equations look different depending on the bioreactor type and processes involved. Mass-balance (change in mass in the system with volume V over time $[d(Vc)/dt]$) can be generalised in the form:

$$\frac{\partial(Vc)}{\partial t} = \sum_{i=1}^{n} F_i(c_{0,i} - c_i) \pm rV \qquad (12)$$

where c is reactor concentration, c_0 feed concentration, F is a mass flow rate (if present) and r a kinetic rate expression (Eq. 11; Fig. 2). There may be various flows F carrying different c_0 into and c out of a reactor system, as well as different rate expressions (zeroth, first, second, fractional orders) for the involved reactions: typically, a number of inter-dependent mass-balance statements with appropriate boundary conditions is required for different solutes of interest. For bioleaching operations targeting sulfidic ores, these include balances for the targeted metal(s) of interest as well as other abundant metals (e.g., Au, Cu, Ni), S and potential intermediary species, other elements of concern, as well as nutrients (N, P) and gaseous species such as oxygen and CO_2 (i.e., carbon and oxygen uptake rates).

Most biometallurgical processes are reactor-operated under mesophilic or moderately thermophilic conditions (20–50 °C; Rawlings et al., 2003). Moderate thermophiles in bioleaching processes may have kinetic advantages over mesophilic conditions, can grow at higher pulp densities than thermophiles (Nemati et al., 2000) and tolerate higher catalysts concentrations such as Ag (Gómez et al., 1999). Because microorganisms have optimum temperatures for their metabolic activity (Norris et al., 2000), temperature is a critical parameter to control bioleaching and biooxidation processes and guarantees high conversion rates. Energy balance equations account for total energy dynamics in a bioreactor system over time (dE/dt) and can be generalised as

$$\frac{\partial E}{\partial t} = (Q_{in} - Q_{out}) + (W_{in} - W_{out}) + (E_{in} - E_{out}) \qquad (13)$$

in which Q is direct heat transfer to and from the reactor, W is work transfer, and E is the energy introduced or removed by mass flow. Energy E is the sum of internal (mU_i), kinetic ($mv_i^2/2$) and potential energy (mgz_i). For open system balances, the kinetic and potential energy components are

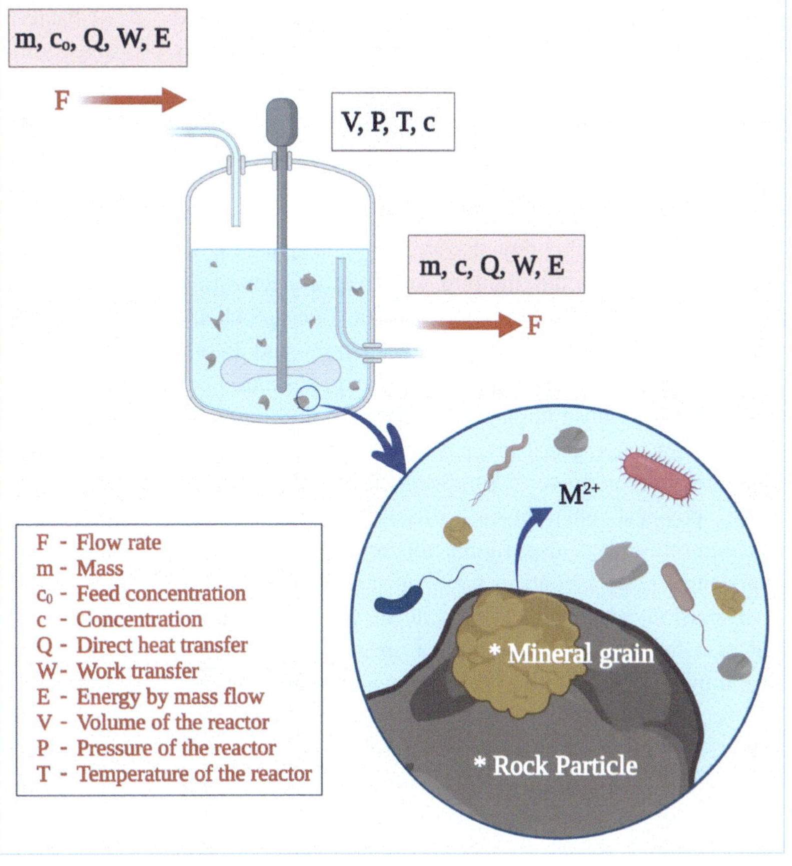

Fig. 2 Schematic illustration of the various mass and energy flows in an aerated stirred tank bioreactor

distinguished from internal energy and flow work and considered as enthalpy H, which allows for facile incorporation of the energies produced or consumed by chemical reactions in the energy balance. Bioreactor design may involve consideration of heat generation or removal by metabolism, agitation, aeration, phase change (e.g., evaporation) and exchange with surroundings (Van't Riet & Tramper, 1991).

Typical operation under steady-state conditions and prioritisation of select major mass and energy transfer processes allow for reactor design with simplified balance statements. However, many simultaneous multiphase transport processes occur in a bioleaching operation: efficient heat transfer is needed to operate the bioreactor at the desired optimal microbial growth temperature, nutrients must reach surface-bound and suspended cells, metabolic products must migrate from the cells to the liquid, and solubilised species must be transported from the mineral surfaces to the bulk liquid. Thus, identification and parameterisation of the relevant mass- and heat transfer processes and the corresponding material properties (e.g., convective heat transfer coefficient of a slurry or effective diffusivity of a solute) is required to optimise bioreactor performance.

Another important consideration in bioreactor design is the autocatalytic nature of microbial growth. In a bioprocess, biomass concentration must remain high enough to achieve a high yield, but not too high when organisms proliferate on the reactor walls or stirrer or cause fouling and clogging. An additional material balance for cell density is often required, deploying, e.g., a Monod-type equation for microbial growth kinetics:

$$\mu = \mu_{\max} \frac{[S]}{K_S + [S]} \tag{14}$$

where μ is a specific growth rate, μ_{max} is a maximum growth rate, $[S]$ is the limiting substrate concentration, and K_S is the Monod constant. Bioreactor design should also consider surface attachment: recalcitrant and poorly soluble sulphides can have low affinity and substantial cell growth, but low substrate-microorganism affinity (corrected Monod's saturation constant) may require the use of material treatment or additives (Crundwell, 2000; Morin, 2007).

The physical environment affects biological performance in other ways also. Bioreactors are usually stirred to homogenise reactor content, maintain microbes and solids in suspension, and increase the rate of oxygen, CO_2 and heat transfer. Under conditions of insufficient agitation, mass transfer processes may become reaction limiting, and the overall performance will decline because segments of bioreactor fluid have insufficient nutrients or inadequate pH or temperature (Bailey & Hansford, 1993). However, excessive stirring results in unnecessary loss of mechanical energy and high shear rates that may induce harm to shear-sensitive organisms and disrupt cell walls (Chong et al., 2002). Overall, mechanical stresses, mass transfer and mixing, pH and temperature are all interrelated to biological growth and activity and affect bioreactor performance.

In summary, key design aspects for metallurgical bioreactors include.

- **Types of ore and particle size**: Bioreactors used to beneficiate polymetallic ores (e.g., Cu, Au, Ni) usually host varied associations of sulphide minerals which can be differently susceptible to bioleaching (Norris & Owen, 1993). Furthermore, the inherent variations within a deposit introduce mineralogical heterogeneity (Dominy et al., 2018). Therefore, the entire mineral assemblage requires careful study. Similarly, physical parameters such as particle size are dependent on the operation mode of the bioreactor. Finer particle sizes can facilitate microbial growth and accelerate reaction kinetics. However, they require increasing energy for milling and increased pulp density (typically 20–30% (Kundu & Kumar, 2014; Mahmoud et al., 2017)), which can negatively affect microbial growth, suspension, homogeneity and downstream settling processes (Acevedo & Gentina, 2007).
- **Microorganisms**: Bioreactors for bioleaching and biooxidation operations often have concurring species of bacteria, archaea or fungi in the same ecosystem and these different species may have variable optimal conditions for growth (Johnson, 2008) and capabilities to adapt to suboptimal conditions (Gomez et al., 1999). The reaction kinetics in bioreactors are strongly controlled by microbial diversity, density, spatial distribution and activity, which in turn are heavily affected by the mineralogical heterogeneity of the ore as well as many other physicochemical factors (temperature, redox potential, particle diameter, solid concentrations, see below). Overall, the bioleaching process efficiency depends on a delicate balance between the bacterial population and controlled growth conditions.
- **Temperature**: The optimal temperature window for bacterial growth depends on the type of microorganisms, often classified as mesophiles (20–40 °C), moderate thermophiles (around 50 °C) and extreme thermophiles (>65 °C) (Rawlings et al., 2003). A challenge for metallurgical bioreactors is that oxidative sulphide dissolution is highly exothermic (up to −1500 kJ/mol) and thus requires adequate control of the system's energy balance. Additional understanding of relevant thermo-hydro-chemical feedbacks is important: e.g., the increasing temperature may be advantageous for improved reaction kinetics but could challenge gas–liquid mass transfer (e.g., O_2 solubility) (Sajjad et al., 2019).

- **pH**: The pH of the bioreactor medium influences the metabolism of microorganisms and geochemistry of the reactor system: different microorganisms have different optimal pH for growth, and the solubility of metals-of-interest and potential secondary mineral products is highly pH-dependent. Most bioleaching reactors exhibit strongly acidic conditions, where transition metals are typically very soluble.
- **Nutrients**: In bioreactions, nutrients may originate from the feed material, but sulfidic metal-ore materials typically contain little carbon, N and P, and supplements are commonly utilised. Reaction rates indirectly depend on a sufficient and effective supply of macro- and micro-nutrients.
- **Oxygen and CO_2**: As most of the metal bacterial-assisted leaching processes are aerobic and chemolithotrophic, an adequate supply of oxygen and CO_2 is necessary. An adequate supply of these gases may be achieved by aeration and stirring reactor content, but oxygen and CO_2 need to be transported from air bubbles to the oxidation sites. Considering the low solubility of oxygen in aqueous solutions, gas mass transfer at the gas–liquid interface is a principal control of bioreactor efficiency (Garcia-Ochoa & Gomez, 2009; van Aswegen et al., 2007). Oxygen and CO_2 supply also present operating challenges for sequential bioleaching reactor systems, with the necessity to balance pressure and consumption in multiple reactors and additional regulation of CO_2 level when acid production is partially offset by carbonate buffering (Acevedo & Gentina, 2007; Kaksonen et al., 2018).

3 Bioreactor Process Monitoring and Control

3.1 Monitoring

As discussed in Sect. 2, various parameters affect the biometallurgical process and require careful monitoring and control. Online monitoring is the logging of measured data directly into a control loop via sensors, whereas offline monitoring is sampling and logging data outside a control loop and usually involves manual entry (Malaguti et al., 2015). Offline methods may produce expired information, which may induce system instability when incorporated into the control loop. Computer-integrated solutions now allow for increasingly rapid (near) real-time data collection and processing and improved sensor technologies enable monitoring of the most critical physical (e.g., temperature, pressure), chemical (e.g., DO, pH) and biochemical (e.g., biomass) parameters within bioreactors at increasing resolution.

Sensors used in bioreactors can be of different types, including optical, electronic, electrochemical, electromagnetic, ultrasonic or bio-based (Busse et al., 2017). A sensor system extracts the parameter or system state of interest and couples it to an actuator framework (e.g., valve, pump) to trigger actions based on the sensors' output signals, thereby regulating the overall process. Numerous intricate subroutines are executed for efficient, robust and high-fidelity operations. Figure 3 illustrates a traditional bioreactor configuration and several parameters that can be measured. Sensors can be

- Direct: directly interact with the bioreactor medium to quantify a magnitude (e.g., temperature)
- Indirect: system not directly probed (e.g., external jacket temperature). Calibrated indirect measurement may be utilised to minimise interference with the process and increase sensor lifespan.
- Inferential: calculates the value of a parameter by measuring other variables that may be easier to probe (e.g., the fluid flow rate through a pipe inferred from a differential pressure transmitter, when the pressure in the pipe is directly related to the flow rate with constant fluid density).

Employing sensors to measure every parameter in a bioreactor process is not always practical: various parameters are challenging to probe at sufficient resolution (e.g., bacterial abundance or enzyme activity), and sensors may be susceptible to time constraints (temporal sensor resolution), and there may also be sensor placement challenges, e.g., due to corrosion/pH concerns or elevated temperatures (Botero & Álvarez, 2011). Considering these restrictions in bioreactor processes, monitoring may be limited to only the most critical parameters. These parameters portray known extensions to other parameters through the bioreactor model.

When sufficient substrate is available, the growth of microorganisms in a bioreactor may be modelled by

$$\frac{dX}{dt} = \mu(S)X \tag{15}$$

Here, X denotes the microorganism mass (or density), with a growth rate constant of μ, which is contingent on the substrate S. In a continuous process, both inflow and outflow affect the growth mechanism and thus Eq. 15 may become

$$\frac{dX}{dt} = \mu(S)X + \frac{F_{in}}{V}X_{in} - \frac{F_{out}}{V}X_{out} \tag{16}$$

where the term $\frac{F}{V}$ describes the dilution coefficient (D), which can be used to describe biomass accretion from inflow and biomass loss through outflow under well-mixed conditions. Furthermore, Eq. 16 relies on a growth coefficient ($\mu(S)$) that can be derived from Eq. 14. Thus, Eq. 16 may also be modified:

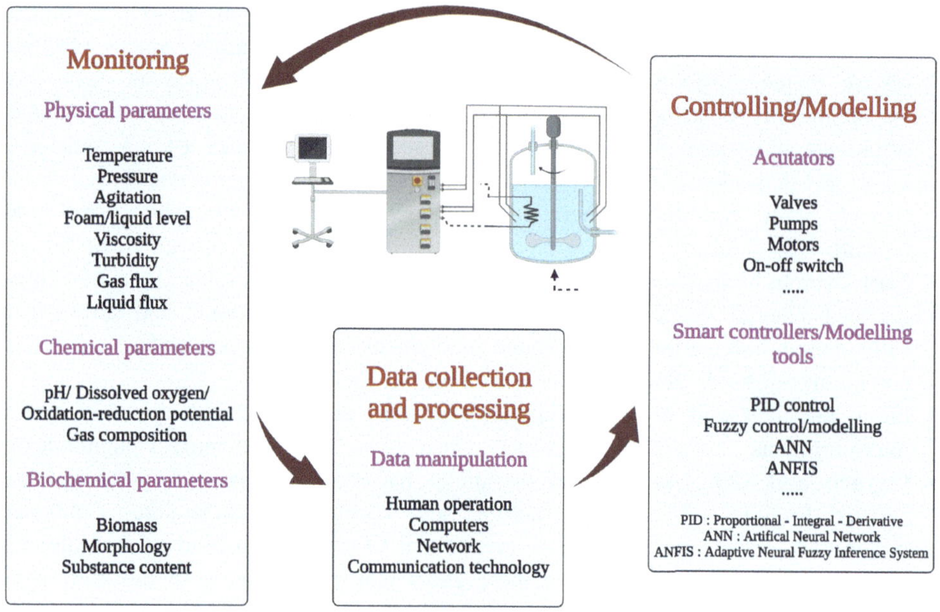

Fig. 3 Monitoring, control and modelling aspects of a conventional bioreactor (modified after Wang et al. (2020))

$$\frac{dX}{dt} = \mu_{\max} \frac{[S]}{K_S + [S]} X + DX_{in} - DX_{out} \tag{17}$$

Bioreactor parameters discussed in Sect. 2 may be incorporated into Eq. 17 to assure an explicit description of microorganism growth. After defining the parameter space $\Omega = f(\text{pH, temperature, DO, agitation speed, etc.})$, the growth coefficient may be converted to

$$\mu(S, \Omega) = \mu_{\max} \frac{[S]}{K_S + [S]} f(pH).f(T).f(DO)\ldots \tag{18}$$

and thereby the mass-balance equation for species X is converted to

$$\frac{dX}{dt} = \mu(S, \Omega)X + DX_{in} - DX_{out} \tag{19}$$

Equation 18 displays the nonlinearity of the microorganism growth process and intricate interdependence of the functional bioreactor parameters. A bioreactor process is highly dynamic and sensitive to subtle changes in parameter fluctuations, and modelling tools are often required to simulate and benchmark process control schemes (Malar & Thyagarajan, 2009). In addition, laboratory studies may disregard unresolved variances and treat dependencies as constants so that models are often poorly generalisable from system-to-system and highly specific instead. Increasingly, detailed monitoring of bioreactor dynamics can facilitate an understanding of reactor behaviour at high precision (subject to validation) but warrants the use of trained computer models to offset computational demands. Artificial Intelligence (AI) and Machine Learning (ML) techniques are gaining precedence in reactor control due to the increased ease of rapid prediction with well-trained models (Babanezhad et al., 2020; Calzolari & Liu, 2021; Mowbray et al., 2021).

3.2 Bioreactor Control/Modelling Strategies

Control of bioprocesses generally requires at least several online measured parameters, including pH, temperature, DO and agitation (Portner et al., 2017). Other online monitoring techniques (e.g., in situ microscopy and impedance spectroscopy for cell concentration, or Infrared and Raman spectroscopy for metabolic compounds) are recent developments in online instrumentation (Abu-Absi et al., 2011; Pacheco et al., 2021). The fluctuation of these process parameters determines the control strategy, which needs to be optimised within the overall process. Observability is a measure of how well internal states can be elucidated by direct measurement (Dahleh et al., 2004): parameters such as temperature, pH, DO are examples of outputs directly probed in a bioreactor, but fully observable systems are scarce in practice (Maes et al., 2019). Most bioreactors are partially observable, and in control theory, a system requires full observability to permit full controllability (Nise, 2020). Controllability is defined as how well the system inputs can manipulate the system state. Observability and controllability are connected: if a system is partially observable, the controllability of the process is limited. Since many bioreactors are partially observable, they present an interesting control challenge for design engineers (Krishna et al., 2018; Wu et al., 2009). The following sections briefly discuss common control schemes for bioreactors.

3.2.1 PID-Based Control

Traditionally, fundamental bioreactor parameters such as pH, temperature or agitation are controlled by proportional-integral-derivative (PID) controllers (Chandra & Samuel, 2010). These controllers compare sensor inputs with a reference value (set value) and determine the error signal to control the system with static or adaptive schemes (Galvanauskas et al., 2019; Oladele & Shaibu, 2019). Regular PID controllers rely on constant tuning parameters (i.e., static), while others attempt to capture time-varying conditions (i.e., adaptive) through gain scheduling, tendency models and more (Akisue et al., 2021; Galvanauskas et al., 2019; Jayachitra & Vinodha, 2014). Further, details on PID control can be found elsewhere (e.g., Johnson & Moradi, 2005; Quiñónez et al., 2019). Although many bioreactor processes are nonlinear (e.g., Eq. 19), PID control is considered linear (Galvanauskas et al., 2019) yet may be applied within a limited (quasi-linear) region at nominal operating conditions of a bioreactor system.

3.2.2 Fuzzy Logic- and Artificial Neural Network-Based Control and Modelling

Fuzzy logic is a branch of Artificial Intelligence (AI) that informs the controller with expert knowledge of the underlying process using fuzzy variables, membership functions, a set of rules and a rule engine (Shaghaghi et al., 2013). Unlike Boolean outputs (i.e., TRUE or FALSE), fuzzy logic permits the controller to emulate human reasoning with scaled values, thereby incorporating control flexibility (Caramihai et al., 2013). Fuzzy logic models are sometimes combined with PIDs to capture nonlinear process features (Butkus et al., 2020). Artificial neural networks (ANNs) are considered universal AI approximators and are used extensively in nonlinear process modelling (Vyas et al., 2020). In contrast to fuzzy controls, ANN application eliminates the requirement of human intervention: ANNs self-learn based on correlations provided by a set of input and output data connected via optimised weights and transfer functions (Abdollahi et al., 2019). Adaptive Neuro-Fuzzy Inference Systems (ANFIS) are an example of combined Fuzzy Systems and ANN, where IF–THEN rules are specified using a feed-forward ANN, allowing the rule engine to self-learn, therefore harnessing the capabilities of both techniques (Xie et al., 2019). Other mixed control schemes include PID controllers with gain scheduling handled by a fuzzy inference system to effectively deal with process nonlinearities (Tamayo et al., 2019) and PID controllers coupled to ANNs that provide gain values (Tronci & Baratti, 2017).

Although fuzzy logic- and ANN-based controls are popular control schemes in many industries, examples of biometallurgical applications of these techniques remain relatively scarce in the published literature. Ahmadi and Hosseini (2015) used fuzzy models to predict Cu bioleaching efficiencies, Bellenberg et al., (2018); Buetti-Dinh et al., (2019) explored ANNs for image analysis to resolve biofilm distribution and evolution (mineral surface attachment) in bioreactors, and Demergasso et al., (2018) and Soto et al., (2013) developed a decision support system for a heap bioleaching plant using historical data on ore mineralogy, solution chemistry and operational parameters. Additional information on potential applications of AI/ML for bioreactor control and modelling may be found in McCoy and Auret (2019), Rabbani et al., (2021) and Mishra (2021).

4 Industrial Applications

The first industrial use of bioreactors for hydrometallurgy dates from 1958 at the Bingham mine, USA (Agioutantis, 2004; Yin et al., 2018). Subsequently, the development of bioreactor technology has been accompanying the steady growth of biometallurgy. The first-ever industrial-scale bioreactor installation was reported in 1986 with the BIOX™ process pioneered by Gencor (now known as Gold Fields), South Africa (van Aswegen et al., 2007). Originally designed for refractory Au ore oxidation, the BIOX™ process has also been explored to recover base metals (Batty & Rorke, 2006). Other noteworthy pilot plants include BHP Billiton's BioCOP™, BioNIC™ and BioZINC™ (Natarajan, 2018; Watling, 2008). Figure 4 presents an overview of metallurgical bioreactor technologies that have reached pilot- or industry-scale. This section discusses select examples of industrial applications of bioreactor-driven recovery of Au, Cu, Ni and Co.

4.1 Gold Industry

Industrial Au extraction often uses cyanidation, and removal of cyanide-consuming gangue is paramount to project economics. Biometallurgical techniques may be employed to biooxidise gangue minerals and remove them from the solid matrix.

The Fairview mine (South Africa) was among the first to commission the BIOX™ process (presently owned by Metso Outotec, Finland—Smart et al. (2017)) for the biooxidation of Au ore. The BIOX™ process has meanwhile found extensive utility as a pre-treatment technique in the Au processing industry, with 13 facilities deployed globally to date (Metso Outotec, 2021). The Fairview mine was designed for a 62 tonnes per day (tpd) feed and is still operational; the Kokpatas mine (Uzbekistan), capable of handling ~ 2,138 tpd of Au ore, is reported as the largest operating BIOX™ facility in the world (Brierley, 2008; Kaksonen et al., 2014; Natarajan, 2019).

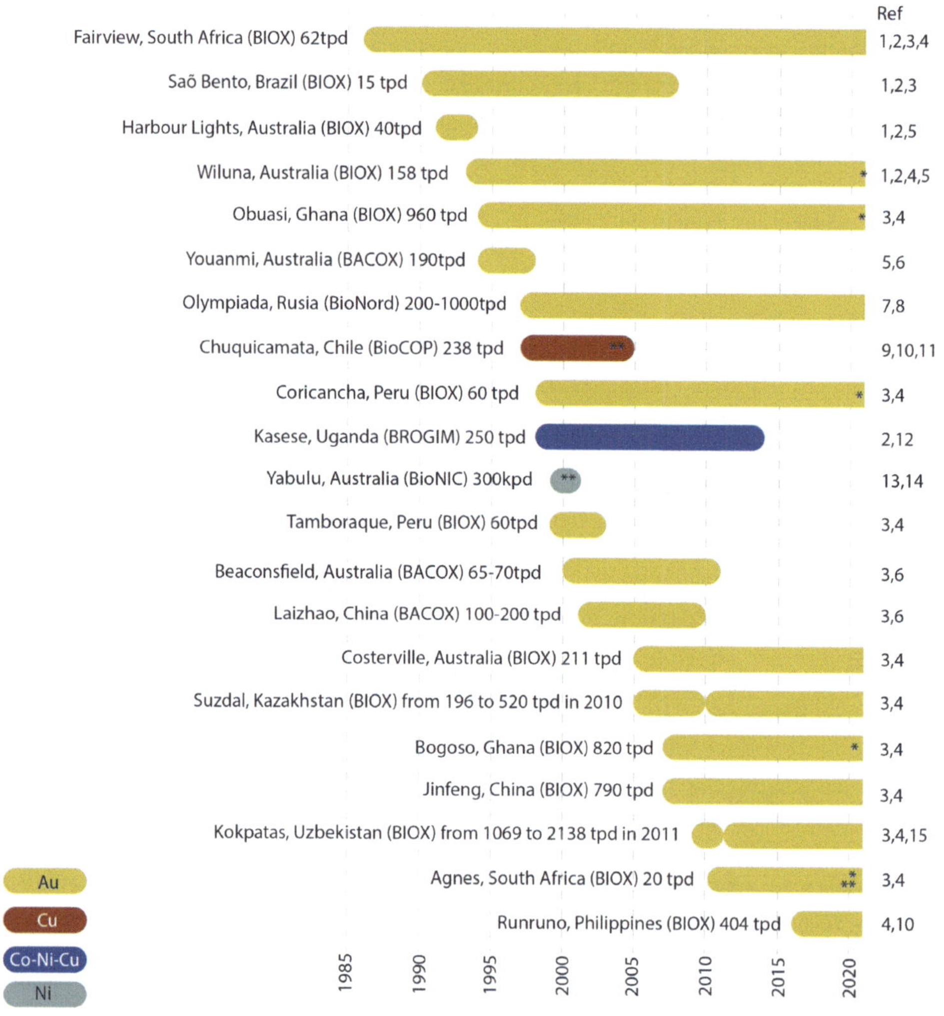

Fig. 4 2021 overview of some bioreactor technologies deployed at industrial and pilot scale for precious and base metals. *: Currently under care and maintenance; **: pilot plant; tpd: tons per day; kpd: kilograms per day. References: [1]: van Aswegen et al. (2007), [2]: Morin (2007), [3]: Mahmoud et al. (2017), [4]: Metso Outotec (2021), [5]: Brierley and Brierley (2001), [6]: Neale et al. (2000), [7]: Sovmen et al. (2009), [8]: Belyi et al. (2018), [9]: Batty and Rorke (2006), [10]: Dreisinger (2006), [11]: Brierley (2016a, 2016b), [12]: BRGM (2021), [13]: Palmer and Johnson (2005), [14]: Natarajan (2018), [15]: Natarajan (2019)

Figure 5 depicts an example flowsheet of the BIOX™ process. The flowsheet typically allows the processing of flotation concentrates with a 75-micron (80% passing) particle size to ensure adequate reaction kinetics along with a recommended 6% sulphide concentration for adequate bacterial activity (van Aswegen et al., 2007). Although further milling may improve Au recovery, finer feeds may also yield negative impacts in the downstream processes due to increased viscosity of the slurry and settling footprints. The milled feed concentrate is mixed with nutrients (i.e., nitrogen, phosphorus and potassium at 1.7, 0.9 and 0.3 kg per tonne, respectively) for microorganism growth (i.e., *Acidithiobacillus ferrooxidans*, *At. thiooxidans* and *Leptospirillum ferrooxidans*, further details in Smart et al. (2017)) and later introduced into three primary BIOX™ reactors arranged in parallel. The total residence time of the process is 4–6 days; the primary reactors account for half of that time to allow for sufficient microorganism growth. A minor fraction of CO_2 is also required for microorganism growth, and carbonate-lacking concentrates require adding limestone/$CO_{2(g)}$ to the primaries. Once microbial growth is established, the overflow of the primary reactors is sequentially treated in secondary reactors with shorter (~1 day) residence times. All secondary reactors are equipped with cooling coils: sulphide oxidation is highly exothermic (Sect. 2.3), and cooling is imperative to regulate the bioreactor conditions to a temperature range of ~30–45 °C for the mixed mesophilic cultures in the BIOX™ process (Brierley, 2008). Moreover, air injection maintains DO at 2 mg/L, and pH is typically kept at 1.2–1.8 (van Aswegen et al., 2007). The oxidation process creates significant loads of dissolved ions and counter-current decantation (CCD) units are employed to reduce these loads before the cyanidation of the final concentrate. The resulting CCD overflow of this process contains Fe and As that require elimination or neutralisation (typically multi-stage—Broadhurst (1994)), whereas the thickened ore product is introduced to the cyanidation step.

Another case study is presented by Mintek and BacTech, who jointly developed the Au recovery technique BACOX™ at the Beaconsfield gold mine (now owned by NQ Minerals), Tasmania, Australia. The Au extraction at

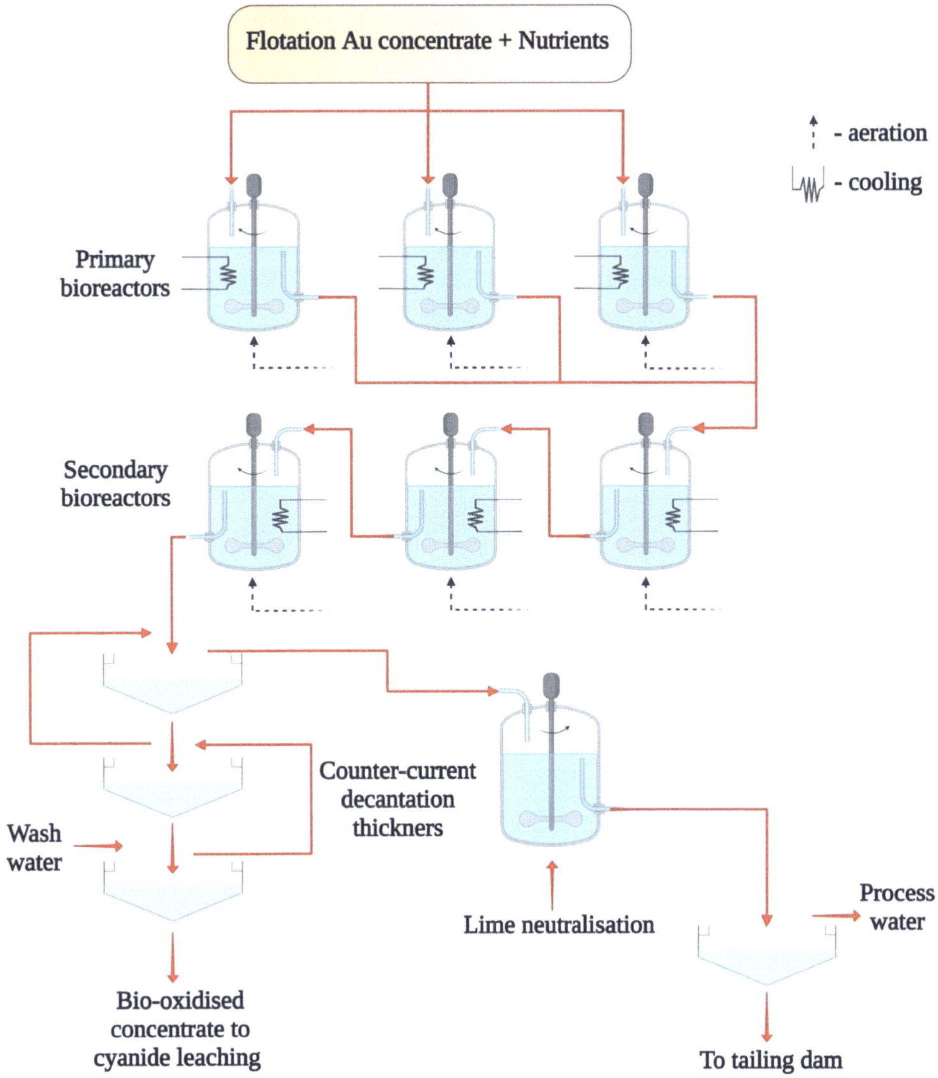

Fig. 5 Schematic overview of the flowsheet of the BIOX™ process. The milled flotation underflow is introduced into the primary (parallel) and secondary (serial) reactors to oxidise Au occluding ore gangue. The oxidised product is washed in the CCD unit; overflow: enriched solution of contaminants such as Fe and As is sent to waste processing, and underflow: Au recovery circuit through cyanidation (adapted from van Aswegen et al. (2007))

the mine has taken place intermittently (i.e., 1877–1914 and 1999–2012) and has produced nearly 1,800 kilo-ounces since 1877 (NQ Minerals, 2021). The BACOX™ process included six reactors, three of them as primaries. The gold ore went through a rigorous crushing, milling and flotation circuit, resulting in a feed size of 75 microns (i.e., d_{80}) with a throughput of 2.1 tonnes per hour. Subsequently, the processed ore was introduced to primary bioreactors (volume 365 m^3), which were regulated at 37–43 °C and contained consortia of mesophilic bacteria (similar to BIOX™; Chingwaru et al. (2017)). The residence time for the entire process was ~ 6 days, and pH values were reported in the range of 1.0–1.4 (Neale et al., 2000). Interestingly, the BACOX™ process did not demand pH control because the deployed bacteria required low pH operations and high Fe concentrations (Neale et al., 2000). Following the biooxidation of the Au ore, the residue was thickened and subject to cyanidation. The Beaconsfield mine reported recoveries of 16–20 g of gold per tonne in the earlier stages (Neale et al., 2000), and by the end of 2012, the average gold recovery was 10.5 g per tonne (NQ Minerals, 2021). Due to these promising results, a new plant is expected to be commissioned in 2022, according to Giles (2021).

4.2 Copper Industry

Despite Cu being the most biometallurgically produced metal in the world, bioreactors are not readily the first choice for Cu processing. Biometallurgy competes against other well-established Cu recovery techniques, e.g., pressure oxidation leaching—which may offer faster reaction kinetics and better economics—even though heap bioleaching is increasingly accepted as a viable Cu ore processing technique (Ghorbani et al., 2016; Schlesinger et al., 2011). There are also advances in the development of bioreactors for Cu processing.

BHP Billiton developed a demonstration-scale bioreactor technology, BioCOP™ (Fig. 6), to treat chalcopyrite ore in the Alliance Cu plant in Chile (Dreisinger, 2006). Production was scaled at 20 kilotonnes per annum (ktpa) from a 77 ktpa Cu concentrate (Havlík, 2008). The process scheme largely resembled that of the BIOX™ process, except for the use of oxygen instead of air (Batty & Rorke, 2006; Dreisinger, 2006, 2016). In addition, thermophilic bacteria (e.g., *Acidimicrobium ferrooxidans*, *Sulfobacillus acidophilus, Sulfolobus metallicus;* Clark et al. (2006)) were employed to oxidise chalcopyrite at temperatures of 65–80 °C (Dreisinger, 2006, 2016). The process housed six bioreactors at 1260 m^3 each, a re-grinding circuit, as well as pre-leach, decantation, and filtration units, a pregnant leach solution storage and one of the world's largest agitators (Batty and Rorke (2006)). The primary reactors performed with > 80% oxygen utilisation, reflecting the process's success at this site (Batty & Rorke, 2006). At this plant, thermophiles presented a significant edge over mesophiles (30–60% recovery) at high concentration chalcopyrite ore that otherwise required very fine milling (Dreisinger, 2016). Consequent to several bench-scale investigations, a prototype Cu plant of 20,000 tonnes per annum (tpa) was successfully operated for 18 months in 2004 and 2005 (Domic, 2007), albeit discontinued due to high operational costs (Knuutila, 2009).

4.3 Nickel and Cobalt Industry

Nickel (Ni) and cobalt (Co) are essential metals in high demand. Even though reported industrial applications of bioreactors for Ni or Co extraction are relatively scarce, biometallurgy is considered a viable option for extracting these metals.

Mondo Minerals assessed a bioreactor technology for recovering Ni and Co from a flotation concentrate based

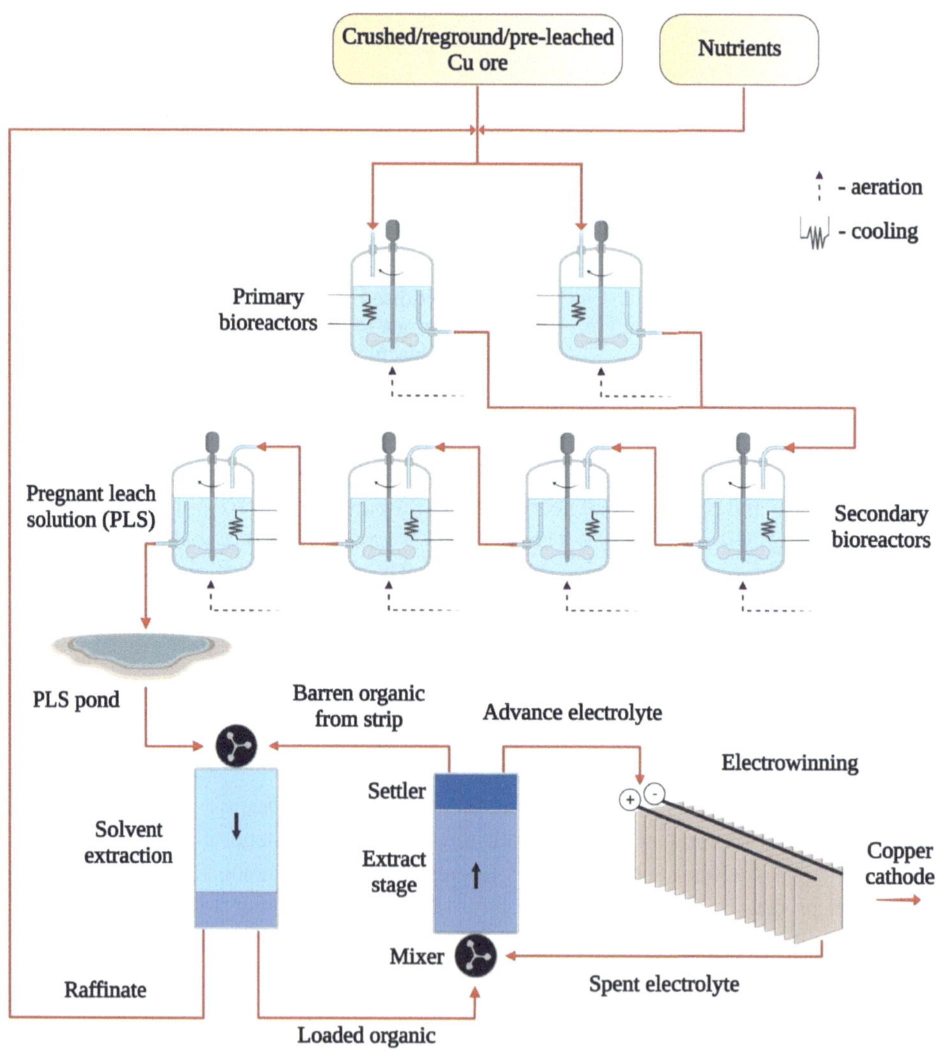

Fig. 6 Schematic overview of the flow sheet of the BioCOP™ process. The process follows a similar route to BIOX™ (Fig. 5), except for collecting the PLS for further processing. The PLS is introduced to the solvent extraction circuit, where a suitable organic solvent is employed to extract Cu. The loaded organic solvent is stripped and transferred to electrowinning. Here, the Cu cathodes are produced, and the rest of the solution is recycled. *Note* the same schematic can be used to recover any other base metal (e.g., Ni, Co) by selecting a suitable organic solvent in the solvent extraction step (modified after Batty and Rorke (2006) and Dreisinger (2006))

on Mintek's biotechnology at their mine sites in Finland (Gericke, 2015). Mondo commenced this project as a route to value-addition for talc mining by-products (Neale et al., 2017). Ni was the main target metal of interest, and the consideration for biometallurgy was partly triggered by high As concentrations (2.18% w/w) in the ore concentrate. The process included pre-processing the feed concentrate ($d_{80}=20$ microns) through magnetic separation, subsequently introducing it to the bioreactor circuit, which encompassed seven tanks. Three tanks were designated primary reactors arranged in parallel, whereas four secondary reactors were placed in series. Every tank was equipped with cooling, airflow and CO_2 supply to accommodate the moderate thermophiles (e.g., *Acidithiobacillus ferrooxidans, At. caldus* or *Leptospirillum ferrooxidans;* Venho (2012)) employed for the bioleaching at reactor temperatures of ~ 45 °C and pH values around 1.7 at both the primary and secondary reactors (Neale et al., 2017). The experimental-scale reactors demonstrated encouraging Ni and Co recoveries > 95%. Construction of a commercial-scale plant immediately followed and is expected to produce 1000 tonnes of Ni (and some Co) annually from a 12,000-tonne concentrate (Gericke, 2015). The sequential development of this Finnish plant at the Vuonos site is well-documented in Neale et al. (2017).

An example of biometallurgical Co recovery is presented by the Kasese pyrite mine in Uganda (BRGM (France)), which housed a bioreactor designed to process waste stockpile material. The initial mining activities from 1956 to 1982 focused on the extraction of Cu, which generated a pyrite-rich waste-concentrate (1.38% w/w Co) containing approximately 11,300 tonnes of Co (Morin & d'Hugues, 2007). BRGM initiated a bioreactor facility in 1998 that included circuits for milling, sizing, thickening, heat control, limestone addition, filtration, solvent extraction and electrowinning. The feed particle size was maintained at < 35 microns (i.e., d_{80}) at a rate of 240 tpd (Morin & d'Hugues, 2007; Schippers et al., 2014). The bioreactor circuit was composed of 5 tanks (each with 1380 m^3 volume) classified as primary, secondary and tertiary. Three tanks in parallel were assigned for the primary stage and two other tanks for each subsequent stage. The average temperature of all tanks was 42 °C with pH values of 1.4–1.7 across the primary and secondary/tertiary stages (Mahmoud et al., 2017). Regarding performance, the primary stage reported Co recoveries of 60%, and the secondary/tertiary stages reported 20–30% (Morin & d'Hugues, 2007). The plant operated until 2014 and produced an average of 800 tonnes of Co per year (Schippers et al. (2014)).

While metallurgical bioreactor applications remain dominated by a few widely employed techniques (e.g., Au biooxidation), the selected industrial case studies discussed in this section and many others, as illustrated in Fig. 4, highlight the diverse potential applications for bioreactor technology. Although a detailed comparison of such reactor performances is of marked importance, it is worth noting the complexity of such analysis. In particular, these performance parameters may drastically vary depending on numerous intricacies, such as locality, geology, mineral type, gangue associated, microorganisms, energy consumption and mineral processing circuits (e.g., comminution and flotation) to name a few. Therefore, it is challenging to set a fixed framework to present an assessment of the reactor operations for different mineral commodities.

5 Opportunities and Future Directions

5.1 Emerging Bioreactor Initiatives

The commencement and success of any mining project are contingent on the socio-economical context and investment opportunities. Critical factors that affect commercial decisions for implementing bioreactor technologies at an industrial scale include commodity price, operating and capital costs of new processing facilities, as well as deposit-specific challenges for metal beneficiation. Bioreactor technologies for Au recovery have shown great success: > 25 million ounces of Au have been produced in the last 30 years through the various BIOX™ plants worldwide (Metso Outotec, 2021). The simplicity of the BIOX™ design results in relatively low operating costs and makes it economically competitive with other technologies (van Aswegen, 2007). Furthermore, high commodity prices have recently encouraged the development of new bioreactor projects in the Au industry. In contrast, only a few technologies have reached industrial-scale operation in the base metal industry (Fig. 4).

The capital cost of developing new facilities is a significant impediment to launching tailor-made bioprocessing facilities. Still, bioreactor technologies may have advantages in specific scenarios, and initiatives continue to emerge. BacTech is seeking investments for their 50 tonnes per day biooxidation plant in the Ponce Enriquez project (Ecuador), reporting initial experimental Au recovery at ~ 96% (BacTech, 2021). The Beaconsfield project is anticipating production in early 2022, reporting promising gold grade estimates at 10.3 g/t, leading to 483,000 oz of gold using bacterial oxidation circuits (Giles, 2021, NQ Minerals, 2021). Further, RioZim, Zimbabwe, has prioritised a US$ 17 million BIOX™ plant at the Cam and Motor mine to process Au ore (The Herald, 2021). Finally, Torres (2021) reports that the Coricancha mine and associated Tamboraque bioreactor plant in Peru are expected to commence processing Au and Ag within years.

5.2 Novel Technologies

Various aspects of hydrometallurgical bioprocesses remain not fully understood, and scientific insights continue to contribute to the further development and application of bioreactor techniques in the industry. For instance, advances in bioengineering and molecular biology allow for genetically engineered microorganisms to be used in biometallurgical technologies (Gumulya et al., 2018), particularly for materials from complex ore deposits or with considerable levels of contaminant impurities (Mahajan et al., 2017). Many microorganisms in acidic environments are yet to be cultured (Brune & Bayer, 2012), and establishing an engineered microbial consortium may require specialised control and monitoring of leaching or oxidation conditions. The ability to control growth conditions in bioreactors beyond what is achievable with ambient operations, e.g., heap leaching (Kaksonen et al., 2020a, 2020b), gives bioreactors a clear advantage. In addition to metal recovery from ore deposits, bioreactors have also been deployed for mine waste management applications, e.g., for the passive treatment of acid mine drainage (AMD) and recovery of residual metal fractions from waste rock and tailings materials (Bailey et al., 2016; Fernando et al., 2018; Habe et al., 2020; Santos & Johnson, 2018; Yildiz et al., 2019). Bioreactors are also being examined to recover critical elements such as Li, Co, In and REEs from mining and even electronics waste (Arshadi et al., 2021; Erust et al., 2021; Hubau et al., 2019, 2020). The operating conditions of such specific applications will vary widely, but bioreactors and the involved microbial consortia have the potential to be tailored to these various requirements.

Bioreactors might even find applications beyond Earth. In pursuit of exploring extra-terrestrial mineral resources, space biomining has become an active field of research (Abbasi et al., 2021; Gumulya et al., 2022; Santomartino et al., 2022). Bioreactors were designed to perform bioleaching of REEs and vanadium from basalt rocks at the International Space Station (ISS) under different gravity conditions (e.g., microgravity, Mars' gravity, and Earth's gravity) with different bacterial species (*Sphingomonas desiccabilis*, *Bacillus subtilis* and *Cupriavidus metallidurans*) (Cockell et al., 2020, 2021). A testimony to the wide range of applications of metallurgical bioreactors, the extreme conditions found in space (irradiation, temperatures, oxygen availability) appear to not prevent the successful operation of bioreactors (Santomartino et al., 2022) despite the high costs and engineering challenges that remain (Klas et al., 2015).

Finally, industrial bioreactor facilities generally include multiple unit processes that require centralised communication systems to regulate the complex flowsheet. The adoption of Industry 4.0 (Fourth Industrial Revolution) technologies—including cloud services, big data, Internet of Things (IoT) and AI (Dalenogare et al., 2018)—can catalyse the transformation of process plant data to digital platforms. This will allow more effective maintenance, control, simulation and waste management through intelligent decision-making (Flevaris & Chatzidoukas, 2021; Gorecky et al., 2014; Vaidya et al., 2018). Due to the increased availability of diverse and large volumes of (historical) data and the increasing ease of data cleaning and curation, the incorporation of Industry 4.0 is an important emerging venue for bioreactor projects (Pereira et al., 2017; Qin et al., 2016).

6 Concluding Remarks

This chapter discussed important aspects of bioreactor design and operation for metallurgical applications. We presented a selection of bioreactor projects to illustrate the wide range of bioreactor applications in the metal extractive industry. While metallurgical bioreactors are currently mostly used in specific industries (e.g., for Au recovery), the technology is actively pursued for various other (base) metal projects. The prospects are promising for bioreactor-based hydrometallurgy, with exciting developments in microbiology, sensors, big data integration, among others, expanding the already wide range of potential applications. Bioreactors can have significant advantages over alternative extraction techniques, including process efficiency, controllability and environmental stewardship. As a result, they will likely play an essential role in the metal supply chain for decades.

References

Abbasi, B., Harper, J., & Ahmadvand, S. (2021). A short critique on biomining technology for critical materials. *World Journal of Microbiology and Biotechnology, 37*(5), 1–6.

Abdollahi, H., Noaparast, M., Shafaei, S. Z., Akcil, A., Panda, S., Kashi, M. H., & Karimi, P. (2019). Prediction and optimization studies for bioleaching of molybdenite concentrate using artificial neural networks and genetic algorithm. *Minerals Engineering, 130*, 24–35.

Abu-Absi, N. R., Kenty, B. M., Cuellar, M. E., Borys, M. C., Sakhamuri, S., Strachan, D. J., Hausladen, M. C., & Li, Z. J. (2011). Real time monitoring of multiple parameters in mammalian cell culture bioreactors using an in-line Raman spectroscopy probe. *Biotechnology and Bioengineering, 108*(5), 1215–1221.

Acevedo, F. (2000). The use of reactors in biomining processes. *Electronic Journal of Biotechnology, 3*(3), 184–194.

Acevedo, F., & Gentina, J. C. (1989). Process engineering aspects of the bioleaching of copper ores. *Bioprocess Engineering, 4*, 223–229.

Acevedo, F., & Gentina, J. C. (2007). Bioreactor design fundamentals and their application to gold mining. *Microbial Processing of Metal Sulfides, 3*, 151–168.

Agioutantis, Z. (2004). 1st international conference on advances in mineral resources management and environmental geotechnology. Heliotopos Conferences.

Ahmadi, A., & Hosseini, M. R. (2015). A fuzzy logic model to predict the bioleaching efficiency of copper concentrates in stirred tank reactors. *International Journal of Nonferrous Metallurgy, 4*(1), 1–8.

Akcil, A., Sun, Z., & Panda, S. (2020). COVID-19 disruptions to tech-metals supply are a wake-up call. Nature Publishing Group.

Akisue, R. A., Harth, M. L., Horta, A. C. L., & de Sousa Junior, R. (2021). Optimized dissolved oxygen fuzzy control for recombinant escherichia coli cultivations. *Algorithms, 14*(11), 1–22.

Arshadi, M., Pourhossein, F., Mousavi, S. M., & Yaghmaei, S. (2021). Green recovery of Cu-Ni-Fe from a mixture of spent PCBs using adapted. A ferrooxidans in a bubble column bioreactor. *Separation and Purification Technology, 272*, 118701.

van Aswegen, P. C., Niekerk, J. van, & Olivier, W. (2007). The BIOX process for the treatment of refractory gold concentrates. In D. E. Rawlings & D. B. Johnson (Eds.), Biomining. Springer.

Aznar-Sánchez, J. A., Velasco-Muñoz, J. F., Belmonte-Ureña, L. J., & Manzano-Agugliaro, F. (2019). Innovation and technology for sustainable mining activity: A worldwide research assessment. *Journal of Cleaner Production, 221*, 38–54.

Babanezhad, M., Taghvaie Nakhjiri, A., Rezakazemi, M., Marjani, A., & Shirazian, S. (2020). Functional input and membership characteristics in the accuracy of machine learning approach for estimation of multiphase flow. *Scientific Reports, 10*(1), 1–15.

BacTech. (2021). Investment presentation. BacTech. Available at: https://bactechgreen.com/investors/. [Last accessed: 30 January 2022].

Bailey, M. T., Gandy, C. J., & Jarvis, A. P. (2016). Reducing life-cycle costs of passive mine water treatment by recovery of metals from treatment wastes. In *Mining Meets Water–Conflicts and Solutions*. Proceedings of the International Mine Water Association Symposium (pp. 1255–1262).

Bailey, A. D., & Hansford, G. S. (1993). Factors affecting bio-oxidation of sulfide minerals at high concentrations of solids: A review. *Biotechnology and Bioengineering, 42*(10), 1164–1174.

Batty, J. D., & Rorke, G. V. (2006). Development and commercial demonstration of the BioCOP™ thermophile process. *Hydrometallurgy, 83*(1–4), 83–89.

Bautista, R. G. (2013). Hydrometallurgical process fundamentals (Vol. 10). Springer.

Bellenberg, S., Buetti-Dinh, A., Galli, V., Ilie, O., Herold, M., Christel, S., Boretska, M., Pivkin, I. v., Wilmes, P., Sand, W., Vera, M., & Dopson, M. (2018). Automated microscopic analysis of metal sulfide colonization by acidophilic microorganisms. *Applied and Environmental Microbiology, 84*(20), e01835–18.

Belyi, A. V., Chernov, D. V., & Solopova, N. V. (2018). Development of BIONORD® technology on Olimpiada deposit refractory arsenic-gold ores treatment in conditions of Extreme North. *Hydrometallurgy, 179*, 188–191.

Blackmore, S., Vriens, B., Sorensen, M., Power, I., Hallam, S., Smith, L., Mayer, K. U., & Beckie, R. D. (2018). Microbial and geochemical controls on waste rock weathering and drainage quality. *Science of the Total Environment, 640*, 1004–1014.

Blowes, D. W., Ptacek, C. J., Jambor, J. L., Weisener, C. G., Paktunc, D., Gould, W. D., & Johnson, D. B. (2014). 11.5—the geochemistry of acid mine drainage. In H. D. Holland, & K.K. Turekian (Eds.) *Treatise on Geochemistry* (pp. 131–190). Elsevier.

Bosecker, K. (1997). Bioleaching: Metal solubilization by microorganisms. *FEMS Microbiology Reviews, 20*(3–4), 591–604.

Botero, H., & Álvarez, H. (2011). Nonlinear state and parameters estimation in chemical processes: Analysis and improvement of three estimation structures applied to a CSTR. *International Journal of Chemical Reactor Engineering 9*.

BRGM. (2021). BRGM and the KCC project. History of the BRGM, the French Geological Survey. Available at: https://histoire.brgm.fr/en/instants/1995-brgm-and-kcc-project. [Last accessed: 30 January 2022].

Brierley, C. L. (2016). Biological processing: Biological processing of sulfidic ores and concentrates-integrating innovations. In Innovative process development in metallurgical industry: Concept to commission (pp. 109–135).

Brierley, C. L. (2008). How will biomining be applied in future? *Transactions of Nonferrous Metals Society of China (english Edition), 18*(6), 1302–1310.

Brierley, C. L. (2010). Biohydrometallurgical prospects. *Hydrometallurgy, 104*(3–4), 324–328.

Brierley, C. L. (2016a). Biological processing: Biological processing of sulfidic ores and concentrates-integrating innovations. In V. Lakshmanan, R. Roy, & V. Ramachandran (Eds.), *Innovative process development in metallurgical industry* (pp. 109–135). Springer.

Brierley, J. A., & Brierley, C. L. (2001). Present and future commercial applications of biohydrometallurgy. *Hydrometallurgy, 59*(2–3), 233–239.

Broadhurst, J. L. (1994). Neutralisation of arsenic bearing BIOX® liquors. *Minerals Engineering, 7*(8), 1029–1038.

Brune, K. D., & Bayer, T. S. (2012). Engineering microbial consortia to enhance biomining and bioremediation. *Frontiers in Microbiology, 3*, 1–6.

Buetti-Dinh, A., Galli, V., Bellenberg, S., Ilie, O., Herold, M., Christel, S., Boretska, M., Pivkin, I. V., Wilmes, P., Sand, W., Vera, M., & Dopson, M. (2019). Deep neural networks outperform human expert's capacity in characterizing bioleaching bacterial biofilm composition. *Biotechnology Reports, 22*, e00321.

Busse, C., Biechele, P., de Vries, I., Reardon, K. F., Solle, D., & Scheper, T. (2017). Sensors for disposable bioreactors. *Engineering in Life Sciences, 17*(8), 940–952.

Butkus, M., Repšytė, J., & Galvanauskas, V. (2020). Fuzzy logic-based adaptive control of specific growth rate in fed-batch biotechnological processes. A simulation study. *Applied Sciences, 10*(19), 6818.

Calzolari, G., & Liu, W. (2021). Deep learning to replace, improve, or aid CFD analysis in built environment applications: A review. *Building and Environment, 206*, 108315.

Caramihai, M., Severin, I. & Matovic, M.D. (2013). Bioprocess modeling and control, In M. D. Matovic (Ed.), Biomass now: Sustainable growth and use (pp. 147–170). Rijeka: InTech.

Chandra, J. A. P., & Samuel, R. D. S. (2010). Modelling, simulation and control of bioreactors process parameters - remote experimentation approach. *International Journal of Computer Applications, 1*(10), 103–110.

Chingwaru, W., Vidmar, J., & Chingwaru, C. (2017). Potential of biotechnology for metals extraction in Zimbabwe: A review. *Journal of the Southern African Institute of Mining and Metallurgy, 117*(4), 381–386.

Chong, N., Karamanev, D. G., & Margaritis, A. (2002). Effect of particle–particle shearing on the bioleaching of sulfide minerals. *Biotechnology and Bioengineering, 80*(3), 349–357.

Clark, M. E., Batty, J. D., van Buuren, C. B., Dew, D. W., & Eamon, M. A. (2006). Biotechnology in minerals processing: Technological breakthroughs creating value. *Hydrometallurgy, 83*(1–4), 3–9.

Cockell, C. S., Santomartino, R., Finster, K., Waajen, A. C., Eades, L. J., Moeller, R., Rettberg, P., Fuchs, F. M., van Houdt, R., Leys, N., Coninx, I., Hatton, J., Parmitano, L., Krause, J., Koehler, A., Caplin, N., Zuijderduijn, L., Mariani, A., Pellari, S. S., & Demets, R. (2020). Space station biomining experiment demonstrates rare earth element extraction in microgravity and Mars gravity. *Nature Communications, 11*(1), 1–11.

Cockell, C. S., Santomartino, R., Finster, K., Waajen, A. C., Nicholson, N., Loudon, C. M., Eades, L. J., Moeller, R., Rettberg, P., Fuchs, F. M., van Houdt, R., Leys, N., Coninx, I., Hatton, J., Parmitano, L., Krause, J., Koehler, A., Caplin, N., Zuijderduijn, L., & Demets, R. (2021). Microbially-enhanced vanadium mining and bioremediation under micro- and mars gravity on the international space station. *Frontiers in Microbiology, 12*, 1–15.

Coward, S., Vann, J., Dunham, S., & Stewart, M. (2009). *The primary-response framework for geometallurgical variables. International Mining Geology Conference*. Australasian Institute of Mining and Metallurgy.

Crundwell, F. K. (2000). Modeling, simulation, and optimization of bacterial leaching reactors. *Biotechnology and Bioengineering, 71*(4), 255–265.

Dahleh, M., Dahleh, M. A. & Verghese, G. (2004). Lectures on dynamic systems and control. Department of electrical engineering and computer science, Massachusetts Institute of Technology.

Dalenogare, L. S., Benitez, G. B., Ayala, N. F., & Frank, A. G. (2018). The expected contribution of Industry 4.0 technologies for industrial performance. *International Journal of Production Economics, 204*, 383–394.

Demergasso, C., Véliz, R., Galleguillos, P., Marín, S., Acosta, M., Zepeda, V., Zeballos, J., Henríquez, F., Pizarro, R., & Bekios-Calfa, J. (2018). Decision support system for bioleaching processes. *Hydrometallurgy, 181*, 113–122.

Domic, E. M. (2007). A review of the development and current status of copper bioleaching operations in chile: 25 years of successful commercial implementation. In D. E. Rawlings & D. B. Johnson (Eds.), *Biomining* (pp. 81–95). Springer.

Dominy, S. C., O'connor, L., Parbhakar-Fox, A., Glass, H. J., & Purevgerel, S. (2018). Geometallurgy—A route to more resilient mine operations. *Minerals, 8*(2), 560.

Dreisinger, D. B. (2016). Case study flowsheets. In Gold ore processing. Elsevier B.V.

Dreisinger, D. (2006). Copper leaching from primary sulfides: Options for biological and chemical extraction of copper. *Hydrometallurgy, 83*(1–4), 10–20.

Edwards, K. J., Bond, P. L., Gihring, T. M., & Banfield, J. F. (2000). A new iron-oxidizing, extremely acidophilic archaea is implicated in acid mine drainage generation. *Science, 287*, 1796–1799.

Ehrlich, H. L. (2004). Beginnings of rational bioleaching and highlights in the development of biohydrometallurgy: A brief history. *The European Journal of Mineral Processing and Environmental Protection, 4*(2), 1303–1868.

Erust, C., Akcil, A., Tuncuk, A., Deveci, H., Yazici, E. Y., & Panda, S. (2021). A novel approach based on solvent displacement crystallisation for iron removal and copper recovery from solutions of semi-pilot scale bioleaching of WPCBs. *Journal of Cleaner Production, 294*, 126346.

Evangelou, V. P. (1998). Pyrite chemistry: The key for abatement of acid mine drainage. *Acidic Mining Lakes, 1982*, 197–222.

Evangelou, V. P., & Zhang, Y. L. (1995). A review: Pyrite oxidation mechanisms and acid mine drainage prevention. *Critical Reviews in Environmental Science and Technology, 25*(2), 141–199.

Evans, C. L., Wightman, E. M., & Yuan, X. (2015). Quantifying mineral grain size distributions for process modelling using X-ray micro-tomography. *Minerals Engineering, 82*, 78–83.

Fagan, M. A., Ngoma, I. E., Chiume, R. A., Minnaar, S., Sederman, A. J., Johns, M. L., & Harrison, S. T. L. (2014). MRI and gravimetric studies of hydrology in drip irrigated heaps and its effect on the propagation of bioleaching micro-organisms. *Hydrometallurgy, 150*, 210–221.

Fernando, W. A. M., Ilankoon, I. M. S. K., Syed, T. H., & Yellishetty, M. (2018). Challenges and opportunities in the removal of sulphate ions in contaminated mine water: A review. *Minerals Engineering 117*.

Fernando, W. A. M., Ilankoon, I. M. S. K., Rabbani, A., & Chong, M. N. (2020a). Applicability of pore networks to evaluate the inter-particle flow in heap leaching. *Hydrometallurgy, 197*, 105451.

Fernando, W. A. M., Ilankoon, I. M. S. K., Rabbani, A., & Yellishetty, M. (2020b). Inter-particle fluid flow visualisation of larger packed beds pertaining to heap leaching using X-ray computed tomography imaging. *Minerals Engineering, 151*, 106334.

Flevaris, K., & Chatzidoukas, C. (2021). Facilitating the industrial transition to microbial and microalgal factories through mechanistic modelling within the Industry 4.0 paradigm. *Current Opinion in Chemical Engineering, 33*, 100713.

Fuerstenau, M. C. & Han, K. N. (2009). Principles of Mineral Processing. Society for Mining, Metallurgy, and Exploration, Inc. (SME Littleton, Colorado, USA).

Galvanauskas, V., Simutis, R., & Vaitkus, V. (2019). Adaptive control of biomass specific growth rate in fed-batch biotechnological processes. A comparative study. *Processes 7*(11).

Garcia-Ochoa, F., & Gomez, E. (2009). Bioreactor scale-up and oxygen transfer rate in microbial processes: An overview. *Biotechnology Advances, 27*(2), 153–176.

Gericke, M. (2012). Review of the role of microbiology in the design and operation of heap bioleaching processes. *Journal of the Southern African Institute of Mining and Metallurgy, 112*(12), 1005–1012.

Gericke, M. (2015). Base metal tank bioleaching: from laboratory test work to commercialization. *Advanced Materials Research, 1130*, 197–200.

Ghassa, S., Boruomand, Z., Abdollahi, H., Moradian, M., & Akcil, A. (2014). Bioleaching of high grade Zn-Pb bearing ore by mixed moderate thermophilic microorganisms. *Separation and Purification Technology, 136*, 241–249.

Ghorbani, Y., Franzidis, J. P., & Petersen, J. (2016). Heap leaching technology—Current State, innovations, and future directions: A review. *Mineral Processing and Extractive Metallurgy Review, 37*(2), 73–119.

Giles. (2021). NQ minerals report 22,000 tonnes moved to Beaconsfield gold mine processing plant. In Directors talk interviews. Available at: https://www.directorstalkinterviews.com/nq-minerals-report-22,000-tonnes-moved-to-beaconsfield-gold-mine-processing-plant/412990799. [Last accessed: 30 January 2022].

Gomez, E., Ballester, A., Blazquez, M. L., & Gonzalez, F. (1999). Silver-catalysed bioleaching of a chalcopyrite concentrate with mixed cultures of moderately thermophilic microorganisms. *Hydrometallurgy, 51*(1), 37–46.

González-Sánchez, A., Meulepas, R., & Revah, S. (2008). Sulfur formation and recovery in a thiosulfate-oxidizing bioreactor. *Environmental Technology, 29*(8), 847–853.

Gorecky, D., Schmitt, M., Loskyll, M., & Zühlke, D. (2014). Human-machine-interaction in the industry 4.0 era. In Proceedings of the 2014 12th IEEE international conference on industrial informatics (pp. 289–294).

Guezennec, A. G., Joulian, C., Jacob, J., Archane, A., Ibarra, D., de Buyer, R., Bodénan, F., & d'Hugues, P. (2017). Influence of dissolved oxygen on the bioleaching efficiency under oxygen enriched atmosphere. *Minerals Engineering, 106*, 64–70.

Gumulya, Y., Boxall, N. J., Khaleque, H. N., Santala, V., Carlson, R. P., & Kaksonen, A. H. (2018). In a quest for engineering acidophiles for biomining applications: Challenges and opportunities. *Genes 9*(2).

Gumulya, Y., Zea, L., & Kaksonen, A. H. (2022). In situ resource utilisation: The potential for space biomining. *Minerals Engineering, 176*, 107288.

Gupta, A. & Yan, D. (2016). Size reduction and energy requirement. In A. Gupta, & D. Yan (Eds.) Mineral processing design and operations, Second ed. Elsevier.

Habashi, F. (1993). Principles of extractive metallurgy: Pyrometallurgy. Gordon and Breach.

Habashi, F. (2009). Recent trends in extractive metallurgy. *Journal of Mining and Metallurgy, Section b: Metallurgy, 45*(1), 1–13.

Habe, H., Sato, Y., Aoyagi, T., Inaba, T., Hori, T., Hamai, T., Hayashi, K., Kobayashi, M., Sakata, T., & Sato, N. (2020). Design, application, and microbiome of sulfate-reducing bioreactors for treatment of mining-influenced water. *Applied Microbiology and Biotechnology, 104*(16), 6893–6903.

Havlík, T. (2008). The current state and prospects of hydrometallurgical processes. *Hydrometallurgy*, pp. 479–534.

The Herald. (2021). Riozim installs US$17m BIOX plant at mine. The Herald. Available at: https://www.herald.co.zw/riozim-installs-us17m-biox-plant-at-mine/. [Last accessed: 30 January 2022].

Herbst, J. A., Lo, Y. C. & Flintoff, B. (2003). Size reduction and liberation. In M. C. Fuerstenau, & K. N. Han (Eds.) Principles of mineral processing. Society for Mining, Metallurgy and Exploration Inc. (SME).

Holmes, P. R., & Crundwell, F. K. (2000). The kinetics of the oxidation of pyrite by ferric ions and dissolved oxygen: An electrochemical study. *Geochimica Et Cosmochimica Acta, 64*(2), 263–274.

Hubau, A., Chagnes, A., Minier, M. & Guezennec, A. G., (2019). Metal recovery from printed circuit boards (PCBs) by acidophilic bioleaching. In International Biohydrometallurgy Symposium-IBS 2019.

Hubau, A., Minier, M., Chagnes, A., Joulian, C., Silvente, C., & Guezennec, A. G. (2020). Recovery of metals in a double-stage continuous bioreactor for acidic bioleaching of printed circuit boards (PCBs). *Separation and Purification Technology, 238*, 116481.

Huminicki, D. M. C., & Rimstidt, J. D. (2009). Iron oxyhydroxide coating of pyrite for acid mine drainage control. *Applied Geochemistry, 24*(9), 1626–1634.

Ilankoon, I. M. S. K., Tang, Y., Ghorbani, Y., Northey, S., Yellishetty, M., Deng, X., & McBride, D. (2018). The current state and future directions of percolation leaching in the Chinese mining industry: Challenges and opportunities. *Minerals Engineering, 125*, 206–222.

Jayachitra, A., & Vinodha, R. (2014). Genetic algorithm based PID controller tuning approach for continuous stirred tank reactor. *Advances in Artificial Intelligence, 2014*, 1–8.

Jia, Y., Tan, Q., Sun, H., Zhang, Y., Gao, H., & Ruan, R. (2019). Sulfide mineral dissolution microbes: Community structure and function in industrial bioleaching heaps. *Green Energy and Environment, 4*(1), 29–37.

Johnson, D. B. (2018). The evolution, current status, and future prospects of using biotechnologies in the mineral extraction and metal recovery sectors. *Minerals 8*(8).

Johnson, D. B. (2008). Biodiversity and interactions of acidophiles: Key to understanding and optimizing microbial processing of ores and concentrates. *Transactions of Nonferrous Metals Society of China (english Edition), 18*(6), 1367–1373.

Johnson, D. B., Grail, B. M., & Hallberg, K. B. (2013). A new direction for biomining: Extraction of metals by reductive dissolution of oxidized ores. *Minerals, 3*(1), 49–58.

Johnson, M. A., & Moradi, M. H. (2005). *PID control new identification and design methods*. Springer-Verlag.

Kaksonen, A. H., et al. (2020b). Prospective direction for biohydrometallurgy. *Hydrometallurgy, 195*, 105376. https://doi.org/10.1016/j.hydromet.2020.105376

Kaksonen, A. H., Boxall, N. J., Gumulya, Y., Khaleque, H. N., Morris, C., Bohu, T., Cheng, K. Y., Usher, K. M., & Lakaniemi, A.-M. (2018). Recent progress in biohydrometallurgy and microbial characterisation. *Hydrometallurgy, 180*, 7–25.

Kaksonen, A. H., Deng, X., Bohu, T., Zea, L., Khaleque, H. N., Gumulya, Y., Boxall, N. J., Morris, C., & Cheng, K. Y. (2020a). Prospective directions for biohydrometallurgy. *Hydrometallurgy, 195*, 105376.

Kaksonen, A. H., Mudunuru, B. M., & Hackl, R. (2014). The role of microorganisms in gold processing and recovery—A review. *Hydrometallurgy, 142*, 70–83.

Kartal, M., Xia, F., Ralph, D., Rickard, W. D. A., Renard, F., & Li, W. (2020). Enhancing chalcopyrite leaching by tetrachloroethylene-assisted removal of sulphur passivation and the mechanism of jarosite formation. *Hydrometallurgy, 191*, 105192.

Keith, M., Smith, D. J., Jenkin, G. R. T., Holwell, D. A., & Dye, M. D. (2018). A review of Te and Se systematics in hydrothermal pyrite from precious metal deposits: Insights into ore-forming processes. *Ore Geology Reviews, 96*, 269–282.

Khaing, S. Y., Sugai, Y., Sasaki, K., & Tun, M. M. (2019). Consideration of influential factors on bioleaching of gold ore using iodide-oxidizing bacteria. *Minerals, 9*(5), 274.

Klas, M., Tsafnat, N., Dennerley, J., Beckmann, S., Osborne, B., Dempster, A. G., & Manefield, M. (2015). Biomining and methanogenesis for resource extraction from asteroids. *Space Policy, 34*, 18–22.

Knuutila, K. (2009). The role of hydrometallurgy in the winning of non-ferrous metals. *World of Metallurgy—ERZMETALL, 62*(3), 142–150.

Krishna, V. S. V., Pappa, N., Rani, S. V. (2018). Implementation of embedded soft sensor for bioreactor on Zynq processing system. In 2018 International conference on recent trends in electrical, control and communication (RTECC) (pp. 297–301).

Kundu, K., & Kumar, A. (2014). Biochemical engineering parameters for hydrometallurgical processes: Steps towards a deeper understanding. *Journal of Mining, 2014*, 290275.

Lee, J., Bazilian, M., Sovacool, B., Hund, K., Jowitt, S. M., Nguyen, T. P., Månberger, A., Kah, M., Greene, S., Galeazzi, C., Awuah-Offei, K., Moats, M., Tilton, J., & Kukoda, S. (2020). Reviewing the material and metal security of low-carbon energy transitions. *Renewable and Sustainable Energy Reviews, 124*, 109789.

Litvinenko, V. S., & Sergeev, I. B. (2019). Innovations as a factor in the development of the natural resources sector. *Studies on Russian Economic Development, 30*(6), 637–645.

Lottermoser, B. G. (2007) Mining environments: The good, the bad, and the ugly. In Sudbury 2007 mining and the environment conference proceedings. From: Sudbury 2007 Mining and the Environment IV International Conference, 19–26 October 2007.

Lund, C., & Lamberg, P. (2014). Geometallurgy–A tool for better resource efficiency. *European Geologist, 37*, 39–43.

Lusty, P. A. J., & Gunn, A. G. (2014). Challenges to global mineral resource security and options for future supply. *Geological Society Special Publication, 393*, 265–276.

Maes, K., Chatzis, M. N., & Lombaert, G. (2019). Observability of nonlinear systems with unmeasured inputs. *Mechanical Systems and Signal Processing, 130*, 378–394.

Mahajan, S., Gupta, A., & Sharma, R. (2017). Bioleaching and biomining. In R. Singh (Ed.), *Principles and applications of environmental biotechnology for a sustainable future* (pp. 393–423). Springer.

Mahmoud, A., Cézac, P., Hoadley, A. F. A., Contamine, F., & D'Hugues, P. (2017). A review of sulfide minerals microbially assisted leaching in stirred tank reactors. *International Biodeterioration and Biodegradation, 119*, 118–146.

Malaguti, A., Mircea, M., la Torretta, T. M. G., Telloli, C., Petralia, E., Stracquadanio, M., Berico, M., et al. (2015). Comparison of online and offline methods for measuring fine secondary inorganic ions and carbonaceous aerosols in the central Mediterranean area. *Aerosol and Air Quality Research, 15*(7), 2641–2653.

Malar, S., & Thyagarajan, T. (2009). Modelling of continuous stirred tank reactor using artificial intelligence techniques. *International Journal of Simulation Modelling, 8*(3), 145–155.

Mambote, R. C. M., Reuter, M. A., Krijgsman, P., & Schuiling, R. D. (2000). Hydrothermal metallurgy: An overview of basic concepts and applications. *Minerals Engineering, 13*(8–9), 803–822.
Metso Outotec (2021). Metso Outotec BIOX process. Available at: https://www.mogroup.com/globalassets/industry-pages/metals-refining/hydrometallurgy/mo-biox-process-brochure_print.pdf. [Last accessed: 30 January 2022].
McCoy, J. T., & Auret, L. (2019). Machine learning applications in minerals processing: A review. *Minerals Engineering, 132*, 95–109.
Midha, V., Jha, M. K., & Dey, A. (2012). Sulfide oxidation in fluidized bed bioreactor using nylon support material. *Journal of Environmental Sciences, 24*(3), 512–519.
Miller, P. C. (1997). The design and operating practice of bacterial oxidation plant using moderate thermophiles (the BacTech Process). In D. E. Rawlings (Ed.), *Biomining: Theory, microbes and industrial processes biomining* (pp. 81–102). Springer.
Mishra, A. K. (2021). AI4R2R (AI for rock to revenue): A review of the applications of AI in mineral processing. *Minerals, 11*(10).
Mishra, D., Kim, D.-J., Ahn, J.-G., & Rhee, Y.-H. (2005). Bioleaching: A microbial process of metal recovery. A Review. *Metals and Materials International, 11*(3), 249–256.
Mooiman, M. B., Sole, K. C., & Kinneberg, D. J. (2005). Challenging the traditional hydrometallurgy curriculum—an industry perspective. *Hydrometallurgy, 79*(1–2), 80–88.
Morin, D. H. R. (2007). Bioleaching of sulfide minerals in continuous stirred tanks. In E. R. Donati & W. Sand (Eds.), *Microbial processing of metal sulfides* (pp. 133–150). Springer.
Morin, D. H. R., & d'Hugues, P. (2007). Bioleaching of a cobalt-containing pyrite in stirred reactors: A case study from laboratory scale to industrial application. In D. E. Rawlings & D. B. Johnson (Eds.), *Biomining* (pp. 35–55). Springer.
Mowbray, M., Savage, T., Wu, C., Song, Z., Cho, B. A., del Rio-Chanona, E. A., & Zhang, D. (2021). Machine learning for biochemical engineering: A review. *Biochemical Engineering Journal, 172*, 108054.
Nancucheo, I., Oliveira, G., Lopes, M., & Johnson, D. B. (2019). Bioreductive dissolution as a pretreatment for recalcitrant rare-earth phosphate minerals associated with lateritic ores. *Minerals, 9*(3), 136.
Natarajan, K. A. (2018). *Biotechnology of metals: Principles, recovery methods, and environmental concerns*. Elsevier.
Natarajan, K. A. (2019). Biotechnology for environmentally benign gold production. In R. Pogaku (Ed.), *Horizons in bioprocessing engineering* (pp. 263–299). Springer.
Neale, J., Seppälä, J., Laukka, A., van Aswegen, P., Barnett, S. & Gericke, M. (2017). The MONDO minerals nickel sulfide bioleach project: from test work to early plant operation. In 22nd international biohydrometallurgy symposium, solid state phenomena (Vol. 262, pp. 28–32). Trans Tech Publications Ltd.
Neale, J. W., Pinches, A., & Deeplaul, V. (2000). Mintek-BacTech's bacterial-oxidation technology for refractory gold concentrates: Beaconsfield and beyond. *Journal of the Southern African Institute of Mining and Metallurgy, 100*(7), 415–421.
Nemati, M., Lowenadler, J., & Harrison, S. T. L. (2000). Particle size effects in bioleaching of pyrite by acidophilic thermophile Sulfolobus metallicus (BC). *Applied Microbiology and Biotechnology, 53*(2), 173–179.
Nise, N. S. (2020). *Control systems engineering*. Wiley.
Nordstrom, D. K. (1982). Aqueous pyrite oxidation and the consequent formation of secondary iron minerals. In J. Kittrick, D. S. Fanning, & L. R. Hossner (Eds.) Soil Science Society of America Special Publications (pp. 37–56).
Norris, P. R., Burton, N. P., & Foulis, N. A. M. (2000). Acidophiles in bioreactor mineral processing. *Extremophiles, 4*(2), 71–76.
Norris, P. R., & Owen, J. P. (1993). Mineral sulphide oxidation by enrichment cultures of novel thermoacidophilic bacteria. *FEMS Microbiology Reviews, 11*(1–3), 51–56.
Northey, S., Mohr, S., Mudd, G. M., Weng, Z., & Giurco, D. (2014). Modelling future copper ore grade decline based on a detailed assessment of copper resources and mining. *Resources, Conservation and Recycling, 83*, 190–201.
Northey, S. A., Mudd, G. M., Werner, T. T., Jowitt, S. M., Haque, N., Yellishetty, M., & Weng, Z. (2017). The exposure of global base metal resources to water criticality, scarcity and climate change. *Global Environmental Change, 44*, 109–124.
NQ Minerals. (2021). Beaconsfield. NQ Minerals Plc. Available at: https://nqminerals.com/beaconsfield/. [Last accessed: 30 January 2022].
Oladele, D. A., & Shaibu, F. E. (2019). An improved performance of fuzzy logic based controller design for bioreactor process plant. *European Journal of Engineering Research and Science, 4*(9), 177–180.
Pacheco, A. D., Delgado-Macuil, R. J., Díaz-Pacheco, Á., Larralde-Corona, C. P., Dinorín-Téllez-Girón, J., & López-Y-López, V. E. (2021). Use of equivalent circuit analysis and Cole-Cole model in evaluation of bioreactor operating conditions for biomass monitoring by impedance spectroscopy. *Bioprocess and Biosystems Engineering, 44*, 1923–1934.
Palmer, C. M., & Johnson, G. D. (2005). The activox® process: Growing significance in the Nickel Industry. *JOM Journal of the Minerals Metals and Materials Society, 57*, 40–47.
Pell, R., Tijsseling, L., Goodenough, K., Wall, F., Dehaine, Q., Grant, A., Deak, D., Yan, X., & Whattoff, P. (2021). Towards sustainable extraction of technology materials through integrated approaches. *Nature Reviews Earth and Environment, 2*, 665–679.
Pereira, T., Barreto, L., & Amaral, A. (2017). Network and information security challenges within Industry 4.0 paradigm. *Procedia Manufacturing, 13*, 1253–1260.
Peters, E. (1992). Hydrometallurgical process innovation. *Hydrometallurgy, 29*(1–3), 431–459.
Petersen, J. (2016). Heap leaching as a key technology for recovery of values from low-grade ores—A brief overview. *Hydrometallurgy, 165*, 206–212.
Popov, K., Grgur, B., & Djokić, S. S. (2007). *Fundamental aspects of electrometallurgy*. Springer.
Pörtner, R., Barradas, O. P., Frahm, B., & Hass, V. C. (2017). Advanced process and control strategies for bioreactors. In C. Larroche, M. A. Sanromán, G. Du, & A. Pandey (Eds.), *Current developments in biotechnology and bioengineering* (pp. 463–493). Elsevier.
Prior, T., Giurco, D., Mudd, G., Mason, L., & Behrisch, J. (2012). Resource depletion, peak minerals and the implications for sustainable resource management. *Global Environmental Change, 22*(3), 577–587.
Qin, J., Liu, Y., & Grosvenor, R. (2016). A categorical framework of manufacturing for Industry 4.0 and beyond. *Procedia CIRP, 52*, 173–178.
Quiñónez, K., Camacho, O., & Chávez, D. (2019). Application of nonlinear PID controllers to bioreactor processes. In 2019 IEEE 4th Colombian conference on automatic control (CCAC) (pp. 1–6).
Rabbani, A., Fernando, A. M., Shams, R., Singh, A., Mostaghimi, P., & Babaei, M. (2021). Review of data science trends and issues in porous media research with a focus on image-based techniques. *Water Resources Research, 57*(10), 1–44.
Radziszewski, P. (2013). Energy recovery potential in comminution processes. *Minerals Engineering, 46–47*, 83–88.
Rasoulnia, P., Barthen, R., & Lakaniemi, A.-M. (2021). A critical review of bioleaching of rare earth elements: The mechanisms and effect of process parameters. *Critical Reviews in Environmental Science and Technology, 51*(4), 378–427.

Rawlings, D. E. (2002). Heavy metal mining using microbes. *Annual Review of Microbiology, 56*, 65–91.

Rawlings, D. E. (2005). Characteristics and adaptability of iron- and sulfur-oxidizing microorganisms used for the recovery of metals from minerals and their concentrates. *Microbial Cell Factories, 4*, 1–15.

Rawlings, D. E., Dew, D., & du Plessis, C. (2003). Biomineralization of metal-containing ores and concentrates. *Trends in Biotechnology, 21*(1), 38–44.

Rawlings, D. E., & Johnson, D. B. (2007). *Biomining*. Springer.

Rohwerder, T., & Sand, W. (2007). Mechanisms and biochemical fundamentals of bacterial metal sulfide oxidation. In E. R. Donati & W. Sand (Eds.), *Microbial processing of metal sulfides* (pp. 35–58). Springer.

Sahinkaya, E., Hasar, H., Kaksonen, A. H., & Rittmann, B. E. (2011). Performance of a sulfide-oxidizing, sulfur-producing membrane biofilm reactor treating sulfide-containing bioreactor effluent. *Environmental Science and Technology, 45*(9), 4080–4087.

Sajjad, W., Zheng, G., Din, G., Ma, X., Rafiq, M., & Xu, W. (2019). Metals extraction from sulfide ores with microorganisms: The bioleaching technology and recent developments. *Transactions of the Indian Institute of Metals, 72*(3), 559–579.

Saldaña, M., Neira, P., Flores, V., Robles, P., & Moraga, C. (2021). A decision support system for changes in operation modes of the copper heap leaching process. *Metals, 11*(7), 1–21.

Santomartino, R., Zea, L., & Cockell, C. S. (2022). The smallest space miners: Principles of space biomining. *Extremophiles 26*(1).

Santos, A. L., & Johnson, D. B. (2018). Design and application of a low pH upflow biofilm sulfidogenic bioreactor for recovering transition metals from synthetic wastewater at a Brazilian copper mine. *Frontiers in Microbiology, 9*, 1–11.

Schippers, A. (2004). Biogeochemistry of metal sulfide oxidation in mining environments, sediments, and soils. *Special Papers-Geological Society of America, 379*, 49–62.

Schippers, A., Breuker, A., Blazejak, A., Bosecker, K., Kock, D., & Wright, T. L. (2010). The biogeochemistry and microbiology of sulfidic mine waste and bioleaching dumps and heaps, and novel Fe(II)-oxidizing bacteria. *Hydrometallurgy, 104*(3–4), 342–350.

Schippers, A., Hedrich, S., Vasters, J., Drobe, M., Sand, W., & Willscher, S. (2014). Biomining: Metal recovery from ores with microorganisms. *Advances in Biochemical Engineering and Biotechnology, 141*, 1–47.

Schippers, A., & Sand, W. (1999). Bacterial leaching of metal sulfides proceeds by two indirect mechanisms via thiosulfate or via polysulfides and sulfur. *Applied and Environmental Microbiology, 65*(1), 319–321.

Schlesinger, M. E., King, M. J., Sole, K. C., & Davenport, W. G. (2011). *Extractive metallurgy of copper* (5th ed.). Elsevier.

Schrenk, M. O., Edwards, K. J., Goodman, R. M., Hamers, R. J., & Banfield, J. F. (1998). Distribution of Thiobacillus ferrooxidans and Leptospirillum ferrooxidans: Implications for generation of acid mine drainage. *Science, 279*(5356), 1519–1522.

Shaghaghi, D., Monir Vaghefi, H., & Fatehi, A. (2013). Generalized predictive control of pH neutralization process based on fuzzy inverse model. In 13th Iranian conference on fuzzy systems (IFSC) (pp. 1–6).

Singer, P., & Stumm, W. (1970). Acid mine drainage: The rate controlling step. *Science, 167*, 1121–1123.

Sklodowska, A., & Matlakowska, R. (2007). Bioleaching of metals in neutral and slightly alkaline environment. In E. R. Donati & W. Sand (Eds.), *Microbial processing of metal sulfides* (pp. 121–129). Springer.

Smart, M., Huddy, R. J., Edward, C. J., Fourie, C., Shumba, T., Iron, J., & Harrison, S. T. L. (2017). Linking microbial community dynamics in BIOX® leaching tanks to process conditions: Integrating lab and commercial experience. *Solid State Phenomena, 262*, 38–42.

Smith, S. L., Grail, B. M., & Johnson, D. B. (2017). Reductive bioprocessing of cobalt-bearing limonitic laterites. *Minerals Engineering, 106*, 86–90.

Soto, P. E., Galleguillos, P. A., Serón, M. A., Zepeda, V. J., Demergasso, C. S., & Pinilla, C. (2013). Parameters influencing the microbial oxidation activity in the industrial bioleaching heap at Escondida mine, Chile. *Hydrometallurgy, 133*, 51–57.

Sovmen, V. K., Belyi, A. v., Danneker, M. Y., Gish, A. A., & Teleutov, A. N. (2009). Biooxidation of refractory gold sulfide concentrate of Olympiada deposit. *Advanced Materials Research* 71–73, 477–480

Stanković, S., Stopić, S., Sokić, M., Marković, B., & Friedrich, B. (2020). Review of the past, present, and future of the hydrometallurgical production of nickel and cobalt from lateritic ores. *Metallurgical and Materials Engineering, 26*(2), 199–208.

Sugai, Y., & Sasaki, K. (2019). Gold dissolution from ore with iodide-oxidising bacteria. *Scientific Reports, 9*(1), 1–11.

Sun, L. X., Zhang, X., Tan, W. S., & Zhu, M. L. (2012). Effects of dissolved oxygen on the biooxidation process of refractory gold ores. *Journal of Bioscience and Bioengineering, 114*(5), 531–536.

Swain, B. (2017). Recovery and recycling of lithium: A review. *Separation and Purification Technology, 172*, 388–403.

Tamayo, E., Zuluaga, C., & Alvarez, H. (2019). A fuzzy-evaluated gain scheduling approach for the PID control of a pressure plant. In 2019 IEEE 4th Colombian conference on automatic control (CCAC) (pp. 1–6).

Torres, S. (2021). Minera great panther mining: Plan de Exploración 2021. Tiempo Minero. Avaiable at: https://camiper.com/tiempominero-noticias-en-mineria-para-el-peru-y-el-mundo/minera-great-panther-mining-plan-de-exploracion-2021/. [Last accessed: 30 January 2022].

Tronci, S., & Baratti, R. (2017). A gain-scheduling pi control based on neural networks. *Complexity, 2017*, 9241254.

Tu, Z., Guo, C., Zhang, T., Lu, G., Wan, J., Liao, C., & Dang, Z. (2017). Investigation of intermediate sulfur species during pyrite oxidation in the presence and absence of Acidithiobacillus ferrooxidans. *Hydrometallurgy, 167*, 58–65.

Vaidya, S., Ambad, P., & Bhosle, S. (2018). Industry 4.0—A glimpse. *Procedia Manufacturing, 20*, 233–238.

Vakylabad, A. B., Nazari, S., & Darezereshki, E. (2022). Bioleaching of copper from chalcopyrite ore at higher NaCl concentrations. *Minerals Engineering, 175*, 107281.

van't Riet, K., & Tramper, J. (1991). Basic bioreactor design. CRC Press.

Vardanyan, N., & Vardanyan, A. (2021). Thermoacidophiles for bioleaching of copper. In D. Egamberdieva, N. K. Birkeland, W. J. Li, & H. Panosyan (Eds.), *Microbial communities and their interactions in the extreme environment* (pp. 177–206). Springer.

Venho, A. (2012). Venho, A. (2013). Immobilization of arsenic through pH control during bioleaching of Nickel Flotation concentrate [Master's thesis, Tampere University].

Vera, M., Schippers, A., & Sand, W. (2013). Progress in bioleaching: Fundamentals and mechanisms of bacterial metal sulfide oxidation—part A. *Applied Microbiology and Biotechnology, 97*(17), 7529–7541.

Vriens, B., Plante, B., Seigneur, N., & Jamieson, H. (2020a). Mine waste rock: Insights for sustainable hydrogeochemical management. *Minerals, 10*(9), 1–38.

Vriens, B., Seigneur, N., Mayer, K. U., & Beckie, R. D. (2020b). Scale dependence of effective geochemical rates in weathering mine waste rock. *Journal of Contaminant Hydrology, 234*, 103699.

Vriens, B., Smith, L., Mayer, K. U., & Beckie, R. D. (2019). Poregas distributions in waste-rock piles affected by climate seasonality and physicochemical heterogeneity. *Applied Geochemistry, 100*, 305–315.

Vyas, S., Das, S., & Ting, Y. P. (2020). Predictive modeling and response analysis of spent catalyst bioleaching using artificial neural network. *Bioresource Technology Reports, 9*, 100389.

Wang, B., Wang, Z., Chen, T., & Zhao, X. (2020). Development of novel bioreactor control systems based on smart sensors and actuators. *Frontiers in Bioengineering and Biotechnology, 8*, 7.

Watling, H. R. (2008). The bioleaching of nickel-copper sulfides. *Hydrometallurgy, 91*(1–4), 70–88.

Watling, H. R. (2015). Review of biohydrometallurgical metals extraction from polymetallic mineral resources. *Minerals, 5*(1), 1–60.

Watling, H. (2016). Microbiological advances in biohydrometallurgy. *Minerals, 6*(2), 49.

Watling, H. R., Shiers, D. W., & Collinson, D. M. (2015). Extremophiles in mineral sulphide heaps: Some bacterial responses to variable temperature, acidity and solution composition. *Microorganisms, 3*(3), 364–390.

Williamson, M. A., & Rimstidt, J. D. (1994). The kinetics and electrochemical rate-determining step of aqueous pyrite oxidation. *Geochimica Et Cosmochimica Acta, 58*(24), 5443–5454. https://doi.org/10.1016/0016-7037(94)90241-0

Wills, B. A. & Finch, J. A. (2016). Chapter 1—introduction. In B. A. Wills, & J. A. Finch (Eds.) Wills' mineral processing technology (Eighth Edition). Butterworth-Heinemann.

Wu, A., Yin, S., Qin, W., Liu, J., & Qiu, G. (2009). The effect of preferential flow on extraction and surface morphology of copper sulphides during heap leaching. *Hydrometallurgy, 95*(1), 76–81.

Wu, A., Yin, S., Yang, B., Wang, J., & Qiu, G. (2007). Study on preferential flow in dump leaching of low-grade ores. *Hydrometallurgy, 87*(3–4), 124–132.

Xie, S., Yu, J., Xie, Y., Jiang, Z., & Gui, W. (2019). A two-layer optimization and control strategy for zinc hydrometallurgy process based on RBF neural network soft-sensor. In 1st international conference on industrial artificial intelligence (IAI), 1–6.

Yang, Y., Ferrier, J., Csetenyi, L., & Gadd, G. M. (2019). Direct and indirect bioleaching of cobalt from low grade laterite and pyritic ores by Aspergillus niger. *Geomicrobiology Journal, 36*(10), 940–949.

Yildiz, M., Yilmaz, T., Arzum, C. S., Yurtsever, A., Kaksonen, A. H., & Ucar, D. (2019). Sulfate reduction in acetate- and ethanol-fed bioreactors: Acidic mine drainage treatment and selective metal recovery. *Minerals Engineering, 133*, 52–59.

Yin, S., Wang, L., Kabwe, E., Chen, X., Yan, R., An, K., Zhang, L., & Wu, A. (2018). Copper bioleaching in China: Review and prospect. *Minerals, 8*(2), 32.

Mineral–Metal Wastes (Bio)/Recycling: Compliance with Circular Economy

Luis Gonzaga Santos Sobral, Louise de Aguiar Sobral, Andriela Dutra Norberto de Oliveira, Ingrid Gomes Silva and Patricia Correia de Araujo

Abstract

This chapter reviews what has been accomplished in metal bio-extraction from so many secondary sources. In some cases, tailings were generated during processing of primary and weathered ores. It has been demonstrated that the use of native microorganisms, naturally present in drainages from mines where sulphide minerals are present, is very attractive in several locations, especially when it comes to the processing of ores with low levels of metals of interest as well as the tailings generated in these processes. In addition, several studies show the possibility of using the mineral substrates remaining in the rocks after such bio-extraction processes, for crops. This is a natural fertilization technique which uses such mineral substrates for augmenting the necessary nutrients for food production, in soils depleted by weathering/leaching or by the inappropriate and intensive use of chemical fertilizers, without affecting the balance of the environment. Following this line of research, there is also a need for technology to recover elements from electronic scraps, bearing in mind that some of them are also used as micronutrients for human beings, especially given the short lifespan of modern electronic equipment. Latterly, electronic waste can encompass dozens of different elements such as base metals, precious metals, rare earth and several heavy metals. Several processes have been used for extracting/recycling those metals. Bearing this in mind, the biotechnological approach plays a pivotal role as a very attractive and cost-effective way for processing such wastes and so many others being dumped in the environment, as natural resources are available in the environment and can be used without much expenses as it is the case of so many microorganisms.

Keywords

Bioleaching · Mining waste · WEEE · Electronic scraps

1 Introduction

Since antiquity, man has been using minerals in the most varied activities. However, more recently, the relentless pursuit for improving the quality of life has contributed a great deal for the technological development of industrial sectors related to mining and metallurgy for producing metallic devices, imposing a progressive demand for the most different types of metals.

According to Luz A. B. et al., "Mineral, or mineral substance, is any inorganic body with defined chemical composition and physical properties, found in the earth's crust. Likewise, ore is the term used to define any rock made up of a mineral or aggregate of minerals containing one or more valuable elements, which can be used economically. The mineral or group of minerals not used in an ore is called gangue".

Considering the processes of formation and transformation that the deposits go through, minerals are classified as primary, secondary and weathered (i.e., the so-called oxidized ore). Figure 1 illustrates, for instance, the transformations likely to occur at the various levels of a copper deposit.

Primary ores of various metals are deposited at great depths. These ores are igneous in origin and consist of sulphide minerals with different characteristics, some more refractory, for example, in the case of copper, chalcopyrite ($CuFeS_2$), and others less refractory, such as covellite

L. G. S. Sobral (✉) · A. D. N. de Oliveira · I. G. Silva · P. C. de Araujo
CETEM—Centre for Mineral Technology, Rio de Janeiro, Brazil
e-mail: Lsobral@cetem.gov.br

L. de Aguiar Sobral
Brazil Federal University of Rio de Janeiro, Rio de Janeiro, Brazil
e-mail: Lsobral@cetem.gov.br

S. Panda et al. (eds.), *Biotechnological Innovations in the Mineral-Metal Industry*,
Advances in Science, Technology & Innovation, https://doi.org/10.1007/978-3-031-43625-3_13

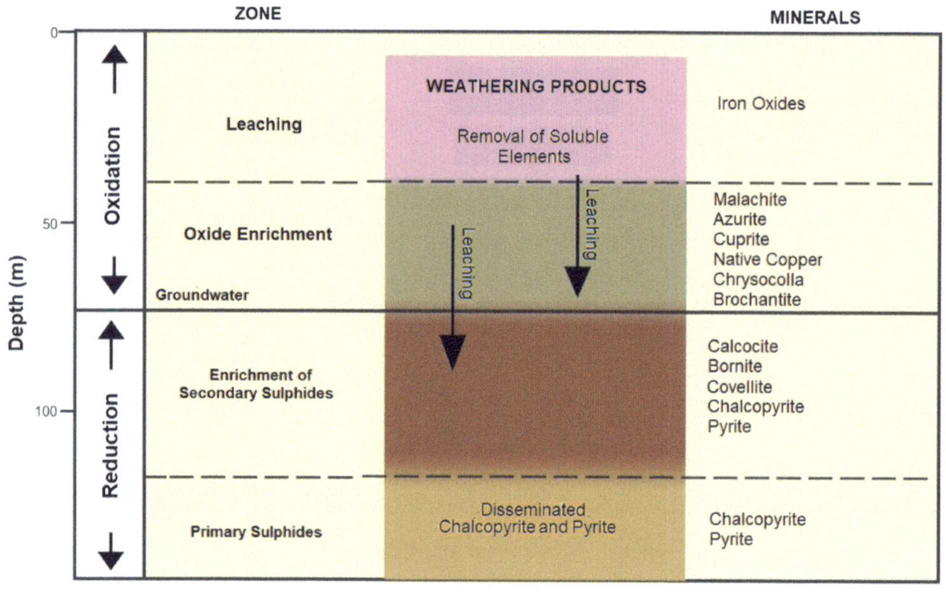

Fig. 1 Typical scheme of a copper deposit with a surface subjected to weather and oxidation phenomena. *Source* Adapted from Domic (2001)

(CuS) and calcosine (Cu_2S). In some deposits, these copper minerals are accompanied by other sulphide minerals, such as pyrite (FeS_2) and molybdenite (MoS_2) (Luz, 2004). Secondary copper minerals, such as calcosine (Cu_2S) and bornite (Cu_5FeS_4), result from oxidative changes that naturally occur in primary minerals.

Finally, one can mention the weathered copper minerals resulting from physical and chemical changes in rocks, which involve fragmentation and decomposition of primary minerals due to various chemical reactions: oxidation, hydration, dissolution, hydrolysis, etc. Among the copper weathered minerals are cuprite (Cu_2O), malachite ($CuCO_3.Cu(OH)_2$, azurite ($2CuCO_3.Cu(OH)_2$), chrysocolla ($CuSiO_3.2H_2O$) and brochantite (Cu_4SO_4 $(OH)_6$ (Domic, 2001).

1.1 Extraction of Metals Out of Weathered Ores/Tailings

According to Domic (2001), some oxidized minerals are quite soluble in acid solution as they do not require an oxidizing agent for their leaching, which is carried out by simple contact with such minerals to release the metals into the aqueous phase (Bingol et al., 2004; Kokes et al., 2014).

The leaching processes are classified into two major groups: static mineral bed and agitated system. The first includes in situ leaching, in heaps (i.e., tailings or ores) or in static tanks (vat leaching). The second group comprises leaching in agitated tanks (open or under pressure) (Brierley, 2008; Lima, 2004; Watling, 2006).

For extracting copper from weathered ore, a static bed (i.e., heap leaching) is used, since the costs of implementation and operation are lower than the costs involved in leaching processes in agitated tanks (Carreteiro, 2010). Figure 2 presents a schematic drawing of a process for heap leaching copper weathered ores. Fundamentally, this technology consists of the deposition of large quantities of ore on a waterproofed base, raising a heap, shaped like a pyramid trunk, the surface of which is irrigated with an acidic solution (i.e., sulphuric acid solution).

The initial heap leaching stage of weathered copper ore consists of a crushing operation to reduce particle size. Both the coarse particles as well as the fines, generated during the crushing operation, are agglomerated and used to raise the heap. This newly constructed mineral bed undergoes an acid curing period prior to finally starting the leaching process.

The beginning of the acid leaching process occurs by irrigating the top of the heap with the diluted sulphuric acid solution that permeates the mineral bed generating the copper pregnant leaching solution—PLS. Sulphuric acid promotes the dissolution of weathered copper minerals, as shown in equations in Table 1.

This leaching process is not 100% efficient due to many problems that occur during the heap operation. One of them is the bed compacting, as the leaching process goes on, caused by the solubilisation of different mineralogical species by using sulphuric acid solution, as previously mentioned. Consequently, percolation of the leaching solution is restricted, and it may accumulate at the top of the heap and run down the sides causing erosion of the slope.

In the case of bioleaching of sulphide mineral-bearing ores or tailings, the heap of such materials also may become clogged when the mineral bed compacts. In this case, not only is leach solution percolation interrupted, but air supply

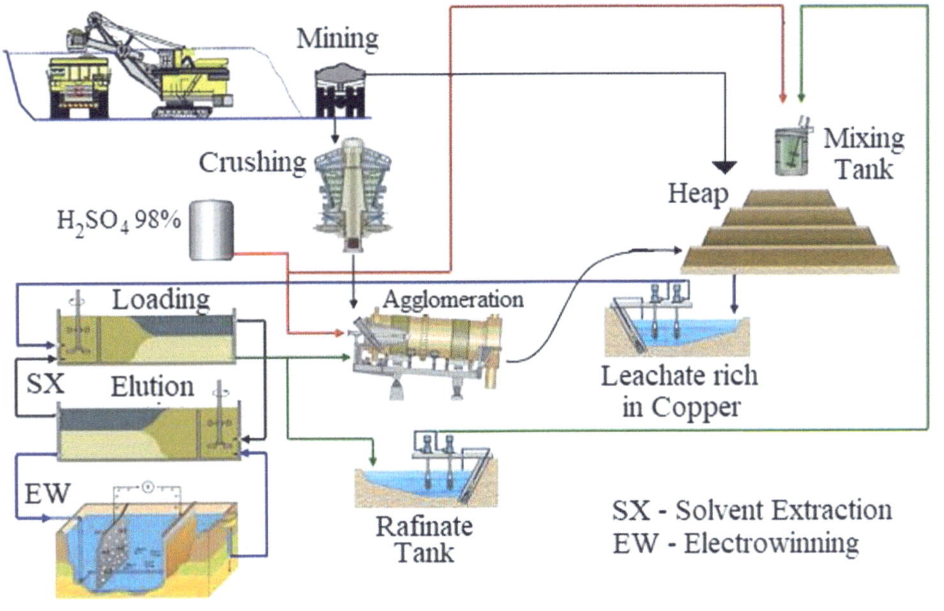

Fig. 2 Typical leaching operation of a heap of copper weathered ore. *Source* Vilca (2013)

Table 1 Leaching reactions of the main copper weathered minerals

Mineral	Chemical dissolution reaction	Equation
Tenorite	$CuO + H_2SO_4 \rightarrow CuSO_4 + H_2O$	(1)
Cuprite	$Cu_2O + 0.5O_2 + 2H_2SO_4 \rightarrow 2CuSO_4 + 2H_2O$	(2)
Azurite	$Cu_3(CO_3)_2(OH)_2 + 3H_2SO_4 \rightarrow 3CuSO_4 + 2CO_2 + 4H_2O$	(3)
Malachite	$Cu_2CO_3(OH)_2 + 2H_2SO_4 \rightarrow 2CuSO_4 + CO_2 + 2H_2O$	(4)
Chrysocolla	$CuSiO_3.2H_2O + H_2SO_4 + 2H_2O \rightarrow CuSO_4.5H_2O + SiO_2$	(5)
Atacamite		
	$2Cu_2(OH)_33Cl + 3H_2SO_4 \rightarrow 2CuSO_4 + 2CuCl_2 + 6H_2O$	(6)
Brochantite	$CuSO_4.3Cu(OH)_2 + 3H_2SO_4 \rightarrow 4CuSO_4 + 3H_2O$	(7)
Antlerite	$CuSO_4.2Cu(OH)_2 + 2H_2SO_4 \rightarrow 3CuSO_4 + 4H_2O$	(8)
Chalcanthite	$CuSO_4.5H_2O \rightarrow CuSO_4 + 5H_2O$ (in the presence of H_2O)	(9)

Source Watling (2006)

is also restricted. As above, this compaction of the mineral bed occurs mainly due to the dissolution of mineralogical species.

Other factors that can influence the clogging of the mineral bed are seismic shocks (at least 3.0 degrees on the Richter scale) and even the movement of trucks during mining operations. These vehicles, due to their robustness, can cause to some extent tremors with sufficient intensities for compaction of mineral beds. One way to solve this problem, while dealing, in particular, with the aforementioned weathered ore tailings, is to turn it over adding elemental sulphur by using proper equipment and machinery, which have high operating cost considering the large volume of such wastes that constitutes the heap. On the other hand, another way of resuming the bio-extractive process consists of adding a layer of elemental sulphur, from a mechanical suspension of fine particles of hydrophilized sulphur in medium containing microorganisms capable of oxidizing such reduced sulphur source, which during its bio-oxidative process produces sulphuric acid allowing for continued extraction of copper still remaining in the heap. Figure 3 shows, schematically, the different ways of using elemental sulphur for processing such tailings.

The potential advantages of generating sulphuric acid from sulphur bio-oxidation in heaps include (1) irrigating sulphuric acid only to the upper part of the heap is depleted as leach solutions percolate the copper ore heaps, (2) addition of big volume of sulphuric acid to a heap may destabilizing the heap, (3) heat is produced from bio-oxidation of sulphur, (4) the sulphuric acid will be less consumed if it is generated and consumed locally at reactive mineral sites and (5) transportation, storage and use of elemental sulphur are safer than sulphuric acid (de Oliviera et al., 2014).

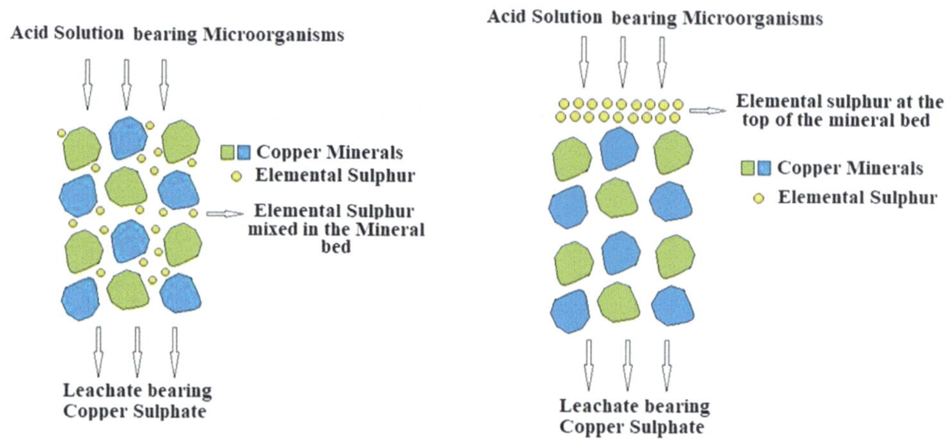

Fig. 3 Different ways of using elemental sulphur for processing weathered ore tailings. *Source* The authors—De Oliviera et al

Figures 4 and 5, as follows, show, respectively, the flow diagram with all unit operations and processes for treating such tailings followed by an schematically way on how to deal with such tailings bearing in mind the directives of circular economy.

Analysing Fig. 4, the mineral processing operations (i.e., crushing, grinding, sampling, etc.) are used so as to prepare the tailing before starting the bioleaching process. Once this mineral substrate is ready, a mechanical suspension of elemental sulphur fine particles bearing microorganisms are mixed with such tailing, in an agglomeration operation, before starting the bioleaching of elemental sulphur for in situ generating sulphuric acid to dissolve the oxidized minerals, as show in reaction 10. According to this reaction

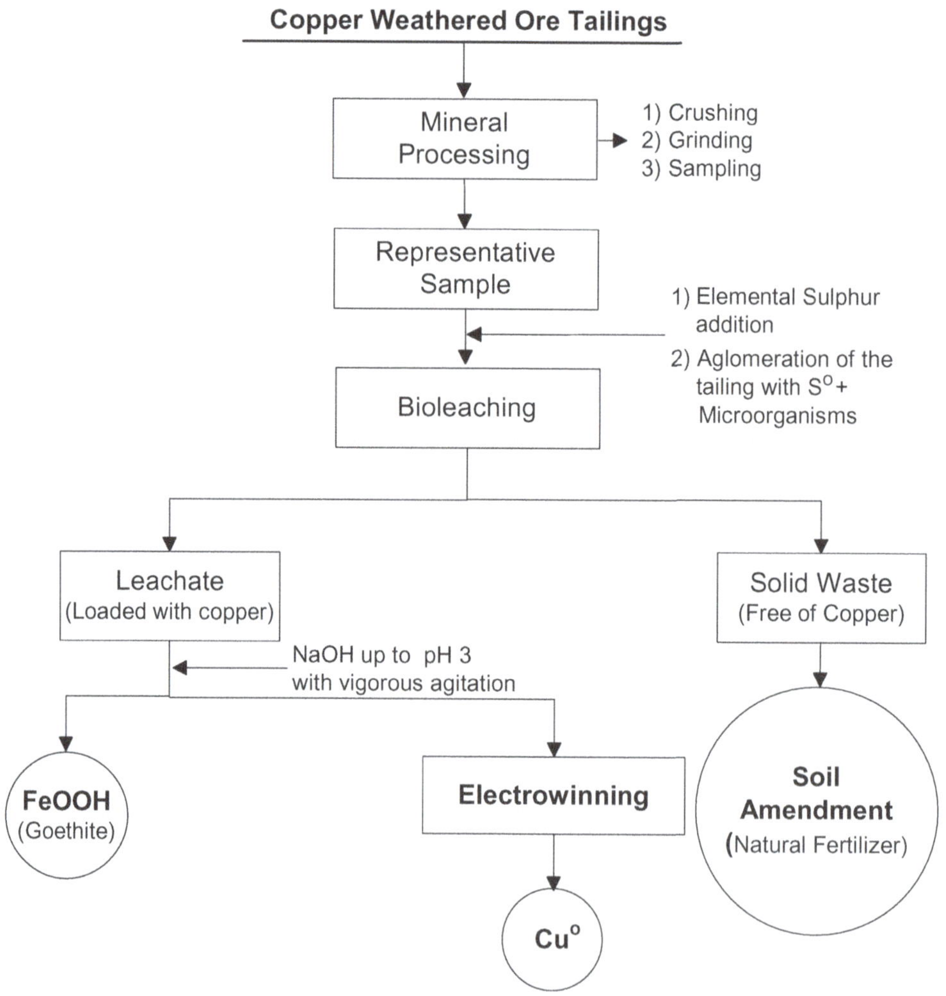

Fig. 4 Flow diagram on how to processing weathered ore tailings. *Source* The authors

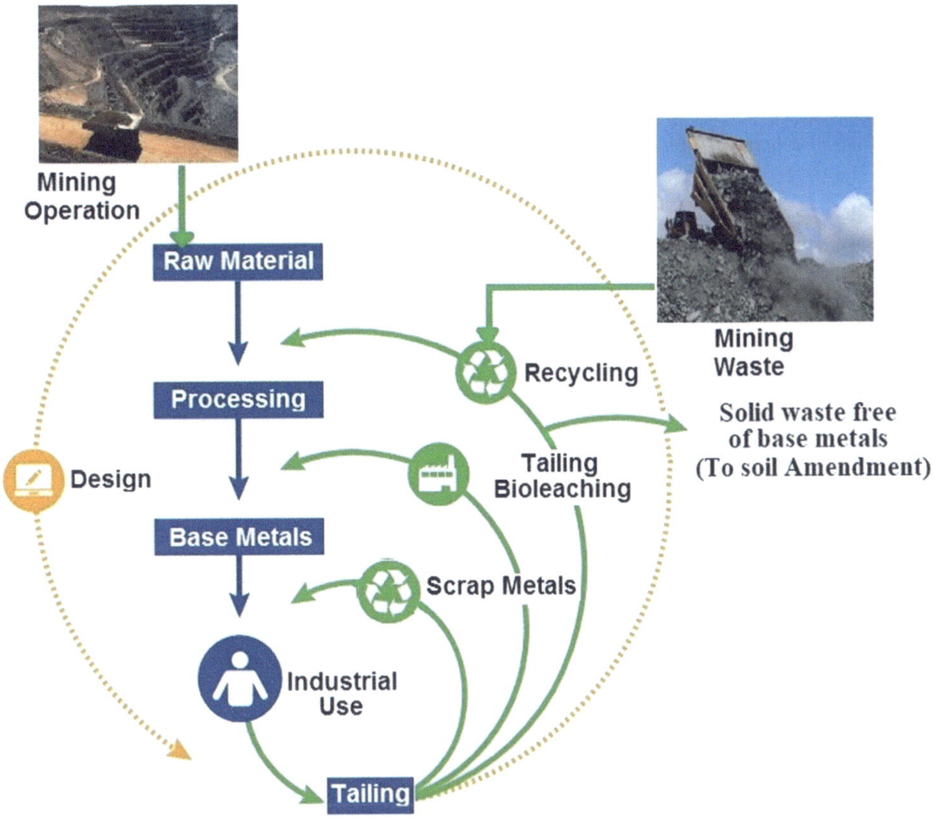

Fig. 5 Bio-extraction of metal values out of tailings under the concepts of circular economy. *Source* The authors

1, ton of elemental sulphur around 3 tons of sulphuric acid are produced.

$$2S^o + 3O_2 + 2H_2O \xrightarrow{Microorganism} 2H_2SO_4. \quad (10)$$

The aqueous phase, out of such leaching process, goes for precipitating iron species as goethite (α-FeOOH) before the copper electrowinning. On the other hand, the solid phase remaining, and free of copper, might be used for soil amendment.

Figure 5 shows, in a generalized way, on how to deal with mining wastes under the directives of circular economy.

Just considering the use of the solid phase after the bioleaching process, we can mention that the vast majority of Brazilian tropical soils are oxisoils that, although with good physical characteristics, present, in almost generalized form, inadequate chemical characteristics, such as high acidity, high levels of exchangeable aluminium and deficiency of nutrients, especially calcium, potassium, magnesium and phosphorus. Soils of this nature, once chemically corrected, have great agricultural potential, enabling improvements in plant nutrition, resulting in good productivity. Thus, several products and methods are used in order to replace nutrients, such as the use of mineral substrates in agriculture, especially basic and ultrabasic rocks, which, in addition to correcting soil acidity, contribute to the replacement of nutrients. This technique called rocks for crops or remineralization of the soil is summed up in the addition of rock dust to the soil, which due to chemical and biological weathering, mediated by the natural local flora, in which water has a solvent role, decomposes the rock dust slowly, thus nutrients are gradually released. In this context, there is the sector of ornamental rocks (i.e., granites, marbles, limestones, slates, quartzite, soapstone, among others), which generates more than 70% of residues from mining to beneficiation, and can be a source of natural fertilizers and soil amendments.

1.2 Bio-extraction of Metals Out of Primary Ores/tailings

In fact, the oxidation of sulphide minerals takes place in the mines and, for a long while, it was defined as just a chemical phenomenon, as the result of the chemical oxidation reactions brought about water and atmospheric oxygen (Pradhan et al., 2008). Pyrite (FeS_2), often found associated with other sulphide minerals, can be oxidized with the generation of ferric sulphate in aerated drainage water, according to Eqs. 11 and 12.

$$2FeS_2 + 7O_2 + 2H_2O \rightarrow 2FeSO_4 + 2H_2SO_4, \quad (11)$$

$$2FeSO_4 + 0.5O_2 + H_2SO_4 \rightarrow Fe_2(SO_4)_3 + H_2O. \quad (12)$$

Equation 12 shows the generation of ferric sulphate, which is an efficient leaching/oxidizing specie, capable of mediating the dissolution of sulphides minerals, such as chalcopyrite ($CuFeS_2$) and pyrite (FeS_2), returning to the ferrous state, as shown in Eq. 13 (Yahya and Johnson 2002).

$$FeS_2 + 7Fe_2(SO_4)_3 + 8H_2O \rightarrow 15FeSO_4 + 8H_2SO_4. \quad (13)$$

However, the reactions shown in Eqs. 11 and 12 occur very slowly in the absence of a catalyst (Hebert, 1999), as the ferrous ion is stable in an acidic medium. Therefore, the oxidation mediated by the ferric ion (Eq. 13) is also slow. Thus, for a significant amount of ferric ions to be generated naturally during the heap leaching process it takes a long while.

After discovering bacteria capable of oxidizing inorganic compounds containing iron and/or sulphur redefined the concept of natural leaching of sulphide minerals as a primarily microbiological process, as the rate of oxidation of various sulphide minerals is significantly accelerated in the presence of those microorganisms. According to Hebert (1999), the reaction catalysed by *Acidithiobacillus ferrooxidans* can be very much faster than by a purely chemical oxidative reaction (i.e., abiotic medium).

Due to the ferric ion formation reaction, as ferric sulphate (Eq. 12), it was believed that bioleaching was an indirect oxidizing process. However, some studies suggest that there is a direct interaction between the microorganism and the sulphide minerals free for such reaction (Crundwell, 2003; Ghauri et al., 2007; Harneit et al., 2006). The microorganisms sticking to the sulphide mineral can also be mediated by extracellular polymeric substances (EPS), which are generated by some microorganisms during the leaching process, such as *At. thiooxidans* (Hugues et al., 2008).

In general terms, according to Crundwell (2003), three types of mechanisms can occur in a bioleaching system, which are schematically represented in Fig. 6. They are direct contact, indirect contact and indirect mechanisms.

In the direct contact mechanism, the cell interacts with the surface of the sulphide minerals, followed by an enzymatic attack on the mineral components that may be oxidized, such as Fe^{2+} (Fig. 6a). In this mechanism, the microorganism extracts electrons directly from the mineral, which, after passing through subsequent biochemical reactions, are transferred to oxygen (i.e., final electron acceptor). The mechanism of direct contact can be described by Eq. 14, for instance, for the oxidation of chalcopyrite.

$$2CuFeS_2 + 8.5O_2 + H_2SO_4 \rightarrow 2CuSO_4 + Fe_2(SO_4)_3 + H_2O. \quad (14)$$

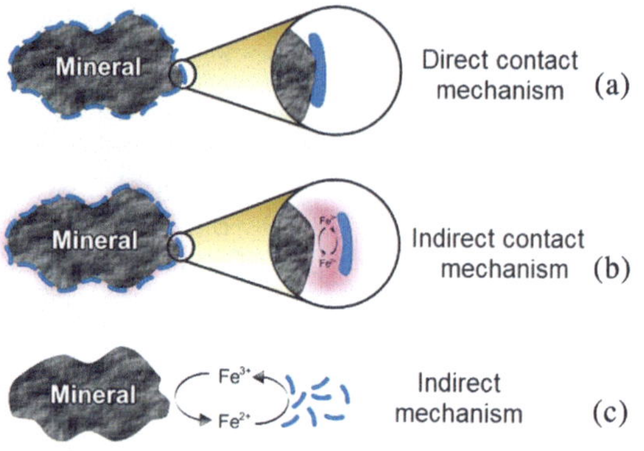

Fig. 6 Microorganism/mineral substrate interaction mechanisms proposed for the bioleaching of sulphide minerals. *Source* The authors

The finding of microbial origin EPS contributed to explain the microorganisms' interaction mechanisms with the sulphide minerals (Fig. 6b). Rodrigués et al. (2003) showed that the release of exopolymeric material is of paramount importance for microbial adhesion, followed by sulphide dissolution, since the complexed iron to the EPS promotes a positive charge to the cell, stablishing an electrostatic attraction between the microbial cell and the negatively charged surface of some sulphide minerals (Sand at. al., 2006). The interaction of microorganism with sulphide minerals, brought about by the EPS, is called the indirect contact mechanism (Crundwell, 2003; Watling, 2006).

In the other mechanism, called the indirect mechanism (Fig. 6C), the sulphide minerals are directly oxidized by ferric ions (Fe^{3+}) in solution (Crundwell, 2003; Watling, 2006).

$$CuFeS_2 + 2Fe_2(SO_4)_3 \rightarrow CuSO_4 + 5FeSO_4 + 2S^o. \quad (15)$$

As the result of the sulphide minerals chemical oxidation, the ferric ion (Fe^{3+}) is reduced to ferrous ion (Fe^{2+}), which are further re-oxidized to ferric ions mediated by the microorganisms (Eq. 12).

Furthermore, when elemental sulphur is generated (Eq. 15), the oxidative action of the microorganism is important (Eq. 10), so as to avoid the elemental sulphur to be deposited on the sulphide minerals surface, a consequent barrier for the oxidant agent diffusion to mineral surface, which would prevent the sulphide minerals oxidation to carry on (Eq. 14).

The metals extraction from sulphide minerals requires the dissolution of these minerals via an oxidative process. The bioleaching, in particular, serves to properly dissolve such minerals by using the natural occurring microorganisms. In the case of running the bioleaching of sulphide minerals bearing ores or tailings the heap of such materials ends up being clogged, as this process goes on, as mentioned for the leaching of weathered ores/tailings, due to the mineral bed packaging obstructing the up-flow of oxygen and carbon dioxide bearing air at the base of that heap or due to the restriction of the leaching solution percolation through such mineral beds, during the bio-extractive process, a fact that leads to the accumulation of this solution at the top of the heap with consequent erosion in the ends of it, which means the interruption of the leaching process. The mineral bed accommodates, in particular due to the dissolution of soluble mineralogical species in sulphuric acid solution bearing microorganisms, causing a rearrangement of the mineral bed making it difficult for the leachate to percolate.

1.3 Extraction of Metals from Electrical-Electronic Waste

Reusing and recirculating products and materials are the basic concept of circular economy (CE). The CE is a new proposal approach resulting in positive impacts such as reduced demand for raw materials, reduced consumption of basic resources, and job creation, as well as preventing negative impacts from the exploitation and processing of natural resources. Mining is notorious for its potential environmental impact, but the mining waste can be treated to recover still valuable material remaining on it, through recycling techniques, which can be practised through urban mining of industrial and post-consumer waste categories (i.e., in the circular economy). Urban mining offers a cost-effective alternative to managing electrical and electronic equipment waste (e-waste) and, at the same time, as a sustainable way to exploit mineral resources, reduces primary material intake and stimulates the circularity in the supply chain.

Waste electrical-electronic equipment, includes end-of-life household appliances such as refrigerators, washing machine, air conditioners and IT equipment including computers, mobile phones and portable electronic devices. WEEE is a mixture of metallic elements, such as Cu, Fe, Al, Pb, Sn, Ni, Cd, Be, Pd, Ag and Au, different types of plastics and ceramics (Tuncuk et al., 2012). The precious metals, in particular, are of paramount importance in the manufacture of electronic appliances due to their high chemical stability and conducting properties. The metal content is around 30% in the PCB (i.e., printed circuit boards) of WEEE (Das et al., 2009). The technologies available for recovering metals from PCB include physical (Cui & Forssberg, 2003), pyrometallurgical (Cui & Zhang, 2008), hydrometallurgical (Tuncuk et al., 2012), and biohydrometallurgical (Liang et al., 2010) processes.

The ever increasing quantities of precious metals and rare earths are required for manufacturing electronic devices—just considering the huge numbers of mobile phones and computers needed now and in future. As an example of such problem, roughly one gram of gold is necessary to manufacture 40 mobile phones (Basel Convention). To get this amount of gold, around one ton of ore has to be mined and processed. This means expenditures in energy and different chemicals—for devices that last roughly two and a half years. The likely alternative is to recover the raw materials from old devices and use them back into the production of new ones, which is known as "urban mining". Figure 7, as follows, gives a quick flavour

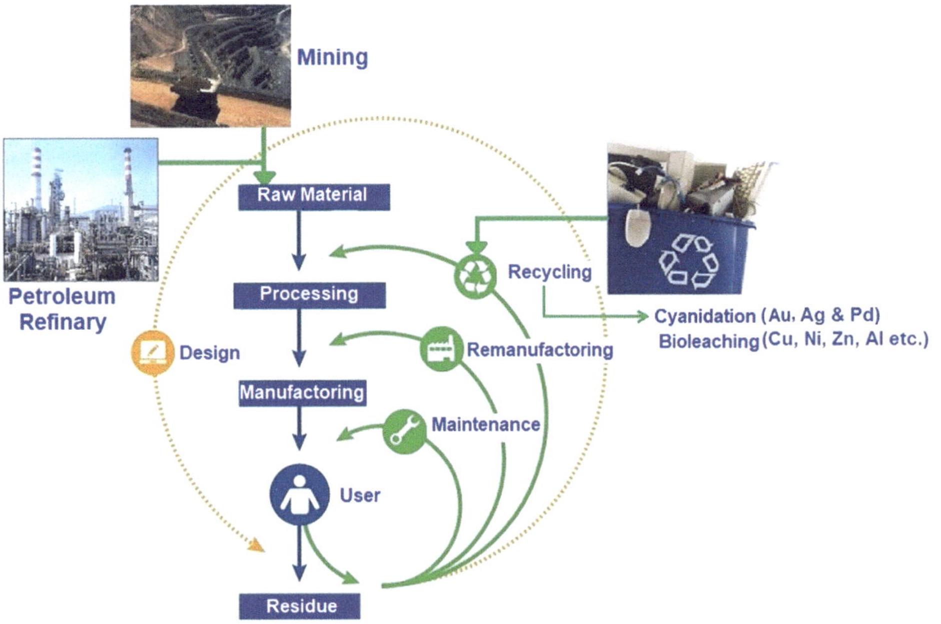

Fig. 7 Production of electro-electronic devices from different natural resources and their post-consume discharge and recycling under the concepts of circular economy. *Source* The authors

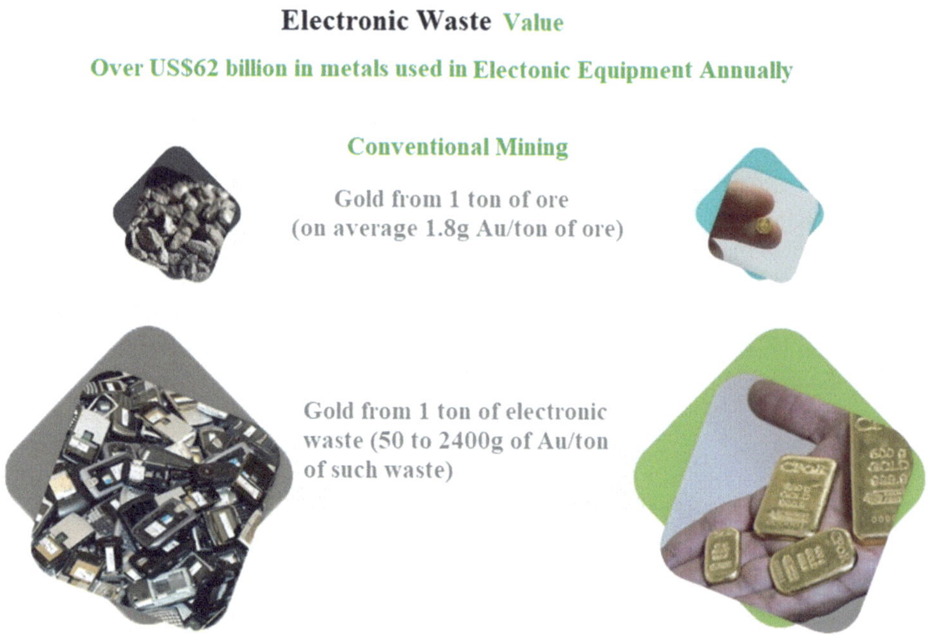

Fig. 8 Gold extraction from 1 ton of gold ore in comparison with the same extraction from 1 ton of useless cell phones. *Source* EnviroLeach.com

on how the electronic equipment is generated, consumed and further recycled so as to recover the metal content.

This figure shows the combination of raw materials from different industrial sectors (i.e., mining, oil industry, etc.) in the manufacture of consumables. Once produced, these goods reach obsolescence in a short period of time and are improperly discarded. However, we are dealing with goods that carry metallic elements in their constitution that come from non-renewable resources. In the circular economy, this obsolete equipment can be repaired during maintenance or can be remanufactured, that is, the parts still serviceable can be reused and, finally, recycling can be done, where, properly processed, the metallic parts are duly extracted in their soluble forms for further separation and obtaining pure metals. The precious metals (i.e., gold, silver and palladium), in particular, always used as a thin layer electrodeposited on other metals surface (i.e., copper, nickel, etc.), to grant them anticorrosive characteristics, can be recovered, in the first place, out of those electronic scraps, by an intensive cyanidation followed by the bioleaching process for extracting base metals. In the particular case of gold, it is interesting to mention that in its extraction process, from 1 ton of gold ore, an average of 1.8 g of gold per ton of processed ore is obtained. However, the gold obtained from 1 ton of useless cell phones can range from 50 to 2400 g of gold, as can be observed in Fig. 8 (Daniel E. Sullivan).

1.3.1 Disposal of Electronic Scrap

The final destination of a WEEE is landfill or incineration; however, when electronic waste is disposed of incorrectly, it causes a series of environmental problems, as it can be naturally leached, producing leachate, resulting in the pollution of groundwater and water bodies. In addition, when these residues are burned, it generates toxic gases (i.e., dioxins and furans) as a result of chemical reactions, due to the flame retardants thermal decomposition (Guo et al., 2010; Jie et al., 2008; Kang et al., 2005).

1.3.2 Electronic Scrap Recycling

One of the alternative routes for disposing e-waste is recycling. Cui and Zhang (2008) mentioned that electronic waste recycling is a relevant issue from the waste treatment point of view in the first place and, additionally, due to the recovery of metals with high intrinsic values, such as gold, silver, platinum and palladium. Therefore, this recycling process is known as a way of transforming solid waste, which would be discarded, with changes in its physical, physical–chemical or biological state, in order to assign characteristics to become, again, raw materials or products.

According to Yamani (2012), other factors that emphasize the importance of electronic equipment waste as an alternative source for metal extraction can be highlighted, given the current mining scenario, with the consequent depletion of ores with high contents in the metals of interest. On the other hand, there is an economy related to the recovery process of these metals, since they are in their metallic forms, in addition to the possibility of a reduction in the exploration of primary mineral reserves to meet the progressive increase in demand for the generation of raw material in order to produce new electronic devices.

Figure 9 shows the unit operations and processes for extracting precious and base metals out of electronic scraps, including pyrometallurgical, chemical leaching and bioextraction of metals.

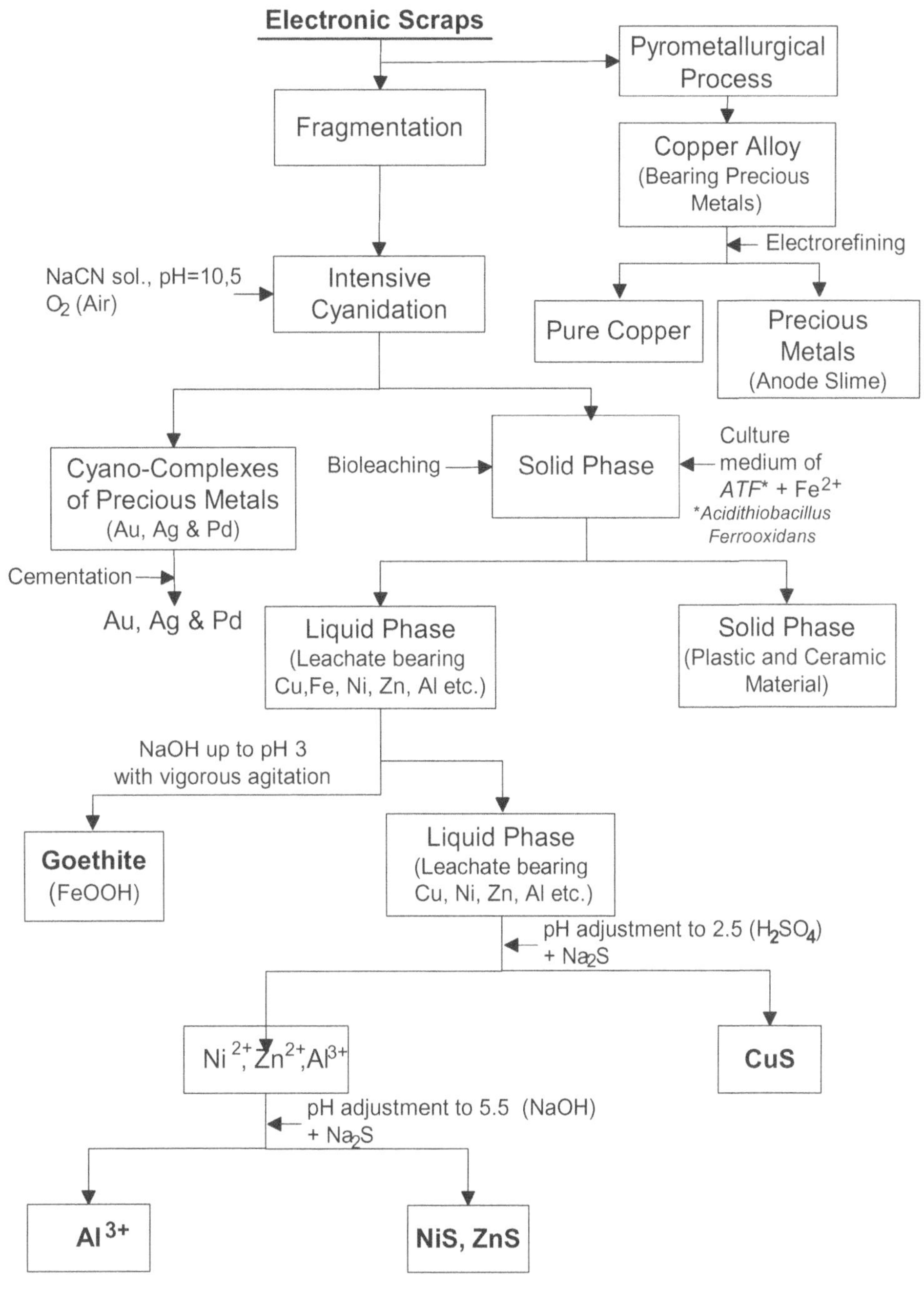

Fig. 9 Proposed unit operations and processes for extracting precious and base metals out of electronic scraps. *Source* The authors

1.3.3 Biotechnological Process for Extracting Base Metals Out of Electronic Scraps

This bioleaching process is a very attractive alternative, bearing in mind that it requires low energy consumption and low implementation and operation costs as such. In this metal extraction process, the *Acidithiobacillus Ferrooxidans* microorganisms are in charge of generating the oxidizing agent (Fe^{3+}), through the Fe^{2+} ions oxidation being previously added to the reaction system as soluble chemical ($FeSO_4.7H_2O$) or as insoluble form (FeS_2).

Brandl et al. (2006), Ilyas et al., (2007) demonstrated using acidophilic and thermophilic microorganisms that metals can be recovered from printed circuit boards through bioleaching. This metals dissolution method from electronic scraps is similar to the one dissolving sulphide minerals, such as copper oxidation (Eq. 16) in which Fe^{3+}, released into the system from the oxidizing action of *Acidithiobacillus Ferrooxidans* bacteria, oxidizes metallic copper to Cu^{2+}. The role of microorganisms is to re-oxidize the Fe^{2+} into Fe^{3+}, thus establishing a reaction cycle.

$$Cu^{o} + 2Fe^{3+} \rightarrow Cu^{2+} + 2Fe^{2+} \rightarrow E^{o} = +0.426\ \text{V versus SHE.} \quad (16)$$

The microorganisms in charge of promoting Fe^{2+} oxidation are acidophilic, chemotrophic (i.e., obtain energy from the oxidation of inorganic compounds), autotrophic (i.e., perform the biosynthesis of all cellular constituents using carbon dioxide (CO_2) as the only source of carbon)) and are classified according to the temperature at which they develop, distinguishing into: mesophilic (up to 40 °C), moderate thermophiles (40–55 °C) and extreme thermophiles (55–80 °C) (Schippers, A. 2006).

Therefore, bacterial leaching of electronic scrap occurs through an indirect mechanism. Microorganisms do not come into contact with the metallic substrates, playing a catalytic role, as they generate chemical species that oxidize most of the metals present in electronic scrap. Ferric ions (Fe^{3+}) act as the main oxidizing agent capable of dissolving such metals, such as metallic copper, the main component of said scrap, as shown in Eq. (16). This equation represents the oxidation of metallic copper to cupric ions by ferric ions, previously generated by oxidizing ferrous ions through microorganism *A. ferrooxidans*, in searching for energy to maintain its metabolic activity. The ferrous ions resulting from this oxidative process are simultaneously re-oxidized to the ferric ions, thus establishing an oxidation–reduction cycle of the iron ions. Thus, the potential for such oxidation reaction to occur is positive (+0.426 V *vs. SHE*), which means that such process is spontaneous. Taking into consideration the vast number of base metals, and so many others present in such electronic scraps, Table 3 shows, as a function of the potentials of the cells formed by the action of ferric ions with the different metals present, that the dissolution reactions are spontaneous. The higher the potential, the more effective are the oxidation/dissolution reaction.

The aforementioned bio-extraction process can be accomplished in a reaction system, outlined in Fig. 10, where the electronic scraps, previously mechanically fragmented, are put in a perforated rotating drum and further immersed in a MKM culture medium leaching solution bearing nutrients and ferrous sulphate, as energy source, and inoculum of *At. ferrooxidans* and *Leptospirillus ferrooxidans* that oxidize Fe^{2+} to Fe^{3+}. The MKM medium is composed of: 0.4 $g.L^{-1}$ of ammonium sulphate; 0.4 $g.L^{-1}$ of magnesium sulphate heptahydrate; 0.04 $g.L^{-1}$ of dihydrogen potassium phosphate.

Table 3 Cell potentials for the metal oxidation reactions

Oxidation reaction	Cell potential, V versus SHE
$2Fe^{3+} + Cu^o \rightarrow Cu^{2+} + 2Fe^{2+}$	$E^o = +0.426V$
$2Fe^{3+} + Ni^o \rightarrow Ni^{2+} + 2Fe^{2+}$	$E^o = +1.024V$
$3Fe^{3+} + Cr^o \rightarrow Cr^{3+} + 3Fe^{2+}$	$E^o = +1.481V$
$2Fe^{3+} + Zn^o \rightarrow Zn^{2+} + 2Fe^{2+}$	$E^o = +1.533V$
$2Fe^{3+} + Al^o \rightarrow Al^{3+} + 3Fe^{2+}$	$E^o = +2.441V$

Source The authors

1.4 Recycling of Aluminium from Waste of the Supply Water Treatment Plant

The treatment of water for a municipality's consumption consists, firstly, of the reaction of this crude water with an aluminium sulphate solution for clarifying the water removing the suspended solids. After adding such chemical to the supply water one or more chemical reactions can take place, generating voluminous aluminium hydroxide whose composition will depend on the pH of the water. If the water has a pH close to neutrality (i.e., with a methyl orange indicator), the floccules will have a likely composition whose formula is: $5Al_2O_3.3SO_3$. However, when the water is sufficiently alkaline, giving a pink coloration with the phenolphthalein indicator, aluminium hydroxide ($Al(OH)_3$) is formed. Should the alkalinity of the water is due to the presence of calcium bicarbonate, the following reaction will occur:

$$Al_2(SO_4)_3 + 3Ca(HCO_3)_2 \rightarrow 3CaSO_4^{\downarrow} + 2Al(OH)_3^{\downarrow} + 6CO_2. \quad (17)$$

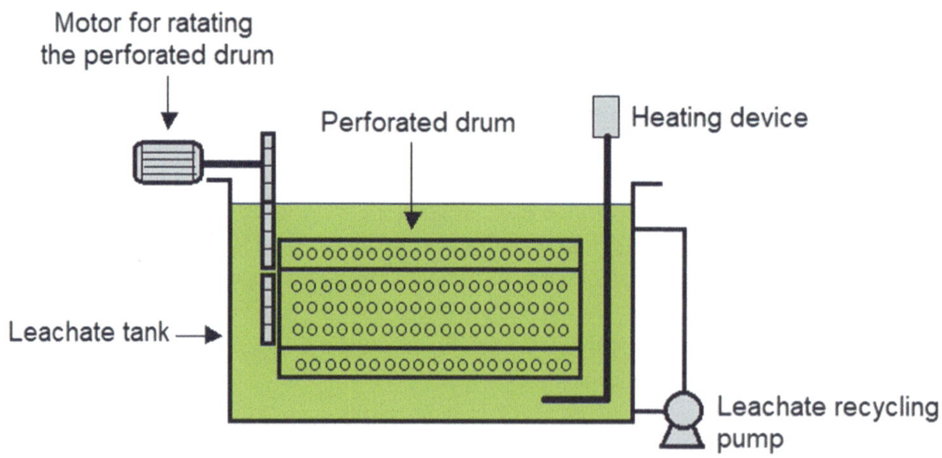

Fig. 10 Reaction system for running the bio-extraction of base metals from electronic scraps. *Source* The authors

When these flocculating aluminium compounds are formed, the result is a suspension of these compounds in the aqueous phase, which, when at rest, decant, dragging with them all suspended solids. The term "solid in suspension" is applied to heterogeneous material contained in water supply, consisting basically of a sludge bearing clay, silica, debris and products originating from industrial activities.

Figures 11 and 12 show, respectively, a schematic flowchart with all the unit operations and processes used for water treatment and recycling of aluminium in the form of sulphate, and how aluminium is extracted out of the sludge generated in a water treatment plant under the concepts of circular economy. As can be observed, in both figures, the sulphuric acid needed for extracting aluminium can be generated by bio-oxidizing elemental sulphur avoiding the transport and handling concentrate sulphuric acid that, frequently, cause so many accidents.

1.5 Bioleaching of Gravity Pyrite Concentrate from a Coal Mine Aiming at Producing Iron-Based Pigments

Coal is a heterogeneous mineral resource from the chemical and physical point of view, consisting basically of carbon and small amounts, by mass, of sulphur, nitrogen and hydrogen. In some countries, the coal produced has high sulphur content, which is the case of the ones produced in Brazilian coal mines. Consequently, during coal combustion, for generating energy, sulphur oxides (SOx) are issued to the atmosphere, being recognized as highly polluting and as acid rain producers. Additionally, coal industries generate large amounts of tailings during the coal processing.

Such tailings contain high iron contents, basically from pyrite (FeS_2), which disposed of inadequately generate acid rock drainage (ARD). As an attempt of minimizing the

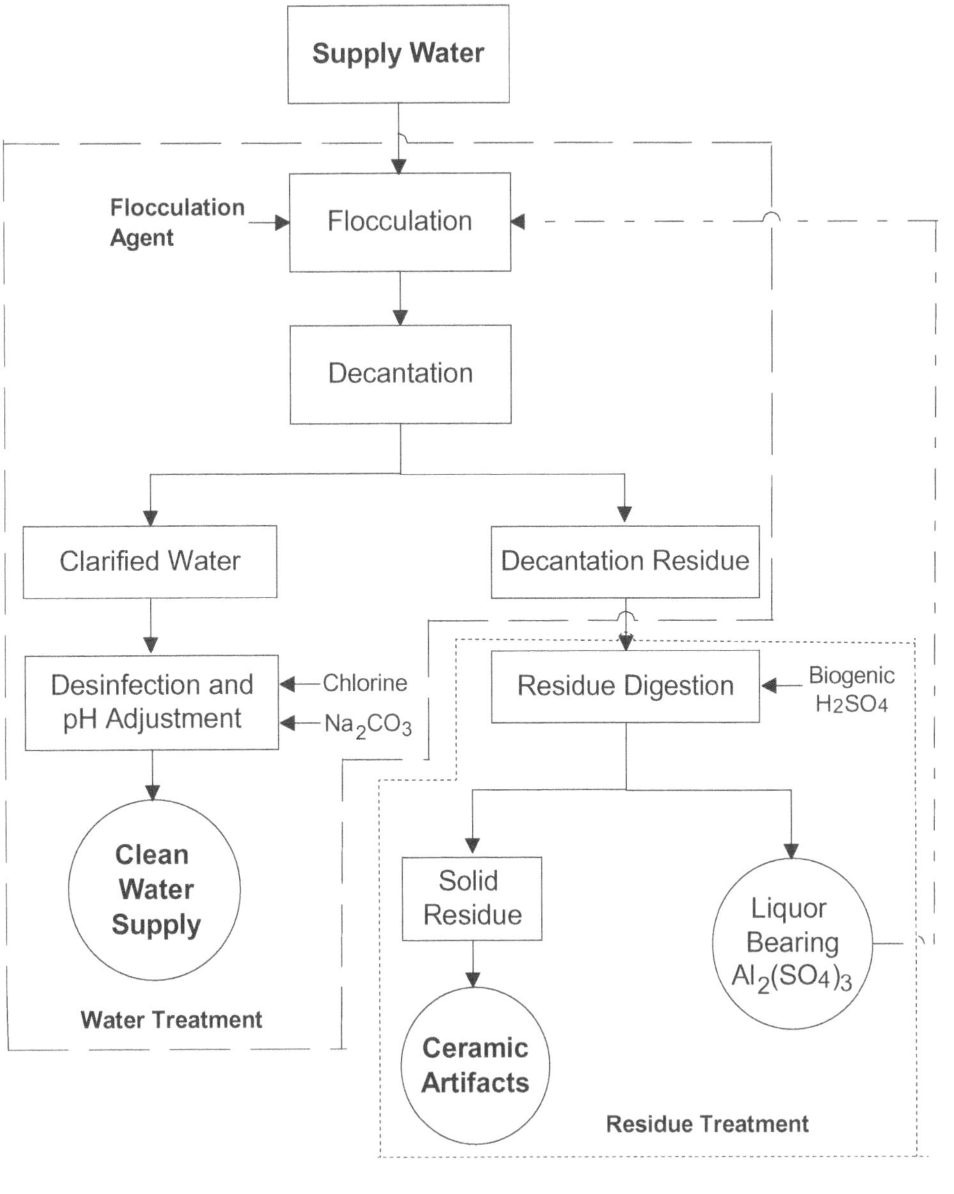

Fig. 11 Unit operations and processes used in water treatment plant *Source* The authors

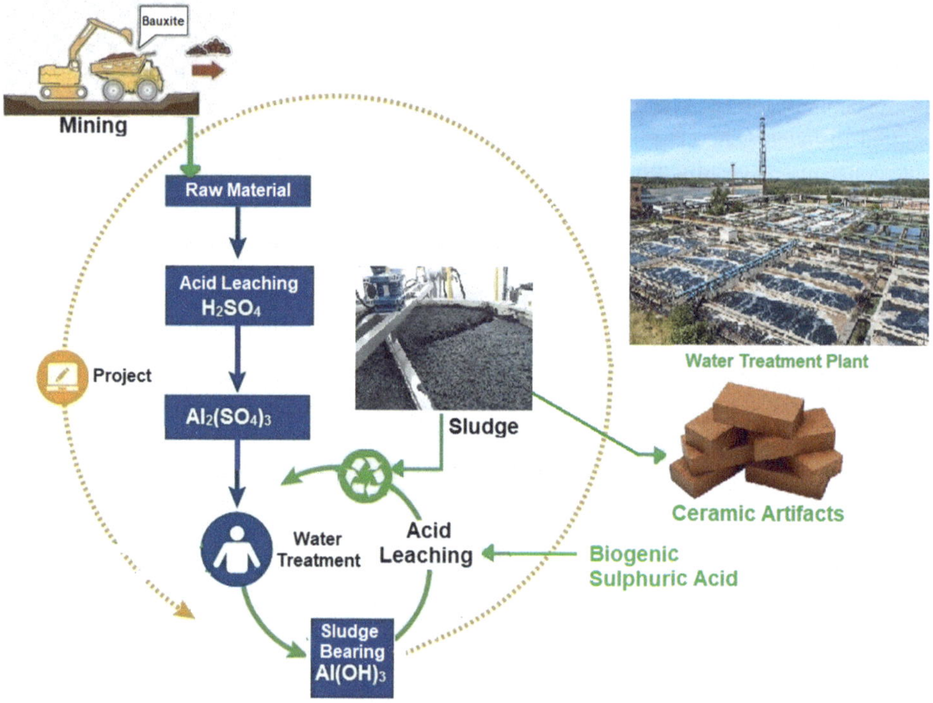

Fig. 12 Extraction of aluminium from the sludge of the water treatment plant under the concepts of circular economy *Source* The authors

environmental impact some coal mines have been producing pyrite concentrates from coal tailings, aiming at producing chemicals with some added value. One of these is the iron oxide-based pigments.

In view of this scenario, it is necessary to develop procedural routes to promote the dissolution of pyrite in order to obtain a leachate bearing ionic iron species to be further used for producing the aforementioned pigments. The attractive alternative way of dissolving such sulphide mineral is the bioleaching of pyrite concentrate at a low operating cost.

1.5.1 Bioleaching of Pyrite Gravity Concentrate

In the column bioleaching test of pyrite gravity concentrate, a sample of the pyrite concentrate, Fig. 13, from a Brazilian coal mine, was used. The iron content in that sample was analysed as being around 30%, after acid digestion of a representative sample of this concentrate.

The pyrite concentrate bioleaching experiment, on a pilot scale, assembled at CETEM—Centre for Mineral Technology, and run by the authors, was carried out in a column made of fibreglass, 4 m high and a diameter of 0.45 m. This column was duly filled with that concentrate. The liquor from the column had its pH adjusted adding, when necessary, 5 M sulphuric acid solution using a dosing pump, which was pumped back to the top of the column for carrying on with the irrigation operation using a spray nozzle. This leachate, in addition to bearing bacteria and archaea capable of oxidizing iron and sulphur, contained nutrients such as nitrogen (N), phosphorus (P) and potassium (K).

Air was supplied at the base of the column, as sources of oxygen and CO_2, whose flow was controlled by rotameters. For the bioleaching process to be established, the availability of dissolved oxygen was essential, which concentration depends on the energetic substrate used and its content. In the case of pyrite, for example, as seen in Eq. 10, four oxygen atoms are needed for each sulphur atom. CO_2 was used by microorganisms as a necessary carbon source, together with other nutrients, in the synthesis of cellular material used to originate new cells and, consequently, increase the microbial population in the reaction system.

The heating of the column was provided by an electrical resistance located around the external area of the column, the same happening, when necessary, with the heating of the liquor in the 200 litter tank. All the above-mentioned parameters were integrated in a Programmable Logic Controller (PLC) for their control and monitoring.

Figure 14 shows a photo of the aforementioned pilot plant for bioleaching pyrite concentrate. This unit was composed of a fibreglass column, with the above-mentioned dimensions, whose body was divided into two modules equipped with removable polycarbonate portholes, with a dual purpose: one for viewing the particles of the mineral bed and for removing the column filling material at the end of the bioleaching process. The reaction system used two metering pumps, one for pumping the leachate from the leaching tank to the top of the column. This solution percolated the mineral bed, providing the oxidative process of sulphide minerals, with the consequent generation of

Fig. 13 Sample of pyrite gravity concentrate from a Brazilian coal mine *Source* The authors

iron sulphate, and returned to the leachate tank, in a closed circuit.

The other pump served to dispense a 5 M sulphuric acid solution into the leachate tank for pH adjustment in the range of 1.0–2.0. This pump was controlled by the signal generated by a combined Eh–pH electrode located in the head of a pneumatic pump responsible for homogenizing the solution in the leachate tank.

The leachate generated in the bio-oxidative process, containing Fe^{2+} and Fe^{3+} ions, was later used in the production of pigment based on iron oxides.

1.5.2 Pigment Production from the Liquor Generated in the Bioleaching of Pyrite Gravity Concentrates

In Brazil, the coal mining has caused some environmental impacts. Currently, one of the most serious environmental problems found in coal mine areas is the acid rock drainage (ARD). The coals from the Brazilian mines have high sulphur contents that can be transformed into ARD by the natural oxidation of their sulphide minerals content, mainly pyrite (FeS_2), by the combined action of native microorganisms (endogenous) with water and oxygen. With the objective of minimizing the environmental impact while prospecting coal, some plants have produced pyrite concentrates from coal tailings, aiming at using them in the production of pigments based on iron oxide, which contain attractive added values, which is the case of goethite (α-FeOOH) (Fig. 15).

The direct precipitation requires preliminary ferrous iron oxidation combined with a partial acid neutralization to a specific pH endpoint. Thus, the process is accomplished in a chemical reactor, such the one depicted in Fig. 11, and air, or sulphur dioxide could be used as a chemical oxidant most frequently. The products of the iron removal process strongly depend on the conditions, under which the process has been carried out, such as iron oxides with a crystalline structure (goethite (αFeOOH).

The goethite precipitation proceeds through the following hydrolysis reaction:

$$Fe^{3+} + 2H_2O \rightarrow FeOOH + 3H^+. \tag{18}$$

The ferric iron concentration in the solution must not exceed 1 g/ L. The process takes place at a higher temperature (80–90 °C) and pH 2–3. Goethite precipitation requires precise monitoring and control between the ferrous iron oxidation (acid consuming process) and ferric iron precipitation (acid generating process) as the net reaction is acid generating:

$$2Fe^{2+} + 0.5O_2 + 3H_2O \rightarrow 2FeOOH + 4H^+. \tag{19}$$

Products similar to goethite were carried out using a leachate from the bio-assisted dissolution of pyrite (FeS_2), out

Fig. 14 Pilot plant for the bioleaching of pyrite gravity concentrate designed and assembled at CETEM—Centre for Mineral Technology/Rio de Janeiro/Brazil. *Source* The authors

of a gravity concentrate generated during the coal processing. This leachate contained 5.0 g.L^{-1} of total iron, where 90% of the iron species were in the form of Fe^{2+} ions. The reaction system used to carry out the synthesis of goethite is composed of a jacketed reactor with a useful capacity of 4 litters, made of refractory glass (borosilicate), equipped with temperature control, variable mechanical agitation, provided by a special impeller, and aeration device within the solution, as seen in Fig. 11. The reactor was properly filled with 3 litters of this leachate, followed by the addition of hydrogen peroxide (H_2O_2), in a dropwise mode, until a Fe^{3+}/Fe^{2+} ratio of 0.9 was obtained, under an insufflation of 5 L.min^{-1} of air and agitation of 1200 rpm. Then, the pH of the reaction system was increased, adding 1 mol.L^{-1} sodium hydroxide (NaOH) solution up to 3.5. After the addition and formation of the hydrolysed material, stirring and air insufflation were continued at the local room temperature of 25°C for 72 h. The solids obtained were filtered, washed with deionized water and dried in an oven.

The production of iron oxide-based pigments from acidic coal mine drainage is possible and represents a potentially attractive route. Such drainages contain high concentrations of iron, coming mainly from the bioleaching of pyrite (FeS_2) and may have a potential environmental

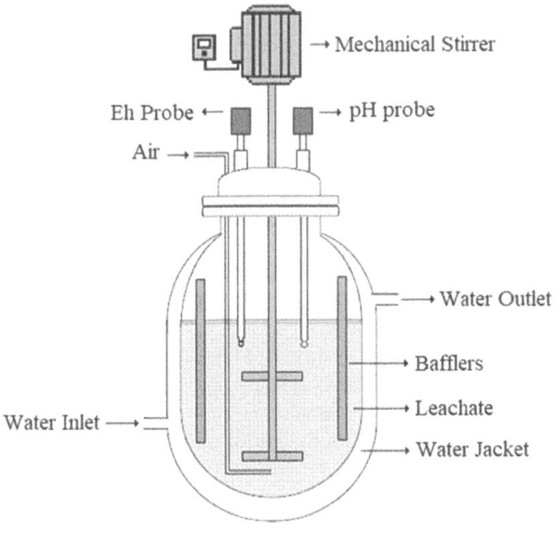

Fig. 15 Schematic drawing of the reaction system for the pigment synthesis. *Source* The authors

impact in regions where its inadequate release occurs. Depending on how pure such pigment is obtained, it can be used in the ceramic industry, for the production of cosmetics and in the pharmaceutic industries.

Final Remarks

A comprehensive study on the bio-extraction of base metals from its mineral residues, resulting from previous heap leaching processes of the respective weathered or primary ore tailings, has to be done, and it is of paramount importance for the proper definition of the experimental conditions aiming at maximizing the extraction of these remaining metallic values. In the case of copper, in vitro and semi-pilot tests carried out, point to high efficiency with regard to the bio-extractive process of this metal, considering the previous definition of operational conditions for the use of elemental sulphur as raw material in biogenic production of sulphuric acid.

Regarding the extraction of base metals from electro-electronic scraps, it is estimated that they will be fully solubilized via the action of ferric ions, generated by the biological oxidation of ferrous ions added to the reaction system, which are continuously maintained in the leaching solution by the action of the microorganisms used. What can be observed is that the action of ferric ions can be intensified by an improvement in the transport of these ions towards the metallic phases, which was observed using the reaction system depicted in Fig. 10. In addition, the use of microorganisms for recovering metals from WEEE could be an economical alternative in comparison to some other hydrometallurgical and pyrometallurgical processes.

The experimental results obtained so far, regarding the processing of acid drainages generated from coal tailings, illustrate the potential for the synthesis of high purity goethite. However, as mentioned above, the synthesis process needs operational adjustments and procedural parameters to obtain a pigment with adequate crystallinity and free from other metallic impurities, a fact that will add greater economic value as well as more noble applications, such as in the industries of cosmetics and pharmaceuticals.

Regarding the recovery of aluminium from the sludge, generated in the treatment of raw water, the chemical process used aimed at contacting this residue with a sulphuric acid solution, generated in the bio-oxidation of elemental sulphur (i.e., biogenic production of H_2SO_4) in a separated reactor. It was experimentally demonstrated that the aluminium and iron contained, in their hydroxide forms, were totally solubilized, generating a liquor with its own characteristics to be used in water treatment, a fact that makes such treatment highly sustainable.

The solid waste generated in the processing of that sludge, consisting, basically, of clay, silica and organic matter, can be used in the production of ceramic artefacts, for constructing popular houses. Additionally, it should be noted that the organic matter present will serve as a source of energy in this production process.

Finally, it is worth to emphasize that there are so many other residues, being generated in the chemical and metallurgical industries, which should be consider as potential metal resources taking into consideration the remarkable advancement of processes using natural microorganisms for recycling the contained metallic values.

Acknowledgements The authors are immensely grateful for the outstanding contribution of the microbiologist Dr. Gregory J. Olson for reviewing this chapter.

References

Basel Convention—UNEP. (30 June 2010). Guidance document on the environmentally sound management of used and end-of-life mobile phones.

Bingol, D., & Canbazoglu, M. (2004). Dissolution kinetics of malachite in sulphuric acid. *Hydrometallurgy, 72*, 159–165.

Brandl, H., & Faramarzi, M. A. (2006). Microbe-metal-interaction for the biotechnological treatment of metal-containing solid waste. *China Particuology, 4*(2), 93–97.

Brierley, J. A. (2008). A perspective on developments in biohydrometallurgy. *Hydrometallurgy, 94*, 2–7.

Carretero, E., Sobral, L. G. S., Oliveira, D. M., Souza, C. E. G. (2010). Bioleaching of metal sulphides ores and concentrates. In L. G. S. Sobral, D. M. Oliveira, C. E. G. Souza (Eds.) Biohydrometallurgical processes: A practical approach. CETEM/MCT.

Crundwell, F. K. (2003). How do bacteria interact with minerals? *Hydrometallurgy, 71*, 75–81.

Cui, J., & Forssberg, E. (2003). Mechanical recycling of waste electrical and electronic equipment: A review. *Journal of Hazardous Materials, 99*, 243–263.

Cui, J., & Zhang, L. (2008). Metallurgical recovery of metals from electronic waste: A review. *Journal of Hazardous Materials, 158*, 228–256.

Das, A., Vidyadhar, A., & Mehrotra, S. P. (2009). A novel flowsheet for the recovery of metal values from waste printed circuit board. *Resources, Conservation and Recycling, 53*, 464–469.

De Oliviera, D. M., Sobral, L. G. S., Olson, G. J., & Olson, S. B. (2014). Acid leaching of a copper ore by sulphur-oxidizing microorganisms. *Hydrometallurgy, 147–148*, 223–227.

Domic, E. M. (2001). Hidrometalurgia: Fundamentos, procesos y aplicaciones, 1a edição (p. 1000). Andros Impresores.

Ghauri, M. A., Okibe, N., & Johnson, D. B. (2007). Attachment of acidophilic bacteria to solid surfaces: The significance of species and strain variations. *Hydrometallurgy, 85*, 72–80.

Guo, Q., Yue, X., Wang, M., & Liu, Y. (2010). Pyrolisis of scrap printed circuit board plastic particles in a fluidized bed. *Powder Technology, 198*, 422–428.

Harneit, K., Göksel, A., Kock, D., Klock, J. H., Gehrke, T., & Sand, W. (2006). Adhesion to metal sulfide surfaces by cells of *Acidithiobacillus ferrooxidans Acidithiobacillus Thiooxidans and Leptospirillum Ferrooxidans*. *Hydrometallurgy, 83*, 245–254.

Hebert Jr, R. B. (1999). MiMi—sulfide oxidation in mine waste deposits: A review with enphasin on dysoxic weathering. Department of Geology and Geochemistry, Stockholm University.

Hugues, P., Joulian, C., Spolaore, P., Michel, C., Garrido, F., & Morin, D. (2008). Continuous bioleaching of a pyrite concentrate in stirred reactors: Population. *Hydrometallurgy, 94*, 34–41.

Ilyas, S., Anwar, M. A., Niazi, S. B., & Ghauri, M. A. (2007). Bioleaching of metals from electronic scrap by moderately thermophilic acidophilic bacteria. *Hidrometallurgy, 88*(1–4), 180–188.

Jie, G., Ying-Shun, L., & Mai-Xi, L. (2008). Product characterization of waste printed circuit board by pyrolysis. *Journal of Analytical and Applied Pyrolysis, 83*, 185–189.

Kang, H. Y., & Shoenung, J. M. (2005). Electronic waste recycling: A review of U. S. infrastructure and technology options. *Resources Conservation and Recycling, 45*, 368–400.

Kokes, H., Morcali, M. H., & Acma, E. (2014). Dissolution of copper and iron from malachite ore and precipitation of copper sulfate pentahydrate by chemical process. *Engineering Science and Technology, 17*, 39–44.

Liang, G., Mo, Y., & Zhou, Q. (2010). Novel strategies of bioleaching netals from printed circuit boards (pcbs) in mixed cutivation of two acidophiles. *Enzyme and Microbial Technology, 47*, 322–326.

Lima L. R. P. A. (2004). A mathematical model for isothermal heap and column leaching. *Brazilian Journal of Chemical Engineering.*

Luz, A. B., Lins, F. A. F. (2004). Introdução ao Tratamento de Minério. In Tratamento de Minérios, 4ª Ed. CETEM/MCT, Cap.1 (pp. 3–14).

Luz, A. B., Sampaio, J. A., França, S. C. A. (2010). Tratamento de minérios, 5ª edição (p. 960).

Pradhan, N., Nathsarma, K. C., Srao, K., Sukla, L. B., & Mishra, B. K. (2008). Heap bioleaching of chalcopyrite: A review. *Minerals Engineering, 21*, 355–365.

Rodriguez, Y., Ballester, A., Blazquez, M. L., Gonzalez, F., & Munoz, J. A. (2003). New information on the chalcopyrite bioleaching mechanism at low and high temperature. *Hydrometallurgy, 71*, 47–56.

Sand, W., & Gehrke, T. (2006). Extracellular polymeric substances mediate bioleaching/biocorrosion via interfacial processes involving iron(III) ions and acidophilic bacteria. *Research in Microbiology, 157*, 49–56.

Schippers, A. (2006). Microorganisms involved in bioleaching and nucleic acid-based molecular methods for their identification and quantification. In E. R. Donati, & W. Sand (Eds.) Microbial processing of metal sulfides. Springer, Cap. 1 (pp. 3–33).

Sullivan, D. E., Recycled cell phones—A treasure trove of valuable metals. USGS—Science for a changing world.

Tuncuk, A., Stazi, V., Akcil, A., Yazici, E. Y., & Deveci, H. (2012). Aqueous metal recovery techniques from e-scrap: Hydrometallurgy in recycling. *Mineral Engineering, 25*, 28–37.

Vilca, A. B. (2013). Studies on the curing and leaching kinetics of mixed copper. Thesis submitted in partial fulfillment of the requirements for the degree of master of applied science in the faculty of graduate studies (materials engineering) the University of British Columbia (Vancouver).

Watling, H. R. (2006). The bioleaching of sulphide minerals with emphasis on copper sulphides: A review. *Hydrometallurgy, 84*, 81–108.

Yamani, L. H. (2012). Recuperação de metais de placas de circuito impresso de computadores obsoletos através de processo biohidrometalúrgico. Tese (Doutorado em Engenharia Metalúrgica e de Materiais)—Universidade de São Paulo (p. 19).

SPRINGER NATURE

GPSR Compliance

The European Union's (EU) General Product Safety Regulation (GPSR) is a set of rules that requires consumer products to be safe and our obligations to ensure this.

If you have any concerns about our products, you can contact us on ProductSafety@springernature.com

In case Publisher is established outside the EU, the EU authorized representative is:

Springer Nature Customer Service Center GmbH
Europaplatz 3
69115 Heidelberg, Germany

The manufacturer's authorised representative in the EU is Springer
Nature Customer Service Centre GmbH, Europaplatz 3, 69115 Heidelberg,
Germany. If you have any concerns regarding our products, please
contact ProductSafety@springernature.com

Printed and bound by CPI Group (UK) Ltd, Croydon, CR0 4YY
28/09/2026
02224097-0001